Advanced Materials for Multifunctional Applications

Advanced Materials for Multifunctional Applications

Editors

Corneliu Munteanu
Ioan Doroftei

MDPI

Basel • Beijing • Wuhan • Barcelona • Belgrade • Novi Sad • Cluj • Manchester

Corneliu Munteanu
Mechanical Engineering,
Mechatronics and Robotics
"Gheorghe Asachi" Technical
University of Iasi
Iasi
Romania

Ioan Doroftei
Mechanical Engineering,
Mechatronics and Robotics
"Gheorghe Asachi" Technical
University of Iasi
Iasi
Romania

Editorial Office
MDPI AG
Grosspeteranlage 5
4052 Basel, Switzerland

This is a reprint of articles from the Special Issue published online in the open access journal *Materials* (ISSN 1996-1944) (available at: www.mdpi.com/journal/materials/special_issues/advance_materials_multifunctional_applications).

For citation purposes, cite each article independently as indicated on the article page online and using the guide below:

Lastname, A.A.; Lastname, B.B. Article Title. *Journal Name* **Year**, *Volume Number*, Page Range.

ISBN 978-3-7258-1540-1 (Hbk)
ISBN 978-3-7258-1539-5 (PDF)
https://doi.org/10.3390/books978-3-7258-1539-5

Contents

About the Editors

Corneliu Munteanu

Munteanu Corneliu, Prof. PhD.Eng. at "Gh. Asachi" Technical University of Iasi, is 65 years old and has expertise in the characterization and production of Mg-based biodegradable materials and thermal plasma-jet depositions, with a PhD which was awarded on 01.02.1995. He has published over 150 ISI/BDI scientific papers as an author/co-author, of which, 98 have an ISI impact factor (22–Q1 (8–first author); 13–Q2 (4–first author)) and 53 of which are ISI Proceedings and Scopus-indexed articles. He is mentioned in over 1200 ISI/BDI citations (h-index ISI: 17; h-index Scopus: 18). He is the author/co-author of 11 OSIM patents and 10 books in the field of Materials Engineering. As a project manager, he has coordinated 11 national projects and has been a member of 10 other contracts.

Ioan Doroftei

Prof. Doroftei Ioan is the head of the Mechanical Engineering, Mechatronics and Robotics Department, "Gheorghe Asachi" Technical University of Iasi. He has published, as the sole author or a co-author, 14 books and manuals and over 260 articles in national and international journals and conference proceedings. He is the editor of 13 volumes (and is the principal editor for 11 of them), including 3 published by Springer and 6 by other international publishers. He is the author or co-author of 13 patents. He has contributed to solving over 24 national and international scientific research grants (4 as the project manager). He is a reviewer for over 30 international journals.

Preface

This Special Issue, entitled "Advanced Materials for Multifunctional Applications", showcases the most recent advancements in the theoretical and experimental study of microstructural aspects, mechanical properties, and corrosion resistance in different metallic materials under various processing methods. Additionally, it highlights the latest progress in utilizing these materials for diverse applications in the medical field, automotive industry, mechatronics, and robotics. This Special Issue focuses on investigating the changes in the microstructure of modern engineering alloys and how these changes affect their mechanical and corrosion characteristics.

The impact of advanced materials and manufacturing is significantly changing a range of developing businesses. These industries are known for their dedication to innovative solutions, efficient operations, and the production of extraordinary customized goods. In these rapidly growing industries, the combination of advanced materials and manufacturing techniques is driving innovation and enabling the creation of complex, high-quality, and personalized products. These innovations not only improve manufacturing processes but also broaden the range of possibilities in these dynamic and exciting industries.

Recently, there have been notable progressions in the fields of materials and production technology. The use of high-throughput technologies and computers has expedited the screening of new materials that exhibit enhanced performance. The use of novel materials and additive manufacturing technology has expanded the potential for creating complex structures with improved functionality.

This editorial summary succinctly captures the core of each contribution to this Special Issue, emphasizing the collaborative effort to push the limits of different types of materials research for advanced applications. The research presented here not only establishes new treatment techniques and biomaterials but also provides the foundation for future investigations in this evolving subject.

Corneliu Munteanu and Ioan Doroftei

Editors

materials

MDPI

Article

Tribological Behavior and Microstructural Analysis of Atmospheric Plasma Spray Deposited Thin Coatings on Cardan Cross Spindles

Corneliu Munteanu [1,2], Viorel Paleu [1,*], Bogdan Istrate [1], Anişoara Dascălu [1], Cornelia Cîrlan Paleu [1,*], Shubrajit Bhaumik [3] and Ana Diana Ancaş [4,*]

1 Mechanical Engineering, Mechatronics and Robotics Department, Mechanical Engineering Faculty, Gheorghe Asachi Technical University of Iaşi, 63 D. Mangeron Blvd., 700050 Iaşi, Romania; cornelmun@gmail.com (C.M.); bogdan_istrate1@yahoo.com (B.I.); naik_6219@yahoo.com (A.D.)

2 Technical Sciences Academy of Romania, 26 Dacia Blvd., 030167 Bucharest, Romania

3 Wagner High Quality Lubricants, Chennai 603202, India; shubrajit.bhaumik@wagner-german-oil.com

4 Building Services Department, Faculty of Civil Engineering and Building Services, Gheorghe Asachi Technical University of Iaşi, 67 D. Mangeron, 700050 Iași, Romania

* Correspondence: vpaleu@tuiasi.ro (V.P.); cpaleu@yahoo.com (C.C.P.); ancas05@yahoo.com (A.D.A.); Tel.: +40-756-797-147 (V.P.)

Abstract: Cardan joints are used in transmissions between misaligned shafts, as in all-wheel-drive (AWD) cars and railway applications. Their functioning is accompanied by heavy cyclical loads, with the cardan cross spindles subjected to intensive abrasive wear and pitting. In this paper, a solution to the mentioned issue is proposed, thin anti-wear coatings of Metco 32 and Metco 72 metallic powders deposited by atmospheric plasma spray (APS) on cylindrical samples cut from spindles of two cardan crosses made of 40Cr10 and RUL2 steel. The morphological analysis of the coated surfaces was realized by scanning electron microscopy (SEM), and the elemental composition of the tested samples was elaborated by energy-dispersive X-ray spectroscopy (EDS). To investigate the wear resistance of the coated samples in dry and grease-lubricated conditions, tests at constant load and constant speed were carried out using an AMSLER tribometer. The results of greased tests proved that the expulsion of the lubricant from the tribological contact occurred no matter the combination of coated or uncoated samples. During grease-lubricated tests of ten minutes, the least coefficient of friction was measured for uncoated specimens with better surface finishing; but in dry friction tests, the lowest values of the mean friction coefficients were obtained for the Metco 72 coatings. The porous coatings may act as lubricant reservoirs in long-lasting tests, providing a solution to the expulsion phenomenon of the lubricant to the boundary outside the area of the larger-diameter roller.

Keywords: coatings; atmospheric plasma spray (APS); grease lubrication; cardan joint; WC Co; AMSLER machine

Citation: Munteanu, C.; Paleu, V.; Istrate, B.; Dascălu, A.; Cîrlan Paleu, C.; Bhaumik, S.; Ancaş, A.D. Tribological Behavior and Microstructural Analysis of Atmospheric Plasma Spray Deposited Thin Coatings on Cardan Cross Spindles. *Materials* **2021**, *14*, 7322. https://doi.org/10.3390/ma14237322

Academic Editor: Pawel Pawlus

Received: 4 November 2021
Accepted: 29 November 2021
Published: 30 November 2021

1. Introduction

Cardan joints, also known as Hook's joints, are parts of the drivelines equipping all-wheel-drive (AWD) vehicles. They are also employed in railway vehicles, marine propulsion systems, and other industrial applications [1,2]. The cardan is one of the most utilized joints, used anytime there is a need to transmit the power and the motion between misaligned shafts. The major problem of a cardan joint is that it converts the constant input speed into a periodically fluctuating one [3]. To ensure a relative steady angular velocity, cardan joints are coupled in pairs, composing the so-called double cardan joint. In AWD cars, this transmission is subjected to the maximum motor torque, high dynamic loadings [2,4], and high torsional and lateral vibrations [5,6]. There are various AWD driveline configurations, presenting permanently engaged cardan joints or engaged "on-demand", as in off-road hikes. The conditions of off-road paths, usually with a lot of dust, subject cardan joints to intense abrasive wear, especially when the seals are broken and dust enters the greased tribological pairs of the cardan. In addition, cardan joints may suffer intensive wear due to severe running conditions: unfavorable lubrication kinematics, heavy loads, shocks, and abrasive contamination of their greased contacts [1,7].

Cardan joint failure can result in very costly consequences. In automotive applications, sudden interruption of power transmission may result in serious problems and failures in other components of the driveline. Vesali et al. [7] presented the problematics of the spindles of the cardan cross, which fail by pitting due to high contact stress with the needles of the roller bearings equipping the hinges between the cardan yokes and its spider (cross). They proposed some design solutions to alleviate the problem. Stamenić et al. [8] emphasized that the spindle of the cardan cross must carry unfavorable distribution of the linear load. Wang and Qi [9] determined the reaction forces in the rolling bearings of drive systems with double cardan joints, revealing the high variation in these forces over time. These multiple causes reflect some of the reasons for intensified wear and premature failure of cardan joints.

An original idea is to prevent or reduce abrasive wear by coating the parts of cardan joints with anti-wear materials. Thermal spray deposition permits the deposition of various coatings (ceramic, metallic, ceramic-metallic, and polymeric) with a thickness of the deposited layer in the range of nanometers to millimeters [10]. Atmospheric thermal spray (APS) is one of the most productive and versatile thermal methods, allowing the deposition of any type of powder at a high rate, good quality, and low costs. This process is characterized by the fact that the particles composing the spraying beam do not interact with each other, the flux of the melted powder directed to hit and coat the material target on imposed trajectory [11]. Thus, after 5 to 20 successive covering layers deposited on the substrate, a multi-layer coating results, its thickness depending on the number of successive passes of the torch over the target. Kowalski [12] reported results on the wear resistance of diamond-like carbon (DLC) coatings in press joint components. Following the results of fatigue tests in rotational bending conditions of the shaft and microscopic analysis of surfaces, wear products in the form of buildups, abrasions and pits were detected. Cimpoeșu et al. [13] presented results on the corrosion resistance of a thin hydroxyapatite (HA) layer, deposited through electrophoresis on the Ti_4Al_4Zr metallic substrate, concluding that the TiO_2 and HA composed layer provided homogeneous and good corrosion-resistant coating. Bita et al. [14] reported results on the adhesion of bio-ceramic coatings, as HA and bioactive glass, on MgCa bio-implants.

WC–Co-based alloys have a similar composition to the Metco powders used for APS deposition in our work, being very hard and wear resistant. Pogrebnyak et al. [15] developed a coating, based on a mixture of powders, 30–70% of Ni Cr B Si Fe (PG-19N-01) and WC–Co (hard alloy), by high-speed plasma jet with detonation, and the plasma detonation method. The fusion of the coating by plasma jet increased the hardness and the wear resistance of the coating several folds over the simple PG-19N-01 coating. Fernandes et al. [16] studied the effect of composition on carbides formation in the WC–M (M = Fe/Ni/Cr) powders, uniaxially pressed at 190 MPa and treated for one hour at 1400 °C in a vacuum of 20 Pa, emphasizing formed phases, binder composition, and the carbon content. Ndumia et al. [17] realized a review on various deposition methods of Fe-based coatings, underlying that the optimization of the spraying parameters conducts to denser and harder coatings with better wear resistance, but concluded that more studies are required to understand the influence of different elements on the properties of Fe-based coatings at high temperatures. Boas et al. [18] reported results on plasma laser-sputtered WC/Co anti-wear coating and found that the low-power laser keeps the substrate melting to a low degree, increasing the abrasive wear resistance of the coating. Guilemany et al. [19] studied the effect of debris formed by WC–Co coatings in contact with different counterparts (Al_2O_3, steel, Si_3N_4, etc.), showing that the main wear mechanism of WC–Co coatings is direct abrasion of joint surfaces (two-body abrasive wear) and abrasion due to wear debris (three-body abrasive wear), followed by carbide pull out. Bolledu et al. [20] performed a comparative study of nanostructured WC-17 wt.% Co coatings deposited by HVOF and APS with nitrogen and highlighted the better wear resistance of HVOF coatings. Nanoindentation tests on WC–Co coatings deposited by APS using H_2 and He as plasma gases proved that decarburization strengthened the coating, and there was a correlation between macroscopic and nanomechanical behavior of the coatings [21]. Rahbar-kelishami et al. [22] examined the effect of the friction stirring process on the wear of thermally sputtered WC-12% Co coatings, finding that friction stirring increased hardness and wear resistance. Simunovic et al. [23] examined the process of patterning and optimization of thermally sputtered Ni-based self-fluxing alloy coatings, these coatings being close in chemical composition to the soft phase of the Metco 32 coating.

Khanna et al. [24] reported that PM 20 alloy (chromium carbide in Ni-Cr powder), WC/Co in Ni-Cr powder were deposited by APS and laser cladding to form hard and friction-resistant coatings, finding optimal coating hardness for 15–35% WC/Co in Ni-25Cr matrix, and arguing for the superiority of laser-cladded coating over APS. Kazamer et al. [25] studied the effect of blending WC–Co with Mo, obtaining the 80WC-16Co 4Mo powder. The results on microstructure, mechanical and tribological performances of this APS-deposited coating on C67 base steel proved that the new coating is less porous and harder, diminishing by 10% the wear rate in comparison with the WC–Co coating.

Bolelli et al. [26] reported that the resistance to abrasive wear of HVOF made Fe–Cr–Ni–Si–B–C (Colferoloy) coatings is less than that of Ni-based alloys and electroplated chromium. Instead, the addition of 20 wt.% and 40 wt.% of a WC–12 wt.% Co powder in Fe–Cr–Ni–Si–B–C composite coatings, deposited by HVOF and high-velocity air fuel (HVAF), mitigated the abrasive wear rate of Fe-based coatings by one order of magnitude [27].

Recent research conducted by Paleu et al. [28] proved that the APS-deposited coatings of AMDRY 1371 (Mo–NiCrFeBSiC) possess a wear intensity of the same order as Colferoloy–NiCrFeSiBC high-velocity oxygen fuel (HVOF)-deposited coating [26], provided that the thickness of the coating is optimized. Similar results were reported by Paleu et al. [29] for APS-deposited Al_2O_3 $40TiO_2$ (AMDRY 6250) coating.

After a thorough survey of the existing literature, we observed that there is no systematic published research regarding the microstructure and tribological behavior of the coatings deposited on cardan joint parts [30], the outcomes on grease lubrication conditions being a novelty. The current paper presents comparative results on the microstructure and tribological behavior of Metco 32 and Metco 72 coatings, deposited by the APS method on substrates obtained from two cardan spindles.

2. Materials and Methods

2.1. Materials

To obtain wear-resistant coatings, two powders were selected from the Oerlikon - Metco catalogue: Metco 32 and Metco 72. Metco 32 contains a hard phase of 80 wt.%; most of it is composed of tungsten carbide. The high concentration of WC confers to the coating extremely wear resistance, but a porous structure after spraying. The chemical symbol of Metco 32 is 20(Ni 17.5Cr 4Fe 4Si 4B 0.5C) 80(WC 12Co). Indicated in wt.%, in the hard phase of Metco 32, there is 88% WC and 12% Co. The matrix of 20% is made of 70% Ni, 17.5% Cr, 4% Fe, 4% B, 4% Si, and 0.5% C. The morphology of the Metco 32 powder particles is spheroidal, with nominal particle size distribution of $-125 + 45$ μm.

Metco 72 powder is 88WC 12Co, the chemical composition indicated by the catalogue as 81% W (min.), 11.5–13.0% Co, 5.25% C (max.), 1.5% Fe (max.), and other elements maximum 1.0%. Metco 72 particles are angular (blocky), with a nominal range $-45 + 11$ μm. As indicated by the manufacturer, Metco 72 produces very dense and well-bonded coatings, with excellent abrasive wear resistance at lower service temperatures (up to 500 °C).

SEM images for the used powders are shown in Figure 1, along with some measured particle sizes. The SEM images show that the Metco 32 particles are spheroidal, with sizes in the range of 6 μm to 28 μm. Due to the spherical shape, the powder provides improved flowability during spraying and gives the coatings better resistance to abrasion, fretting and erosion. The manufacturer recommends such coatings for wear protection. APS, Combustion Powder Thermospray (CPT) and HVOF are recommended as deposition processes of coatings for automotive, turbomachinery, petrochemical, agricultural and hydroelectric applications.

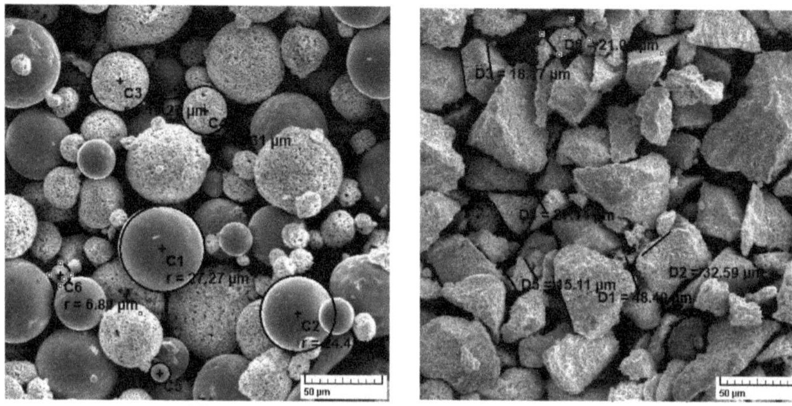

(a) (b)

Figure 1. SEM images of Metco 32 (**a**), and Metco 72 (**b**), 500×.

Metco 72 particles are irregular in shape, obtained by sintering and crushing, with measured sizes ranging from a few microns to a maximum of 50 µm. The material contains fine-grained carbides, also providing abrasion, erosion and fretting wear resistance, but only in dry, non-corrosive environments, our subsequent dry friction results proving this aspect.

The base material was obtained by cutting the spindles of cardan crosses equipping Mercedes Benz transporters and ARO Romanian AWD vehicles, denoted hereafter as Cardan 1 and Cardan 2, respectively. Cardan 1 is made of 40Cr10 steel (SR EN ISO 683-17 standard), containing 98.26% Fe, 0.39% C, 0.21% Si, 0.62% Mn, 0.03% P, 0.04% S, and 0.45% Cr. Cardan 2 is made of 100CrMnSi6-4 (according to SR EN ISO 683-17) or RUL2 steel (according to Romanian standards), with 96.1% Fe, 0.98% C, 0.42% Si, 1.03% Mn, 0.03% P, 0.02% S, and 1.42% Cr.

Substrates for deposition were prepared as follows: samples were cut to size, sand-blasted with casting sand, cleaned of ethyl alcohol and positioned on a support plate for the deposition of coatings.

The obtained testing samples by combining the different cardan steels as base materials and the Metco 32 and 72 coatings are figured in Table 1.

Table 1. Materials and tested samples.

Test Number	Lower Disk	Upper Disk	
		Substrate	Coating
TEST 1	AISI52100	40Cr10	without
TEST 2	AISI52100	RUL2	without
TEST 3	AISI52100	40Cr10	Metco 32
TEST 4	AISI52100	40Cr10	Metco 72
TEST 5	AISI52100	RUL2	Metco 32
TEST 6	AISI52100	RUL2	Metco 72

The employed grease is Mobilux EP2, recommended for antifriction applications (rolling bearings, sleeve bearings, needle roller bearings), meeting the requirements of standard DIN 51825:2004-06-KP 2 K-20. The grease grade is NLGI2, and the thickener type is lithium. The base oil viscosity of the grease @ 40 °C is 160 mm^2/s; penetration, 60×, 0.1 mm, ASTM D217; the Timken OK Load, according to ASTM D2509 is 40 kg; Four-Ball Extreme Pressure Test, Weld Point, kgf, ASTM D2596, is 250 Kgf; and Four-Ball Wear Test, Scar Diameter, mm, ASTM D2266, is 0.4 mm.

2.2. *Methods*

2.2.1. Coating Deposition

The deposition of the coatings by the APS method was realized by Sulzer Metco 9MCE plasma jet equipment, made Sulzer & Metco, Pfäffikon, Switzerland. The process parameters are: spraying distance, 150 mm; current, 500 A; electric voltage, 630 V; argon flow, 50 m^3/h; and hydrogen flow, 70 m^3/h. For each coating, five successive passes of the deposition gun were accomplished.

2.2.2. Microhardness Measurements and Scratch Tests

The Rockwell microhardness and the scratch tests of each coating carried out by using the equipment of the CETR UMT-2 micro-tribometer (Luleå, Sweden). Five similar tests were performed for each coating, and a mean value was obtained. The Rockwell diamond indenter has an opening angle of 120° ± 0.35°, radius 200 ± 10 μm, deviation from profile ±2 μm. The micro-indentation procedure consisted of a progressive increase in the indentation force from 0 to 5 N and back to nil value. The micro scratch tests of the coatings were realized at a constant force of 5 N and constant speed of 10 mm/min. The software equipping the tribometer is the CETR-UMT Test Viewer.

2.2.3. Morphological and Structural Analyses

Scanning electron microscopy (SEM) equipment, model Quanta 200 3D Dual Beam (Waltham, MA, USA) was used to get images of the surface and cross-section of the coatings and substrates. The chemical characterization was made by energy-dispersive X-ray spectroscopy elemental mapping (EDS), model XFlash (Bruker, Billerica, MA, USA), [31].

For the cross-section SEM analysis, the coated specimens were cut and embedded in a BAK-R-type resin produced by Metkon. The samples were ground using seven sand papers of 180 to 1200 grit. The metallographic attack is performed by immersing the surface of the sample to be analyzed in a chemical reagent (Nital reagent, having a concentration of 5% of HNO_3), each phase and constituent reacting differently; therefore, under the microscope, it will be possible to distinguish their shape and distribution as well as their nature.

2.2.4. Surface Topography Measurement

A stylus profilometer model Form Talysurf 50 (made by Taylor Hobson, Leicester, UK), and the μltra Intra Form Talysurf software were employed for roughness measurements. Each measurement on the circumferential direction of the cylindrical samples was repeated three times, and mean values are reported.

2.2.5. Friction and Wear Tests

The AMSLER machine is a bi-disc tribometer, type A135, made by Schaffhausen, Switzerland, the same as fabricated by Alfred J. AMSLER & Co., allowing friction and wear tests for various tribological pair configurations [32].

The cylindrical test specimen, 30 mm in diameter and 10 mm thick, was mounted in the upper position and the lower rotating disc was made of AISI 52,100 bearing steel (hardness 60–64 HRC) with equal radii of 29.5 mm in both circumferential and axial directions and a thickness of 10 mm. The kinematic chains of the AMSLER machine provide a creepage (sliding to rolling ratio) of approximately 10% for equal diameters of the rollers, the ratio of the speeds realized by the upper and lower rollers being 0.906. In our case, the diameter of the upper disc is 30 mm, and the diameter of the lower disc is 59 mm, resulting in a sliding-to-rolling ratio of approximately 74%. The testing machine is fully described in our previous published papers [33–35], including the kinematic drivelines, data acquisition system, data acquisition chain calibration, and description of the main parts. A general view of the AMSLER machine is provided in Figure 2a. Figure 2b presents the tribological arrangement, including the upper coated testing specimen and the lower rolling bearing steel disk. To assure good repeatability, each test was repeated thrice, for both dry and grease-lubricated tests. According to Table 1, a total number of 36 friction tests were carried out for dry and grease-lubricated conditions, that is 18 dry tests and 18 lubricated tests. Mean values of the friction torque and coefficients of friction (CoF) were reported, besides the corresponding confidence limits. All the tests were carried out at a constant speed of 100 rpm, for 600 s each one. For dry friction, a constant load of 20 N was applied by the dead weights method. To obtain observable wear traces in rapid tests, the grease-lubricated tests were realized at a twofold constant load of 40 N. The friction torque was monitored by strain gauges measurements with the aid of a Vishay P3 tensometric bridge. To compute the mean friction torque and mean friction coefficient, and also to apply smoothening filtering to the acquired data and to estimate the statistical parameters (arithmetic mean, standard deviation, and signal to noise ratios), the acquired data were post-processed by the mean of a virtual instrument realized in LabVIEW 7.1 software, made National Instruments [33]. Figure 3 presents an image of the developed LabVIEW interface.

(**a**) (**b**)

Figure 2. General view of the AMSLER machine (**a**), and bi-disc testing arrangement (**b**).

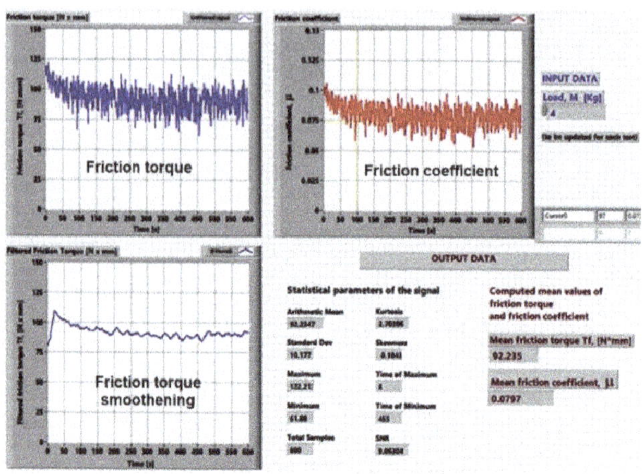

Figure 3. The LabVIEW interface for data post-processing.

3. Results and Discussions

3.1. Hardness and Elasticity Modulus

When the procedure for Rockwell microhardness HR measurements was adopted, it was considered the maximum displacement of the indenter (approximately 15 μm), far below the minimum coating thickness (over 55 μm). The adopted $HR_{0.5}$ procedure consisted of applying a preload of 0.5 N and a continuous increase up to 5 N. The values of the microhardness and elastic modulus were computed from five similar measurements (Figure 4). The results for microhardness and Young modulus are presented in Table 2, with the confidence limits.

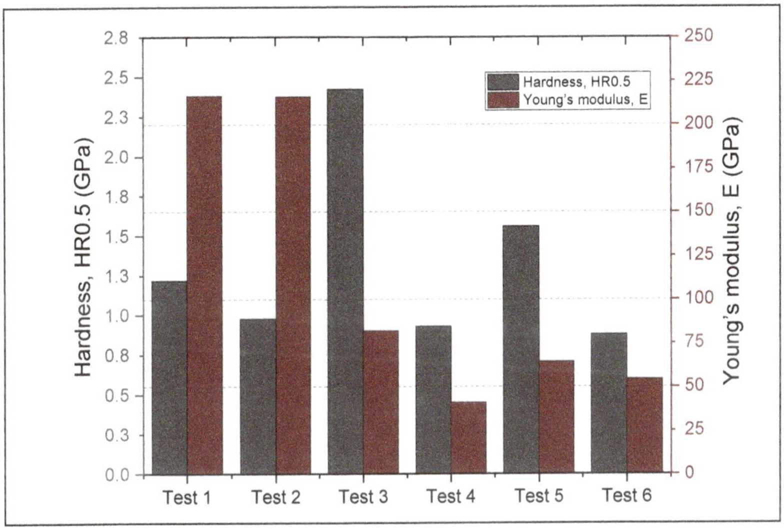

Figure 4. HR0.5 microhardness and Young's modulus.

Table 2. Young's elasticity modulus, E, and Rockwell microhardness HR0.5.

Parameter	Metco 32/40Cr10	Metco 72/40Cr10	Metco 32/RUL2	Metco 72/RUL2	40Cr10	RUL2
Microhardness, $HR_{0.5}$ [GPa] (and confidence limits)	2.42 (−0.37; +0.48)	0.93 (±0.08)	1.56 (−0.28; +0.31)	0.88 (−0.15; +0.3)	1.22 (±0.3)	0.98 (±0.2)
Young modulus, E [GPa] (and confidence limits)	81.8 (−7.51; +11.43)	40.88 (±3.25)	64.21 (−8.68; +4.80)	54.34 (−6.93; +7.18)	216.33 (±5.11)	215.93 (±1.04)
Coating thickness, hc [μm] (and confidence limits)	59.64 (−4.47; +2.52)	91.70 (−5.05; +8.94)	126.47 (−8.74; +9.32)	49.74 (−7.96; +9.52)	0 -	0 -

As can be observed from the measured mean values of the coatings, Metco 32 powder has better affinity for RUL2 substrate, with an average thickness of the coating of 126.47 μm, while Metco 72 proved better for 40Cr10 substrate (91.7 μm of coating). The worst deposition rate was obtained for Metco 72 on RUL2 substrate, the average thickness of the coating being approximately 49.74 μm. As subsequent friction tests will prove, all layers underwent slight plastic deformation and abrasion, with a final steady frictional torque evolution, the SEM analysis proving that no layer was completely removed from the base material.

The best microhardness values were obtained for Metco 32 coatings regardless of the substrate. The confidence limits presented in Table 2 show that the microhardness and Young's modulus values were influenced by the rough and porous structure of the coating. Mrdak et al. [36] found that the microhardness of APS coatings depends on spraying distance. In our research, the spraying distance was kept constant (150 mm). At the microhardness level, each indentation point presents different roughness, porosity and chemical composition, the spread of the indentation results being of the same order as measured by Vashishtha et al. [37] for HVOF coatings of WC-12Co and WC-10Co-4Cr. The corresponding hardness values of the base materials are in good agreement with the existing materials databases [38], with near microhardness and Young modulus values obtained. However, the results of our indentation tests show a slightly better hardness of 40Cr10 steel of 1.22 GPa, while for RUL2 steel it was 0.99 GPa.

3.2. *Surface Topography*

For all the samples used in dry and lubricated tests, the arithmetic mean roughness, Ra, of each sample was measured in the circumferential (rolling) direction. The results of the profilometer measurements are presented in Table 3. For the roughness analysis, the adopted parametric values are: cutoff (Lc) 0.25 mm, cutoff (Ls) 0.008 mm, Gaussian filter, bandwidth 30:1. The lower rotating disc, made of AISI 52,100 steel, employed in friction and wear tests on the AMSLER machine, was polished each time by using sandpaper grit 320. In such a way, a similar roughness of the AISI 52,100 disc was kept constant during all the friction experiments. The obtained circumferential roughness of the lower disc was 1.0 ± 0.1 μm, and the transversal (axial) roughness was 1.2 ± 0.1 μm.

Table 3. Roughness for the batches of tested samples (upper discs), R_a.

Sample	Testing Conditions	
	Dry	Grease Lubricated
	Roughness Ra, [μm]	Roughness Ra, [μm]
40Cr10 (no coating)	1.96 ± 0.11	1.04 ± 0.07
RUL 2 (no coating)	0.37 ± 0.04	0.28 ± 0.03
Metco 32/40Cr10	7.54 ± 0.29	4.61 ± 0.14
Metco 72/40Cr10	5.97 ± 0.26	5.22 ± 0.21
Metco 32/RUL2	5.55 ± 0.24	4.31 ± 0.17
Metco 72/RUL2	4.44 ± 0.19	5.19 ± 0.24

The two batches of tested rollers in dry and grease-lubricated conditions confirmed roughness of the same magnitude, proving the quality of the APS deposition technique on cylindrical rotating surfaces. It must be mentioned that the obtained surface quality was maintained during tests, with no further polishing of the coatings. The surface of the uncoated samples was obtained by turning. As the results of the friction tests will show, the roughness values are greatly influencing the friction torque and the friction coefficient in grease-lubricated conditions. The variation in the roughness values is typical to the nature of the APS deposition process, the successive passes of the deposition gun over the base material conducting to erratic splat-on-splat melted powder structures.

3.3. *Microstructural Characterization of the Coatings by SEM Analysis*

3.3.1. Surface SEM Analysis

Surface SEM images of Metco 32 and Metco 72 coatings are presented in Figure 5. Metco 32 spheroidal particles resulted in a multi-layer splatted-like flakes structure. Correlated with the roughness measurements, one can observe that the resulting roughness of both Metco 32 and Metco 72 is in the range of 4.0 μm up to 8.0 μm, the boron and silica elements from Metco 32 assuring good melting of the powder and the same surface roughness as obtained for the smaller particles with irregular form of Metco 72 powder.

In addition, the coated surfaces possess a morphology characteristic to the APS deposition process, with splats, pores, and rare partially melted powder particles.

3.3.2. Cross-Section SEM Analysis of Metco 32 and Metco 72 Coatings on 40Cr10 Substrate

Figures 6 and 7 present the cross-section SEM images of Metco 32 and Metco 72 powders, deposited by APS on 40Cr10 substrate (Figure 6) and RUL2 substrate (Figure 7). A relatively homogeneous structure is observed for all the coatings, with the presence of partially melted particles and the appearance of a few micro-cracks. These microcracks at the interface were generated due to internal stresses arising from non-uniform shrinkage and high particle temperature during the cooling process. These aspects are characteristic of the APS deposition process. It is obvious that the coatings adhered well to the substrate, but some micropores appeared at the substrate–coating interface. Furthermore, the average thickness of the deposited layers differs from a coating to another and from a substrate to another. The differences in layer thicknesses may result from the different sizes of the constituent particles of the powders and the different affinity of the melted particles to the substrates. Microstructure in Metco 32 coatings shows a spall-like structure evidenced by the formation of Ni-based compounds with a light-colored lamellar appearance.

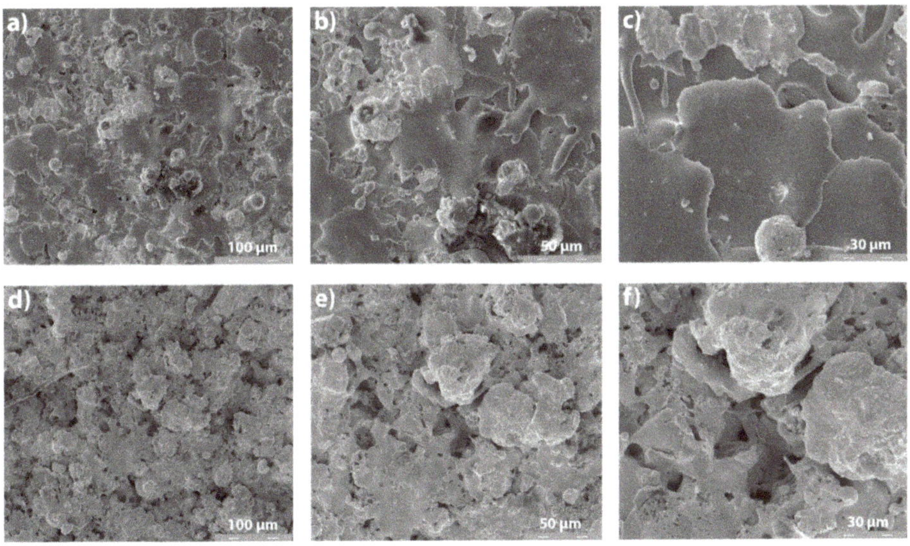

Figure 5. Surface SEM analysis of the coatings: (**a**) Metco 32, 500×; (**b**) Metco 32, 1000×; (**c**) Metco 32, 2000×; (**d**) Metco 72, 500×; (**e**) Metco 72, 1000×; (**f**) Metco 72, 2000×.

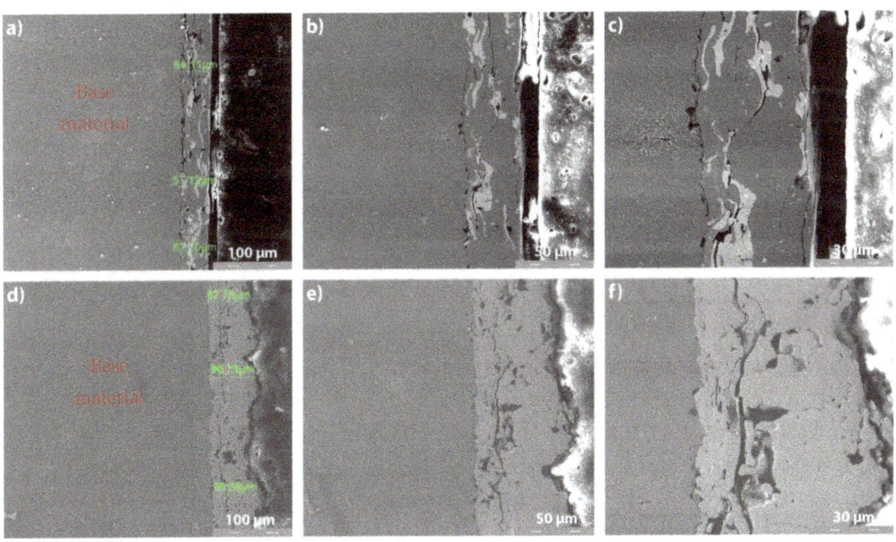

Figure 6. Cross-section SEM analysis of METCO 32 and Metco 72 coatings deposited on 40 Cr10 substrate: (**a**) Metco 32/40Cr10, 500×; (**b**) Metco 32/40Cr10, 1000×; (**c**) Metco 32/40Cr10, 2000×; (**d**) Metco 72/40Cr10, 500×; (**e**) Metco 72/40Cr10, 1000×; (**f**) Metco 72/40Cr10, 2000×.

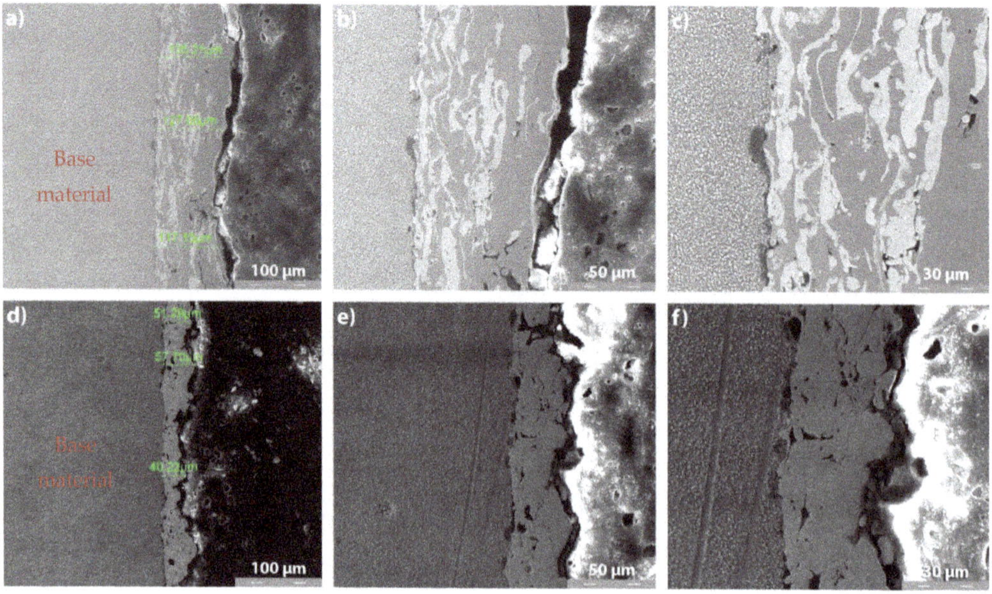

Figure 7. Cross-section SEM analysis of METCO 32 and Metco 72 coatings deposited on RUL2 substrate: (**a**) Metco 32/RUL2, 500×; (**b**) Metco 32/RUL2, 1000×; (**c**) Metco 32/RUL2, 2000×; (**d**) Metco 72/RUL2, 500×; (**e**) Metco 72/RUL2, 1000×; (**f**) Metco 72/RUL2, 2000×.

The base materials have a predominantly pearlite structure, with the occurrence of primary and secondary carbides, carbides that impart hardness and wear resistance. The APS coatings showed uniform, compact layers with the presence of micropores and micro-cracks and some semi-melted particles. Regarding the microstructural analysis of APS-deposited Ni alloys coatings, Zhang et al. [39] underline that the darker color on the SEM images represents unmelted particles and pores (see Figure 6c,f and Figure 7c,f at 2000×). The unmelted particles and partially melted particles are produced due to a low rate of gas flow or high powder feed rate. The presence of close-placed spherical-shape pores into the coating has as the main source of the entrapped gases generated in the spraying process. The intra-lamellar cracks occur within the splats and are formed because of the thermal shrinking of the splats during rapid solidification.

3.4. EDS Analysis

The EDS results of the coatings provide the elemental chemical composition, in weight percent, wt.%. The measurements were obtained in the middle cross-section area of each coating, as presented in Figure 8. A comparison between the composition of the deposited powders (see Section 2.1) and the values obtained by the EDS analysis (Figure 9), shows the tendency of iron (Fe) to diffuse in coatings. Metco 32 powder already contains some percent of Fe into its soft phase (matrix). The EDS spectra of the coatings shown in Figure 9 correspond to the area marked in Figure 8. Figure 10 presents the EDS maps of the distribution of elements in coatings.

In Metco 32 powder, nickel and chromium have the role of sustaining oxidation and corrosion resistance at high temperatures, increasing the hardness of the coating. The low percent elements from the Metco 32 powder, as boron and Si, representing 4.0 wt.% from 20 wt.% of the matrix (soft phase), are not detectable by EDS. Anyway, boron has the role of lowering the melting temperature, helping the formation of hard phases, especially with Ni-Ni_3B [40], and supplementary augmenting the hardness of the coating. During the melting process, this element formed Ni compounds. Silicon is added to increase the self-melting properties of the exposed alloy. The presence of WC–Co in Metco 72 significantly contributes to increased hardness. Powdered tungsten can create so-called "self-bonding" coatings because it belongs to the category of metals with high metallurgical or diffusion bonding capabilities. These diffusion bonds are produced locally at shallow depths of max. 0.5 µm, [41].

3.5. Friction and Wear Results

3.5.1. Friction Results

As specified in Section 2.2.5, the friction characteristics of the Metco 32 and Metco 72 coatings, deposited on both Cardan 1 (40Cr10) and Cardan 2 (RUL2) base materials, were tested on the AMSLER tribometer in dry and grease-lubricated conditions. The mean values of the coefficients of friction (CoF) computed for three identical tests are indicated in Table 4.

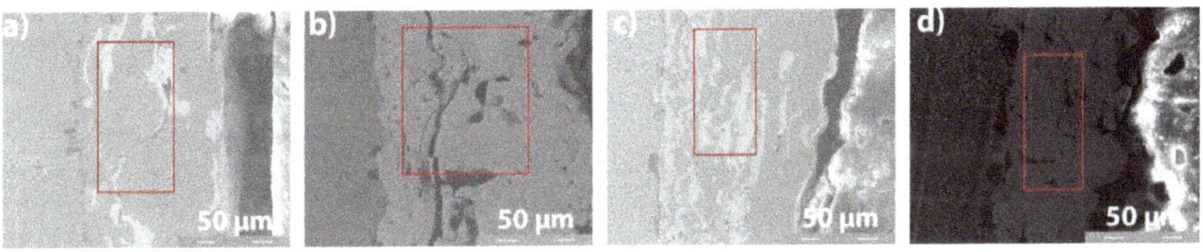

Figure 8. Selected cross-section middle area (2000× SEM images) of coatings: (**a**) Metco 32/40Cr10; (**b**) Metco 72/40Cr10; (**c**) Metco 32/RUL2; (**d**) Metco 72/RUL2.

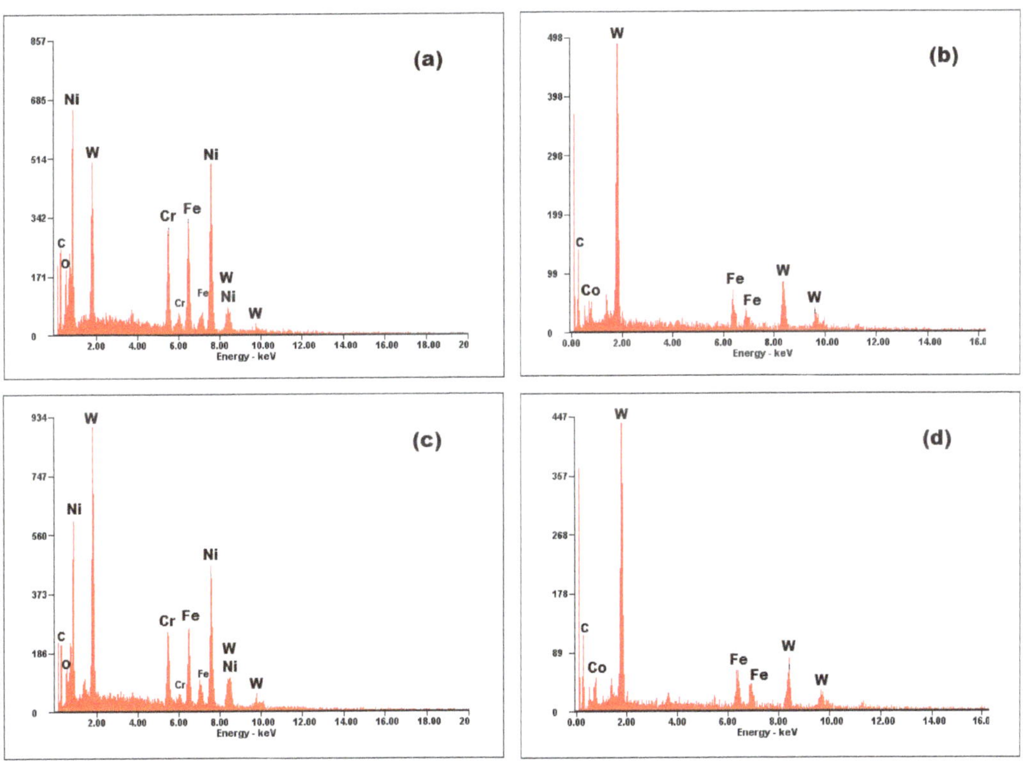

Figure 9. EDS spectra of coatings: (**a**) Metco 32/40Cr10; (**b**) Metco 72/40Cr10; (**c**) Metco 32/RUL2; (**d**) Metco 72/RUL2.

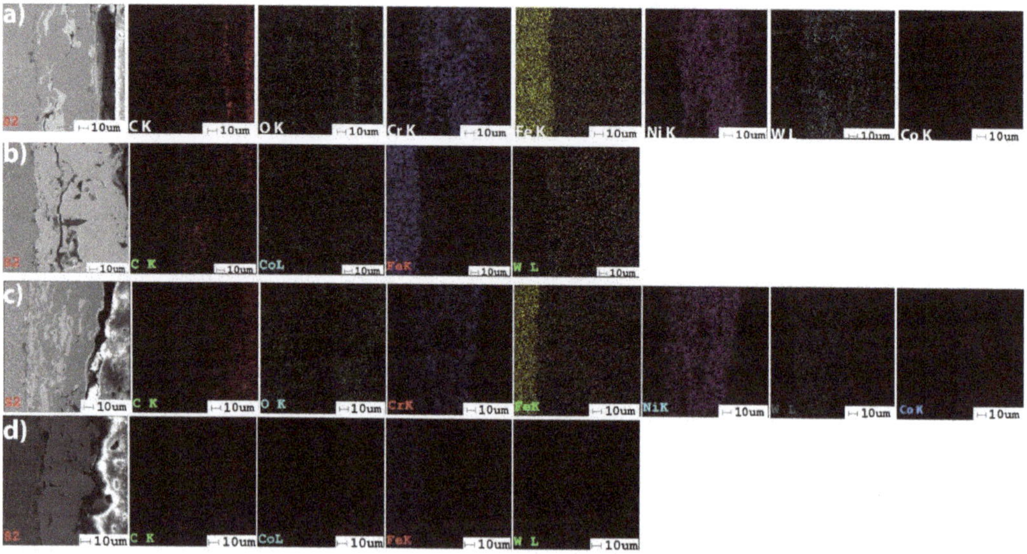

Figure 10. EDS maps of the distribution of elements in coatings: (**a**) Metco 32/40Cr10; (**b**) Metco 72/40Cr10; (**c**) Metco 32/RUL2; (**d**) Metco 72/RUL2.

Table 4. Average coefficient of friction for dry and grease-lubricated tests.

Test No. [1]	Test 1	Test 2	Test 3	Test 4	Test 5	Test 6
Dry CoF (and confidence limits)	0.228 (±0.021)	0.216 (±0.019)	0.256 (±0.027)	0.184 (±0.016)	0.260 (±0.03)	0.182 (±0.019)
Greased CoF (and confidence limits)	0.0556 (±0.0041)	0.0496 (±0.0037)	0.079 (±0.0057)	0.0801 (±0.0062)	0.0797 (±0.0059)	0.0795 (±0.0054)

[1] Test number corresponds to Table 1.

The batches 1 and 2 of testing specimens were tested in dry and grease-lubricated conditions, respectively. The variation in the CoF for both tested batches is represented in Figure 11a.

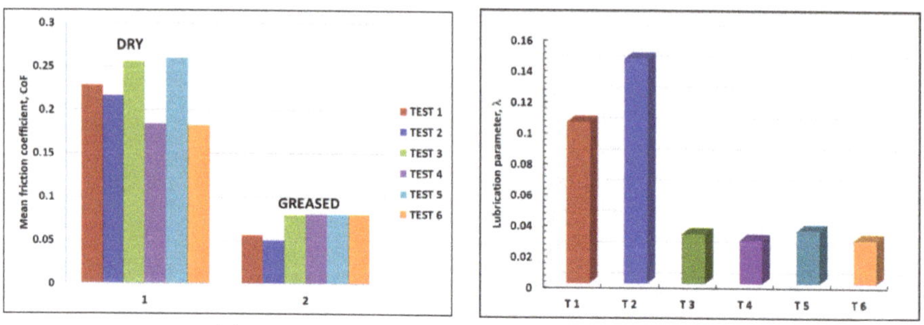

(**a**) (**b**)

Figure 11. (**a**) Coefficient of friction mean values for the two batches of tested specimens, 1 and 2, and various tribological pairs in dry and grease lubrication (Test 1 to Test 6); (**b**) lubrication parameter, λ, in grease-lubricated tests.

It can be seen that even though the load was doubled from 20 to 40 N, the average CoF values decreased by an order of magnitude in the grease tests compared to the dry tests.

For dry and grease-lubricated tests, the corresponding Hertz pressure value, computed according to [28], is 0.574 and 0.723 GPa, respectively. It seems that during rapid tests the lubricant (Mobilux EP2 grease) successfully separated most of the mating surfaces asperity contacts between the lower AISI52100 steel roller and the upper discs testing specimens. The better surface finishing of the uncoated specimens, Test 1 and Test 2, assured lower friction coefficients, as the lubrication regime is dictated by the λ lubrication parameter, given by Equation (1), [42]:

$$\lambda = \frac{h_{min}}{\sqrt{R_{a\_lower}^2 + R_{a\_upper}^2}} \quad (1)$$

where h_{min} is the minimum lubricant film thickness, in μm, computed according to the Hamrock and Dowson formula, presented in [43], and R_{a_lower} and R_{a_upper} are the measured arithmetic mean roughness values of the lower and upper discs, respectively. Neglecting the influence of the coating micrometric layer on the overall Young modulus of the coated discs, the corresponding mean values for the λ lubrication parameter in grease lubrication conditions are indicated in Table 5, taking into account the real roughness values of the tested specimens. It can be seen that the better roughness is very important, as the composite surface roughness term, $\sqrt{R_{a\_lower}^2 + R_{a\_upper}^2}$, plays a crucial role and dictates the type of the lubrication regime.

Table 5. The lubrication parameter, λ, for grease-lubricated tests.

Test No. [1]	Test 1	Test 2	Test 3	Test 4	Test 5	Test 6
λ parameter	0.104	0.145	0.032	0.028	0.034	0.028
Lubrication regime	Mixed			Boundary		

[1] Test number corresponds to Table 1.

The variation in lubrication parameter, λ, in grease-lubricated tests is presented in Figure 11b. The CoF variation should follow the inverse variation in the lubrication parameter in different tests because better lubrication conditions mean lower friction. In conclusion, the higher the λ parameter, the lower the CoF values that must be obtained. The values of the lubrication parameter correspond for uncoated surfaces with better roughness to the mixed lubrication regime, that is preponderant grease separation of the mating roughness of lower and upper discs, with occasional asperity contacts. For the "as obtained" unpolished coatings, the λ parameter indicates a boundary lubrication regime, with preponderant direct contacts between the mating asperities of the rough surfaces and occasional grease separator lubricating film. From this combined theoretical and experimental interpretation of the results, the conclusion is that after the APS deposition process, the roughness of coatings should be reduced by proper polishing to ensure better lubrication and lower friction.

For the dry tests (Figure 11a), the results showed that tests 4 and 6, i.e., Metco 72/40CR10 and Metco 72/RUL2, gave lower CoF values. This can be also explained by the decreased microhardness of Metco 72-deposited coatings obtained by the APS process, which is approximately half of the microhardness for Metco 32 coatings (see Table 2).

3.5.2. Wear Mechanism

As the needle roller bearings equipping the cardan spindles are always grease lubricated and sealed, only the wear SEM images of the lubricated tests are presented herein.

To establish the wear modes of the uncoated and coated samples, the variation in the mean friction torque T (Figure 12) must be correlated with SEM images of the wear paths at different magnifications (Figure 13), mean roughness and hardness values of mating surfaces, and mean CoF values. In Test 1 (uncoated sample of 40Cr10), the friction was intensified at the beginning of the test (Figure 12, Test 1), indicating a running-in period accompanied by intense abrasion and plastic deformation of the roughness (Figure 13a).

Figure 13b shows rare crushed abrasion particles entrapped in the surface of the testing specimen. The evolution of the friction torque in Test 2 (uncoated RUL2 specimen) was constant and the sporadic damages look like abrasion wear (Figure 13c), and small wear particles attached to the surface of the tested sample (Figure 13d). The friction torque in Test 3 (Metco 32/40Cr10) and Test 5 (Metco 32/RUL2) is similar (Figure 12), and also the wear mode is the same. At a lower magnification of 100×, the ceramic coating of Metco 32 looks intact (Figure 13e,i), excepting a few regions with deformed splatted particles (Figure 13f,j). From the SEM images, it is clear that the Metco 32 coating wears by plastic deformation of the top roughness and abrasion. The microtopography of the two Metco 32 coatings is almost the same with near roughness values, but the hardness of Metco 32/40Cr10 (Test 3) is higher than that of Metco 32/RUL2 (Test 5), from here an intensified wear resulted for the sample from Test 5 (Figure 13j). For the Metco 72 coatings, more abrasive damage can be observed for Test 6 (Metco 72/RUL2), Figure 13l, its hardness a little lower than that of the coating from Test 4 (Metco 72/40Cr10), Figure 13h. For this reason, even if the roughness of the two coatings is almost the same, the friction torque is a little intensified at the beginning of Test 6. The running-in period of all the coated samples is the same, approximately 200 s (Figure 12), but for the coating of Metco 72, the friction was more intense at the beginning of the tests.

It should be noted that all coatings remained on the substrate, with only superficial plastic deformation of roughness and mild abrasive wear. In the dry friction tests, the lowest values of the mean friction coefficients were obtained for the Metco 72 coating, but in the lubricated tests, the CoF values of both coatings were almost the same. In any case, wear tests and SEM images proved that Metco 32 performs well with grease lubrication, with mild abrasive wear and no significant damage to the coating surface observed. Considering the hardness of the coatings, which will dictate the wear resistance in the long-lasting functioning of cardan joints, a good coating could be Metco 32 on 40Cr10 substrate, with the highest micro-hardness and low CoF in grease lubrication. Anyway, for the dry running of cardan joint bearings, the best coating is Metco 72 on 40Cr10, with the lowest CoF.

The results of the AMSLER tribometer friction tests should be correlated with the results of the scratch tests performed on the four combinations of coatings and substrates shown in Table 6. Additionally, some significant SEM images of the scratch tests are presented in Figure 14.

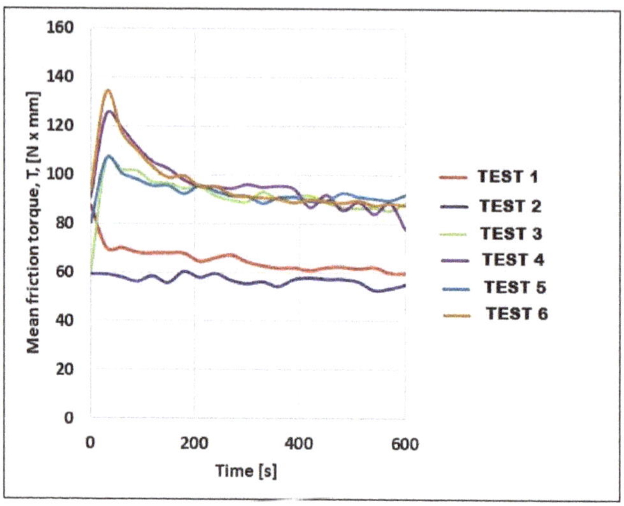

Figure 12. Evolution of mean friction torque in grease-lubricated tests.

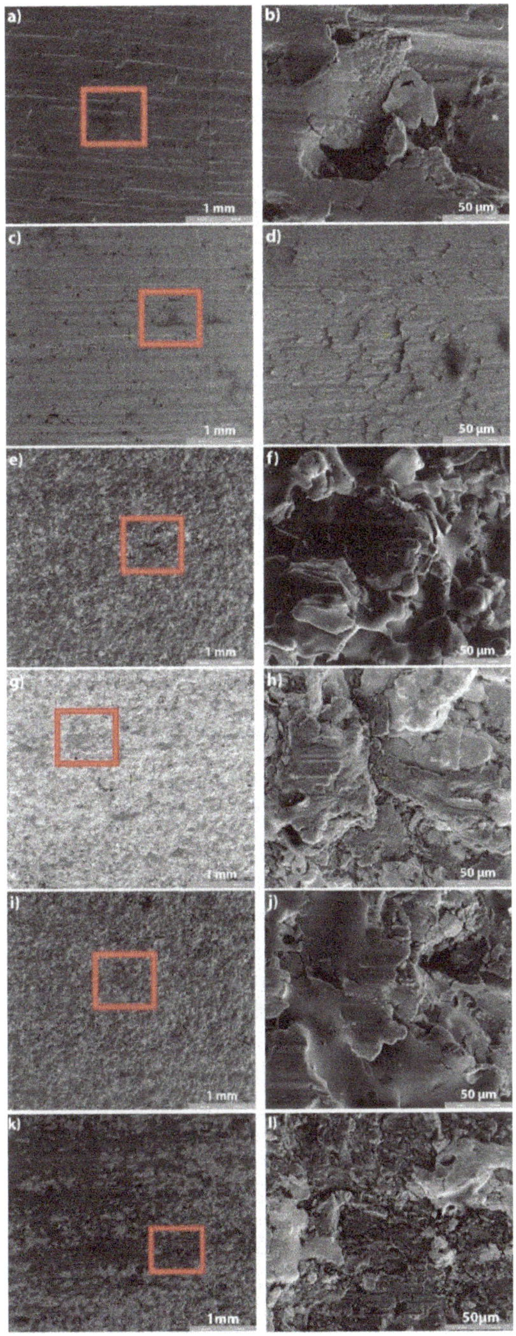

Figure 13. SEM images of tested samples in grease-lubricated environment: (**a**) Test 1 (100×); (**b**) Test 1 (2000×); (**c**) Test 2 (100×); (**d**) Test 2 (2000×); (**e**) Test 3 (100×); (**f**) Test 3 (2000×); (**g**) Test 4 (100 ×); (**h**) Test 4 (2000×); (**i**) Test 5 (100×); (**j**) Test 5 (2000×); (**k**) Test 6 (100×); (**l**) Test 6 (2000×).

Table 6. Average coefficient of friction during the scratch tests.

Coating/ Substrate	Metco 32/ 40Cr10	Metco 72/ 40Cr10	Metco 32/ RUL2	Metco 72/ RUL2
Average dry CoF	0.481	0.550	0.604	0.626
Standard deviation	0.190	0.205	0.240	0.225

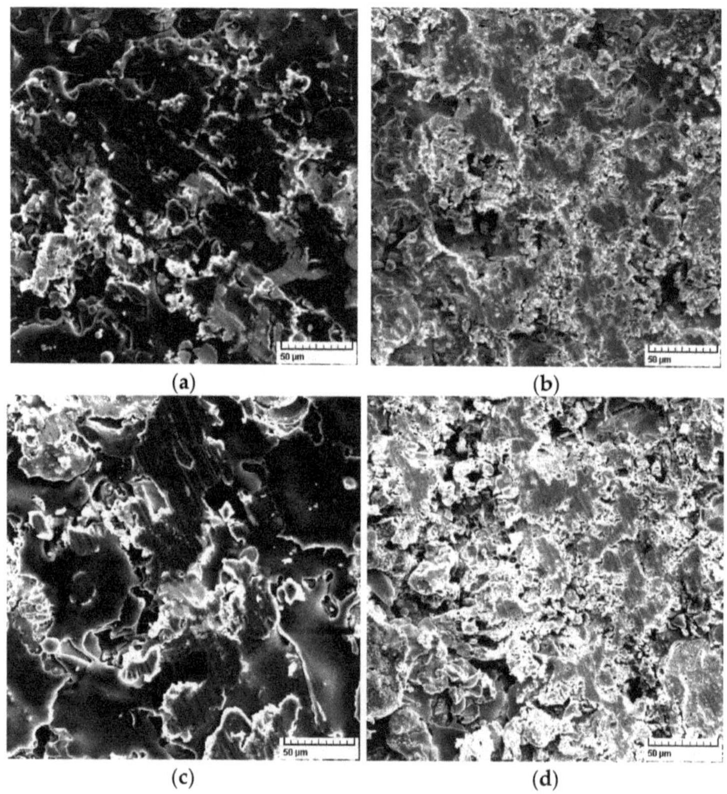

Figure 14. SEM images of scratch tests (500×): (**a**) Metco 32/40Cr10; (**b**) Metco 72/40Cr10; (**c**) Metco 32/RUL2; and (**d**) Metco 72/RUL2.

It can be seen that the average CoF in the scratch tests is higher for the RUL2 substrate regardless of the applied coating, suggesting better adhesion of the coating to the substrate. The best adhesion was found for the Metco 72 layer on the RUL 2 substrate, interestingly with the lowest thickness of all layers (approximately 49 µm). In terms of the wear modes manifested in the scratch tests, the results correlate well with the results of the friction tests (Figure 14). The worn surfaces of the coatings subjected to scratch tests show cut particles. The lower hardness of the Metco 72 coating resulted in an extended worn surface, with the largest worn surface for Metco 72/40Cr10 (Figure 14b). The smallest area with deformed particles was obtained for Metco 32/RUL2 (Figure 14c), the result consistent with those obtained from friction tests.

These new results, correlated with the previous ones obtained from the friction torque evolution during AMSLER tests (Figure 12), recommend Metco 32/RUL 2 (test 5) as the best coating and substrate solution, the friction coefficient evolution during the running-in period being more favorable than that obtained for the Metco 72/RUL 2 coating. In addition, it was found that the microhardness of the Metco 32 coating was higher than that of Metco 72 (see Table 2). There was good correlation between the results of microhardness, friction and scratch tests, and SEM analysis.

However, an unexpected phenomenon of grease expulsion from the contact occurred in all lubricated tests, the grease was trapped on the lower roller of AISI521000, due to its higher tangential speed, and was removed from the contact towards the edge of the lower roller (Figure 15).

Figure 15. Grease expulsion from the contact.

Figure 15 shows that the larger-diameter disc entrains most of the grease and expels it outwards, the contact being lubricated at the end of the tests with a tiny amount of grease. Correlating this information with friction and wear results for both dry and lubricated tests, it results that the coatings ensure better protection to wear than the bare steel 40Cr10 and RUL2. In any case, the expulsion of grease from the contact indicates that, in actual application, the grease is attracted to the spindle of the cardan cross and pushed against the seals, leaving the needle rollers poorly lubricated. This fact explains why the spindles of the cardan crosses fail by abrasive wear and pitting. In needle bearing cardan joint operation, grease will tend to be expelled towards the seals. Any cracking or deterioration of the shaft and bearing seals, for example, due to salts deposited on the road surface in winter, will lead to a change to dry friction.

From the graph of the variation in the average friction torque (Figure 12) under grease lubrication conditions, it can be seen that the roughness of the sample dictates the magnitude of the friction torque. One solution to reduce friction and reduce wear is to polish Metco coatings to a lower roughness, and future tests will focus on this as well.

4. Conclusions

The Metco 32 and Metco 72 coatings on 40Cr10 and RUL 2 substrates were prepared by APS. The indentation tests showed that the microhardness of the Metco 32 coating is higher than that of Metco 72. The surface and cross-section SEM analysis of the coatings revealed their splatted morphology with some partially melted particles, micro-cracks and micropores, specific to the APS process. The deposition parameters in the APS process were kept constant and the number of passes of the deposition gun over the samples was the same; coatings with uniform thickness were obtained, their thickness depending on the substrate and coating type.

Two batches of six samples each, obtained by combining the substrates and the coatings and including also the uncoated specimens, were tested on AMSLER machine in friction and wear tests at a constant speed and constant load for both dry and grease-lubricated conditions, the obtained results correlating with the results of the scratch tests. Following the results of the friction tests and SEM analysis, it was noticed that all coatings remained on the substrate with only superficial plastic roughness deformation and slight abrasive wear. In the dry friction tests, the lowest mean CoF values were obtained for the Metco 72 coating, but in the lubricated tests, the CoF values of both coatings were almost identical irrespective of the substrate.

The average CoF in scratch tests is higher for the RUL2 substrate regardless of the applied coating, suggesting better adhesion of the coating to the substrate. Even if the best adhesion was found for the Metco 72 coating on RUL 2 substrate (test 6), the better hardness of the Metco 32 coatings, the uniform evolution of the CoF during running-in period of the friction tests and the results of the SEM analysis on worn surfaces are better for Metco 32/RUL2 (test 5), recommending Metco 32/RUL2 as the best coating and substrate solution.

In lubricated tests, it was observed that most of the grease is transferred from the smaller-diameter disc to the larger-diameter disc due to the different peripheral velocities, with the lower disc with higher peripheral velocity taking up the grease. In addition, the grease is expelled to the outside of the lower disc, the contact remaining with a minute amount of grease. It is possible that over a longer period, the more porous coatings may form micro-lubricant reservoirs, and supplementary future long-term testing is needed to elucidate this.

Author Contributions: Conceptualization, C.M. and V.P.; methodology, B.I., C.C.P. and S.B.; software, V.P., B.I. and A.D.; validation, C.M., A.D.A. and V.P.; formal analysis, B.I., S.B. and C.C.P.; investigation, C.M., V.P. and B.I.; resources, C.M., A.D.A., A.D., B.I. and C.C.P.; data curation, S.B., B.I., C.C.P., A.D.A., A.D. and V.P.; writing—original draft preparation, C.M., V.P. and B.I.; writing—review and editing, V.P. and B.I.; visualization, A.D.A., C.C.P., A.D. and S.B.; supervision, C.M.; project administration, C.M.; funding acquisition, C.M. and A.D.A. All authors have read and agreed to the published version of the manuscript.

Funding: This research was funded by the Internal Grants—Publications program of the "Gheorghe Asachi" Technical University of Iasi, grant number GI/P8/2021.

Conflicts of Interest: The authors declare no conflict of interest.

References

1. Krause, H.; Hammel, C. The wear behaviour of copper alloy-steel and polyamide-steel sliding pairs for heavily loaded cardan joints. *Wear* **1984**, *93*, 127–143. [CrossRef]
2. Bulut, G.; Parlar, Z. Dynamic stability of a shaft system connected through a Hooke's joint. *Mech. Mach. Theory* **2011**, *46*, 1689–1695. [CrossRef]
3. Chaban, A.; Łukasik, Z.; Popenda, A.; Szafraniec, A. Mathematical Modelling of Transient Processes in an Asynchronous Drive with a Long Shaft Including Cardan Joints. *Energies* **2021**, *14*, 5692. [CrossRef]
4. Bulut, G. Dynamic stability analysis of torsional vibrations of a shaft system connected by a Hooke's joint through a continuous system model. *J. Sound Vib.* **2014**, *333*, 3691–3701. [CrossRef]
5. Xia, Y.; Pang, J.; Yang, L.; Zhao, Q.; Yang, X. Nonlinear numerical and experimental study on the second-order torsional and lateral vibration of driveline system connected by cardan joint. *J. Vib. Control* **2020**, *26*, 540–551. [CrossRef]
6. Song, M.-H.; Nam, T.-K.; Lee, J.-U. Self-Excited Torsional Vibration in the Flexible Coupling of a Marine Propulsion Shafting System Employing Cardan Shafts. *J. Mar. Sci. Eng.* **2020**, *8*, 348. [CrossRef]
7. Vesali, F.; Rezvani, M.A.; Kashfi, M. Dynamics of universal joints, its failures and some propositions for practically improving its performance and life expectancy. *J. Mech. Sci. Technol.* **2012**, *26*, 2439–2449. [CrossRef]
8. Stamenić, Z.; Tasic, M.B.; Mitrović, R. Influence of the Geometry Parameters of Cardan Joint Rolling Parts on the Load Distribution. *FME Trans.* **2012**, *40*, 135–143.
9. Wang, G.; Qi, Z. Approximate determination of the joint reaction forces in the drive system with double universal joints. *Proc. Inst. Mech. Eng. Part C J. Mech. Eng. Sci.* **2017**, *232*, 1191–1207. [CrossRef]
10. Pawlowski, L. *The Science and Engineering of Thermal Spray Coatings*, 2nd ed.; Wiley: New York, NY, USA, 2008. [CrossRef]

11. Niranatlumpong, P.; Koiprasert, H. The effect of Mo content in plasma-sprayed Mo-NiCrBSi coating on the tribological behavior. *Surf. Coat. Technol.* **2010**, *205*, 483–489. [CrossRef]
12. Kowalski, S. Influence of diamond-like carbon coatings on the wear of the press joint components. *Wear* **2021**, *486–487*, 204076. [CrossRef]
13. Cimpoeșu, R.; Vizureanu, P.; Știrbu, I.; Sodor, A.; Zegan, G.; Prelipceanu, M.; Cimpoeșu, N.; Ioanid, N. Corrosion-Resistance Analysis of HA Layer Deposited through Electrophoresis on Ti4Al4Zr Metallic Substrate. *Appl. Sci.* **2021**, *11*, 4198. [CrossRef]
14. Bita, A.I.; Stan, G.E.; Niculescu, M.; Ciuca, I.; Vasile, E.; Antoniac, I. Adhesion evaluation of different bioceramic coatings on Mg-Ca alloys for biomedical applications. *J. Adhes. Sci. Technol.* **2016**, *30*, 1968–1983. [CrossRef]
15. Pogrebnyak, A.D.; Bratushka, S.N.; Ilyashenko, M.V.; Makhmudov, N.A.; Kolisnichenko, O.V.; Tyurin, Y.N.; Uglov, V.V.; Pshik, A.V.; Kaverin, M.V. Tribological and physical-mechanical properties of protective coatings from Ni-Cr-B-Si-Fe/WC-Co-Cr before and after fission with a plasma jet. *J. Frict. Wear* **2011**, *32*, 84–90. [CrossRef]
16. Fernandes, C.M.; Popovich, V.; Matos, M.; Senos, A.M.R.; Vieira, M.T. Carbide phases formed in WC–M (M = Fe/Ni/Cr) systems. *Ceram. Int.* **2009**, *35*, 369–372. [CrossRef]
17. Ndumia, J.N.; Kang, M.; Gbenontin, B.V.; Lin, J.; Nyambura, S.M. A Review on the Wear, Corrosion and High-Temperature Resistant Properties of Wire Arc-Sprayed Fe-Based Coatings. *Nanomaterials* **2021**, *11*, 2527. [CrossRef]
18. Boas, M.; Bamberger, M.; Revez, G. Laser-alloying of a plasma-sprayed WC/Co layer to enhance wear properties. *Surf. Coat. Technol.* **1990**, *42*, 175–186. [CrossRef]
19. Guilemany, J.M.; Miguel, J.M.; Vizcaino, S.; Climent, F. Role of three-body abrasion wear in the sliding wear behaviour of WC–Co coatings obtained by thermal spraying. *Surf. Coat. Technol.* **2001**, *140*, 141–146. [CrossRef]
20. Bolleddu, V.; Racherla, V.; Bandyopadhyay, P.P. Comparative study of air plasma sprayed and high velocity oxy-fuel sprayed nanostructured WC-17wt.%Co coatings. *Int. J. Adv. Manuf. Technol.* **2015**, *84*, 1601–1613. [CrossRef]
21. Rayón, E.; Bonache, V.; Salvador, M.D.; Roa, J.J.; Sánchez, E. Hardness and Young's modulus distributions in atmospheric plasma sprayed WC–Co coatings using nanoindentation. *Surf. Coat. Technol.* **2011**, *205*, 4192–4197. [CrossRef]
22. Rahbar-kelishami, A.; Abdollah-zadeh, A.; Hadavi, M.M.; Banerji, A.; Alpas, A.; Gerlich, A.P. Effects of friction stir processing on wear properties of WC–12%Co sprayed on 52100 steel. *Mater. Des.* **2015**, *86*, 98–104. [CrossRef]
23. Simunovic, K.; Havrlisan, S.; Saric, T.; Vukelic, D. Modeling and Optimization in Investigating Thermally Sprayed Ni-Based Self-Fluxing Alloy Coatings: A Review. *Materials* **2020**, *13*, 4584. [CrossRef] [PubMed]
24. Khanna, A.S.; Kumari, S.; Kanungo, S.; Gasser, A. Hard coatings based on thermal spray and laser cladding. *Int. J. Refract. Met. Hard Mater.* **2009**, *27*, 485–491. [CrossRef]
25. Kazamer, N.; Pascal, D.T.; Marginean, G.; Şerban, V.A.; Codrean, C.; Uţu, I.D. A Comparison between Hardness, Corrosion and Wear Performance of APS Sprayed WC-CoMo and WC-Co Coatings. *Solid State Phenom.* **2016**, *254*, 71–76. [CrossRef]
26. Bolelli, G.; Bonferroni, B.; Laurila, J.; Lusvarghi, L.; Milanti, A.; Niemi, K.; Vuoristo, P. Micromechanical properties and sliding wear behaviour of HVOF-sprayed Fe-based alloy coatings. *Wear* **2012**, *276–277*, 29–47. [CrossRef]
27. Bolelli, G.; Börner, T.; Milanti, A.; Lusvarghi, L.; Laurila, J.; Koivuluoto, H.; Vuoristo, P. Tribological behavior of HVOF- and HVAF-sprayed composite coatings based on Fe-Alloy + WC–12%Co. *Surf. Coat. Technol.* **2014**, *248*, 104–112. [CrossRef]
28. Paleu, C.C.; Munteanu, C.; Istrate, B.; Bhaumik, S.; Vizureanu, P.; Bălţatu, M.S.; Paleu, V. Microstructural Analysis and Tribological Behavior of AMDRY 1371 (Mo–NiCrFeBSiC) Atmospheric Plasma Spray Deposited Thin Coatings. *Coatings* **2020**, *10*, 1186. [CrossRef]
29. Paleu, V.; Cîrlan Paleu, C.; Istrate, B.; Bhaumik, S.; Munteanu, C. Friction and wear resistance of Al_2O_3 $40TiO_2$ (AMDRY 6250) coating of a pump shaft sleeve bearing. In Proceedings of the IOP Conference Series: Materials Science and Engineering, Cluj-Napoca, Romania, 19–21 September 2019; IOP Publishing: Bristol, UK, 2020; Volume 724, p. 012064. [CrossRef]
30. Dascălu, A.; Istrate, B.; Munteanu, C.; Paleu Cîrlan, C.; Paleu, V. Morphological and tribological studies of thermal plasma jet deposited coatings used in cardan joints. In Proceedings of the IOP Conference Series: Materials Science and Engineering, Iași, Romania, 4–5 June 2020; IOP Publishing: Bristol, UK, 2020; Volume 997, p. 012022. [CrossRef]
31. Istrate, B.; Rau, J.V.; Munteanu, C.; Antoniac, I.V.; Saceleanu, V. Properties and in vitro assessment of ZrO2-based coatings obtained by atmospheric plasma jet spraying on biodegradable Mg–Ca and Mg–Ca–Zr alloys. *Ceram. Int.* **2020**, *46*, 15897–15906. [CrossRef]
32. Doxey, R.C. Use of an AMSLER Wear Testing Machine to Investigate the Wear of Steel. Master's Thesis, United States Naval Postgraduate School, Monterey, CA, USA, 1956. Available online: https://core.ac.uk/download/pdf/36708066.pdf (accessed on 27 October 2020).
33. Paleu, V.; Georgescu, S.; Baciu, C.; Istrate, B.; Baciu, E.R. Preliminary experimental research on friction characteristics of a thick gravitational casted babbit layer on steel substrate. In Proceedings of the IOP Conference Series: Materials Science and Engineering, Iasi, Romania, 9–10 June 2016; IOP Publishing: Bristol, UK, 2016; Volume 147, p. 012028. [CrossRef]
34. Paulin, C.; Chicet, D.; Paleu, V.; Benchea, M.; Lupescu, Ş.; Munteanu, C. Dry friction aspects of Ni-based self-fluxing flame sprayed coatings. In Proceedings of the IOP Conference Series: Materials Science and Engineering, Sibiu, Romania, 14–17 June 2017; IOP Publishing: Bristol, UK, 2017; Volume 227, p. 012091. [CrossRef]
35. Paleu, C.C.; Paleu, V.; Istrate, B.; Cimpoesu, N.; Munteanu, C. Thin coatings for pumping station mechanical components. In Proceedings of the IOP Conference Series: Materials Science and Engineering, Iasi, Romania, 19–22 June 2019; IOP Publishing: Bristol, UK, 2019; Volume 591, p. 012007. [CrossRef]

36. Mrdak, M.; Vencl, A.; Ćosić, M. Microstructure and mechanical properties of the Mo–NiCrBSi coating deposited by atmospheric plasma spraying. *FME Trans.* **2009**, *37*, 27–32.
37. Vashishtha, N.; Khatirkar, R.K.; Sapate, S.G. Tribological behaviour of HVOF sprayed WC-12Co, WC-10Co-4Cr and Cr3C2−25NiCr coatings. *Tribol. Int.* **2017**, *105*, 55–68. [CrossRef]
38. Total Materia Database. Available online: https://portal.totalmateria.com (accessed on 29 October 2021).
39. Zhang, X.C.; Xu, B.S.; Xuan, F.Z.; Tu, S.T.; Wang, H.D.; Wu, Y.X. Porosity and effective mechanical properties of plasma-sprayed Ni-based alloy coatings. *Appl. Surf. Sci.* **2009**, *255*, 4362–4371. [CrossRef]
40. Neikov, O.D.; Naboychenko, S.S.; Murashova, I.B. Production of Nickel and Nickel-Alloy Powders. In *Handbook of Non-Ferrous Metal. Powders*, 2nd ed.; Neikov, O.D., Naboychenko, S.S., Dowson, G., Eds.; Elsevier: Amsterdam, The Netherlands, 2019; pp. 633–668. [CrossRef]
41. Longo, F.N. Coating Processing. In *Handbook of Thermal Spray Technology*; Davis, J.R., Ed.; ASM International: Materials Park, OH, USA, 2004; pp. 108–119.
42. Yan, Y. Tribology and tribocorrosion testing and analysis of metallic biomaterials. In *Metals for Biomedical Devices*, 2nd ed.; Niinomi, M., Ed.; Elsevier: Amsterdam, The Netherlands; Woodhead Publishing: Duxford, UK, 2019; pp. 218–234. [CrossRef]
43. Bhaumik, S.; Kamaraj, M.; Paleu, V. Tribological analyses of a new optimized gearbox biodegradable lubricant blended with reduced graphene oxide nanoparticles. *Proc. Inst. Mech. Eng. Part J J. Eng. Tribol.* **2020**, *235*, 901–915. [CrossRef]

materials MDPI

Article

Pulsed TIG Cladding of a Highly Carbon-, Chromium-, Molybdenum-, Niobium-, Tungsten- and Vanadium-Alloyed Flux-Cored Wire Electrode on Duplex Stainless Steel X2CrNiMoN 22-5-3

Daniel Mutașcu [1], Olimpiu Karancsi [2], Ion Mitelea [1], Corneliu Marius Crăciunescu [1], Dragoș Buzdugan [1] and Ion-Dragoș Uțu [1,*]

[1] Department of Materials and Fabrication Engineering, Politehnica University Timisoara, Bulevardul Mihai Viteazul nr.1, 300222 Timisoara, Romania; daniel.mutascu@student.upt.ro (D.M.); ion.mitelea@upt.ro (I.M.); corneliu.craciunescu@upt.ro (C.M.C.); dragos.buzdugan@upt.ro (D.B.)

[2] Department of Oral Implantology and Prosthetic Restorations on Implants, Victor Babeș University of Medicine and Pharmacy Timișoara, Eftimie Murgu Square, No. 2, 300041 Timișoara, Romania; karancsi.olimpiu@umft.ro

* Correspondence: dragos.utu@upt.ro

Abstract: The hardfacing process aims to increase the life span of structural components in the petrochemical, mining, nuclear and automotive industries. During operation, these components are subject to demands of abrasion wear, cavitation erosion and corrosion. Duplex stainless steels are characterized by high mechanical characteristics and corrosion resistance, but poor behavior to abrasive wear and cavitation erosion. The improvement in wear resistance is possible by selecting and depositing a special alloy on the surface using a joining technique that ensures a metallurgical bonding between the layer and the substrate. The experimental investigations carried out in this work demonstrate the ability of the TIG pulsed welding process to produce layers with good functional properties for engineering surfaces. The "Corodur 65" alloy was deposited on a duplex-stainless-steel substrate, X2CrNiMoN22-5-3, using a series of process parameters that allowed for the control of the cooling rate and heat input. The properties of the deposited layers are influenced not only by the chemical composition, but also by the dilution degree value. Since the deposition of layers through the welding operation can be considered as a process with several inputs and outputs, the control of the input parameters in the process aims at finishing the granulation and the structure in the fusion zone as well as limiting the segregation phenomena. The aim of this work is to investigate the microstructural characteristics of the iron-based alloy layer, Corodur 65, deposited via pulsed current TIG welding on duplex X2CrNiMoN22-5-3 stainless-steel substrates.

Keywords: pulsed current TIG welding; Corodur 65 alloy; duplex stainless steel

Citation: Mutașcu, D.; Karancsi, O.; Mitelea, I.; Crăciunescu, C.M.; Buzdugan, D.; Uțu, I.-D. Pulsed TIG Cladding of a Highly Carbon-, Chromium-, Molybdenum-, Niobium-, Tungsten- and Vanadium-Alloyed Flux-Cored Wire Electrode on Duplex Stainless Steel X2CrNiMoN 22-5-3. *Materials* **2023**, *16*, 4557. https://doi.org/10.3390/ma16134557

Academic Editors: Corneliu Munteanu, Ioan Doroftei and Francesco Iacoviello

Received: 24 May 2023
Revised: 21 June 2023
Accepted: 21 June 2023
Published: 24 June 2023

1. Introduction

In most cases, the removal of a mechanical part from operation is mainly caused by the following three phenomena: wear, corrosion and fatigue [1–3]. They are localized in mechanical constructions, in the automobile and aeronautical industry, in the railway and weapon industry, etc. Estimates of the past 20 years made in the USA, Germany, France and Canada show that the direct expenses related to the replacement of parts degraded by these three phenomena represent approx. 5% of the national industries' turnover [3].

Surface layers play an important role in the behavior of mechanical parts; in recent decades, we have been witnesses to a real revolution in the field of surface engineering techniques [4].

To limit the harmful effects of mechanical wear and cavitation erosion, a number of possible solutions were suggested without significant success [5,6]. Thus, some research focused on the deposition of protective coatings using thermal spraying, electric arc welding, PVD and CVD processes, thermochemical treatments, etc. [4–6].

Panadda Niranatlumpong et al. [7] studied the microstructure and wear resistance of NiCrBSi–WC and NiBSi–WC coatings deposited on the 304 SS substrate using electric arc spraying. It was highlighted that the microstructure of the NiCrBSi–WC coating consists predominantly of WC/W_2C phases together with NiCr and NiCrW solid solutions. The microstructure of the NiBSi–WC coating consisted of WC/W_2C phases together with Ni and NiW solid solutions. The absence of hard chromium carbide precipitates contributed to the lower hardness of the NiBSi–WC coating compared to the NiCrBSi–WC coating.

Thermal spraying processes are usually easy to conduct, carried out in a short time and, through the reduced thermal shock, they allow for the obtainment of layers with minor changes in the properties of the substrate. However, these layers show a series of drawbacks related to the presence of porosities, low adhesion to the substrate and inhomogeneous microstructure [8].

On the other hand, the deposition processes via electric arc welding ensure a good adhesion of the layer to the substrate, but the thermal shock induced in the material can cause unfavorable microstructural changes. Additionally, the dilution degree of the filler material with the base material can worsen the performance properties of the deposited layer [9–12].

The hardfacing techniques recommended to be performed using welding are plasma transferred arc welding (PTA), TIG welding, welding in shielding gas atmosphere (SMA) and laser beam welding [13,14]. Generally, Fe-, Ni- and Co-based alloys are selected for hardfacing.

Qian M. and colleagues [15] studied the effects of the powder feed rate (PFR) and translation speed (TS) of a laser beam being investigated by the following three nickel-based laser-clad hardfacing alloys: Colmonoy 6, Colmonoy 88 and AI-1236. The width-to-height ratio of the single clad pass and the dilution rate were found to decrease with the increasing PFR and reducing TS.

Rojacz and colleagues [16] analyzed the microstructural changes caused by the processing of complex-alloyed hardfacings and the resulting wear performance under impact–abrasive conditions. Beyond the chemical composition of welded hardfacings, processing parameters, such as heat input, interpass temperature, etc., have a crucial influence on the final tribological properties of the deposited coatings.

Singh S. and colleagues [17] analyzed the hardness and wear behavior of the surface hardfacing process using EN 353 alloy steel deposited via thermal spraying, TIG welding and electric arc welding. Due to the high cost of remanufacturing engineering components via welding, they recommend the hardfacing process.

The scientific paper [18] found that the toughness of tungsten carbides was a predominant factor in determining the high-stress abrasion resistance of Ni-WC hardfacing coatings. Cemented tungsten carbide and spherical tungsten carbide contribute to improved wear resistance due to their toughness.

The research work [19] recommends the hardfacing process using Ni-Cr-B-Si alloy on 316 LN stainless steel components in sodium-cooled fast reactors to enhance wear resistance and to prevent self-welding. The following three coating processes were comparatively analyzed: plasma transferred arc, TIG arc and laser.

Bembenek M. and colleagues [20] investigated the microstructure and mechanical properties of hardfacing alloys based on the Fe-Mo-B-C system. The obtained results show that the addition of Ti and Mn favor a significant increase in abrasion and impact–abrasion wear resistance by 1.2 and 1.3 times, respectively.

Gurumoorthy et al. [21] analyzed NiCrB alloy deposited on 316 LN SS substrate using the PTA process for nuclear applications. Aging studies were performed at 650 °C for 250 h on the deposited layer to estimate its microstructural stability. It was observed that the microstructure of the deposited layer, after artificial aging at 650 °C for 250 h, consisted of a solid solution γ, with the Ni base having a dendritic shape, and the Cr carbides and borides ensured good wear resistance.

Hemmati et al. [22] explained the role of Fe content in changing the microstructure of Colmonoy 69 layers deposited via laser cladding on S355 steel. It was observed that as the Fe content of the steel substrate increases to 25%, chromium borides with higher Fe fractions are formed. The precipitation of primary chromium borides is suppressed by increasing the Fe content above 40%. The Ni-B-Si eutectics led to a higher dilution. Due to these changes, the hardness decreased from 800 HV to 500 HV.

The present paper analyzes the parameters of the TIG welding deposition process using Corodur 65 electrode wire (EN 14,700 DIN 8555) on X2CrNiMoN22-5-3 duplex-stainless-steel substrates, having potential applications in the repairment of mechanical systems that work under conditions of mechanical wear and cavitation erosion. The parameters of the process are presented together with a diffusion analysis of the alloying elements from the base metal in the deposited metal, the hardness gradient and the macro- and microstructure of the layer–substrate system.

2. Experimental Procedure

The filler metal selected for the experiments was Corodur 65 (Corthal 65), DIN EN 14,700; its chemical composition is shown in Table 1.

Table 1. Chemical composition of Corodur 65.

C%	Si%	Mn %	Cr%	Mo%	Nb%	V%	W%	Fe%
5.08	0.94	0.42	21.1	6.94	6.72	1.02	1.88	Rest

This alloy was chosen because it has special characteristics in terms of high resistance to abrasive wear due to its chemical composition that combines a high content of C with that of Cr, Mo, Nb, W and V, maintaining a lower content of Fe. The base metal was stainless steel X2CrNiMoN22-5-3 (AISI 2205), which is used in the manufacturing of some components from the petrochemical industry, rotors and motor shafts of mining equipment, desalination plants, etc. Its chemical composition is displayed in Table 2.

Table 2. Chemical composition of X2CrNiMoN22-5-3 stainless steel.

C%	Si%	Mn%	Cr%	Ni%	Mo%	P%	S%	N%	Fe%
0.017	0.72	1.80	22.08	5.02	2.90	0.021	0.012	0.16	Rest

The TIG surface modification process provides a quality weld and a good metallurgical bonding between the deposited layer and the substrate. The intense heat of the electric arc progressively melts the base metal and the filler metal (Figure 1).

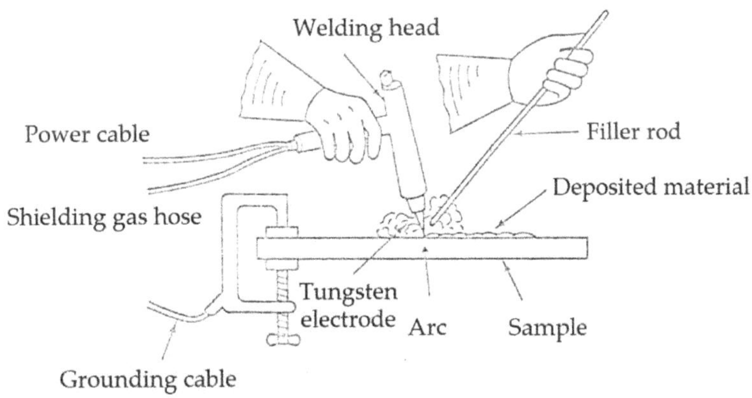

Figure 1. Schematic of the TIG welding process.

The technological variant in the pulsed current involves the rapid variation of the welding current between two levels, the higher current state, known as the pulse current (I_p), and the lower current state, called the base current (I_b), with a certain frequency. During the pulse current period, the welding area is heated and melted, and by lowering to the base current, the stable electric arc is maintained, and the cooling and partial crystallization of the molten metal pool occurs. Figure 2 describes the parameters of the deposition process via pulsed current TIG welding.

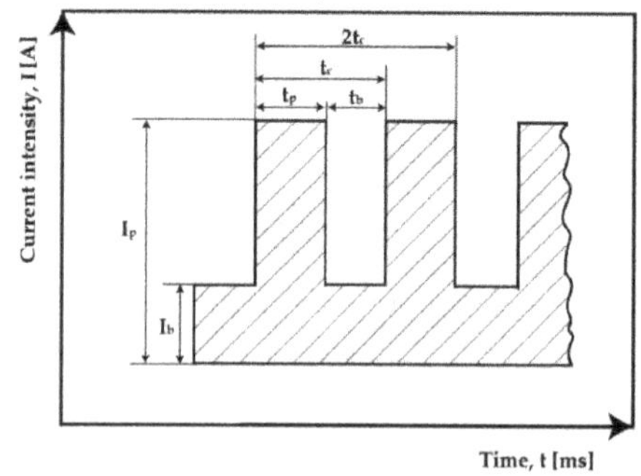

I—current t—time

I_p—pulse current I_b—base current

t_p—pulse time t_b—base time

t_c—cycle time f—pulse frequency ($f = \frac{1}{t_c}$)

$t_a = \frac{t_p}{t_p + t_c} \cdot 100$, % (active pulse time)

Figure 2. The parameters of pulsed current TIG welding process.

By properly adjusting the level of the specific parameters I_p, I_b, t_p and t_b, respectively, and the frequency f, it is possible to precisely control the heat input introduced into the components. At the same average welding current, the seam penetration is greater than in classic TIG welding, and the heat input is lower, which leads to reduced welding stresses and deformations. As general recommendations, the pulse current has values that are 1.5–2 times higher than the classic constant current. The base current is usually taken as 25% of the pulse current, and the pulse time is between 0.02–1 s. depending on the welding technique. In practice, the so-called "one third rule" is frequently used according to which, to obtain an average current in the pulsed current that is equivalent to a constant reference current, the following values for the pulsed current are recommended:

$$I_p = 1.8\, I_{ref.}$$

$$Ib = \frac{I_p}{3} \text{ (one third rule)}$$

$$t_p = 0.2 \ldots 0.3 \text{ s}$$

$$t_b = 2 \cdot t_p$$

In this case, the following is obtained:

$$I_{ref.} = (I_p \cdot t_p + I_b \cdot t_b)/(t_p + t_b)$$

For example,

$$\text{for } I_{ref.} = 50 \text{ A},$$

the results are as follows:

$$I_p = 90 \text{ A}$$

$$Ib = \frac{I_p}{3} = 30 \text{ A}$$

$$t_p = 0.3 \text{ s}$$

$$t_b = 2 \cdot t_p = 0.6 \text{ s}$$

$$I_m = I_{ref.} = (I_p \cdot t_p + I_b \cdot t_b)/(t_p + t_b) = (90 \times 0.3 + 30 \times 0.6)/(0.3 + 0.6) = (27 + 18)/0.9 = 45/0.9 = 50 \text{ A}$$

From the operator mode's point of view, the welding is carried out as follows: the torch is held in place during the pulse time and is withdrawn during the base time, and the filler material is inserted during the pulse time and removed during the base time.

The experimental tests were carried out in the following welding conditions:

- Base metal—X2CrNiMoN22-5-3 (AISI 2205);
- Sample sizes—ø20 × 20 mm;
- Filler material—Corodur 65 (Corthal 65), DIN EN 14700;
- Type of filler material—tubular wire with self-protection;
- Wire diameter—1.6 mm;
- Shielding gas—Ar 100%, I1, EN 14,175 (Ar 4.8—LINDE);
- Shielding gas flow rate—10 l/min;
- Gas nozzle no. 7 (inner diameter 11.2 mm);
- Non-fusible electrode—EWC20 EN 26,848;
- Electrode diameter—2.4 mm;
- How to arrange the passes—on the frontal surface of the sample;
- Pass number—2 per layer, from outside to inside, as shown in Figure 3;
- Number of layers:
    - Sample 1—1 layer;
    - Sample 2—2 layers;
    - Sample 3—3 layers.

It is specified that the selected wire is intended for welding deposition using the process with self-protection tubular wire. From this electrode, rods with dimensions ø1.6 × 200 mm were taken. The characteristic of this wire is that the deposition is free slag, making it suitable for TIG welding.

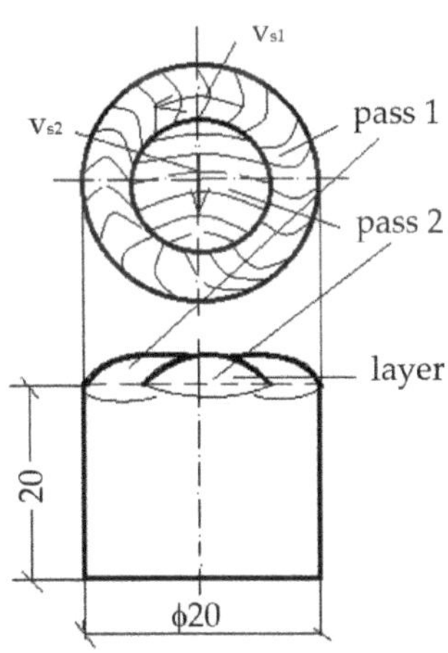

Figure 3. The order of layer deposition.

The process parameters were selected based on the results of several preliminary experiments and data obtained from the specialized literature. They have the following values:

- Impulse current—130 A;
- Base current—45 A;
- Average welding current—49 A; [$I_m = (I_p \cdot t_p + I_b \cdot t_b)/(t_p + t_b)$]—rectangular wave;
- Arc voltage—11 V;
- Pulse time—0.3 s;
- Base time—0.6 s ($t_b = 2\ t_p$);
- Pulse frequency—f = 1 Hz;
- Arc length—l_{arc} = 2.0–2.5 mm;
    - Layer I, pass 1:
        - Welding time—10 s;
        - Weld length—43 mm;
        - Welding speed—26 cm/min;
        - Linear energy—2540 J/cm.
    - Layer I, pass 2:
        - Welding time—5 s;
        - Weld length—10 mm;
        - Welding speed—12 cm/min;
        - Linear energy—5500 J/cm.
    - Layer II, pass 1:
        - Welding time—8 s;
        - Weld length—40 mm;
        - Welding speed—30 cm/min;
        - Linear energy—2200 J/cm.
    - Layer II, pass 2:
        - Welding time—4 s;
        - Weld length—10 mm;
        - Welding speed—15 cm/min;
        - Linear energy—4400 J/cm.
    - Layer 3, pass 1:
        - Welding time—8 s;
        - Weld length—40 mm;
        - Welding speed—30 cm/min;
        - Linear energy—2420 J/cm.
    - Layer III, pass 2:
        - Welding time—3 s;
        - Weld length—10 mm;
        - Welding speed—20 cm/min;
        - Linear energy—3300 J/cm.

Figure 4 shows the settings of the used welding equipment, namely, the source with inverter MW—300 from the Fronius company.

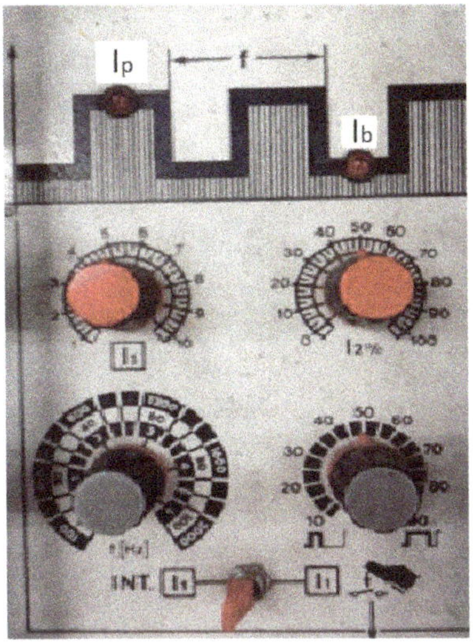

(**a**)

(**b**)

Figure 4. The settings of the used equipment: (**a**) tp = tb = 50% (lower right potentiometer); (**b**) f = 1 Hz.

It can be seen that the welding parameters were kept constant except for the welding speed, which changed in value from one layer to another, respectively, from one pass to another, due to the sample's heating conditions. It is specified that between the layers, each sample was allowed to cool down to a temperature of 100–150 °C in order to limit its overheating and to have a good control over the metal bath.

The choice of the pulse current (I_p = 130 A) was made according to the energy required for the stable melting of the rod, taking into account the chemical composition of the core, characterized by chemical elements with a high melting temperature. This involves the need for a higher heat delivered by the electric arc, specifically, the use of a welding current higher than that claimed by the 1.6 mm rod diameter. The distance between the electrode tip and the samples to be processed was 10 mm.

Although this value is unusual for conventional TIG welding, in the case of making multilayer welds, a previous study [23] showed that this distance ensures the least possible melting of the base metal without compromising the strong bond between it and the deposited metal. The distance from the wire tip to the samples to be processed was set at 3 mm to obtain a good correlation between the geometry of the welded joint and the dilution [24].

The dilution was maintained at acceptable values by changing the volume of the filler material introduced into the molten metal bath, which allowed for the regulation of the bath temperature and, therefore, the dilution with the base metal [25]. For example, when depositing the first layer, approx. 90 mm of filler material was used with 6 current pulses, which corresponds to 15 mm of molten rod at each pulse, which is relatively a lot for a 1.6 mm rod in normal welding conditions.

3. Evaluation of the Experimental Data

3.1. Chemical Heterogeneities and Degree of Dilution

For the chemical and metallographic analyses, the samples were cross sectioned followed by their conventional preparation; thus, after cutting, they were grounded with abrasive paper using silicon carbide and polished with diamond paste, and finally, they were etched via immersion for 30 s in the V_2A reagent (20% hydrogen chloride, 5% nitric acid and 75% distilled water).

The dilution phenomenon is caused by the difference between the melting points of the base metal and the hardfacing alloy. The process of hard loading a substrate is accompanied by the penetration of its component elements into the deposited metal, affecting, to a certain degree, the properties of the hard coating alloy. Consequently, the properties of the alloy are diluted to a certain region at the interface, known as the dilution region. The evaluation of the dilution degree was made based on the chemical composition of the three distinct areas (base metal, deposited metal and filler metal) using an X-ray energy dispersion analysis. The calculation relationship is as follows [3]:

$$G_d = (Fe_{wm} - Fe_{fm})/(Fe_{bm} + Fe_{fm})\cdot 100 = (65.02 - 55.90)/(67.27 + 55.90) \times 100 = 7.4\%$$

where Fe_{wm}, Fe_{fm} and Fe_{bm} represent the iron contents in weight percent of the weld metal, filler metal and base metal.

It is known that in the first moments of its formation, the molten metal bath is the place where important movements occur due to the thermal and dynamic action of the heat source. The superheated liquid undergoes a complex peripheral movement starting from the melting front to the rear area of the bath, through which a matter transfer occurs immediately. The duration of which the atoms are kept in the liquid phase is variable. Some of them are blocked by the solidification front movement, and others reach this front later, depending on the trajectory that is imposed on them. The rest of the atoms leave the environment in which they are via volatilization or chemical reactions. Consequently, the chemical composition of the molten zone varies both continuously and discontinuously.

Its continuous evolution is caused by the following [26]:

- Physical processes—volatilization of alloying elements;
- Chemical reactions between the components of the alloy;
- Reactions between the molten metal and the environment;
- Dilution;
- The fluctuation of the solidification speed.

Its discontinuous evolution is caused by the following [26]:

- Constitutional subordination;
- Dendritic segregation;
- Pollution;
- Irregular gas protection.

These aspects justify the changes in the chemical composition of the deposited metal layers from the Corodur 65 alloy. In Figure 5a, the measurement direction of the main alloying elements' concentrations in the base metal and in the first zone of the dilution region is marked, and in Figure 5b, their linear variations are shown. If, in the base metal, there are limited variations in the concentrations of Fe, Cr, Ni and Mo, in the deposited metal, there are wide variations of Fe (from approx. 40% wt. to approx. 70% wt.), of Cr (from approx. 20% wt. to approx. 54% wt.), of Mo (from approx. 1% wt. to approx. 14.5% wt.) and of Ni (from 1 to 5% wt.) (Figure 5b).

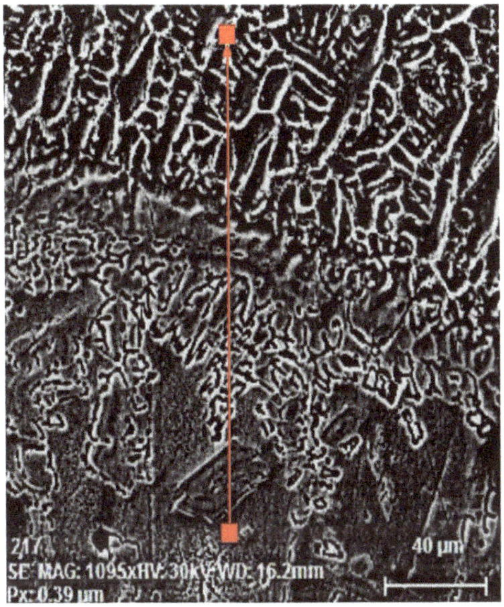

(a)

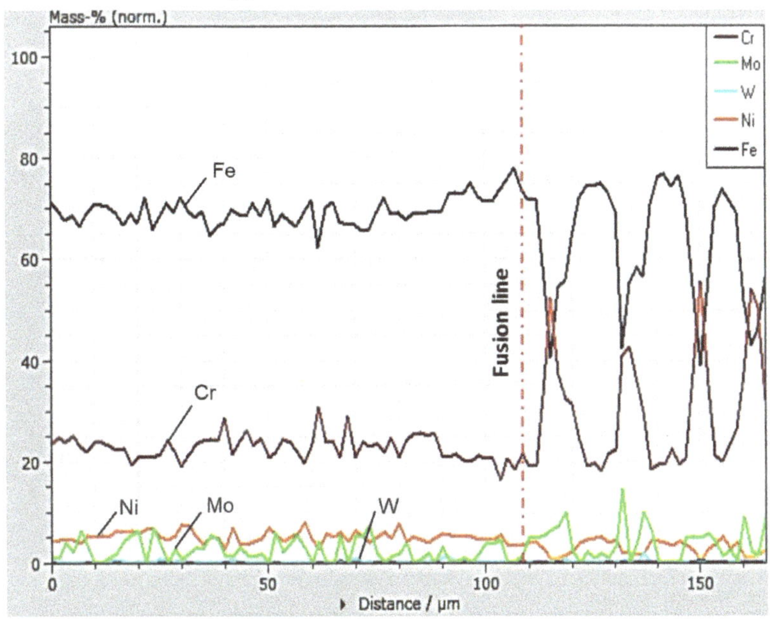

(b)

Figure 5. The linear variation of the concentrations of alloying elements in the region of the base metal–deposited metal interface (layer 1): (**a**) measurement direction; (**b**) evolution curves.

The X-ray energy dispersion analyses carried out on the areas of deposited metal in the immediate vicinity of the fusion line with the base metal (Figure 6) highlight important changes in the chemical composition determined by the volatilization phenomena of chemical elements, by the previously specified chemical reactions, dilution, etc.

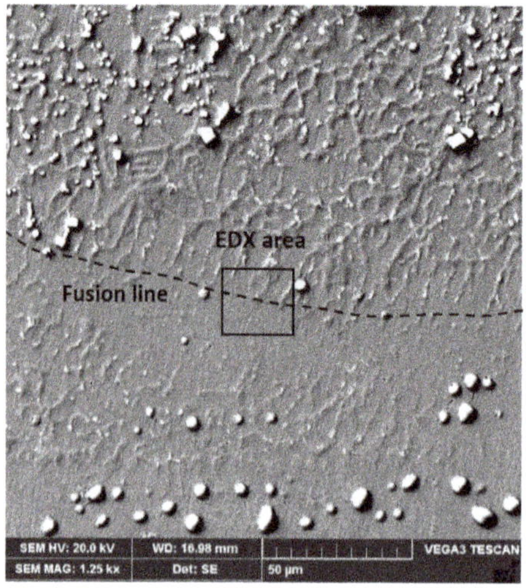

(**a**)

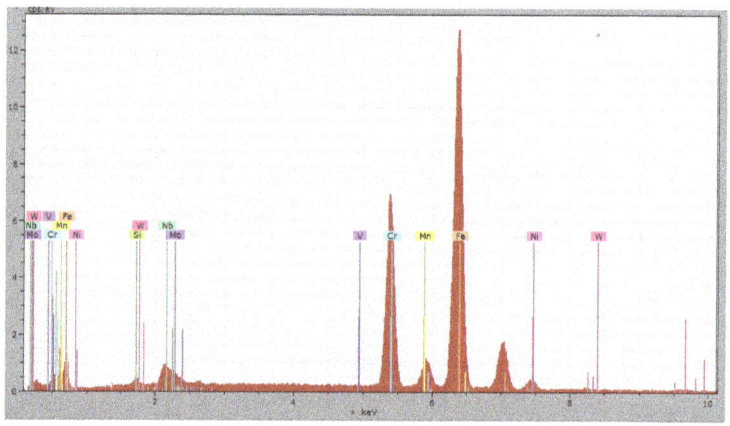

(**b**)

Element	Series	Mass C. [wt.%]	norm. C. [wt.%]	Atom C. [at.%]	Error (1 Sigma) [wt.%]
Si	K series	0.75	0.80	1.60	0.07
Cr	K series	21.01	22.41	24.05	0.58
Mn	K series	0.97	1.03	1.05	0.06
Fe	K series	63.44	67.68	67.61	1.64
Mo	K series	1.75	1.86	1.08	0.24
Ni	K series	2.49	2.66	2.53	0.11
Nb	K series	3.11	3.32	1.99	0.27
V	K series	0.02	0.02	0.02	0.03
W	L series	0.20	0.21	0.06	0.04
	Total	**93.74**	**100.00**	**100.00**	

(**c**)

Figure 6. SEM micrograph (**a**), the energy dispersion spectrum of X-rays (**b**) and the chemical composition of a microregion from the base metal–deposited metal interface (**c**).

During the molten metal bath solidification, the cooling rates specific to TIG welding are so high that they cause a segregation of the alloying elements. Since the alloy deposited via welding has a high content of Cr, which has a strong potential for segregation and is, at the same time, an element that forms complex carbides $(FeCr)_{23}C_6$, then one can expect wide variations in the concentrations of Fe and Cr in the entire volume of deposited metal (Figure 7).

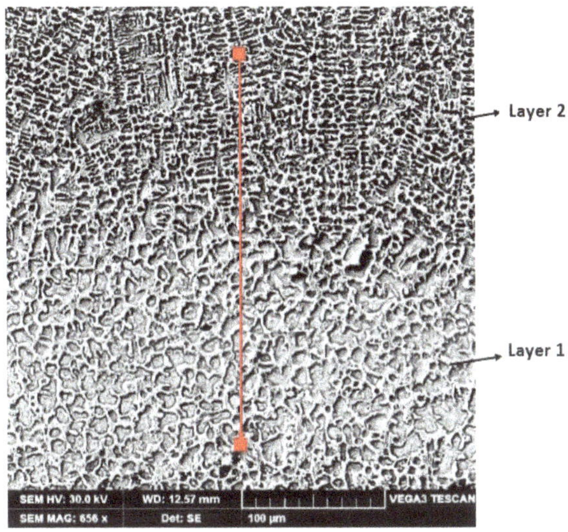

(**a**)

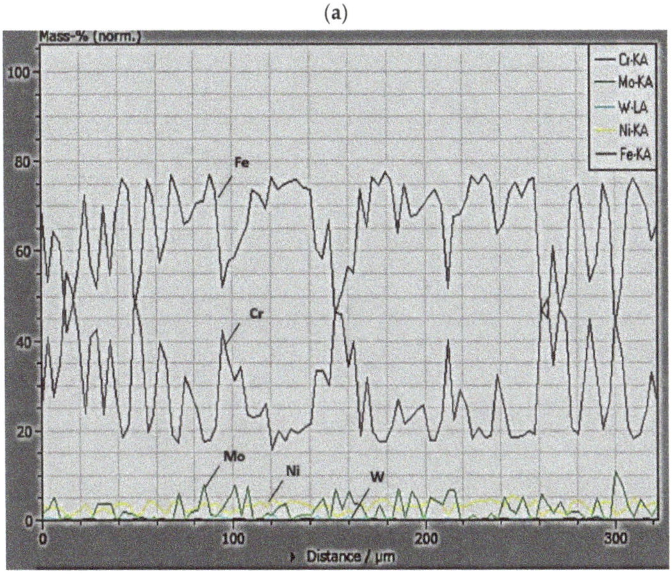

(**b**)

Figure 7. The microstructure (**a**) and the linear variation of the concentrations of alloying elements (**b**) in the region of the interface between layer 1 and layer 2 of the deposited metal.

Figure 8 exemplifies the microstructure and the profile of the chemical composition in a direction transverse to that of cellular and dendritic growth. This shows an increase in the Cr concentration in the intercellular region, and for the same region, there is a depletion of Fe concentration.

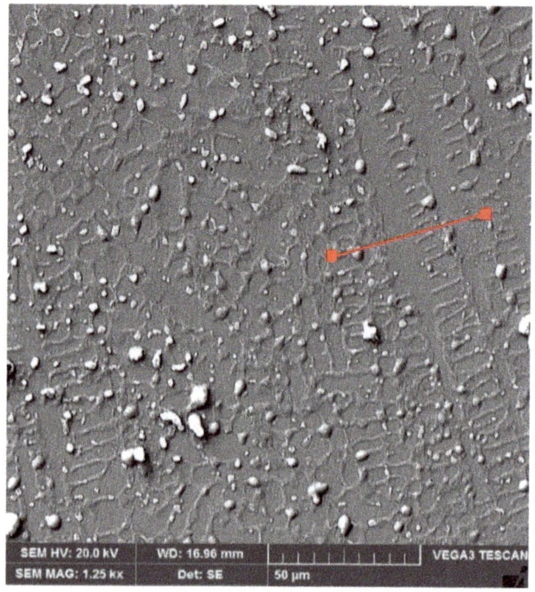

(**a**)

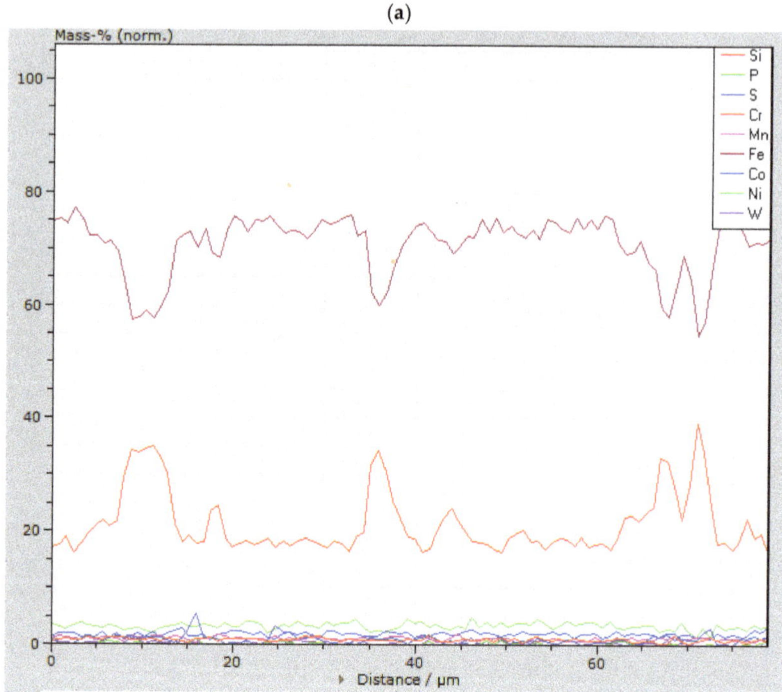

(**b**)

Figure 8. The microstructure (**a**) and the linear variation of the concentrations of alloying elements (**b**) in a cellular–dendritic region.

3.2. The Microstructure of the Layer–Substrate System

A microscopic analysis provides the knowledge that allows for the quality evaluation of the deposited layers. The microstructural examinations were conducted using a Leica DM 2700 M optical microscope and a TESCAN VEGA 3 LMU Bruker scanning electron microscope equipped with an X-ray energy dispersive analysis system (EDX).

Figures 9–13 analyze the structural domains found in the deposited metal (DM), in the base metal (BM) and in the heat-affected influenced zone (HAZ) of the BM. Thus, in the deposited metal appears a microstructure composed of primary and eutectic Cr carbides, plus Nb-Mo-W-V carbides incorporated in an austenite matrix (Figure 9). These results are consistent with the existing data in other works from the specialized literature [27,28].

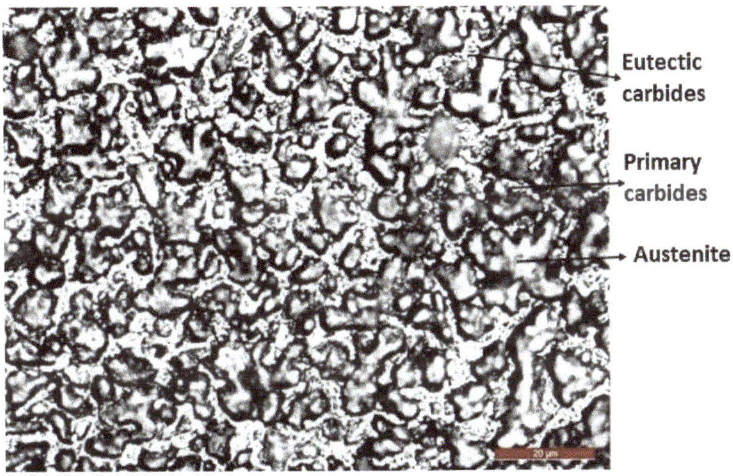

Figure 9. Optical microscopy (×500)—microstructure of the deposited metal.

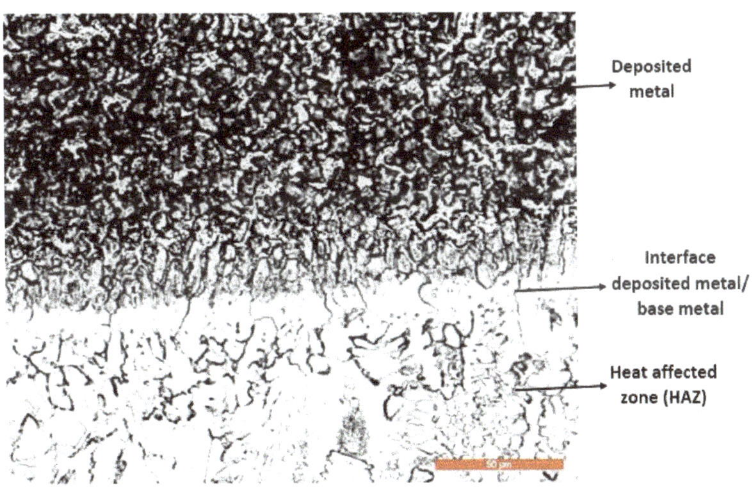

Figure 10. Optical microscopy (×500)—the microstructure of the interface between the deposited metal and the base metal.

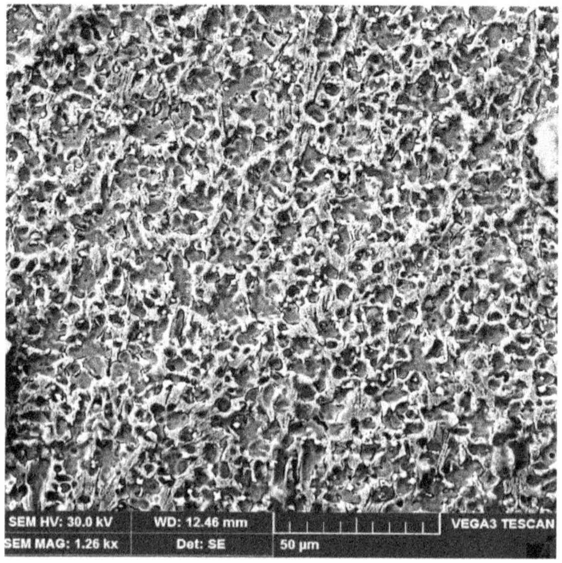

Figure 11. SEM micrograph (×1260) microstructure of the first deposited layer.

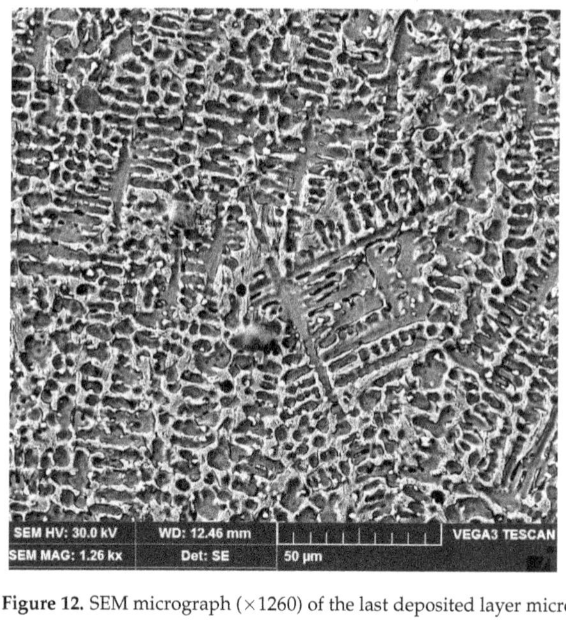

Figure 12. SEM micrograph (×1260) of the last deposited layer microstructure.

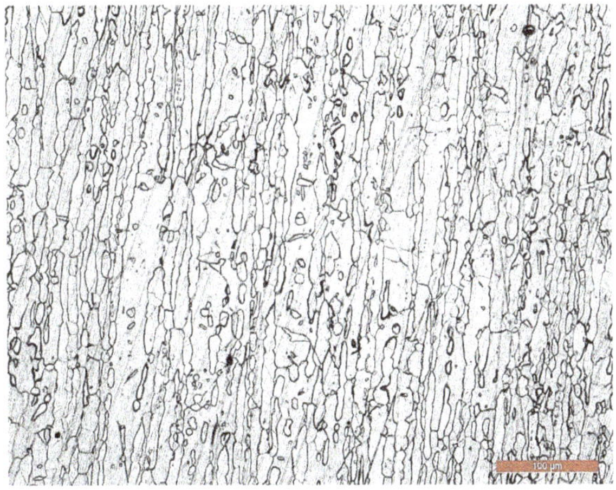

(**a**)

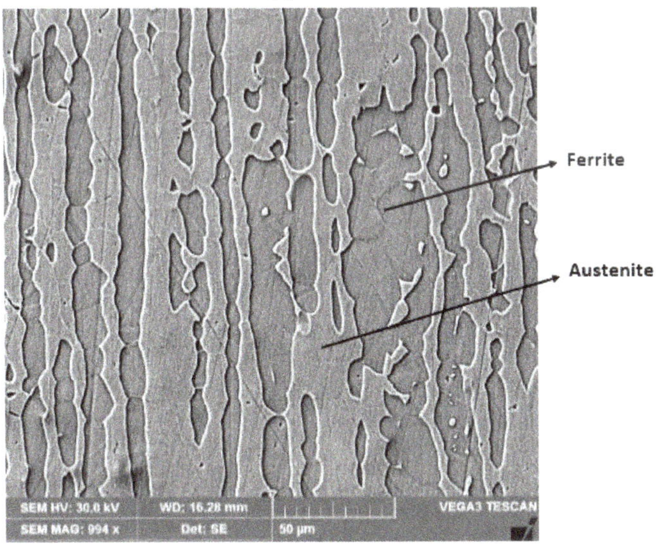

(**b**)

Figure 13. Microstructure of the BM: (**a**)—M.O. × 200; (**b**)—SEM × 1000.

The investigations carried out using the scanning electron microscope highlight the fact that the secondary phases are precipitated both in the intercellular and interdendritic regions. The cell and cell-dendritic structures are located at the bottom of the fusion zone, immediately after the line of its interface with the base metal (Figure 11).

The interface between the first layer of the deposited metal and the heat-affected zone (HAZ) of the base metal contains a strip with a predominantly ferritic microstructure (Figure 10) without the appearance of cracking phenomena or other metallic continuity defects such as the lack of melting, pores, cracks and crevices. It is known that duplex stainless steels can tolerate relatively high linear energies and that their solidification structure is much more resistant to hot cracking than that of austenitic stainless-steel welds. However, if the linear energy is too high, the danger of forming a high proportion of intermetallic phases in the HAZ increases [3]. To avoid problems in the HAZ, the welding process must allow for a high-speed cooling of this region after welding. For these reasons, the maximum temperature between two successive passes was limited to 150 °C, as this temperature has the greatest effect on the HAZ cooling.

The growth of the dendritic zone comprises the largest part of the deposited metal volume. In the upper part of the fusion zone, a dendritic structure was also observed with a certain proportion of secondary phases that precipitated along all the interdendritic boundaries (Figure 12).

The base metal is delivered in the form of bars, sheets, pipes, wires, cast or forged pieces, which are either in a hyper-hardened state (solution treatment), or in a mechanically quenched state. It presents a microstructure consisting of approx. 52% austenite and 48% ferrite (determined using a Fischer ferritoscope) (Figure 13), which is typical for a duplex stainless steel. These results are in concordance with those found in other works from the specialized literature [3].

3.3. X-ray Diffraction Analyses

X-ray diffraction investigations were used to characterize secondary phases precipitated during solidification and/or solid-state transformations. The equipment used is the X Panalytical XPert Pro MPD spectrometer, and the X'Pert HighScore software and its database were used for post-processing and interference peak indexing. The scan details applied in these analyses were a 2Theta range from 20 to 100 degrees, with a step size of approximately 0.02 degrees, and counting at each step for 4 s. Cu Ka radiation was used, and the X-ray power generator was set at 45 kV and 30 mA.

Figures 14–16 exemplify the diffractograms of the deposited metal, the heat-affected zone of the base metal and the unaffected base metal via the heat welding cycle. Their analysis confirms the results of the microstructural examinations, highlighting the following aspects:

- The structural phases detected in the deposited metal are the particles of complex carbides dispersed in the solid solution γ with the fcc crystalline lattice.
- The high temperatures reached in the area adjacent to the fusion line promote, in the HAZ of the base metal, a predominantly ferritic microstructure with a reduced proportion of austenite resulting from the transformation into a solid state, initiated in the cooling stage of the deposited metal bath.
- The base metal in the solution treatment state has a microstructure made up of solid solutions, ferrite and austenite.

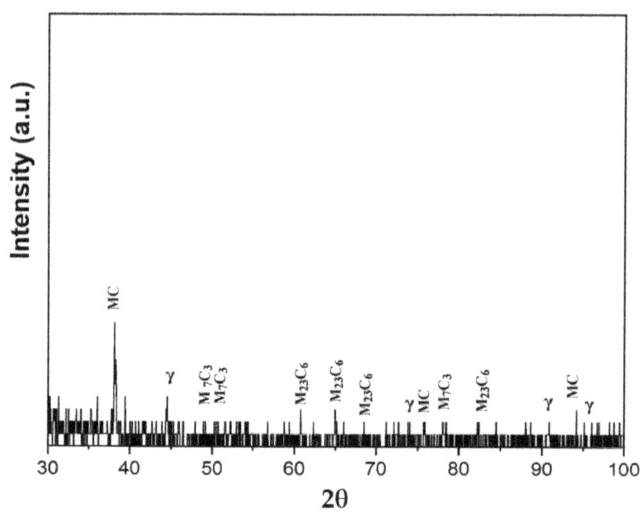

Figure 14. XRD pattern of the deposited metal Corodur 65.

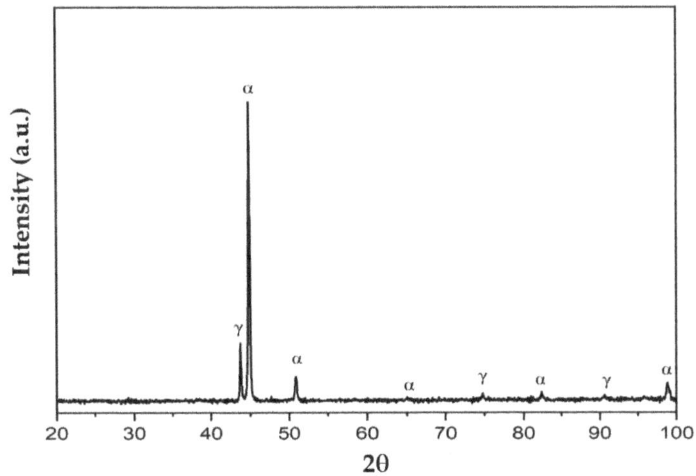

Figure 15. XRD pattern for the HAZ of the BM X2CrNiMo22-5-3.

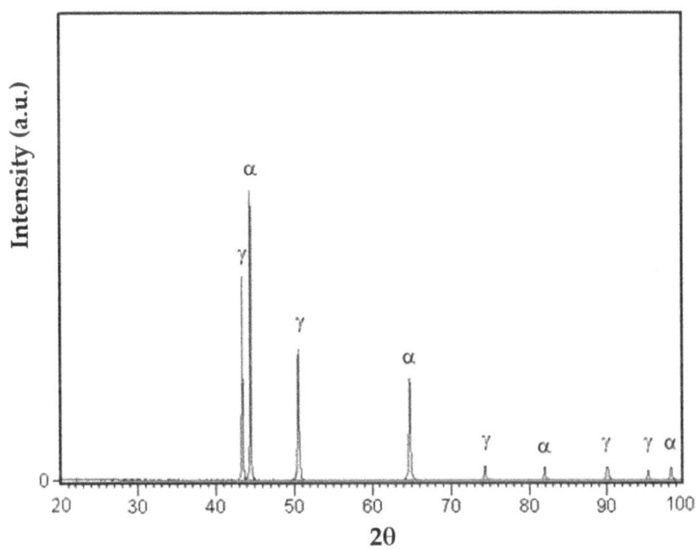

Figure 16. XRD pattern of the base metal X2CrNiMo22-5-3.

3.4. Hardness Measurements

Since hardness is the most sensitive mechanical property to the structural changes in a metallic material, the samples coated with the Corodur 65 alloy and those from the solution-treated base metal were subjected to such examinations. On the front surface of these samples, seven Vickers hardness measurements were made with a load of 50 N/mm^2 using the Zwick/Roell (YHV-S, Ulm, Germany) device. Based on the obtained values, the histogram shown in Figure 17 was drawn. The presented data demonstrate that there is a full agreement between the hardness and the microstructure of the layer–substrate system. The lowest hardness values are specific to duplex stainless steel (274–282 HV5). Instead, the hardfacing process using the Corodur 65 alloy ensures a high hardness (798–825 HV5), which favors a decrease in the wear rate. The high hardness values are attributed to the complex carbide particles in the layer deposited on the surface.

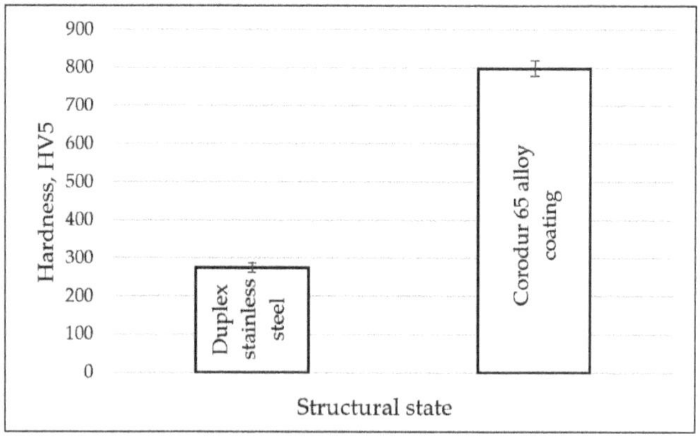

Figure 17. Hardness measurements of the investigated materials.

4. Conclusions

Based on the results obtained in this study regarding the microstructural characteristics of the Corodur 65 alloy layers deposited via TIG pulsed current welding on a duplex-stainless-steel substrate, X2CrNiMoN 22-5-3, the following conclusions are made:

- The most important process parameter is the welding current, as it influences the linear energy and the melting of the materials selected for the layer and for the substrate.
- The base metal was covered with three layers of the Corodur alloy—single, double and triple layer. The obtained hardness is remarkable, at approx. 810 HV5. Due to the low degree of dilution, there were no differences in the hardness between the deposited metal layers.
- The hardfacing process promotes, in the deposited layer, precipitation reactions of complex carbides finely dispersed in a tenacious austenite matrix.
- The precipitation of the carbide phases takes place both in the intercellular and interdendritic regions from the fusion zone located in close proximity to the base metal.
- The profile of the chemical composition in a direction transverse to that of cellular and dendritic growth indicates an increase in the Cr concentration and a decrease in the Fe concentration in the intercellular region.

Author Contributions: Conceptualization, I.M. and I.-D.U.; methodology, I.M., O.K., D.M., D.B., I.-D.U. and C.M.C.; formal analysis, I.M., C.M.C. and I.-D.U.; investigation, I.M., D.M., O.K., D.B., I.-D.U. and C.M.C.; resources, I.M., I.-D.U. and C.M.C.; data curation, D.M. and I.-D.U.; writing—original draft preparation, I.M.; writing—review and editing, I.M. and I.-D.U.; visualization, I.M., O.K., D.M., D.B., I.-D.U. and C.M.C.; supervision, I.M. All authors have read and agreed to the published version of the manuscript.

Funding: This research received no external funding.

Institutional Review Board Statement: Not applicable.

Informed Consent Statement: Not applicable.

Data Availability Statement: Not applicable.

Conflicts of Interest: The authors declare no conflict of interest.

References

1. Madadi, F.; Ashrafizadeh, F.; Shamanian, M. Optimization of pulsed TIG cladding process for stellite alloy on carbon steel using RSM F. *J. Alloys Compd.* **2012**, *510*, 71–77. [CrossRef]
2. Rajeev, G.P.; Kamaraj, M.; Srinivasa, R. Bakshi. Hardfacing of AISI H13 tool steel with Stellite 21 alloy using cold metal transfer welding process. *Surf. Coat. Technol.* **2017**, *326*, 63–71.

3. Messler, R.W., Jr. *Principles of Welding—Processes, Physics, Chemistry, and Metallurgy*; John Wiley & Sons.: Hoboken, NJ, USA, 2004; pp. 49–154.
4. Dobrzański, L.A.; Mikuła, J. Structure and properties of PVD and CVD coated Al_2O_3 + TiC mixed oxide tool ceramics for dry on high speed cutting processes. *J. Mater. Process. Technol.* **2005**, *15*, 822–831. [CrossRef]
5. Koseki, S.; Inoue, K.; Morito, S.; Ohba, T.; Usuki, H. Comparison of TiN-coated tools using CVD and PVD processes during continuous cutting of Ni-based superalloys. *Surf. Coat. Technol.* **2015**, *283*, 353–363. [CrossRef]
6. Thakur, A.; Gangopadhyay, S. Influence of tribological properties on the performance of uncoated, CVD and PVD coated tools in machining of Incoloy 825. *Tribol. Int.* **2016**, *102*, 198–212. [CrossRef]
7. Niranatlumpong, P.; Koiprasert, H. Phase transformation of NiCrBSi–WC and NiBSi–WC arc sprayed coatings. *Surf. Coat. Technol.* **2011**, *206*, 440–445. [CrossRef]
8. Daram, P.; Banjongprasert, C. The influence of post treatments on the microstructure and corrosion behavior of thermally sprayed NiCrMoAl alloy coating. *Surf. Coat. Technol.* **2020**, *384*, 125166. [CrossRef]
9. Suraj, R. Hardfacing and its effect on wear and corrossion performance of various ferrous welded mild steels. *Mater. Today Proc.* **2021**, *42*, 842–850. [CrossRef]
10. Özen, F.; Onar, V.; Çil, G.; Gel, M. Wear resistance and microstructural evaluation of a hardfacing welded S355J2 steel pipe piles. *Mater. Test.* **2022**, *64*, 800–808. [CrossRef]
11. Sönmez, M.; Topuz, P. Effect of TIG welding parameters in joining grade 2 pure titanium. *Mater. Test.* **2021**, *63*, 420–425. [CrossRef]
12. Di Schino, A.; Testani, C. Corrosion behavior and mechanical properties of AISI 316 stainless steel clad Q235 plate. *Metals* **2020**, *10*, 552. [CrossRef]
13. Balaguru, S.; Vela, M.; Chellapandi, P. Measurement of the residual stresses and investigation of their effects on a hardfaced grid plate due to thermal cycling in a pool type sodium-cooled fast reactor. *Sci. Technol. Nucl. Install.* **2016**, *2016*, 1–8. [CrossRef]
14. Xi, B.; Liu, B.; Li, S.; Wang, D.; Zhang, Y.; Szakálos, P.; Ejenstam, J.; Wallenius, J.; He, G.; Zhang, W. Influence of TIG and Laser Welding Processes of Fe-10Cr-4Al-RE Alloy Cracks Overlayed on 316L Steel Plate. *Materials* **2022**, *15*, 3541. [CrossRef] [PubMed]
15. Qian, M.; Lim, L.C.; Chen, Z.D. Laser Cladding of nickel-based hardfacing alloys. *Surf. Coat. Technol.* **1998**, *106*, 174–182.
16. Rojacz, H.; Katsic, C.; Kirchgaßner, M.; Kirchmayer, R.; Badisch, E. Impact-abrasive wear of martensitic steels and complex iron-based hardfacing alloys. *Wear* **2022**, *492–493*, 204183. [CrossRef]
17. Singh, S.; Kumar, R.; Goel, P.; Singh, H. Analysis of wear and hardness during surface hardfacing of alloy steel by thermal spraying, electric arc and TIG welding. *Mater. Proc.* **2022**, *50*, 1599–1605. [CrossRef]
18. Liu, W.; Gao, D. Microstructure and wear of Ni-WC hardfacing used for steel-body PDC bits. *Int. J. Refract. Met. Hard Mater.* **2021**, *101*, 105683. [CrossRef]
19. Balaguru, S.; Mohammad, A.; Gupta, M. Investigations on different hardfacing processes for High temperature applications of Ni-Cr-B-Si alloy hardfaced on austenitic stainless steel components. *J. Mater. Res. Technol.* **2020**, *9*, 10062–10072. [CrossRef]
20. Bembenek, M.; Prysyazhnyuk, P.; Shihab, T.; Machnik, R.; Ivanov, O.; Ropyak, L. Microstructure and Wear Characterization of the Fe-Mo-B-C-Based Hardfacing Alloys Deposited by Flux-Cored Arc Welding. *Materials* **2022**, *15*, 5074. [CrossRef]
21. Gurumoorthy, K.; Kamaraj, M.; Prasad Rao, K.; Sambasiva Rao, A.; Venugopal, S. Microstructural aspects of plasma transferred arc surfaced Ni-based hardfacing alloy. *Mater. Sciece Eng. A* **2007**, *456*, 11–19. [CrossRef]
22. Hemmati, I.; Ocelik, V.; De Hosson, J.T.M. Dilution effects in laser cladding of Ni-Cr-B-Si-C hardfacing alloys. *Mater. Lett.* **2012**, *84*, 69–72. [CrossRef]
23. Silva, C.C.; Miranda, E.C.; Motta, M.F.; Miranda, H.C.; Farias, J.P. Influence of arc length on dilution and weld bead geometry of Ni-based alloy using GTAW process with cold wire feed. In *Proceedings, 20th International Congress of Mechanical Engineering*; Brazilian Society of Mechanical Science and Engineering: Gramado, Brazil, 2009.
24. Miranda, E.C.; Silva, C.C.; Miranda, H.C.; Motta, M.F.; Farias, J.P. Adjustment of the wire feeder parameters in GTAW process for coatings applications. In *Proceeding, XXXV CONSOLDA—National Welding Congress*; Brazilian Welding Society: São Paulo, Brazil, 2009.
25. Anik, S.; Dorn, L. *Schweißeignung Metallischer Werkstoffe*; Deutscher Verlag fuer Schweisstechnik, DVS-Verlag: Frankfurt am Main, Germany, November 1995; pp. 303–311.
26. Mitelea, I.; Budau, V. *Materiale si Tratamente Termice Pentru Structuri Sudate (Materials and Heat Treatments for Welded Structures)*; West Publishing House Timisoara: Timisoara, Romania, 1992; pp. 225–236.
27. Klimpel, A. Industrial surfacing and hardfacing technology, fundamentals and applications. *Weld. Technol. Rev.* **2020**, *91*, 33–42. [CrossRef]
28. Pelz, A. Microstructure and Properties of New Developed Fe-Cr-C-B Powders for Wear-Protection Purposes. In Proceedings of the Thermal Spray 2009: Proceedings from the International Thermal Spray Conference, Las Vegas, NV, USA, 4–7 May 2009; pp. 849–854.

materials

Article

Analysis of Incremental Sheet Forming of Aluminum Alloy

Costel Catalin Coman, Simona-Nicoleta Mazurchevici *, Constantin Carausu and Dumitru Nedelcu

Faculty of Machine Manufacturing and Industrial Management, Machine Manufacturing Technology Department, "Gheorghe Asachi" Technical University of Iasi, Blvd. Mangeron 59A, 700050 Iasi, Romania; catalin.coman06@gmail.com (C.C.C.); c_carausu@yahoo.com (C.C.); nedelcu1967@yahoo.com (D.N.)
* Correspondence: simona-nicoleta.mazurchevici@academic.tuiasi.ro

Abstract: Recent developments in incremental sheet forming have resulted in the creation of novel manufacturing processes that are highly adaptable and could bring significant economic benefits for advanced technologies and low-volume production. In this manuscript, the following variables were examined: the variation in the deformation forces for a part with a pyramidal trunk shape; the variation in the deformations and thinning of the Al 3003 material during the incremental forming process; and the variation in the accuracy of the incrementally formed part and the quality of the surfaces (surface roughness). The components of the forces in the incremental forming have increasing values from the beginning of the process to the maximum value due to the hardening process. The TiN-coated tool ensures lower values of the forming components. Due to the kinematics of the forming process, deviations, especially in shape, from the part in the drawing are observed, which are shown by the radius of curvature of the side wall of the part, the appearance of a radius of connection between the wall and the bottom of the part, as well as dimensional deviations that are expressed by the variation in the forming depth. Concerning the smoothness of the surfaces, it was observed that the best roughness results were obtained in the case of the TiN-coated tool.

Keywords: incremental forming; sheet; Al 3003; ANOVA analysis; deformation forces

Citation: Coman, C.C.; Mazurchevici, S.-N.; Carausu, C.; Nedelcu, D. Analysis of Incremental Sheet Forming of Aluminum Alloy. *Materials* **2023**, *16*, 6371. https://doi.org/10.3390/ma16196371

Academic Editor: Daniela Kovacheva

Received: 25 August 2023
Revised: 15 September 2023
Accepted: 21 September 2023
Published: 23 September 2023

1. Introduction

Incremental sheet forming (ISF) is a novel and adaptable technology for prototyping and small-lot production.

A classification of the ISF process is conventional and hybrid methods with a brief description as follows:

- Single-point incremental forming—one shaping or deforming tool passes on a sheet through a series of tiny progressive deformations until the desired shape is obtained [1–6];
- Two-point incremental forming—two forming tools are used: one deforms the sheet into the appropriate geometry, while the other one supports the sheet from the rear [7–9];
- Multi-point incremental forming (flexible 3D manufacturing process)—used for big sheets and has two opposed hard dice (lower and upper) with systems consisting of multiple punches with the highly precise movements that are necessary to obtain the desired shape being carried out with linear actuators [10–12];
- Single-point incremental forming with hydraulic fluid—the forming tool is on one side, and the hydraulic fluid pumped to a high pressure is on the opposite side [13–15];
- Two-point incremental forming with partial die—the sheet is formed using a partial die on one side and a moving forming tool on the other [16–18];
- Two-point incremental forming with full die—a complete die is applied to one side of the sheet, while the forming tool rotates on the other side [19–22].

The process parameters involved in ISF encompass various factors, such as tool path, wall angle and tilting angle, incrementation/step depth, speed of rotation, tool diameter, temperature, and lubrication. These parameters interact with each other, influencing the overall outcome of the ISF process. Other key output parameters are also taken into consideration, such as the following [23–26]:

- forming force—There are three types of forces that contribute to sheet deformation: those derived through prediction, analysis, and experimentation. Finally, the forming force shifts when different inputs are used [27];
- formability—The evaluation of a material's deformation characteristics can be conducted using various methods, such as the swift cup drawing test and the Erichsen cupping test [27,28];
- surface roughness—The tool radius, the initial roughness of the active part of the tool, the thickness of the sheet, the flow rate, the incremental depth, and the various tool designs are all process characteristics that might affect surface roughness [29];
- spring back—There are two types of spring back: local and global. The first one produces poor accuracy, and the second produces residual stresses [30];
- failure to fracture—During the process of sheet material deformation, it undergoes multiple stages that ultimately lead to the fracture stage [31].

The vast majority of incremental sheet-forming applications have been carried out in the metalworking industry, and the most significant raw materials that are used currently in the industry are mild steel (small-batch production or rapid prototyping); aluminum/aluminium alloys (frequently used due to small forming loads); stainless steel/high-strength steels (limited by difficulties related to significant elastic spring back); copper, brass, and titanium (heat up during forming—blister appears); magnesium; platinum alloys; polymers (not really suitable—short life cycles but, by using single-point incremental forming, this raw material can be considered); and even sandwich panels (sheets from different materials as steel/polymer/steel or aluminum/polymer/aluminum with a lightweight core that improves buoyancy, impact absorption, thermal insulation, vibration absorption, and sound-deadening) [32–39].

Single-point incremental forming is an improvement over traditional sheet forming because it does not require the use of dies. For low-volume production items, the reduction in die can save tooling costs while increasing turnaround time. Die for the two-point incremental forming procedure can be constructed from cheap materials like wood, allowing for a low total tooling cost. However, this method is inherently time-consuming, and it often results in diminished dimensional accuracy—especially for intricate components. When it comes to the construction of automobile prototypes and the manufacture of parts in small quantities, incremental sheet forming is a method that is both practical and cost-effective [40–43].

The researchers are conducting different studies in the ISF area in order to explain some behaviors or to observe, for instance, the influence of process parameters on the obtained part characteristics.

The mechanism of sheet deformation is associated with the shearing and stretching that appear in the plane perpendicular to the direction of the tool and the shear parallel to it [1,44]; also, other researchers [25] consider that this is produced with a mixture of spinning and stretching.

The reduction in part size to the micro domain is another goal, achieved by using devices installed on a scanning electron microscope [45], a conventional CNC machine modified to function as a micro incremental sheet forming system [46], different types of thin sheets from classic alloys (aluminum, cooper), stainless steels, and silver or titanium [47–51].

The physical and mechanical characteristics of the sheet are significantly influenced by the microscale and macroscale forming processes [52,53]. Also, classic determination of mechanical properties was realized on deformed sheets: tensile test at different temperatures [54,55], yield strength, ultimate tensile strength, total elongation [56,57], spring back defect [58], the phenomenon of shape distortion resulting from supplementary heat treatment [59], and the effect of lubrication [60,61].

From a simulation point of view, the studies are focused on the sequential limit model for thickness distribution [62], influence of the thickness and average grain size ratio on the yield stress and tangent modulus [63], finite element model (FEM) to observe deformations and elasto-plastic behavior material [64], FEM that simulate the impact of plastic deformation on the force precision [40], and three-dimensional FEM that incorporates both elastic and plastic material behavior for three completely different materials (wood, polymer, and rubber) [65–67].

In order to highlight even more, the studies carried out so far on single-incremental forming (SPIF) have been selected from the specialized literature (Table 1).

Table 1. Studies from the technical literature regarding the SPIF.

No. crt.	Material	Analysis	References
1	polymer sheets	optimization of forming forces	[68]
2	polycarbonate	formability limits (necking, fracture)	[69]
3	titanium and magnesium	surface roughness, design and production of parts, maximum forming angle	[70]
4	AZ31B-H24 (magnesium sheet)	temperature effect on roughness, microstructure and hardness, coating of parts using electrospun polycaprolactone	[71]
5	AA 1050 (aluminium alloy sheet)	role of heat in formability, comparation between elevated temperature incremental forming and conventional method	[72]
6	Ti-6Al-4V (titanium sheet)	analysis of forming force, surface roughness, geometric accuracy, thickness profile, micro-hardness for induction heating SPIF; finite element analysis—FEA (strain and stresses), Arrhenius model	[73]
7	Ti-6Al-4V (titanium sheet)	tensile test, stretch forming experiments (at 400 °C), FEA	[74]
8	Al alloy 5083 (aluminium alloy sheet)	surface roughness, deformation of thin-walled, mathematical models, optimization	[75]
9	Al2219-O, AA2219-T6 (aluminium alloy sheet)	tool rotational speed influence on tensile strength, microstructure analysis	[76]
10	AA 2014-T6 (aluminium alloy sheet)	surface roughness, Taguchi's analysis, input parameters influence	[77]
11	AA1050-H111 (aluminium alloy sheet)	fracture forming limit, failure mechanism, ductile fracture, tensile stress, process parameter influence	[78]
12	AA 5052 (aluminium alloy sheet)	formability, optimization, variable wall angle, orthogonal array	[79]
13	AA1200 H14 (aluminium alloy sheet)	formability analysis, ANOVA analysis, microstructure	[80]
14	AA 1050 (aluminium alloy sheet)	residual stresses, hole drill test, Taguchi analysis	[81]
15	St12 (mild steel sheet)	surface roughness, Taguchi analysis, ANOVA methodology	[82]
16	AA1050-H14 (aluminium alloy sheet)	surface profile, influence of process parameters	[83]
17	Monolithic geometry	performance, strategies of experimental conditions, applications	[84]
18	AA1050 (aluminium alloy sheet)	microstructure, texture based analysis, structural and morphological analysis	[85]
19	AA2024, AA6061 (aluminium alloy sheet)	void coalescence and cluster, ductile fracture, GTN model, FEA	[86]
20	Al 6063-T6 (aluminium alloy sheet)	residual stress, nanoindentation fractal geometry	[87]
21	LITECOR® composite material (metal-plastic)	residual stresses, structural and morphological analysis	[88]
22	AA1100 (aluminium alloy sheet)	hardness, artificial neural network, relative importance	[89]
23	TA1 (titanium sheet)	FEM, tensile test, thinning ratio, forming strain	[90]
24	titanium and titanium alloy sheets	microstructure, friction, lubrication	[91]
25	AL1060 (aluminium alloy sheet)	ultrasonic vibration, forming force, experimental verification, analytical modelling	[92]
26	2024-T3, 7075-T6 (aluminium alloy panels)	surface finish, surface roughness, surface topography, artificial neural networks modelling	[93]

The main problems that are not solved by the previous researchers are the particularities of incremental forming in a single point for obtaining parts in the pyramid trunk form (without revolution movements) and incremental forming with titanium nitride-coated tools in order to observe their characteristics.

No research has been carried out on forming processing performed with tools coated with titanium nitride led by industrial robots. This type of processing is different from that performed with CNC milling machines due to the lack of rotation movement of the tool as well as a lower rigidity of the technological system.

The original contribution of this study is that tools coated with TiN (used in the forming process) allow for an improvement of the results of the incremental forming process of a part with a pyramid trunk shape.

Extensive experimental investigations and numerical simulations have been conducted to study the incremental sheet forming (ISF) process. These studies have contributed to establishing a comprehensive understanding of the deformation mechanism involved in macro ISF. In this manuscript, we analyzed the variation in the deformation forces for a part with a pyramid trunk shape; performed ANOVA analysis of the experimental results regarding the maximum components of force in the incremental forming process; and analyzed the variation in the accuracy of the incrementally forming part and the quality of the obtained surfaces (surface roughness).

2. Materials and Methods

2.1. Materials

The semi-finished products used to obtain the parts with incremental stamping were constructed of sheet materials of EN AW 3003 aluminium alloys with a thickness of 1.8 mm. The main mechanical and physical properties of EN AW 3003 aluminium alloy (AlMnCu DIN 3.0517) used for incremental drawing are the following [94]: tensile strength (120–160) MPa; yield strength—min 85 MPa; elongation at break (sheet thickness 1.2–6.3 mm)—6% (out of 50 mm); hardness—36 HB; density—2.73 (g/cm^3); modulus of elasticity—69 GPa; thermal conductivity at 25 °C—193W/m.K; melting point—655 °C; and coefficient of thermal expansion (20–100 °C)—23.2 µm/m/°C. The geometric features of the sheet were the following: thickness—1.8 mm; length and width before cutting (6000 × 2000) mm; length and width after cutting (240 × 200) mm; forming depth—200 mm; large base of the pyramid trunk—a square with 85 mm side; and pyramid wall inclination angle—55°.

The material used for the execution of the incremental forming tool was X153CrMoV12 steel with the following mechanical and physical properties [95]: tensile strength—860 MPa; yield strength—420 MPa; density 7.7 (g/cm^3); modulus of elasticity—210 GPa; thermal conductivity at 350 °C—21 W/m·K; coefficient of thermal expansion (at 20 °C)—13 µm/m/°C; and hardness—54–60 HRC.

Following the processing with spirit, the produced parts are presented in Figure 1.

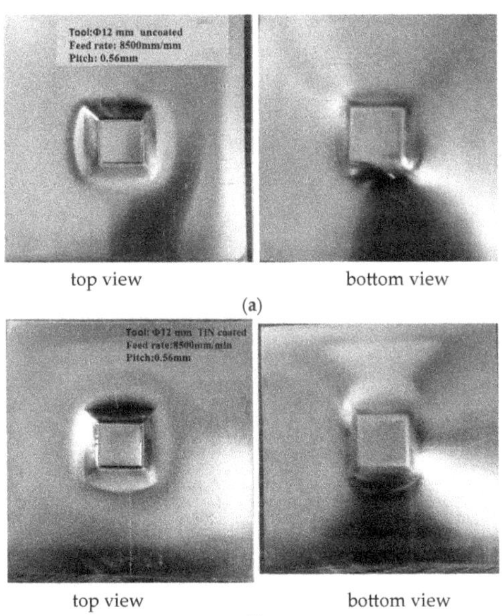

Figure 1. Images of the semi-finished products after processing. (**a**) uncoated tool. (**b**) TiN-coated tool.

2.2. Machine Tools, Equipment, and Tools

The machines, equipment, and tools used in the incremental ambush were as follows:

- ✓ Kuka KR210-2 6-axis robot, developing a maximum load of 2100 N with a positioning repeatability of ±0.06 mm. It is equipped with a KR C2 controller (KUKA Roboter GmbH, Augsburg, Germany), and the numerical programming software is SprutCam (X version, SprutCAM Tech, Limassol, Cyprus) [96,97]. This robot, together with the Aramis measurement system, was used for measuring deformations and wall thinning;
- ✓ Incremental forming force measurement device. A measuring system, consisting of the PCB261A13 sensor (PCB Piezotronics, Inc., NY, USA), the CMD600 signal amplifier (KUKA Roboter GmbH, Augsburg, Germany), and the Quantum X MX840B acquisition system (Hottinger Brüel & Kjær, Naerum, Denmark), was used to measure the forces during the incremental forming process. The force sensor makes it possible to acquire forces in dynamic and quasi-static format, and the amplifier creates the signal voltages required for acquisition with the Quantum X system. The force transducer consists of a piezoelectric sensor assembled between 2 plates.

The CMD600 amplifier was used to obtain ±10 V voltages on the signal acquired from the force sensor PCB261A13 [98]. Three such amplifiers were used for the experimental research, one on each channel (for each of the forces F_z, F_x, and F_y).

The Quantum X MX840B acquisition system [99] has 8 individually configurable and electrically isolated measurement channels. For each channel, 16 types of transducers can be connected.

The OGP SMARTSCOPE 600 FLASH (OGP Messtechnik GmbH, Wallau, Germany) and Formtracer Avant FTA-H8 D4000-D (Mitutoyo, Sakado, Japan) measuring machine were used for the measurement of the acquired parts.

The power tools used for the preparation of the incremental forming process were the following: angle grinder Panzer AP125-12 (Panzer, Germany); and the drilling and screwing machine MAKITA DF347DWE (Makita Corporation, Anjo, Japan);

A 10W40 engine oil (MOTUL, Aubervilliers, France) was chosen as the lubricant for better stretching of the material during the incremental forming process.

The tools used were of two types as follows: simple ball-tip tool (uncoated tool) and a ball-tip tool coated with TiN titanium nitride (coated tool). In the case of both tools, the ball had a diameter of Ø12 mm and a hardness of (53–58) HRC. The TiN coating has the advantage of good wear and corrosion resistance. The hard material applied without a metallic binder has a high hardness, i.e., it has an extremely fine grain size that allows for very good friction resistance [100]. The hardness of the TiN layer is approximately 2300 HV, and the coating thickness is (2–4) μm.

To determine the microstructure of the tools, SEM and EDS analyses were performed using VegaTescan LMHII equipment (TESCAN ORSAY HOLDING, Kohoutovice, Czech Republic) with an EDX detector X Flash 6I10 from Bruker, Germany, Esprit 2.2 software. Figure 2 shows the two tools, the uncoated and the coated one. The SEM image of the uncoated tool (after use) highlights visible traces of wear. The sample coated with titanium nitride, according to the SEM and EDS analyses, reveals a uniform deposition of the layer but also an insensible wear of the tool after use.

The Solidworks 2018 software package was used for the 3D modeling of the parts as well as for the execution drawings.

In order to perform the finite element modelling, the software package Abaqus/CAE version 6.14 was used.

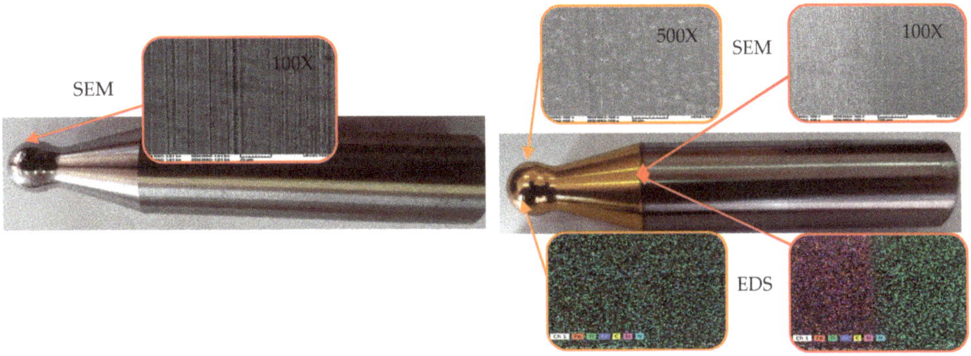

(a) (b)

Figure 2. Tools used for SPIF. (**a**) uncoated tool. (**b**) TiN-coated tool.

2.3. Methods

The Erichsen method was used for tensile/compression testing of the semi-finished product. The specimens used for the punching test were rectangular with one side 50 mm and the other side 300 mm. The condition of the test sample is prescribed by the respective material standards. No machining of the samples was carried out prior to the test because machining may affect the material's ability to be drawn and influence the test results. Sample deformation in the Erichsen equipment was performed with radial and circumferential stretching [101].

The experiments were conducted following a full factorial experimental design with three influencing factors, each on 2 levels, of type $2^3 = 8$ experiments. The factors considered were the tool (−1—uncoated tool; +1—TiN-coated tool), feed rate (−1—5500 mm/min; +1—8500 mm/min), and increment step (−1—0.28 mm; +1—0.56 mm).

Table 2 shows the experimental results of the maxima of the force components in the full factorial design.

Table 2. Experimental data with maximum values of force components.

No.	Tool	Feed Rate, Vs [mm/min]	Forming Step, Pi [mm]	F_z [kN]	F_x [kN]	F_y [kN]
1.	uncoated	5500	0.28	1.2	0.4	0.35
2.	uncoated	5500	0.56	1.39	0.38	0.4
3.	uncoated	8500	0.28	1.0	0.21	0.17
4.	uncoated	8500	0.56	1.41	0.4	0.47
5.	coated with TIN	5500	0.28	1.25	0.37	0.35
6.	coated with TIN	5500	0.56	1.38	0.42	0.44
7.	coated with TIN	8500	0.28	1.26	0.4	0.41
8.	coated with TIN	8500	0.56	1.41	0.37	0.41

The flowchart of the conducted experiments is presented synthetically in Figure 3.

The ANOVA method with the MiniTab software (17.1.0 version, Minitab, LLC, State College, PA, USA) package was used for statistical processing and determining the influence of factors, and the LS-DYNA software package (3.2 version, Livermore Software Technology Corporation (LSTC), Livermore, California, USA) was used for numerical simulation.

The type of problem required for the selected simulation from the LS-DYNA program was Forming. The created model was composed of the following elements: semi-finished product, active plate, and withholding plate and tool—modeled in the form of a sphere.

Figure 4 shows the model used to define the problem in two situations: at the beginning of the incremental forming process (Figure 4a) and at the end of the processing (Figure 4b).

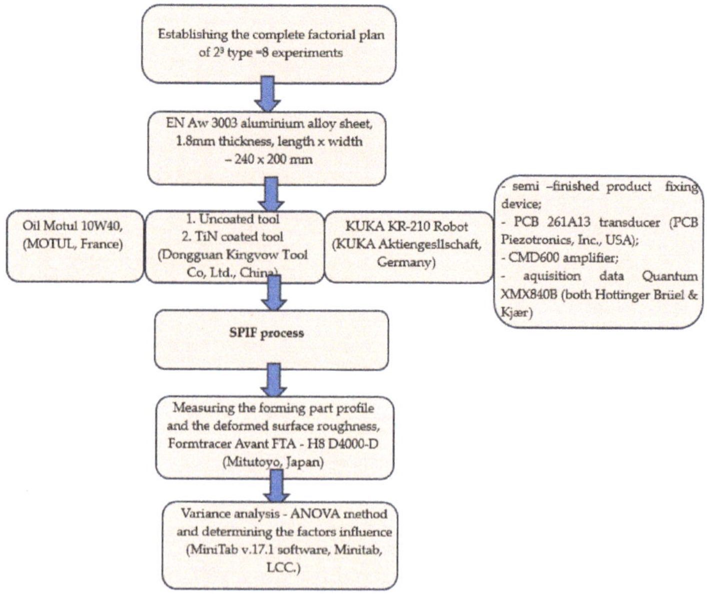

Figure 3. Flow chart of the experimental procedure.

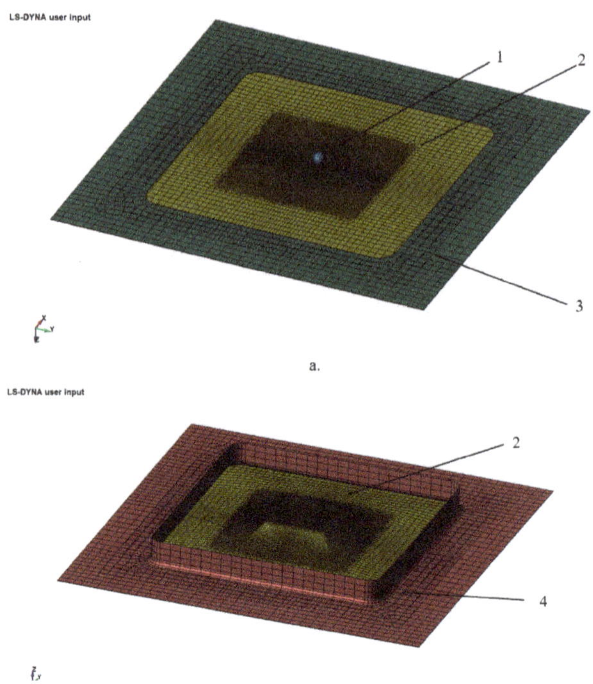

Figure 4. Simulation model: (**a**) at the beginning of the incremental forming process; (**b**) at the end of the process; 1—tool; 2—semi-finished product; 3—restraint ring; 4—active plate.

The semi-finished product was discreet with elements of Shell type (with the help of the Elform parameter) for which the explicit formulation #16 was chosen, forms that ensure the calculation of the deformation forces but also an increase in the solving time. The number of integration points by thickness, nip, was chosen as 7, which allowed for a high accuracy to the detriment of increasing the running time.

For the other two boards and for the sphere, the same type of element (Shell) was chosen but with two integration points.

For discretization, the method of adaptive refinement was used, starting from a network of coarse finished elements. At each iterative step, each element from the previous step was divided into four smaller finished elements.

Three contacts of the type "forming-other-wayfaceosurface" were generated between the electric elements: between the sphere (tool) and the semi-finished product, between the active and semi-finished plate, respectively, and between the withholding and semi-finished plate. In order to solve the contact problem, the penalty method was applied.

The kinematics of the tool was defined with the key word *Boundery_Prescriebed_Motion_Rigid, which allowed for the direction of movement and speed by specifying the movement curves after the three axes. The action of the retention plate was modeled by applying a 2000 N press force in the Z direction.

The rubbing between the tool and the semi-finished product was modeled by adopting a friction coefficient, which allowed for the differentiation of the use of the two types of tools. Thus, for the uncoated tool, the coefficient of friction was 0.09, while for the coated tool, the coefficient was 0.075. The running time on the computer that was required to solve the problem (either with an uncoated or coated tool) was 68 h.

3. Results

3.1. Numerical Simulation of the Variation in the Components of the Ambulation Force

In the numerical simulation, two cases were considered as follows: case 1 corresponds to the uncoated tool (Figures 5 and 6), and case 2 corresponds to the tool coated with TIN (Figures 7 and 8). Thus, the F_z component (Figures 1 and 3) increases from zero towards the maximum value, reaching the maximum towards the end of the working stroke but also having a series of local maxima at each change in direction of the plane of the tool. The F_x component (Figures 2 and 4) and F_y have the same type of variation, lagging each other, going from negative to positive values depending on the direction the tool is moving at a given time. For both the F_x and F_y components, the maximum value is obtained at the end of the tool's working tool.

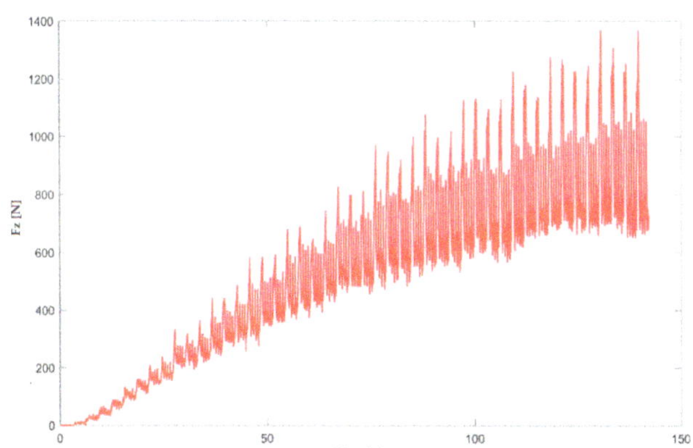

Figure 5. Variation in F_z force for case 1—uncoated tool—obtained with numerical simulation with LS-Dyna.

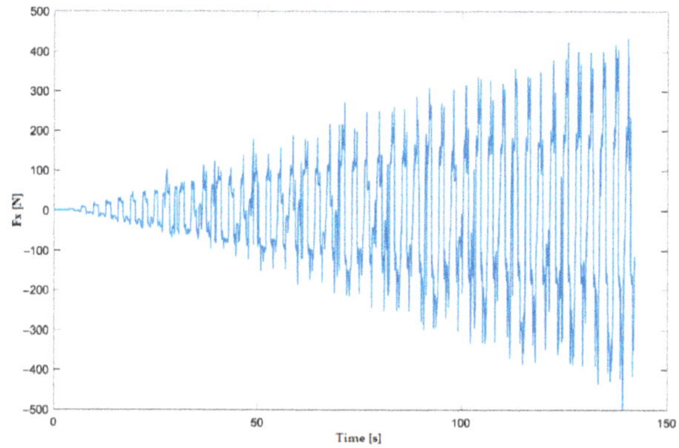

Figure 6. Variation in F_x force for case 1—uncoated tool—obtained with numerical simulation with LS-Dyna.

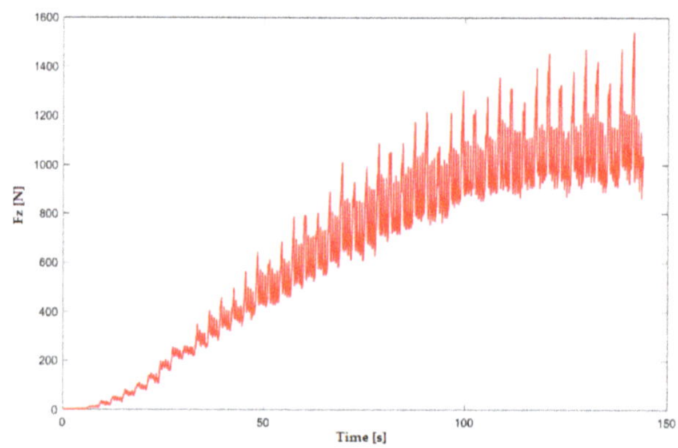

Figure 7. Variation in F_z force for case 2—coated tool with TIN—obtained with numerical simulation with LS-Dyna.

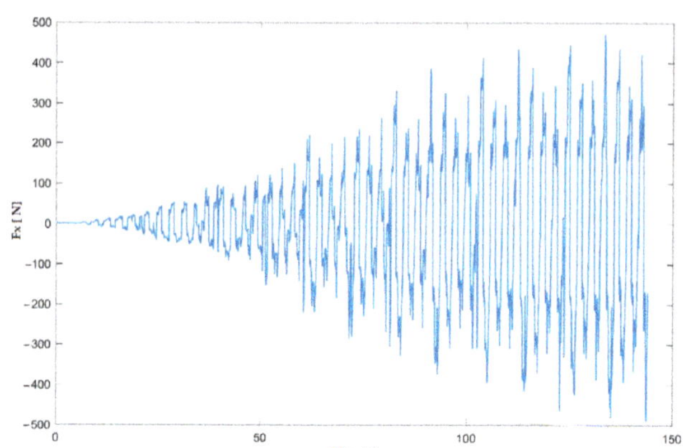

Figure 8. Variation in F_x force for case 2—coated tool with TiN— obtained with numerical simulation with LS-Dyna.

3.2. Influence of Working Parameters on the Components of the Forming Force

The deformation force is an important process parameter during the incremental forming process, which has a major influence on the machining results of the parts. The deformation forces in incremental forming are relatively small compared to those in classical forming but are an important factor when using CNC machining centers or industrial robots. The forming forces used in the incremental forming process depend largely on the depth of forming, the tool diameter, the angle of the part being formed, the blank used and its thickness, and the lubricants used; the axial force F_z is not influenced by the rotational speed of the tool and the lubrication of the blank during the forming process—only by the radial force F_x [64].

For comparison with the results obtained from the numerical simulation, the experimental results of experiments 2 and 6 in the experimental design will be presented.

In Figures 9 and 10, the variation in the components F_z and F_x along the direction of the two coordinate axes XZ in experiment 2 is shown graphically. The forming time was 157.4 s; this was lower because the incremental step was double, and the number of passes was lower. The maximum forming force F_z obtained in experiment 2 is 1.39 kN; the maximum value obtained for the F_x force is 0.38 kN.

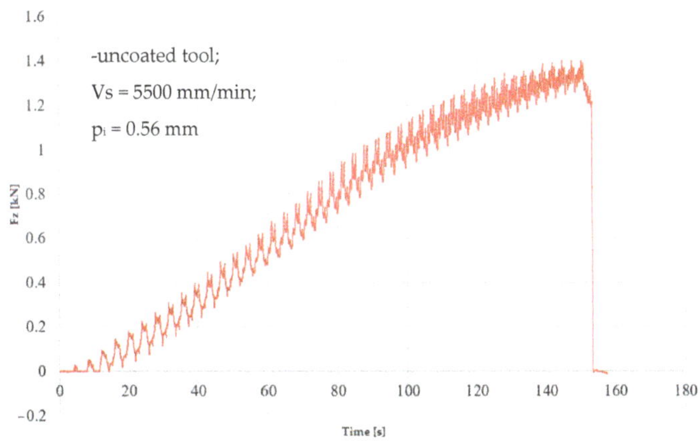

Figure 9. Variation in F_z force in experiment 2.

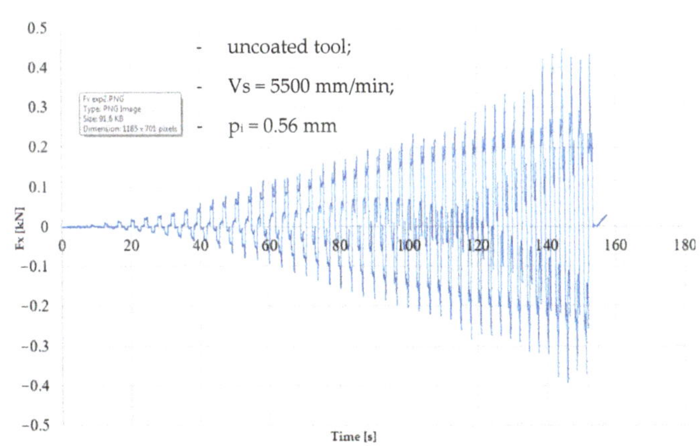

Figure 10. Variation in F_x force in experiment 2.

In Figures 11 and 12, the variation in the force components F_z and F_x along the direction of the three XYZ coordinate axes for experiment 6 (case 2 of the coated tool) is plotted. Thus, the maximum values of the forces are as follows: F_z reaches the maximum value of 1.38 kN, and F_x reaches the maximum value of 0.42 kN.

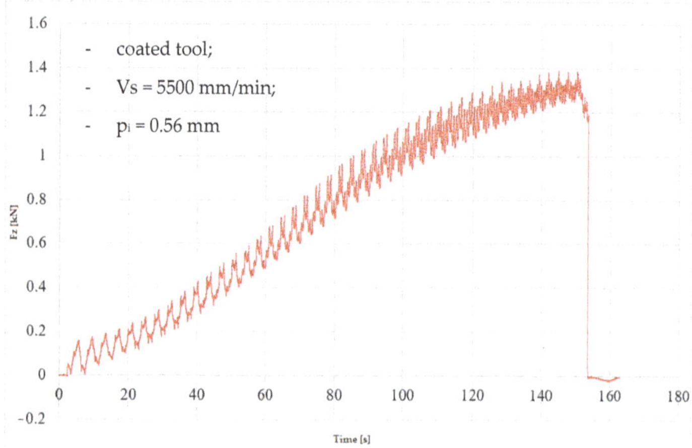

Figure 11. Variation in F_z force in experiment 6.

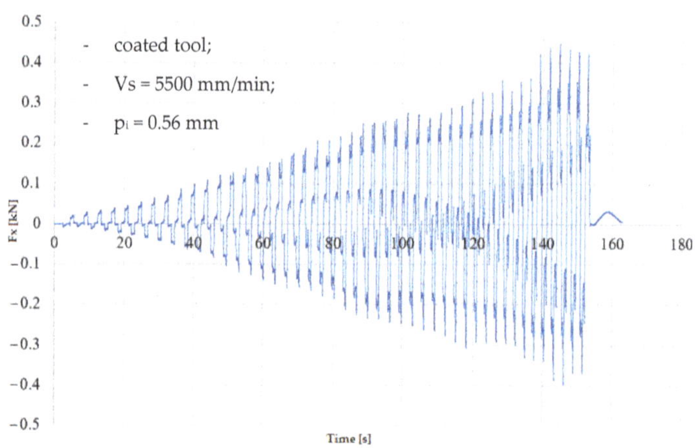

Figure 12. Variation in F_x force in experiment 6.

After running the numerical analyses with the finite element method, it was possible to compare the results obtained theoretically (numerically) with those obtained experimentally. A summary of these results and, especially, of the differences obtained between the numerical and experimental results is presented in Table 3 (comparative results for the uncoated tool, case 1 of the simulation and experiment 2) and Table 4 (comparative results for the TIN-coated tool, case 2 of the simulation and experiment 6).

It should be mentioned at the outset that the results obtained for the force components in the numerical simulations have a higher percentage than those obtained for the deformations due to the fact that the program calculates these components as being those of the reaction occurring between the nodes of the tool and those of the blank. However, as can be observed from Tables 3 and 4, the differences are acceptable with the highest values being 7.97% and 7.89%, respectively. From the analysis of the data presented in Tables 3 and 4, it can be observed that the values for case 2 are higher in the OX and OY directions and lower for the force in the OZ direction than those for the forces obtained in case 1. In conclusion, it can be stated that the simulation using the finite element method leads to results with a relatively low percentage of error and can be successfully used for numerical simulation of the incremental buckling process.

Table 3. Numerical and experimental results for forces—case 1.

Case	Feed Rate [mm/min]	Forming Step [mm]	Maximum Force Fz [N]	Maximum Force Fx [N]	Maximum Force Fy [N]
Experimental	5500	0.56	1390	380	400
Simulation	5500	0.56	1395	410	420
Diference [%]			0.36	7.89	5

Table 4. Numerical and experimental results for forces—case 2.

Case	Feed Rate [mm/min]	Forming Step [mm]	Maximum Force Fz [N]	Maximum Force Fx [N]	Maximum Force, Fy [N]
Experimental	5500	0.56	1380	420	440
Simulation	5500	0.56	1490	440	450
Diference [%]			7.97	4.76	2.27

The influence on the maximum force component F_z is shown in Table 5, and the influence graph is shown in Figure 13.

Table 5. Analysis of variance of factors on the maximum force component F_z.

Factor	No. Degrees of Freedom (DF)	Adj. MS	F Value	Threshold of Significance, *p*	Statistical Significance *
Tool	1	0.0968	11.90	0.026	S
Feed rate, V_s	1	0.005	0.61	0.477	NS
Forming step, P_i	1	0.0072	0.88	0.400	NS

* S—statistically significant; NS—not statistically significant.

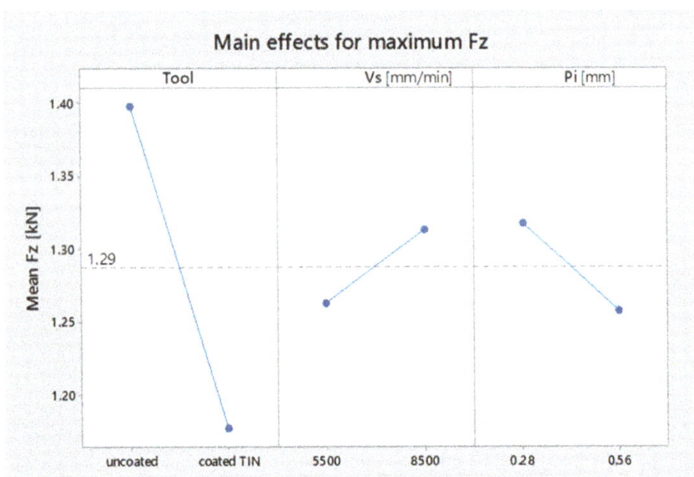

Figure 13. Influence of factors on the maximum force component F_z.

Tool is the only factor that has a statistically significant influence on the magnitude of the F_Z force (significance threshold $p < 0.026$). The other factors have no statistically significant influence ($p > 0.03$).

The TIN-coated tool provides a lower maximum force F_Z during incremental forming than when using the uncoated tool.

The influence on the maximum force F_X is shown in Table 6, and the influence graph is shown in Figure 14. According to Table 6, none of the factors has a statistically significant influence.

Table 6. Analysis of variance of factors on the maximum force component F_X.

Factor	No. Degrees of Freedom (DF)	Adj. MS	F Value	Threshold of Significance, *p*	Statistical Significance *
Tool	1	0.004513	1.31	0.316	NS
Feed rate, Vs	1	0.010513	3.06	0.155	NS
Forming step, Pi	1	0.002113	0.61	0.477	NS

* S—statistically significant; NS—not statistically significant.

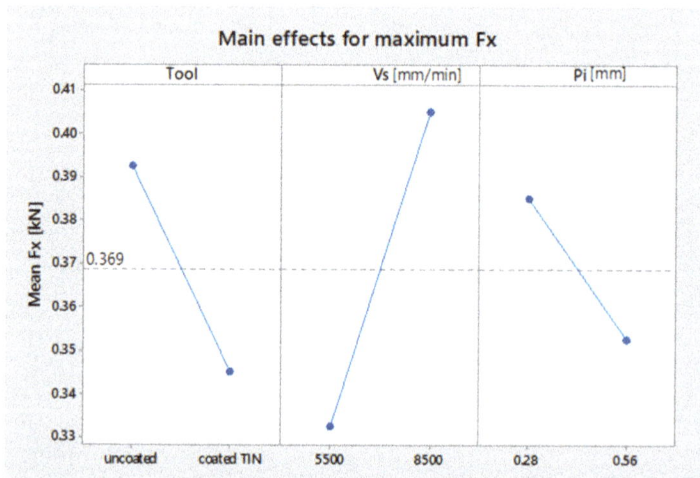

Figure 14. Influence of factors on the maximum force component F_X.

It is found that the value of the maximum force component F_X increases with increasing forward speed.

As in the case of the maximum value of the F_X force, a similar influence of the analyzed factors is found.

Due to the large variation in the disturbances influence in the case of maximum values of the forces components (obtained at the end of the forming process), the experimental results have variations that have a high dispersion, which determines that the advance speed influences and the incremental step are not statistically insignificant in the case of the F_Z component. In the case of the X and Y directions (the tool movement directions in the perpendicular plane to the forming direction), no parameter is statistically significant, and the variability of these components is even greater. One solution would be to increase the distance between the two levels (−1 and +1) of these parameters but only for quantitative parameters (speed advance and increment). Thus, in the study carried out by [102] with the same type of material as in the present paper, by the statistical processing of the experimental results with the MiniTab software 17.1.0 version, the increment step influence on the F_Z component came out strongly significant ($p = 0.01$). The result was obtained due to the fact that the difference between the two levels was seven times (0.1 mm and 0.7 mm, respectively).

Disturbance can also occur due to the variability of the parameters of the technological system.

3.3. Execution Accuracy

After the parts are produced with the incremental forming process, the measurement of the parts obtained is carried out. For the measurement of the depth of the part, the measuring equipment took, as a basis, a theoretical straight area as can be observed in Figures 15 and 16.

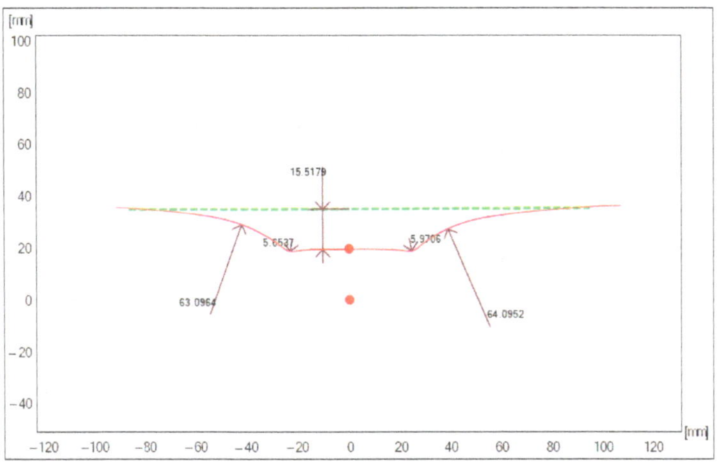

Figure 15. Measurement of the part profile obtained in experiment 2.

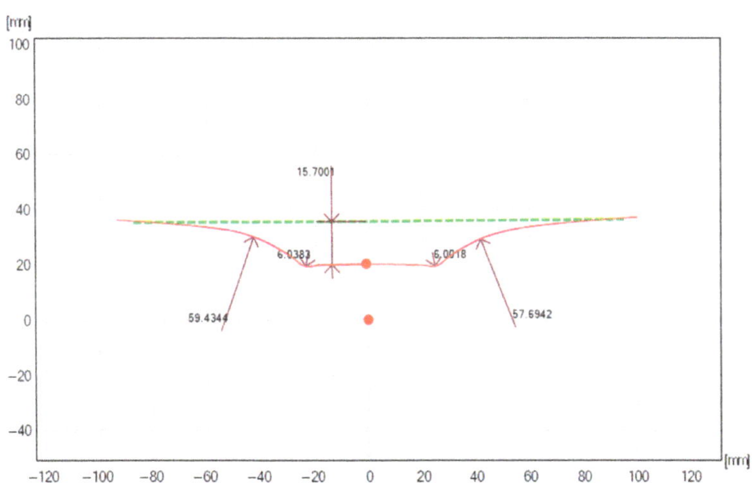

Figure 16. Measurement of the part profile obtained in experiment 6.

In the study [103], the geometric accuracy of the pieces processed by using SPIF is analyzed, underlining that the geometric and form deviations are produced as a result of unwanted plastic deformations, spring back, and the pillow effect. The study does not signal the deviation from the rectilinearity of the wall of the part obtained with forming.

The present study signals the appearance of a wall form deviation of convex shape of wall curvature. This form deviation occurs due to the forming force variation during the process, the increase in the cold hardening (due to the spring back reduction), and the material thinning; thus, at the beginning of processing, the F_z is smaller, and the deformation of the material is lower, which leads to a smaller forming angle [104]. At the maximum forming depth, the wall thinning is maximum, and the forming force is maximum, which determines a higher deformation degree, thus, obtaining a maximum forming angle.

The deviation from the forming depth is due to spring back and has relatively high values due to the SPIF (one tool). The reduction solutions are the use of the counter-tool and the introduction in the deformation area of a controlled temperature of 200–300 °C.

Another cause of deviations is the use of the robot (compared to milling CNC) that, according to [105], induces position errors of the tool center that are caused by the elastic deformations to which the robot is subjected to during the process.

3.4. Quality of the Surfaces Obtained

The quality of the surfaces obtained with the incremental forming process is influenced by a number of factors, such as the surface quality of the blank and the die and their shape deviations. Another key factor is the process parameters used in the incremental forming process. If all these factors have been checked, the next step is to determine how the process is to be carried out with or without lubrication, as friction between the tool and the blank causes a temperature rise; for this reason, we can have significant influences on the roughness of the machined surface. A very important thing is the choice of the type of lubricant that meets all the conditions to allow the tool to stretch the material so that there will be influence on the quality of the incrementally ground surface. In this work, an aluminium alloy blank was used, where it was taken into account to avoid, as much as possible, material deposits on the tool surface by choosing 10W40 engine oil, which could have a significant influence on the roughness of the incrementally ground surface. In order to check the quality of the surface obtained, the roughness R_a (μm) and R_z (μm) were measured (Figures 17 and 18) for all the parts obtained experimentally. It can be observed that, in the case of the TiN-coated tool, a better smoothness of the surfaces of the obtained parts was obtained.

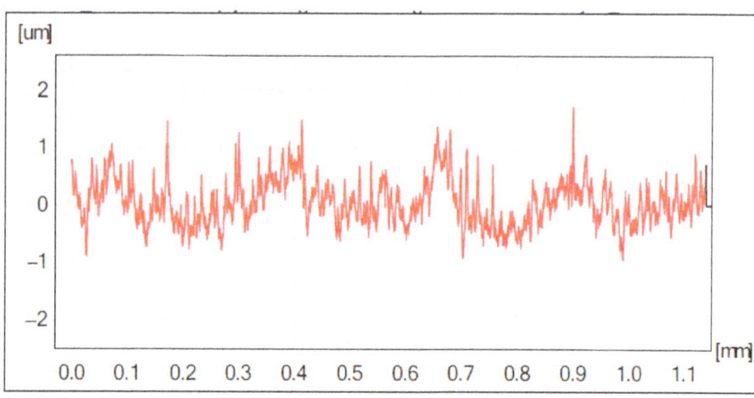

Parameter Summary Sheet

Parameter	Profile=R_VDA - Section= 1	Average Value
Ra (um)	0.3129	0.3129
Rz (um)	2.2088	2.2088
RSm (mm)	0.0375	0.0375

Figure 17. Roughness values for the part in experiment 2.

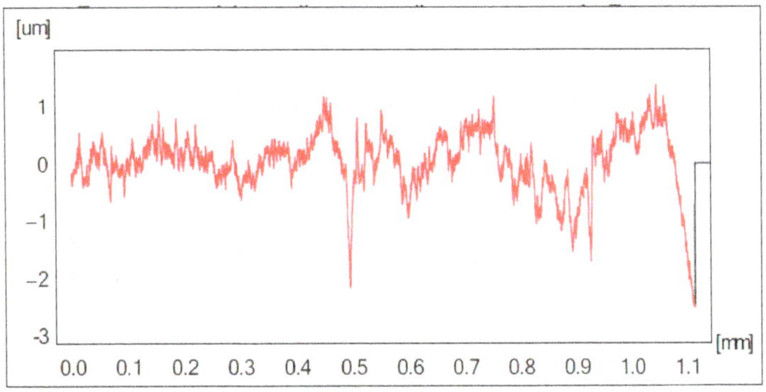

Parameter Summary Sheet

Parameter	Profile=R_VDA - Section= 1	Average Value
Ra (um)	0.4377	0.4377
Rz (um)	2.7188	2.7188
RSm (mm)	0.0549	0.0549

Figure 18. Roughness values for the part in experiment 6.

An improved smoothness of the surfaces was attained in the tool case with the TiN coating due to a superior superficial hardness that reduced the local deformation in the contact area, tool semi-finished product, and, as a consequence, the contact area. On the other hand, the initial roughness of the tool coated with TiN was lower (R_a = 0.2 μm) compared to that of the uncoated tool that had the parameter of the roughness R_a = 0.3 μm.

4. Conclusions

This manuscript is a comparative study on the SPIF process with the use of coated and uncoated tools led by an industrial robot.

The variation in the deformation forces for a part with a pyramidal trunk shape, the ANOVA analysis for experimental research on the maximum deformation force in the incremental forming process, numerical simulation with LS-DINA software, 3.2 version, of incremental deformation process, the variation in the accuracy of the incrementally formed part, and the quality of the surfaces (surface roughness) were all examined in this manuscript. The forces in incremental forming have components with rising values from the start of the procedure to the maximum value as a result of the hardening process. The F_x and F_y components oscillate between positive and negative values as a result of the change in the direction of the tool, and the F_z component oscillates with a certain amplitude and frequency due to the modification of the feed rate when changing the direction of the formation. Lower values of the forming components are ensured with the TiN-coated tool. Due to the kinematics of the forming process, deviations from the part in the drawing are observed, particularly in shape.

Based on the data obtained, the study limitations consist of the following:

- the temperature is not controlled in the deformed area (with the possibility of improving the material deformability);
- reduced precision of the process (need to use the counter-pouch (tool)/counter-matrice);
- the inclination angle of the wall cannot exceed 70° (under special conditions) because breaks and cracks occur;
- the process is limited and of a lower precision due to the use of a robot for traveling the tool, which has low rigidity. On the other hand, industrial robots can be used in the processing with incremental forming of areas of large assemblies that cannot be placed on the CNC milling machines' table;
- long processing time compared to classic forming, which recommends it only for small and unique series production.

The practical recommendation that can be made is the use of tools coated with hard alloys (as titanium nitride) that, among others, improve the contact from the tool and the piece (deformed area). Also, it should be used counter-tool with a correlated movement with that of the tool (the kinematics of the equipment used, programming the movement of both tools and others).

Author Contributions: Conceptualization, D.N. and C.C.; methodology, C.C.; software, C.C.; validation, D.N. and S.-N.M.; formal analysis, S.-N.M.; investigation, C.C.C.; resources, C.C.C.; data curation, C.C., C.C.C. and D.N.; writing—original draft preparation, S.-N.M.; writing—review and editing, S.-N.M., D.N. and C.C.C.; visualization, D.N. and S.-N.M.; supervision, D.N.; project administration, D.N.; funding acquisition, C.C.C. All authors have read and agreed to the published version of the manuscript.

Funding: "Gheorghe Asachi" Technical University of Iasi, Bd. Dimitrie Mangeron, no. 53A, 700050 Iasi, Romania, under funding number.

Institutional Review Board Statement: Not applicable.

Informed Consent Statement: Not applicable.

Data Availability Statement: Not applicable.

Conflicts of Interest: The authors declare no conflict of interest.

References

1. Emmens, W.C.; van den Boogaard, A.H. An overview of stabilizing deformation mechanisms in incremental sheet forming. *J. Mater. Process. Technol.* **2009**, *209*, 3688–3695. [CrossRef]
2. Button, S.T. Introduction to Advanced Forming Technologies. In *Comprehensive Materials Processing*; Hashmi, S., Batalha, G.F., Van Tyne, C.J., Yilbas, B., Eds.; Elsevier: Amsterdam, The Netherlands, 2014; pp. 1–5. [CrossRef]
3. Zheng, K.; Politis, D.J.; Wang, L.; Lin, J. A review on forming techniques for manufacturing lightweight complex—Shaped aluminium panel components. *Int. J. Lightweight Mater. Manuf.* **2018**, *1*, 55–80. [CrossRef]
4. Kumar, Y.; Kumar, S. *Advances in Material Forming and Joining*; Springer India: New Delhi, India, 2015; pp. 29–47. [CrossRef]
5. Duflou, J.R.; Habraken, A.-M.; Cao, J.; Malhotra, R.; Bambach, M.; Adams, D.; Vanhove, H.; Mohammadi, A.; Jeswiet, J. Single point incremental forming: State-of-the-art and prospects. *Int. J. Mater. Form.* **2018**, *11*, 743–773. [CrossRef]
6. Behera, A.K. Ricardo Alves de Sousa, Giuseppe Ingarao, Valentin Oleksik, Single point incremental forming: An assessment of the progress and technology trends from 2005 to 2015. *J. Manuf. Process.* **2017**, *27*, 37–62. [CrossRef]
7. Tisza, M. General overview of sheet incremental forming. *Manuf. Eng.* **2012**, *55*, 113–120.
8. Peng, W.; Ou, H.; Becker, A.; Forming, D.-S.I.; Eng, J.M.S. Double-Sided Incremental Forming: A Review. *J. Manuf. Sci. Eng.* **2019**, *141*, 050802. [CrossRef]
9. Mostafanezhad, H.; Menghari, H.G.; Esmaeili, S.; Shirkharkolaee, E.M. Optimization of two-point incremental forming process of AA1050 through response surface methodology. *Measurement* **2018**, *127*, 21–28. [CrossRef]
10. Li, M.Z.; Cai, Z.Y.; Sui, Z.; Yan, Q.G. Multi-point forming technology for sheet metal. *J. Mater. Process. Technol.* **2002**, *129*, 333–338. [CrossRef]
11. Bhaskar, R.S. Analysis of Various Sheet Incremental Metal Forming Processes: A Review. *Int. J. Innov. Sci. Res. Technol.* **2017**, *2*, 128–131.
12. Ramkumar, K.; Baskar, N.; Elangovan, K.; Narayanan, C.S.; Selvarajan, K.A.; Jesuthanam, C. Comparison of Multi Point Incremental Forming Tool with Single Point Incremental Forming Tool Using FLD, Fractography and 3D-Surface Roughness Analysis on Cr/Mn/Ni/Si Based Stainless Steel. *Silicon* **2021**, *13*, 487–494. [CrossRef]
13. Kumar, Y.; Kumar, S. Experimental and analytical evaluation of Incremental Sheet Hydro-Forming strategies to produce high forming angle sheets. *Heliyon* **2019**, *5*, e01801. [CrossRef] [PubMed]
14. Shamsari, M.; Mirnia, M.J.; Elyasi, M.; Baseri, H. Formability improvement in single point incremental forming of truncated cone using a two-stage hybrid deformation strategy. *Int. J. Adv. Manuf. Technol.* **2018**, *94*, 2357–2368. [CrossRef]
15. Shang, M.; Li, Y.; Yang, M.; Chen, Y.; Bai, L.; Li, P. Wall Thickness Uniformity in ISF of Hydraulic Support: System Design, Finite Element Analysis and Experimental Verification. *Machines* **2023**, *11*, 353. [CrossRef]
16. Bhatt, D.K.; Rana, J.; Shah, K.; Patel, K.J. Incremental Sheet Metal Forming Process: A Review. *Int. J. Eng. Res. Technol.* **2016**, *4*, 1–9.
17. Silva, M.B.; Martins, P.A.F. Two-Point Incremental Forming with Partial Die: Theory and Experimentation. *J. Mater. Eng. Perform.* **2013**, *22*, 1018–1027. [CrossRef]
18. Lu, H.; Kearney, M.; Wang, C.; Liu, S.; Meehan, P.A. Part accuracy improvement in two point incremental forming with a partial die using a model predictive control algorithm. *Precis. Eng.* **2017**, *49*, 179–188. [CrossRef]
19. Young, D.; Jeswiet, J. Wall thickness variations in single-point incremental forming. *Proc. Inst. Mech. Eng. Part B J. Eng. Manuf.* **2004**, *218*, 1453–1459. [CrossRef]

20. Reddy, N.V.; Lingam, R. Double Sided Incremental Forming: Capabilities and Challenges. *J. Phys. Conf. Ser.* **2018**, *1063*, 012170. [CrossRef]
21. Nourmohammadi, A.A.; Elyasi, M.; Mirnia, M.J. Flexibility improvement in two-point incremental forming by implementing multi-point die. *Int. J. Adv. Manuf. Technol.* **2019**, *102*, 2933–2952. [CrossRef]
22. Bhasker, R.S.; Kumar, Y. Process capabilities and future scope of Incremental Sheet Forming (ISF). *Mater. Today Proc.* **2023**, *72*, 1014–1019. [CrossRef]
23. Gatea, S.; Ou, H.; McCartney, G. Review on the influence of process parameters in incremental sheet forming. *Int. J. Adv. Manuf. Technol.* **2016**, *87*, 479–499. [CrossRef]
24. Patel, D.; Gandhi, A. A review article on process parameters affecting Incremental Sheet Forming (ISF). *Mater. Today Proc.* **2022**, *63*, 368–375. [CrossRef]
25. Kim, Y.H.; Park, J.J. Effect of process parameters on formability in incremental forming of sheet metal. *J. Mater. Process. Technol.* **2002**, *130–131*, 42–46. [CrossRef]
26. Maaß, F.; Hahn, M.; Tekkaya, A.E. Interaction of Process Parameters, Forming Mechanisms, and Residual Stresses in Single Point Incremental Forming. *Metals* **2020**, *10*, 656. [CrossRef]
27. Xiao, X.; Kim, C.I.; Lv, X.D.; Hwang, T.S.; Kim, Y.S. Formability and forming force in incremental sheet forming of AA7075-T6 at different temperatures. *J. Mech. Sci. Technol.* **2019**, *33*, 3795–3802. [CrossRef]
28. Shi, X.; Hussain, G.; Zha, G.; Wu, M.; Kong, F. Study on formability of vertical parts formed by multi-stage incremental forming. *Int. J. Adv. Manuf. Technol.* **2014**, *75*, 1049–1053. [CrossRef]
29. Gandla, P.K.; Inturi, V.; Kurra, S.; Radhika, S. Evaluation of surface roughness in incremental forming using image processing based methods. *Measurement* **2020**, *164*, 108055. [CrossRef]
30. Edwards, W.L.; Grimm, T.J.; Ragai, I.; Roth, J.T. Optimum Process Parameters for Springback Reduction of Single Point Incrementally Formed Polycarbonates. *Procedia Manuf.* **2017**, *10*, 329–338. [CrossRef]
31. Ai, S.; Long, H. A review on material fracture mechanism in incremental sheet forming. *Int. J. Adv. Manuf. Technol.* **2019**, *104*, 33–61. [CrossRef]
32. Heggemann, T.; Homberg, W.; Sapli, H. Combined Curing and Forming of Fiber Metal Laminates. *Procedia Manuf.* **2020**, *47*, 36–42. [CrossRef]
33. Ding, Z.; Wang, H.; Luo, J.; Li, N. A review on forming technologies of fibre metal laminates. *Int. J. Lightweight Mater. Manuf.* **2021**, *4*, 110–126. [CrossRef]
34. Magrinho, J.P.G.; Silva, M.B.; Martins, P.A.F. Incremental sheet forming. In *Reference Module in Earth Systems and Environmental Sciences*; Elsevier: Amsterdam, The Netherlands, 2022. [CrossRef]
35. Zhang, Q.; Li, G.; Liu, Z.; Zhao, A.; Xiong, X.; Lei, Y. Friction Stir Incremental Forming of AA7075-O Sheets: Experimental Investigations on Performance Evaluation of Formed Parts. *Rare Met. Mater. Eng.* **2018**, *47*, 3659–3665. [CrossRef]
36. Heinz, A.; Haszler, A.; Keidel, C.; Moldenhauer, S.; Benedictus, R.; Miller, W. Recent development in aluminium alloys for aerospace applications. *Mater. Sci. Eng. A* **2000**, *280*, 102–107. [CrossRef]
37. Weblet Importer, (n.d.). Available online: https://european-aluminium.eu/resource-hub/aluminium-automotive-manual/ (accessed on 7 January 2018).
38. Harrison, N.R.; Luckey, S.G. Hot stamping of a B-Pillar outer from high strength aluminum sheet AA7075, 2014-01e0981. *SAE Int. J. Mater. Manuf.* **2014**, *7*, 567–573. [CrossRef]
39. Hirsch, J. Recent development in aluminium for automotive applications. *Trans. Nonferrous Met. Soc. China* **2014**, *24*, 1995–2002. [CrossRef]
40. Sousa, R. Incremental Sheet Forming Technologies. In *Reference Module in Materials Science and Materials Engineering*; Elsevier: Amsterdam, The Netherlands, 2016; ISBN 9780128035818. [CrossRef]
41. Kridli, G.T.; Friedman, P.A.; Boileau, J.M. Chapter 7—Manufacturing processes for light alloys. Updated by Boileau, J.M.; Mallick, P.K.; Woodhead Publishing in Materials; In *Materials, Design and Manufacturing for Lightweight Vehicles*, 2nd ed.; Mallick, P.K., Ed.; Woodhead Publishing: Sawston, UK, 2021; pp. 267–320. ISBN 9780128187128. [CrossRef]
42. Cheng, Z.; Li, Y.; Xu, C.; Liu, Y.; Ghafoor, S.; Li, F. Incremental sheet forming towards biomedical implants: A review. *J. Mater. Res. Technol.* **2020**, *9*, 7225–7251. [CrossRef]
43. Emmens, W.C.; Sebastiani, G.; van den Boogaard, A.H. The technology of Incremental Sheet Forming—A brief review of the history. *J. Mater. Process. Technol.* **2010**, *210*, 981–997. [CrossRef]
44. Jackson, K.; Allwood, J. The mechanics of incremental sheet forming. *J. Mater. Process. Technol.* **2009**, *209*, 1158–1174. [CrossRef]
45. Saotome, Y.; Okamoto, T. An in-situ incremental microforming system for three-dimensional shell structures of foil materials. *J. Mater. Process. Technol.* **2001**, *113*, 636–640. [CrossRef]
46. Obikawa, T.; Satou, S.; Hakutani, T. Dieless incremental micro-forming of miniature shell objects of aluminum foils. *Int. J. Mach. Tool Manu.* **2009**, *49*, 906–915. [CrossRef]
47. Dejardin, S.; Thibaud, S.; Gelin, J.C. Experimental and Numerical Investigations in Single Point Incremental Sheet Forming For Micro parts. *AIP Conf. Proc.* **2007**, *907*, 145–150.
48. Obikawa, T.; Hakutani, T.; Sekine, T.; Numajiri, S.; Matsumura, T.; Yoshino, M. Single-Point Incremental Micro-Forming of Thin Shell Products Utilizing High Formability. *J. Adv. Mech. Des. Syst.* **2010**, *4*, 1145–1156. [CrossRef]

49. Sekine, T.; Obikawa, T. Micro Incremental Forming Characteristics of Stainless Foil. *Key Eng. Mater.* **2010**, *447–448*, 346–350. [CrossRef]
50. Otsu, M.; Hiroshi, T.; Kazuki, T. Micro-incremental forming of Ti and Au foils by indentation method. *Key Eng. Mater.* **2007**, *345–346*, 1101–1104. [CrossRef]
51. Song, X.; Zhang, J.; Zhai, W.; Taureza, M.; Castagne, S.; Danno, A. Numerical and Experimental Study of Micro Single Point Incremental Forming Process. *Procedia Eng.* **2017**, *207*, 825–830. [CrossRef]
52. Vollertsen, F. Categories of size effects. *Prod. Eng.* **2008**, *2*, 377. [CrossRef]
53. Geiger, M.; Vollertsen, F. Fundamentals on the Manufacturing of Sheet Metal Microparts. *CIRP Ann. Manuf. Technol.* **1996**, *45*, 277–282. [CrossRef]
54. Iadicola, M.A.; Foecke, T.; Banovic, S.W. Experimental observations of evolving yield loci in biaxially strained AA5754-O. *Int. J. Plast.* **2008**, *24*, 2084–2101. [CrossRef]
55. Ramesh, R.; Bhattacharya, R.; Williams, G. Effect of ageing on the mechanical behaviour of a novel automotive grade AleMgeSi alloy. *Mater. Sci. Eng. A 541* **2012**, 128–134. [CrossRef]
56. Abedrabbo, N.; Pourboghrat, F.; Carsley, J. Forming of AA5182-O and AA5754-O at elevated temperatures using coupled thermo-mechanical finite element models. *Int. J. Plast.* **2007**, *23*, 841–875. [CrossRef]
57. Aerospace Aluminum Distributor Supplier. Available online: http://www.aerospace-metals.com/aluminum-distributor.html (accessed on 15 June 2023).
58. Wang, A.; Zhong, K.; El Fakir, O.; Liu, J.; Sun, C.; Wang, L.-L.; Lin, J.; Dean, T.A. Springback analysis of AA5754 after hot stamping: Experiments and FE modelling. *Int. J. Adv. Manuf. Technol.* **2017**, *89*, 1339–1352. [CrossRef]
59. Fan, X.; He, Z.; Zheng, K.; Yuan, S. Strengthening Behavior of Al-Cu-Mg Alloy Sheet in Hot Forming-Quenching Integrated Process With Colda^V"Hot Dies. *Mater. Des.* **2015**, *83*, 557–565. [CrossRef]
60. Jawale, K.; Duarte, J.F.; Reis, A.; Silva, B. Microstructural investigation and lubrication study for single point incremental forming of copper, in press. *Int. J. Solids Struct.* **2018**, *21*, 145–151, in press. [CrossRef]
61. Azevedo, N.G.; Farias, J.S.; Bastos, R.P.; Teixeira, P.; Davim, J.P.; De Sousa, R.J.A. Lubrication aspects during single point incremental forming for steel and aluminum materials. *Int. J. Precis. Eng. Manuf.* **2015**, *16*, 589–595. [CrossRef]
62. Mirnia, M.J.; Dariani, B.M.; Vanhove, H.; Duflou, J.R. Thickness improvement in single point incremental forming deduced by sequential limit analysis. *Int. J. Adv. Manuf. Tech.* **2014**, *70*, 2029–2041. [CrossRef]
63. Yang, C.; Wen, T.; Liu, L.; Zhang, S.; Wang, H. Dieless incremental hole-flanging of thin-walled tube for producing branched tubing. *J. Mater. Process. Technol.* **2014**, *214*, 2461–2467. [CrossRef]
64. Bahloul, R. A study on optimal design of process parameters in single point incremental forming of sheet metal by combining Box- Behnken design of experiments, response surface methods and genetic algorithms. *Int. J. Adv. Manuf. Technol.* **2014**, *74*, 163–185. [CrossRef]
65. Li, J.; Bai, T. Numerical simulation and experimental investigation of incremental sheet forming with an elastic support. *Int. J. Adv. Manuf. Technol.* **2014**, *74*, 1649–1654. [CrossRef]
66. Echrif, S.B.M.; Hrairi, M. Materials and Manufacturing Processes Research and Progress in Incremental Sheet Forming Processes. *Mater. Manuf. Process.* **2011**, *26*, 1404–1414. [CrossRef]
67. Araghi, B.T.; Manco, G.L.; Bambach, M.; Hirt, G. Investigation into a new hybrid forming process: Incremental sheet forming combined with stretch forming. *CIRP Ann.* **2009**, *58*, 225–228. [CrossRef]
68. Formisano, A.; Boccarusso, L.; Durante, M. Optimization of Single-Point Incremental Forming of Polymer Sheets through FEM. *Materials* **2023**, *16*, 451. [CrossRef] [PubMed]
69. Rosa-Sainz, A.; Centeno, G.; Silva, M.B.; López-Fernández, J.A.; Martínez-Donaire, A.J.; Vallellano, C. On the Determination of Forming Limits in Polycarbonate Sheets. *Materials* **2020**, *13*, 928. [CrossRef] [PubMed]
70. Jeswiet, J.; Adams, D.; Doolan, M.; McAnulty, T.; Gupta, P. Single point and asymmetric incremental forming. *Adv. Manuf.* **2015**, *3*, 253–262. [CrossRef]
71. Cusanno, A.; Negrini, N.C.; Villa, T.; Farè, S.; Garcia-Romeu, M.L.; Palumbo, G. Post Forming Analysis and In Vitro Biological Characterization of AZ31B Processed by Incremental Forming and Coated With Electrospun Polycaprolactone. ASME. *J. Manuf. Sci. Eng.* **2021**, *143*, 011012. [CrossRef]
72. Singh, S.A.; Priyadarshi, S.; Tandon, P. Comparative Study of Incremental Forming and Elevated Temperature Incremental Forming Through Experimental Investigations on AA 1050 Sheet ASME. *J. Manuf. Sci. Eng.* **2021**, *143*, 064501. [CrossRef]
73. Li, W.; Attallah, M.M.; Essa, K. Experimental and numerical investigations on the process quality and microstructure during induction heating assisted incremental forming of Ti-6Al-4V sheet. *J. Mater. Process. Technol.* **2022**, *299*, 117323. [CrossRef]
74. Kotkunde, N.; Gupta, A.K.; Paresi, P.R.; Singh, S.K. Experimental and Finite Element Studies of Stretch Forming Process for Ti-6Al-4V Alloy at Elevated Temperature. *Mater. Today Proc.* **2017**, *4 Pt D*, 5266–5273. [CrossRef]
75. Cheng, D.-J.; Xu, F.; Xu, S.-H.; Zhang, C.-Y.; Zhang, S.-W.; Kim, S.-J. Minimization of Surface Roughness and Machining Deformation in Milling of Al Alloy Thin-Walled Parts. *Int. J. Precis. Eng. Manuf.* **2020**, *21*, 1597–1613. [CrossRef]
76. Riaz, A.A.; Hussain, G.; Ullah, N.; Wei, H.; Alkahtani, M.; Khan, M.N. An investigation on the effects of tool rotational speed and material temper on post-ISF tensile properties of Al2219 alloy. *J. Mater. Res. Technol.* **2021**, *10*, 853–867. [CrossRef]
77. Kumar, V.; Kumar, R. Investigation of Surface Roughness in Incremental Sheet Forming of AA 2014-T6 using Taguchi's Method. *J. Phys. Conf. Ser.* **2020**, *1519*, 012009. [CrossRef]

78. Wu, R.; Hu, Q.; Li, M.; Cai, S.; Chen, J. Evaluation of the forming limit of incremental sheet forming based on ductile damage. *J. Mater. Process. Technol.* **2021**, *287*, 116497. [CrossRef]
79. Pandivelan, C.; Jeevanantham, A.K.; Sathiyanarayanan, C. Optimization Study on Incremental Forming of Sheet Metal AA5052 for Variable Wall Angle using CNC Milling Machine. *Mater. Today Proc.* **2018**, *5 Pt 2*, 12832–12836. [CrossRef]
80. Kumar Singh, A.; Agrawal, A. Formability Analysis of AA1200 H14 Aluminum Alloy Using Single Point Incremental Forming Process. *Trans. Indian Inst. Met.* **2020**, *73*, 1975–1984. [CrossRef]
81. Rashid, H.; Hussain, G.; Rehman, K.; Khan, S.; Alkahtani, M.; Abidi, M.H. Characterization of residual stresses in an asymmetrical shape produced through incremental forming. *CIRP J. Manuf. Sci. Technol.* **2020**, *31*, 478–491. [CrossRef]
82. Shojaeefrd, M.; Khalkhali, A.; Shahbaz, S. Sensitivity analysis and optimization of the surface roughness in the incremental forming of mild steel sheets. *Sci. Iran.* **2021**, *28*, 316–325. [CrossRef]
83. Dabwan, A.; Ragab, A.E.; Saleh, M.A.; Anwar, S.; Ghaleb, A.M.; Rehman, A.U. Study of the Effect of Process Parameters on Surface Profile Accuracy in Single-Point Incremental Sheet Forming of AA1050-H14 Aluminum Alloy. *Adv. Mater. Sci. Eng.* **2020**, *2020*, 7265941. [CrossRef]
84. Mandaloi, G.; Nagargoje, A.; Gupta, A.K.; Banerjee, G.; Shahare, H.Y.; Tandon, P. A Comprehensive Review on Experimental Conditions, Strategies, Performance, and Applications of Incremental Forming for Deformation Machining, ASME. *J. Manuf. Sci. Eng.* **2022**, *144*, 110802. [CrossRef]
85. Shrivastava, P.; Tandon, P. Microstructure and texture based analysis of forming behavior and deformation mechanism of AA1050 sheet during Single Point Incremental Forming. *J. Mater. Process. Technol.* **2019**, *266*, 292–310. [CrossRef]
86. Chang, Z.; Chen, J. A new void coalescence mechanism during incremental sheet forming: Ductile fracture modeling and experimental validation. *J. Mater. Process. Technol.* **2021**, *298*, 117319. [CrossRef]
87. Nirala, H.K.; Agrawal, A. Reprint of: Residual stress inclusion in the incrementally formed geometry using Fractal Geometry Based Incremental Toolpath (FGBIT). *J. Mater. Process. Technol.* **2021**, *287*, 116623. [CrossRef]
88. Kubit, A.; Al-Sabur, R.; Gradzik, A.; Ochał, K.; Slota, J.; Korzeniowski, M. Investigating Residual Stresses in Metal-Plastic Composites Stiffening Ribs Formed Using the Single Point Incremental Forming Method. *Materials* **2022**, *15*, 8252. [CrossRef] [PubMed]
89. Najm, S.M.; Paniti, I.; Trzepieciński, T.; Nama, S.A.; Viharos, Z.J.; Jacso, A. Parametric Effects of Single Point Incremental Forming on Hardness of AA1100 Aluminium Alloy Sheets. *Materials* **2021**, *14*, 7263. [CrossRef] [PubMed]
90. Li, R.; Wang, T. Research on Single Point Incremental Forming Characteristics of Perforated TA1 Sheet. *Metals* **2022**, *12*, 1944. [CrossRef]
91. Oleksik, V.; Trzepieciński, T.; Szpunar, M.; Chodoła, Ł.; Ficek, D.; Szczęsny, I. Single-Point Incremental Forming of Titanium and Titanium Alloy Sheets. *Materials* **2021**, *14*, 6372. [CrossRef] [PubMed]
92. Bai, L.; Li, Y.; Yang, M.; Lin, Y.; Yuan, Q.; Zhao, R. Modeling and Analysis of Single Point Incremental Forming Force with Static Pressure Support and Ultrasonic Vibration. *Materials* **2019**, *12*, 1899. [CrossRef]
93. Trzepieciński, T.; Kubit, A.; Dzierwa, A.; Krasowski, B.; Jurczak, W. Surface Finish Analysis in Single Point Incremental Sheet Forming of Rib-Stiffened 2024-T3 and 7075-T6 Alclad Aluminium Alloy Panels. *Materials* **2021**, *14*, 1640. [CrossRef]
94. Available online: https://www.atlassteels.com.au/documents/Atlas_Aluminium_datasheet_3003_rev_Oct_2013.pdf (accessed on 26 May 2023).
95. Available online: https://www.theworldmaterial.com/din-en-1-2379-steel-x153crmov12-material/ (accessed on 26 May 2023).
96. Available online: https://www.robots.com/robots/kuka-kr-210 (accessed on 26 May 2023).
97. Available online: https://www.kuka.com/-/media/kuka-downloads/imported/8350ff3ca11642998dbdc81dcc2ed44c/pf0031_kr_2102_k_en.pdf (accessed on 3 May 2023).
98. 2823-12. Available online: https://www.sensor-hbm.com (accessed on 4 May 2023).
99. QuantumX MX840B/MX440B: Universal Data Acquisition Module. Available online: https://www.hbm.com/en/2129/quantumx-mx840b-8-channel-universalamplifier/ (accessed on 23 September 2022).
100. Available online: https://www.archimina.ro/articole/Articles-Detail/52-Placarea-sculelor-aschietoare-cu-Nitrura-de-Titan{-}{-}-TiN/0 (accessed on 6 March 2021).
101. Available online: http://magnum.engineering.upm.ro/~gabriela.strnad/Tehnologia%20materialelor%20I%20-%20curs%20licenta%20an%20I/2%20LABORATOR/saptamana%205-6/Incercarea%20de%20ambutisare%20prin%20metoda%20Erichsen.pdf (accessed on 11 March 2021).
102. Oraon, M.; Sharma, V. Predicting Force in Single Point Incremental Forming by Using Artificial Neural Network. *Int. J. Eng.* **2018**, *31*, 88–95.
103. Harfoush, A.; Haapala, K.R.; Tabei, A. Application of Artificial Intelligence in Incremental Sheet Metal Forming: A Review. *Procedia Manuf.* **2021**, *53*, 606–617. [CrossRef]
104. Arshad, S. Single Point Incremental Forming. A Study of Forming Parameters, Forming Limits and Part Accuracy of Aluminium 2024, 6061 and 7475 Alloys. Ph.D. Thesis, KTH Royal Institute of Technology, Stockholm, Sweden, June 2012.
105. Trzepiecinski, T.; Oleksik, V.; Pepelnjak, T.; Najm, S.M.; Paniti, I.; Maji, K. Emerging Trends in Single Point Incremental Sheet Forming of Lightweight Metals. *Metals* **2021**, *11*, 1188. [CrossRef]

materials

Article

Technological Processes for Increasing the Cavitation Erosion Resistance of Nimonic 80A Superalloys

Cosmin Belin [1], Ion Mitelea [1], Ilare Bordeașu [2], Corneliu Marius Crăciunescu [1] and Ion-Dragoș Uțu [1,*]

[1] Department of Materials and Fabrication Engineering, Politehnica University Timisoara, Bulevardul Mihai Viteazul nr.1, 300222 Timisoara, Romania; cosmin.belin@student.upt.ro (C.B.); ion.mitelea@upt.ro (I.M.); corneliu.craciunescu@upt.ro (C.M.C.)

[2] Department of Mechanical Machines, Equipment and Transports, Politehnica University Timisoara, Bulevardul Mihai Viteazul nr.1, 300222 Timisoara, Romania; ilare.bordeasu@upt.ro

* Correspondence: dragos.utu@upt.ro

Abstract: Nickel-based superalloys are frequently used to manufacture the components that operate under cavitation erosion conditions, such as aircraft gas turbine construction, nuclear power systems, steam turbine power plants, and chemical and petrochemical industries. Their poor performance in terms of cavitation erosion leads to a significant reduction in service life. This paper compares four technological treatment methods to improve cavitation erosion resistance. The cavitation erosion experiments were carried out on a vibrating device with piezoceramic crystals in accordance with the prescriptions of the ASTM G32—2016 standard. The maximum depth of surface damage, the erosion rate, and the morphologies of the eroded surfaces during the cavitation erosion tests were characterized. The results indicate that the thermochemical plasma nitriding treatment can reduce mass losses and the erosion rate. The cavitation erosion resistance of the nitrided samples is approximately 2 times higher than that of the remelted TIG surfaces, approximately 2.4 times higher than that of the artificially aged hardened substrate, and 10.6 times higher than that of the solution heat-treated substrate. The improvement in cavitation erosion resistance for Nimonic 80A superalloy is attributed to the finishing of the surface microstructure, graining, and the presence of residual compressive stresses, factors that prevent crack initiation and propagation, thus blocking material removal during cavitation stresses.

Keywords: Nimonic 80A; cavitation erosion; alternative technologies

Citation: Belin, C.; Mitelea, I.; Bordeașu, I.; Crăciunescu, C.M.; Uțu, I.-D. Technological Processes for Increasing the Cavitation Erosion Resistance of Nimonic 80A Superalloys. *Materials* **2023**, *16*, 3206. https://doi.org/10.3390/ma16083206

Academic Editors: Corneliu Munteanu and Ioan Doroftei

Received: 29 March 2023
Revised: 13 April 2023
Accepted: 17 April 2023
Published: 18 April 2023

1. Introduction

Nickel-based superalloys are engineering materials whose mechanical properties can be improved by alloying, by the kinetics of solidification processes, and by thermal and thermochemical treatments [1,2]. They are intended for the manufacturing of components of aircraft gas turbines, steam turbine power plants, components of the nuclear energy production industry, the chemical and petrochemical industry, and other hydromechanical equipment [2].

Such components are frequently exposed during operation to cavitation erosion phenomena. Cavitation is the process of the formation, growth, and implosion of bubbles containing vapor, gas, or vapor–gas mixtures due to cyclic pressure changes in the flowing liquid [2–8].

The cavitation phenomenon involves the interaction of hydrostatic, mechanical, metallurgical, and chemical processes. This notion includes the formation and subsequent rupture of voids in liquids, which are caused by the local drop in the hydrostatic pressure below a critical value, which is the saturation vapor pressure. The momentary decreases of the local pressure can be caused, on the one hand, by the spontaneous increase in the flow speed, and on the other hand, by an expansion of the shock waves during the low-pressure phase.

The voids contain the vapor phase of the liquid with or without parts of the gases that have been dissolved in it [1].

The research carried out so far has shown that the level of cavitation damage in metallic alloys depends, on the one hand, on the cavitation intensity, which is specific to the hydro-dynamics of the cavitation current, and, on the other hand, on the nature and physical condition of the material that produces a certain influence [1–4]:

- Chemical composition, respectively the nature and amount of alloying elements;
- Semi-finished product development technology (cast, laminated, etc.);
- Surface hardening treatment (thermal, thermomechanical, thermo-chemical, etc.);
- The values of the mechanical properties (Rm, Rp0.2, HV, KCU);
- The morphology of the microstructural constituents.

In order to improve the cavitation erosion resistance, research efforts are oriented towards the application of volume or surface thermal treatments [2,5,7,8], as well as some techniques of physical surface modification (welding deposition [9,10], cathodic arc plasma deposition [6], laser shock peening [11], and laser or TIG surface remelting [12,13], etc.). In many cases, the coatings developed by these techniques show a series of drawbacks, among which the presence of porosities, low adhesion, dilution, and inhomogeneous microstructure are mentioned. Due to these defects, the performances of the deposited layers in terms of cavitation erosion are limited. The technical literature contains numerous data on the cavitation erosion behavior of stainless steels, alloyed cast irons, cobalt-based alloys, and shape memory alloys, but few data on the cavitation erosion of nickel-based alloys [2,9,14,15]. Mesa D.H. et al. [8] investigated the effect of a composite surface treatment, which consisted of a high-temperature gas nitriding (HTGN) followed by a low-temperature plasma nitriding (LTPN), applied to a Duplex stainless steel UNS S31803, on the strength of cavitation erosion. The composed treatment caused an almost 9-fold increase in the incubation time, a phenomenon due to the mechanical resistance and higher hardness of the surface layer. Chang J.T. et al. [3] demonstrated that the deposition of a Ni-Al film on AISI 1045 carbon steel using the cathodic arc plasma (CAP) ion plating process is beneficial for cavitation erosion resistance. Chollet S. et al. [5] investigated the microstructure of Ni-based superalloys, either monocrystalline or polycrystalline, subjected to plasma nitriding, highlighting the variation in the thickness of the nitrogen-enriched layer depending on the chemical composition of the alloy and the duration of the thermochemical treatment.

The present contribution investigates the cavitation erosion response of a nickel base superalloy subjected to different hot manufacturing processes (plasma nitriding, TIG surface remelting, and solution treatment with or without artificial aging).

2. Researched Material

The research material is Nimonic 80A alloy (EN NiCr20TiAl; UNS No 7080; W. Nr. 24952& 24631), with the chemical composition listed in Table 1.

Table 1. Chemical composition of Nimonic 80A alloy, wt.%.

Chromium	20.5
Titanium	1.98
Aluminum	1.54
Iron	2.14
Cobalt	1.61
Manganese	0.63
Silicon	0.45
Copper	0.12
Carbon	0.08
Sulfur	0.014
Zirconium	0.11
Nickel	Balance

The main alloying element is chromium, which improves the resistance to hot oxidation. Additions of aluminum and titanium further favor the strengthening of the alloy through aging. The semi-finished products made of this alloy are delivered either in a solution heat-treated state or in an artificial aged state. The best mechanical properties of the Nimonic 80A alloy are obtained by applying the solution treatment followed by artificial aging.

The microstructure given in Figure 1 is made up of a solid solution matrix γ with a Ni base, of primary carbides, MC, with dimensions up to 20–30 µm, arranged intragranularly, that are rich in Cr of the $M_{23}C_6$ type with dimensions of a few 100 nm, arranged intergranularly and from precipitations of intermetallic phases γ' of the $Ni_3(Al, Ti)$ type, whose size is ≈40–60 nm), inside the matrix grains. These observations fully agree with the results of other research works [5,16].

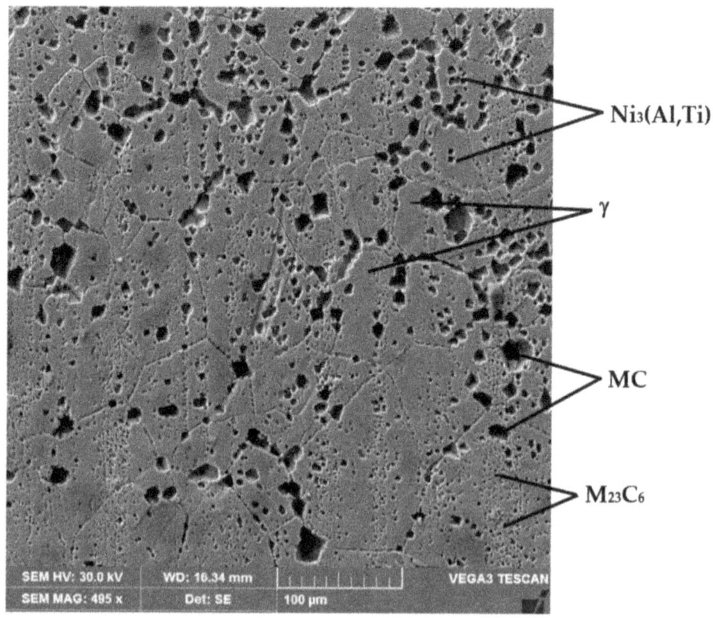

Figure 1. The microstructure of the Nimonic 80A alloy solution heat-treated followed by artificial aging.

3. Experimental Stand and Work Procedure

From Nimonic 80A superalloy delivered in the form of cylindrical bars with a diameter of 20 mm, cavitation samples were made (Figure 2), which were subjected to the following technological processing:

- Solution treatment, 1080 °C/8 h/air;
- Solution treatment followed by artificial ageing, 700 °C/16 h/air (the cyclogram shown in Figure 3);
- Solution treatment followed by surface local remelting using the TIG technique;
- Solution treatment followed by plasma nitriding at 530 °C/14 h.

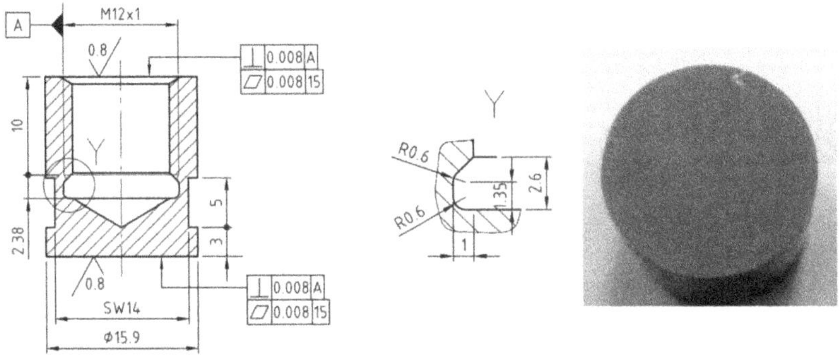

(**a**) (**b**) (**c**)

Figure 2. Geometry of cavitation samples: (**a**) shape and sizes; (**b**) manufacturing detail; (**c**) surface before cavitation.

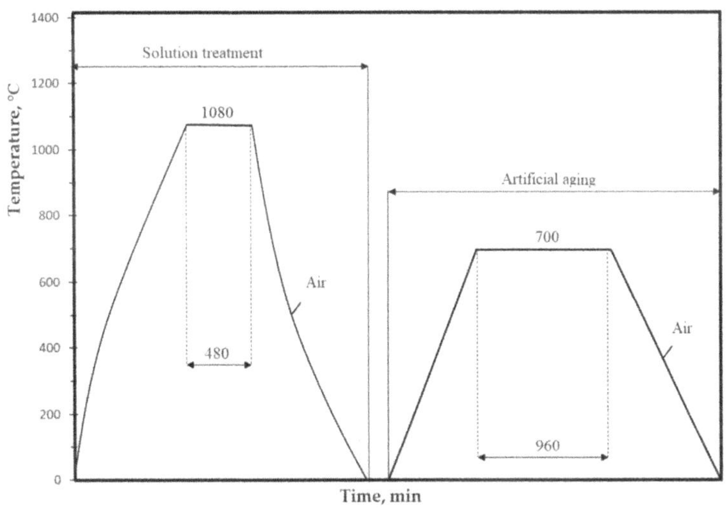

Figure 3. Cyclogram of the solution heat treatment followed by artificial aging.

The components of the experimental stand used for the local surface TIG remelting are shown in Figure 4. The mobility, which provided to the welding source in order to allow the correct positioning of the samples on the work table was achieved with the help of a movement-pendulum device built for this purpose.

Previous research [17] established that the optimal values of the melting current of the heat input that ensure the highest performance of cavitation erosion resistance are as follows: Is = 80 A, El = 4704 J/cm. The parameters of the local TIG surface remelting process together with the macroscopic image of the sample in the two stages (remelted surface, respectively later ground and polished) are shown in Figure 5.

The thermochemical plasma nitriding treatment was carried out in a furnace equipped with a PROTHERM 500 (Mississauga, Ontario L5L 5Y7 Canada) controller. The cystogram of this treatment is shown in Figure 6 [17]. The samples were introduced into the retort of the installation and preheated to 350 °C for 30 min. This was followed by the introduction of ammonia and the continuation of the heating—holding phases up to the operating temperature of 530 °C. From 480 °C, the dissociation of ammonia practically started with the release of active nitrogen atoms. The cooling of the samples was carried out in the treatment oven up to a temperature of 150 °C, at which time they were taken out into the air.

Figure 4. Overview of the TIG remelting stand: 1 metal frame; 2 piece positioning plate; 3 displacement device; 4 oscillation device; 5 power supplies of displacement–oscillation devices; 6 TIG torch; 7 welding source ESAB ET 300 iP.

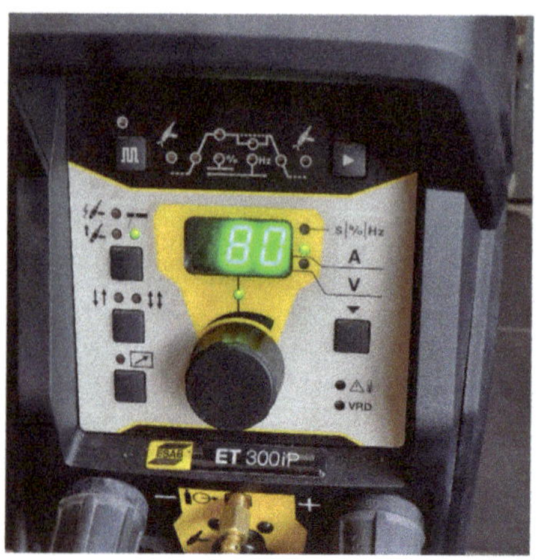

(a)

Figure 5. *Cont.*

(**b**) (**c**)

Current (Is)	80 A
Arc voltage (Ua)	9.8 V
Displacement speed (V)	10 cm/min
The distance between piece electrode (L)	1.5 mm
The step between the lines (P)	2 mm
Linear energy (El)	$El = \frac{Ua \times Is}{V} \times 60 = 4704$ [J/cm]

Figure 5. The control panel of the welding source (**a**), the aspect of the cavitation sample with the TIG remelted surface (**b**), with the surface prepared for testing (**c**), and the TIG remelting parameters.

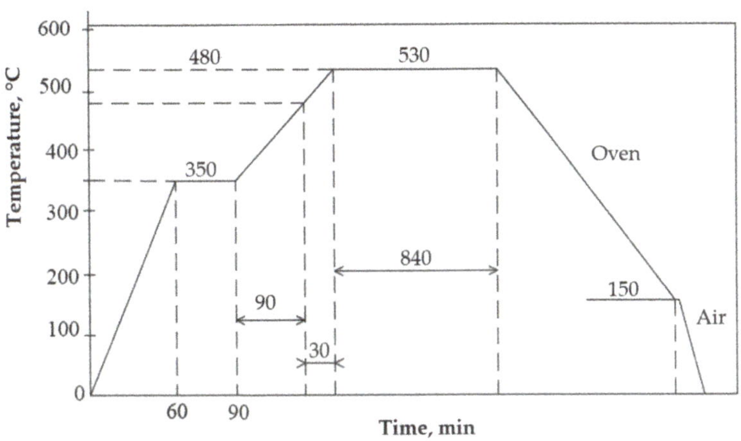

Figure 6. Cyclogram of thermochemical plasma nitriding treatment [18].

The acoustic cavitation of these treated samples was performed using an ultrasonic emitter according to the G32-2016 standard of the American Society for Testing and Materials (ASTM) [19]. The oscillating pressure waves generated by the ultrasonic vibrations cause numerous bubbles to rise and collapse. The bubbles cluster in the cavity, and the sample can oscillate with a frequency lower than the excitation frequency, i.e., subharmonic oscillation. Previous studies on this material [17,18] have shown that ultrasonic cavitation erosion results in plastic deformation, initiation and crack propagation, partial material removal, and ultimately part failure.

Figure 7 shows the detailed image of a vibrating device with piezoceramic crystals used in the experimental program. The functional parameters of the device are as follows:

- Electronic ultrasonic generator power, 500 W;
- Vibration frequency, 20,000 ± 2% Hz;
- Vibration amplitude, 50 μm;
- Sample diameter: 15.8 ± 0.05 mm;
- Power supply: 220 V/50 Hz;
- Working media: drinking water from the industrial network, with a temperature of 22 ± 1 °C.

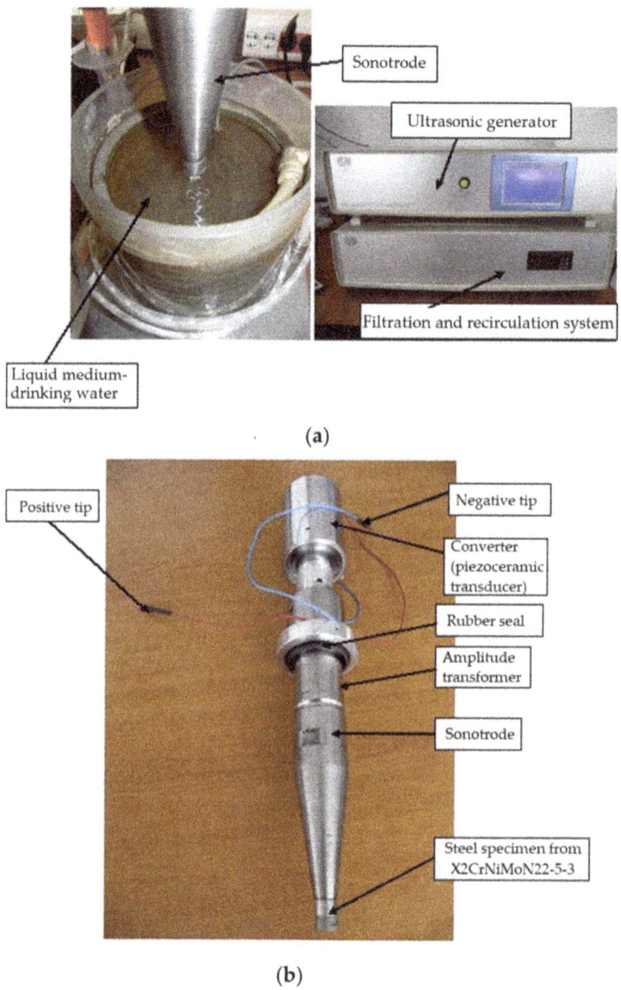

Figure 7. Image of the sample in the cavitation process (**a**) and of the mechanical resonator assembly (**b**).

Before cavitation testing, the exposed surface of each sample was polished on a Buehler Phoenix Beta machine to a roughness of Ra = 0.2 ÷ 0.8 μm.

The total testing time of each sample was 165 min, which was divided into 12 periods (5 min, 10 min and 10 periods of 15 min). At the end of each test period, the samples were washed under a jet of water, with alcohol, acetone, and dried under a jet of warm air. This was followed by their weighing with an analytical balance whose precision was 5 significant decimals (up to 0.00001 g), type Zatklady Mechaniki Precyzyjnej WP 1, in order to determine material losses through cavitation erosion.

After each test period, the surface of the samples was photographed using the Canon PowerShot Sx200 IS, 12× Optical Zoom, whose resolution additionally allowed highlighting the extent of damage over the entire surface. At the end of the 165 min of attack, the roughness of degraded surfaces was investigated using the Mitutoyo SJ210 (Japan) device. The topography of the surfaces was examined with a TESCAN VEGA 3 LMU Bruker EDX Quantax scanning electron microscope.

Based on the mass losses Δm_1, Δm_2, Δm_3, from each period "i", for the set of three samples related to each structural state, the average mass loss was determined, according to the following relationship:

$$\Delta m_i = \sum_{j=1}^{3} \frac{\Delta m_j}{3}, \text{ where j = 1, 2, 3- is the sample number} \tag{1}$$

The mean cumulative mass loss over a certain cavitation attack time, until completion (165 min), was determined with the relationship:

$$M_i = \sum_{i=1}^{12} \Delta m_i \tag{2}$$

Based on these losses, following the procedure described in [20,21], the values of the cumulative mean depths MDE_i and of the erosion penetration rates $MDER_i$ were determined with the relations:

$$MDE_i = \frac{4 \cdot M_i}{\rho \cdot \pi \cdot d_p^2} \; [\mu m] \tag{3}$$

$$MDER_i = \Delta MDE_i / \Delta t_i \quad [\mu m/min.] \tag{4}$$

where

i is the testing time;

Δm_i is the mass of material, lost through erosion, in period "i", in grams;

ρ material density, in grams/mm^3;

Δt_i cavitation time corresponding to period "i" (5 min, 10 min, or 15 min);

d_p diameter of the sample surface, exposed to cavitation (d_p= 15.8 mm);

ΔMDE_i value of the mean penetration depth of erosion, achieved by cavitation during the period Δt_i.

4. Results and Discussions

4.1. Cavitation Curves

Figures 8 and 9 show the variation curves of these two parameters depending on the cavitation attack duration of the four technological options for modifying the microstructure of the considered alloy.

The main observations derived from the analysis of these curves are as follows:

- The highest values of the mean penetration depth of erosion, MDE, are characteristic of the solution treatment;
- The lowest values of MDE were obtained after applying the thermochemical nitriding treatment;
- The local modification by TIG remelting (Is = 80 A) of the material surface led to lower cavitation losses compared to the heat treatment composed of a solution followed by artificial aging (MDE = 6865 μm).

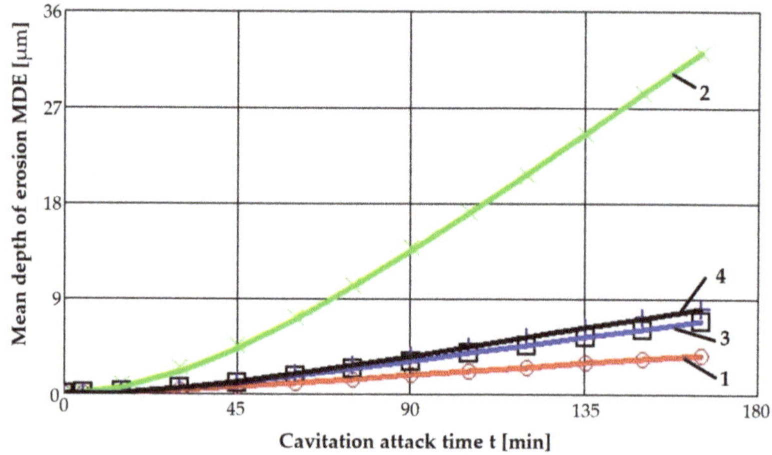

Figure 8. The variation of the mean penetration depth of erosion with the cavitation attack time: 1 nitriding; 2 solution treatment; 3 TIG remelting with Is = 80 A; 4 solution treatment followed by artificial aging.

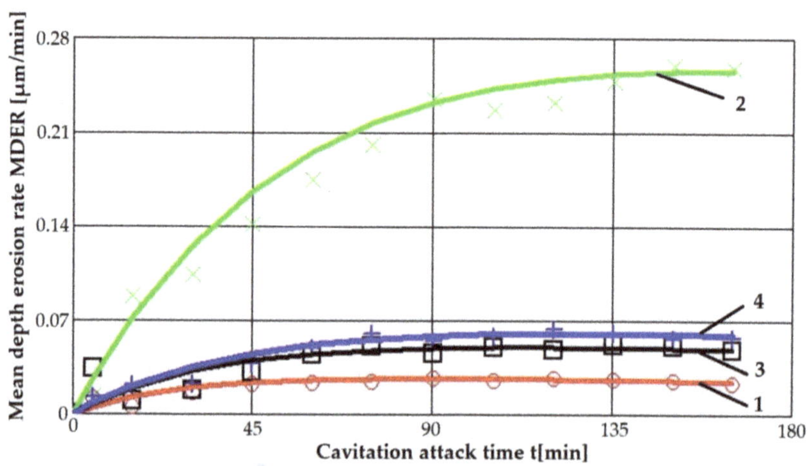

Figure 9. The variation in the mean erosion penetration rate with the cavitation attack time: 1 nitriding; 2 solution treatment; 3 TIG remelting with Is = 80 A; 4 solution treatment followed by artificial aging.

According to the international norm ASTM G32—2016, the inverse of the other cavitation parameter towards which it tends to stabilize, MDERs, defines the resistance to cavitation, R_{cav}.

Table 2 shows its values, presented comparatively for the processing conditions applied to the material.

The data from this table highlights an increase of approximately 4…10 times the cavitation resistance of the strengthened surfaces using the applied techniques.

Table 2. The effect of treatment processes on the cavitation erosion parameters.

Structural State	Cavitation Erosion Resistance Parameter		Variation Compared with Solution Treated State
	$MDER_s$ (μm/min)	$R_{cav.}$ (min/μm)	
Solution treatment	0.255	3.92	-
Solution treatment followed by artificial aging	0.059	16.94	Increase of 4.3 times
Solution treatment followed by local TIG remelting, Is = 80 A	0.049	20.4	Increase of 5.2 times
Solution treatment + nitiding	0.024	41.67	Increase of 10.6 times

4.2. Surface Hardness Measurements

Since hardness is the mechanical property most sensitive to structural changes in a metallic material, the samples processed by the four technological variants were subjected to such measurements. On the front surface of these samples, eight Vickers hardness measurements were made with a load of 50 N. Based on the obtained results, the histogram shown in Figure 10 was drawn. These data demonstrate that there is a full correspondence between the hardness and the material's resistance cavitation erosion degradation. The lowest hardness values are specific to solution heat treatment (approx. 181 HV5), and they correspond to the highest erosion rate (0.255 μm/min). Instead, the thermochemical plasma nitriding treatment ensures a high hardness (approx. 800 HV5), which favors a pronounced decrease in the erosion rate (0.024 μm/min). The high values of the hardness are attributed to the microstructure generated in the nitrided layer, consisting of carbonitride particles embedded in a Ni-based γ solid solution matrix.

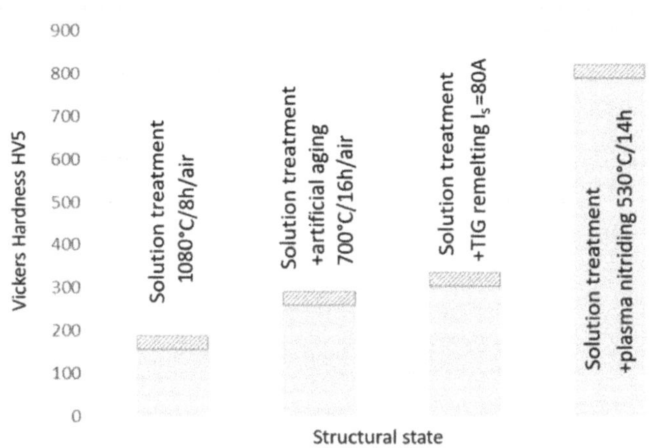

Figure 10. Histogram of surface hardness values.

4.3. Roughness of the Cavitation Tested Surfaces

The samples surfaces subjected to the four technological variants of processing and tested for cavitation erosion for 165 min were roughness measured using the Mitutoyo SJ210 device. Figure 11 exemplifies the three-way measurement of the roughness Ra, Rz, and Rt for the state of nitriding thermochemical treatment. Table 3 summarizes their mean roughness results together with the values of mean erosion penetration depth for the 4 surface ennobling techniques of the Nimonic 80A alloy. The comparative analysis of the results obtained (Table 3) highlights the differences between the solution-treated state and the structural state obtained after applying the other three thermal processing processes.

They prove that by applying the thermochemical plasma nitriding, the greatest increase in cavitation erosion resistance is obtained.

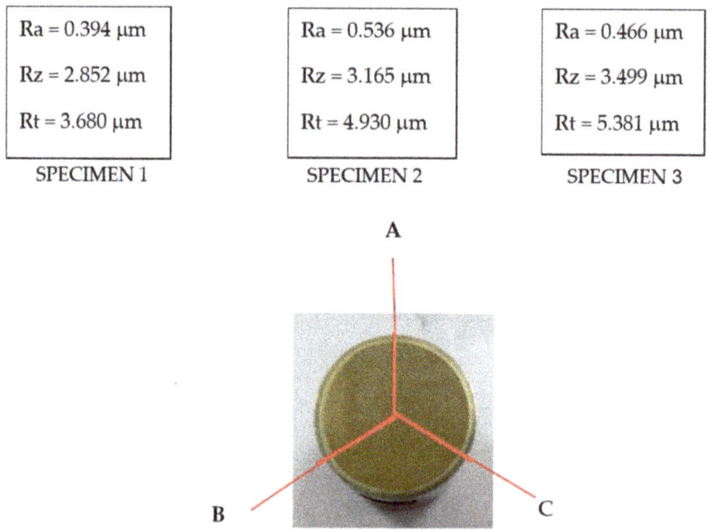

Figure 11. Roughness measurement mode with the Mitutoyo SJ210 device at minute 165.

Table 3. Mean values for MDE and roughness parameters.

Structural State	MDE (165 min.)	Ra, μm	Rz, μm	Rt, μm
Solution treatment	32.11	5.539	31.598	48.026
Solution treatment + artificial aging	8.068	1.975	12.504	17.306
TIG remelting, Is = 80 A	6.86	1.659	10.507	14.543
Nitriding	3.71	0.465	3.172	4.663

4.4. *Structural Analyses*

Figure 12 shows images of some cross-sections through the surface layer of the samples processed by the four technological variants and tested by cavitation for 165 min. In the unetched state, the differences are revealed regarding the intensity of the surface cavitation erosion phenomenon, which are specific to each processing process. The greatest irregularities (Figure 12a) are identified on the material surface subjected to the solution treatment, which leads to the lowest hardness values, respectively, to the highest values of the cavitational parameters, MDE and MDER. On the other hand, the lowest surface degradation (Figure 12d) occurs on surfaces strengthened by plasma nitriding.

The same observations can be derived from the macroscopic images of the degraded surfaces by ultrasonic cavitation for 45 min and 165 min (Figure 13). It should be noted that after the application of the solution treatment, the surface degradation is visible even after the first 45 min of cavitation attack and is pregnant at the end of the 165 min test period (Figure 13a). Instead, the other three processes of the material structure modification through which the hardness increases more and more are manifested by a less intense surface degradation throughout the cavitation erosion attack (Figure 13b–d). The morphology of the structural transformations involved in the layer-substrate system before and after the cavitation erosion tests have been analyzed in scientific works [17,18].

Scanning electron microscope examination of the topography of the surfaces treated differently and tested for cavitation (Figure 14a–d) proves that the material removal mechanisms, during cavitation erosion, are controlled by a plastic deformation process, accompanied by the formation of sliding steps, characteristic of the solid solution matrix with a nickel base (γ), which has an fcc crystalline lattice. From their analysis, it can be seen that in the samples subjected to the solution treatment, the intensity of the surface degradation phenomenon is at a maximum, with the boundaries between the grains being preferentially attacked by microcracking followed by the expulsion of the crystalline grains. The topographic image of the samples surface from the reference material (Figure 14a) highlights the formation of craters whose depths are greater than 20 μm, determined by the preferential cavities of the boundaries between the grains of the solid solution γ, which are more fragile. It is emphasized that the boundaries between the grains are less able to absorb the deformation energy due to the stresses induced in the material by the cavitation impact waves, and as a result, they erode faster than the solid solution microstructure γ.

(a)

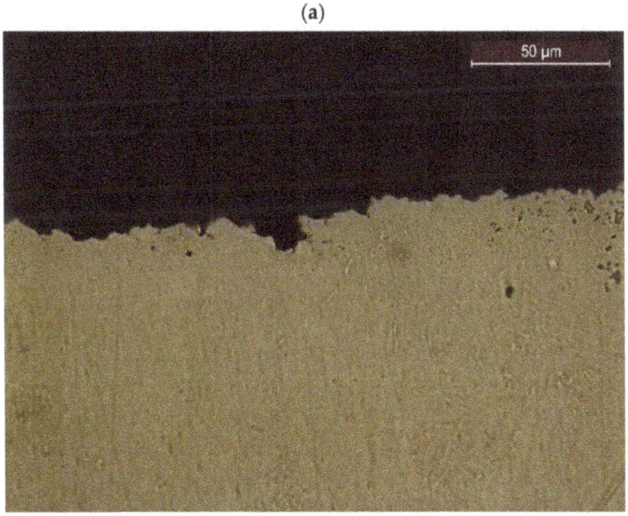

(b)

Figure 12. *Cont.*

(c)

(d)

Figure 12. The images of some cross sections through the cavitated layers after 165 min: (**a**) solution treatment; (**b**) solution treatment followed by artificial aging; (**c**) solution treatment followed by TIG remelting; (**d**) solution treatment followed by nitriding.

In the case of the samples treated by the solution followed by artificial aging, the nucleation of fatigue cracks occurred on the boundaries between the solid solution grains γ and the precipitated secondary phases, i.e., in the areas with stress concentrators. Interfaces cannot undergo mechanical strengthening and consequently become brittle, and they fragment due to cavitation attack. At the same time, the localized strength of the structural matrix increased with the degree of cold deformation during cavitation erosion. This explains why the boundaries between the grains were attacked more compared to the γ solid solution grains in the studied alloy. The pinches that initially formed underwent the phenomenon of coalescence over time and gave rise to microcracks. As the test time increased, the attacked area of the samples became larger, with more micro-pinching, pinching, and subsequent deformation of the solid solution matrix. The removal of the material through the coalescence of these pinches led to the formation of microcraters, as can be seen in Figure 14b.

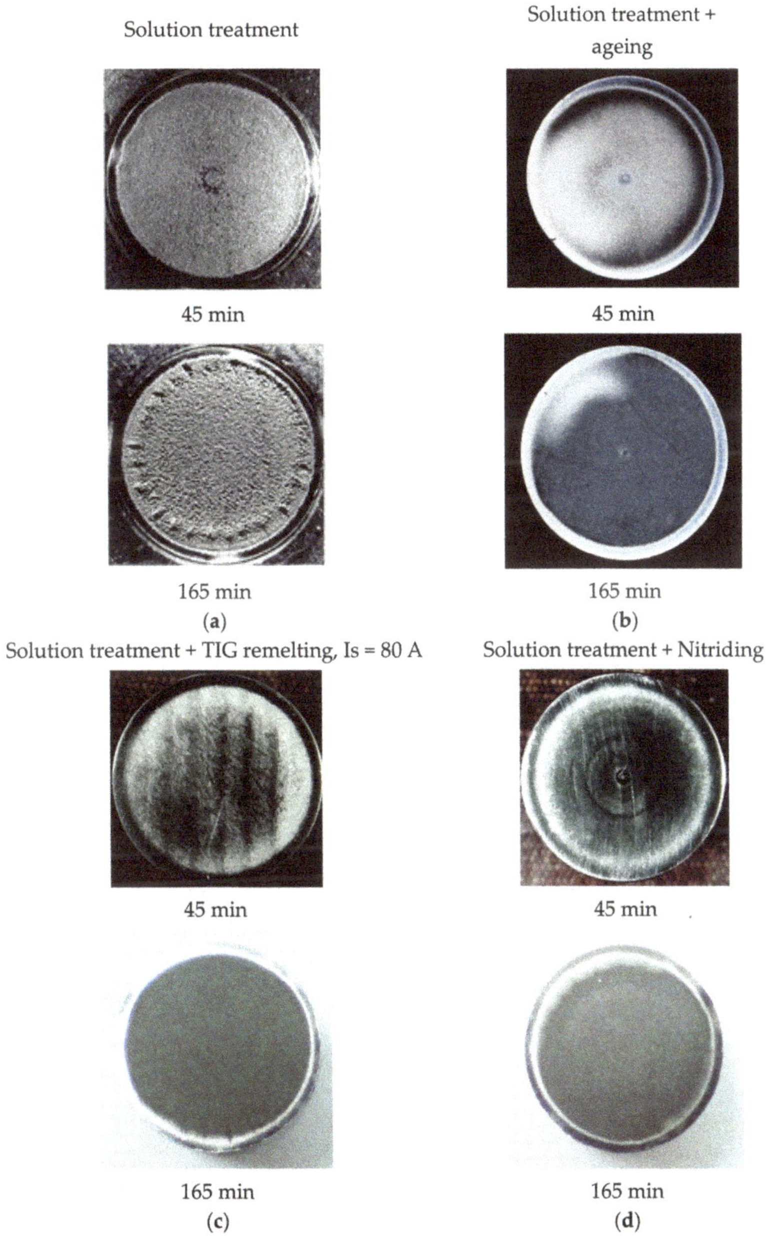

Figure 13. The macrographic images of the degraded surfaces after 45 min and 165 min of cavitation: (**a**) solution treatment; (**b**) solution treatment followed by artificial aging; (**c**) solution treatment followed by TIG remelting with Is = 80 A; (**d**) solution treatment followed by nitriding.

A similar situation from a qualitative point of view also occurs in the samples whose surface was locally remelted using the TIG technique, in the sense that the initiation of cracks occurred in the area of the interfaces between the solid solution grains γ and the particles of chemical combinations (Figure 14c), with poor mechanical resistance characteristics. Finishing the grain and limiting the precipitation of secondary phases after TIG surface remelting increased the energy absorption capacity of the cavitation impact wave, which delayed the nucleation of cracks.

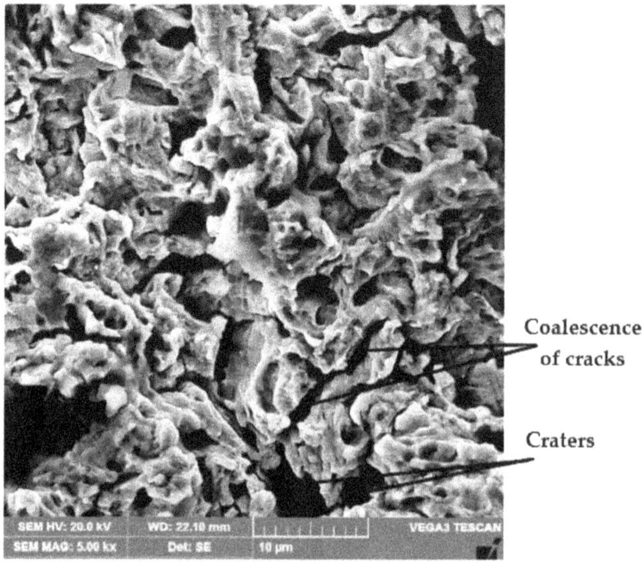

(a)

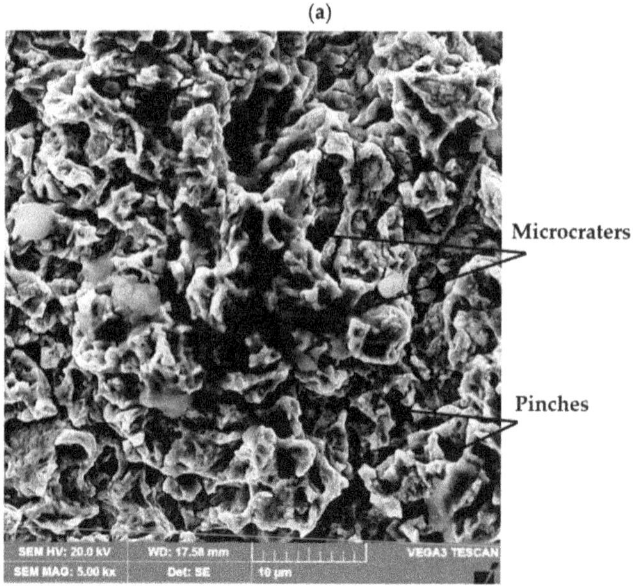

(b)

Figure 14. *Cont.*

(c)

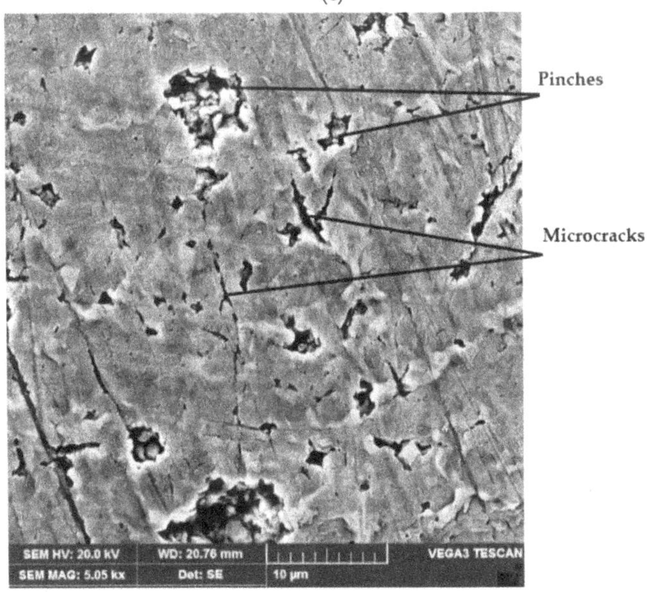

(d)

Figure 14. The microfractographic images of the degraded surfaces at 165 min of cavitation: (**a**) solution treatment; (**b**) solution treatment followed by artificial aging; (**c**) solution treatment followed by TIG remelting with Is = 80 A; (**d**) solution treatment followed by nitriding.

By plasma-nitriding samples, the crack initiations were determined by the nitride particles and the separation limits between them, as well as the separation limits between the grains of solid solution alloyed with nitrogen. The high-hardness microstructure of the marginal layer causes small and uniform wear with fine nicks without the appearance of deep craters. The pinches present in the eroded surface, especially on the boundaries of the grains of solid solution γ are the places of the former nitride particles of the alloying elements. Increasing the nitrogen content in the surface layer increases the elastic energy returned to the cavitation medium and decreases the amount of plastic energy absorbed by the alloy in the impact points of the cavitation. The cavitation depths do not exceed 10 μm, and the values of the roughness parameters decrease by approx. 10 times. The microcraters present on the eroded surface are the places of the former nitride particles of the alloying elements.

5. Conclusions

Compared to the solution treatment, the application of artificial aging caused a 4.3 times increase in the cavitation erosion resistance.

Ennobling the surface by local TIG remelting (Is = 80 A) or by thermochemical plasma nitriding treatment ensured an increase in cavitation erosion resistance by 5.2 and 10.6 times, respectively.

The low hardness values (approx. 181 HV5), which are specific to the solution treatment, are responsible for the highest erosion rates (0.255 μm/min.) and the high hardness values (approx. 800 HV5), characteristic of the thermochemical plasma nitriding treatment, which favors a pronounced decrease in the erosion rate (0.024 μm/min.).

The microstructure resulting from nitriding, which has a high hardness, suffers a slower and uniform degradation with extremely fine pinches.

Although the nitriding treatment involves additional costs, it is fully justified due to the extension of the life of the components that work in cavitation conditions and the reduction in maintenance times and costs during the operation period.

The topography of the cavitation erosion surfaces highlights the presence of caverns arranged towards the periphery of the solution-treated samples, which have a lower hardness and a slower and uniform degradation with the finishing of the crystalline grains and with the increase in the strengthening degree through the applied techniques (TIG and plasma nitriding).

Author Contributions: Conceptualization, I.M. and I.B.; methodology, I.M., I.B., C.B., I.-D.U. and C.M.C.; formal analysis, I.M., I.B. and I.-D.U.; investigation, I.M., C.B., I.B., I.-D.U. and C.M.C.; resources, I.M., I.B. and C.M.C.; data curation, C.B. and I.-D.U.; writing—original draft preparation, I.M. and I.B.; writing—review and editing, I.M., I.B., I.-D.U. and C.M.C.; visualization, I.M., I.B., C.B., I.-D.U. and C.M.C.; supervision, I.M. and I.B. All authors have read and agreed to the published version of the manuscript.

Funding: This research received no external funding.

Institutional Review Board Statement: Not applicable.

Informed Consent Statement: Not applicable.

Data Availability Statement: Not applicable.

Conflicts of Interest: The authors declare no conflict of interest.

References

1. Franc, J.-P.; Michel, J.M. *Fundamentals of Cavitation*; Kluwer Academic Publishers: Dordrecht, The Netherlands; Boston, MA, USA; London, UK, 2004.
2. Eliaseny, K.M.; Christiansen, T.; Somers, M. Low temperature gaseous nitriding of Ni based superalloys. *Surf. Eng.* **2010**, *26*, 248–255. [CrossRef]
3. Chang, J.T.; Yeh, C.H.; He, J.J.; Chen, K.C. Cavitation erosion and corrosion behavior of Ni–Al intermetallic coatings. *Wear* **2003**, *255*, 162–169. [CrossRef]
4. Krella, A.K.; Krupa, A. Effect of cavitation intensity on degradation of X6CrNiTi18-10 stainless steel. *Wear* **2018**, *408–409*, 180–189. [CrossRef]
5. Chollet, S.; Pichont, L.; Cormier, J.; Dubois, J.B.; Villechaise, P.; Drouet, M.; Declemy, A.; Templier, C. Plasma assisted nitriding of Ni-based superalloys with various microstructures. *Surf. Coat. Technol.* **2013**, *235*, 318–325. [CrossRef]

6. Han, S.; Lin, J.H.; Kuo, J.J.; He, J.L.; Shih, H.C. The cavitation-erosion phenomenon of chromium nitride coatings deposited using cathodic arc plasma deposition on steel. *Surf. Coat. Technol.* **2002**, *161*, 20–25. [CrossRef]
7. Lo, K.H.; Kwok, C.T.; Wang, K.Y.; Ai, W. Implications of solution treatment on cavitation erosion and corrosion resistances and synergism of austenitic stainless steel. *Wear* **2017**, *392–393*, 159–166. [CrossRef]
8. Mesa, D.H.; Pinedo, C.E.; Tschiptschin, A.P. Improvement of the cavitation erosion resistance of UNS S31803 stainless steel by duplex treatment. *Surf. Coat. Technol.* **2010**, *205*, 1552–1556. [CrossRef]
9. Santa, J.F.; Blanco, J.A.; Giraldo, J.E.; Toro, A. Cavitation erosion of martensitic and austenitic stainless steel welded coatings. *Wear* **2011**, *271*, 1445–1453. [CrossRef]
10. Mitelea, I.; Bordeasu, I.; Mutascu, D.; Buzdugan, D.; Craciuneescu, C.M. Cavitation resistance of Stellite 21 coatings tungsten inert gas (TIG) deposited onto duplex stainless steel X2CrNiMoN22-5-3. *Mater. Test.* **2022**, *64*, 967–976. [CrossRef]
11. Tong, Z.; Jiao, J.; Zhou, W.; Yang, Y.; Chen, L.; Liu, H.; Sun, Y.; Ren, X. Improvement in cavitation erosion resistance of AA5083 aluminium alloy by laser shock processing. *Surf. Coat. Technol.* **2019**, *377*, 124799. [CrossRef]
12. Alabeedi, K.F.; Abboud, J.H.; Benyounis, K.Y. Microstructure and erosion resistance enhancement of nodular cast iron by laser melting. *Wear* **2009**, *266*, 925–933. [CrossRef]
13. Mitelea, I.; Bordeasu, I.; Riemschneider, E.; Utu, I.D.; Craciunescu, C.M. Cavitation erosion improvement following TIG surface-remelting of gray cast iron. *Wear* **2022**, *496–497*, 204282. [CrossRef]
14. Godoy, C.; Mancosu, R.D.; Lima, M.M.; Brandão, D.; Housden, J.; Avelar-Batista, J.C. Influence of plasma nitriding and PAPVD $Cr_{1-x}N_x$ coating on the cavitation erosion resistance of an AISI 1045 steel. *Surf. Coat. Technol.* **2006**, *200*, 5370–5378. [CrossRef]
15. Mitelea, I.; Dimian, E.; Bordeasu, I.; Craciunescu, C. Ultrasonic cavitation erosion of gas nitrided Ti-6Al-4V alloys. *Ultrason. Sonochem.* **2014**, *21*, 1544–1548. [CrossRef] [PubMed]
16. Manova, D.; Hirsch, D.; Gerlach, J.W.; Mändl, S.; Neumann, H.; Rauschenbach, B. In situ investigation of phase formation during low energy ion nitriding of Ni80Cr20 alloy. *Surf. Coat. Technol.* **2014**, *259*, 434–441. [CrossRef]
17. Belin, C.; Mitelea, I.; Bordeasu, I.; Craciunescu, C.M. Ultrasonic cavitation erosion mechanism in TIG remelted surfaces of Nimonic 80 A alloy. *Mater. Today Proc.* **2021**, *45*, 4207–4210. [CrossRef]
18. Mitelea, I.; Bordeasu, I.; Belin, C.; Utu, I.D.; Craciunescu, C.M. Cavitation Resistance, Microstructure, and Surface Topography of Plasma Nitrided Nimonic 80 A Alloy. *Materials* **2022**, *15*, 6654. [CrossRef] [PubMed]
19. *ASTM G 32 Standard*; Standard Test Method for Cavitation Erosion Using Vibratory Apparatus. ASTM International: West Conshohocken, PA, USA, 2016.
20. Bordeaşu, I.; Patrascoiu, C.; Badarau, R.; Sucitu, L.; Popoviciu, M.; Balasoiu, V. *New Contributions in Cavitation Erosion Curves Modeling*; FME Transactions Faculty of Mechanical Engineering, New Series, vol. 34, Nr.1/2006, YU ISSN 1451-2092; University of Belgrade: Belgrade, Serbia, 2006; pp. 39–44. ISSN 1451-2092.
21. Bordeasu, I. *Eroziunea Cavitationala a Materialelor*; Editura Politehnica Timisoara: Timișoara, Romania, 2006.

materials
MDPI

Article

Investigating the Efficiency of a Heat Recovery–Storage System Using Heat Pipes and Phase Change Materials

Robert Stefan Vizitiu [1,*], Andrei Burlacu [1,*], Chérifa Abid [2], Marius Costel Balan [1], Stefanica Eliza Vizitiu [1], Marius Branoaea [1] and Nicoleta Elena Kaba [3]

1 Faculty of Civil Engineering and Building Services, "Gheorghe Asachi" Technical University of Iasi, 700050 Iasi, Romania
2 Aix-Marseille Université, CNRS, IUSTI, 13453 Marseille, France
3 Faculty of Civil Engineering, Politehnica University of Timisoara, 300223 Timișoara, Romania
* Correspondence: robert-stefan.vizitiu@academic.tuiasi.ro (R.S.V.); andrei.burlacu@academic.tuiasi.ro (A.B.)

Abstract: This study presents an experimental and numerical investigation into the efficiency of a two-stage heat recovery–storage system for reducing the thermal energy losses in the industry. The system is designed to recover and store waste thermal energy from residual fluids using heat pipes for recovery and an environmentally friendly phase change material for heat storage. Experimental investigation was conducted using water as the primary agent and varying the temperature between 60 °C, 65 °C, and 70 °C at a constant flow rate of 24 L/min. The secondary agent, also water, was used at an initial temperature of 10 °C and the flow rate was varied between 1 L/min, 2 L/min, and 3 L/min. The results show that the system had a peak efficiency of 78.1% and was able to recover a significant amount of thermal energy. This study demonstrates the potential of this system to reduce the thermal energy losses in the industry and highlight the importance of further research and development in this field, as the industry is responsible for approximately 14% of the total thermal energy losses and finding efficient ways to recover and store waste thermal energy is crucial to achieving sustainable energy consumption.

Keywords: heat recovery; heat storage; heat pipe; PCM

Citation: Vizitiu, R.S.; Burlacu, A.; Abid, C.; Balan, M.C.; Vizitiu, S.E.; Branoaea, M.; Kaba, N.E. Investigating the Efficiency of a Heat Recovery–Storage System Using Heat Pipes and Phase Change Materials. *Materials* **2023**, *16*, 2382. https://doi.org/10.3390/ma16062382

Academic Editors: Corneliu Munteanu and Ioan Doroftei

Received: 24 February 2023
Revised: 10 March 2023
Accepted: 14 March 2023
Published: 16 March 2023

1. Introduction

The negative impact of fossil fuels on the climate is well known and humanity is seeking non-polluting, renewable resources to replace them. The Paris Agreement of 2017 aims to keep the global temperature rise below 2 °C by the end of the century, which would require USD 3.5 trillion in annual investments in renewable and sustainable energy by 2050, according to a study by the International Energy Agency and International Renewable Energy Agency. In 2015, USD 1.8 trillion was invested in energy, with 12% going towards energy efficiency [1].

Heat recovery is a technique that has the potential to contribute to the reduction in greenhouse gas emissions. The process involves the capture and conversion of residual thermal energy into useful forms of energy, such as thermal, electrical, or mechanical energy. The most common forms of residual thermal energy are steam, hot air, or hot water, and the efficiency of the heat recovery process increases with the temperature of the residual heat. There are various heat recovery technologies available, including energy recovery heat exchangers, such as waste heat recovery units. Among these, heat pipe heat exchangers are considered particularly efficient due to the high thermal conductivity of heat pipes, which allows for a minimal temperature drop during heat transfer over long distances.

A number of studies have been conducted to evaluate the various types of heat recovery systems, primarily utilizing residual heat from industrial sources. The sources of energy that can be recovered in this manner include the heat generated from industrial equipment, heat generated from combustion processes, and the heat lost through radiation, conduction, or convection during industrial processes [2].

Jouhara et al. [3] designed and studied a radiative heat recovery system for recovering waste heat radiated during the steel manufacturing process. Another study, conducted by Delpech et al. [4], examined a heat recovery system that captures waste heat from the cooling of ceramic parts. The results of this study indicate that, at temperatures below 200 °C, the heat transfer is primarily convective, while at higher temperatures, the heat transfer is a combination of radiation and convection. Other studies have also been conducted on heat pipe heat recovery systems utilizing flue gases [5] and hot air [6,7] as primary sources. The waste heat recovery from steam is a well-established method and there have been studies investigating the use of steam generated from nuclear power plants as a primary source of seawater desalination [8].

The heat recovery process can be further optimized by incorporating latent heat storage systems. These systems consist of three main components: a storage medium, typically a phase change material, a heat transfer mechanism, and a containment system [9]. However, one of the main limitations of phase change materials is their low thermal conductivity, which has led to numerous studies aiming to improving their performance [10–13].

There are some experiments combining heat pipe technology with phase change materials [14,15]. A study [16] performed in 2017 by Amini et al. investigated a heat recovery system that uses steam as a primary agent and a salt hydrate PCM which melts at 89 °C as the heat storage medium. The results showed the good efficiency of the systems but it can be improved by increasing the thermal conductivity of the PCM.

Fan et al. [17] investigated a two-stage latent thermal energy storage tank that is integrated with an air purification pilot plant. A test was then carried out under actual operating conditions. The tank has units that are filled with commercial paraffin with a melting point of 70 °C and 48 °C, respectively. For a better heat transfer performance, heat pipes are combined with vertical fins for this research. According to the results, adding a latent thermal energy storage tank to an air separation system allows for a 51,369.5 kJ heat savings during a charging–discharging session. The waste heat requirement of 40 °C allowed for the recovery of 64.7% of the excess heat in the exhaust air and annually, 25,784.3 kg less CO_2 is emitted. In contrast with previous known latent thermal energy storage for industrial purposes, the latent thermal energy storage exhibits great results.

In another study [18] from 2015, a high-temperature latent heat thermal energy storage system assisted by finned heat pipes was subjected to numerical simulations, to check the charging process of PCM with a different configuration of embedded heat pipes. It was found that the bigger the number of heat pipes, the smaller the thermal resistance within the system, resulting a faster charging process.

This previous study is relevant to the recent research, as the use of finned storage units was also explored by [19], as a method to enhance the thermal performance of a vertically configured-cylindrical copper LHTES in combination with nano-enhanced PCM (NEPCM) and both outer and inner longitudinal fins. The study's outcomes demonstrate that the inclusion of six longitudinal fins on the internal and external surfaces of the LHTES wall leads to a notable decrease in the duration needed for discharging and charging by 71% and 62%, correspondingly, compared to the pure PCM reference condition lacking fins.

In their study, Nie et al. [20] also investigated the aspects of successively charging and discharging the phase change material in a thermal energy storage unit with fins. According to the results, it was observed that the heat transmission was enhanced when using a straight fin design, particularly when there are fewer fins used and when the thermal conductivity of the outer tube material is lower. Additionally, it was discovered that, when continuously charging and discharging, heat transfer was improved by using longer fins rather than shorter fins, while the fin volume remained the same.

In-depth experimental and computational investigation by Ruiz et al. [21] resulted in the determination of the optimum design for a water heat recovery and storage unit based on phase change material plates. The phase change material storage unit is designed to store approximately 75% of the thermal energy during the heat charging phase and some of it is recovered to provide approximately 50% of the thermal energy needed for the subsequent process. The performance of heat transfer for the latent thermal energy storage unit appeared to be strongly linked to the working parameters.

In order to enhance the efficiency of latent heat thermal energy storage (LHTES) with embedded heat pipes, Ladekar et al. [22] experimentally investigated the charging and discharging process in the case of using a heat pipe and a copper pipe. The results show that the capability to store heat for the charging process phase increases when majoring the volume flow rate, and early heat extraction when discharging becomes achievable. The charging and discharging of latent LHTES with integrated heat pipes performed well during testing with a flow rate of 10 L/min. When compared to systems with copper pipes, LHTES with integrated heat pipes has greater and more consistent efficiency and efficacy. It was also concluded that, as the volume flow drops, the overall melting time of the phase-changing material reduces.

The novelty of this research lies in the combination of gravitational heat pipes and phase change materials in a heat recovery–storage system (HRSS) for recovering and storing waste thermal energy from residual fluids. The system utilizes gravitational heat pipes to recover the low-temperature residual heat and store it as latent heat within a phase change material.

Previous research has separately focused on either the use of heat pipes or phase change materials, indicating the lack of experimental studies using both the quick heat transfer ability of heat pipes and the latent heat storage property of phase change materials. Therefore, it is essential to give more attention to the experimental part in order to reveal methods for improving the heat transfer performance of phase change materials and heat pipes [23].

This study proposes a new approach that combines both technologies to increase the efficiency of the heat recovery process and store thermal energy for future use. Heat pipes are used in the device's modular design to increase the efficiency of the heat recovery process, and phase change materials are used to store the thermal energy. The apparatus can recover the waste thermal energy from used water at low and medium temperatures. The thermal energy can then be transformed into additional thermal energy and used to pre-heat or heat water for domestic use or other heating systems.

This research effort seeks to examine the potential of the suggested HRSS in recovering and storing waste thermal energy from low- to medium-temperature residual fluids. The study intends to provide experimental proof of the HRSS' efficiency and highlight the technology's potential benefits in reducing thermal energy losses and achieving sustainable energy management.

In the following sections of this paper, there are descriptions of the design of the equipment and its testing parameters, followed by the presentation of results and conclusions.

2. Materials and Methods

2.1. *The Design of the Equipment*

The two-stage heat recovery–storage system consists of the connection of a water-to-water heat exchanger (heat exchanger A) to a water-to-water heat exchanger with integrated PCM (heat exchanger B). Heat exchanger A was previously tested as a stand-alone heat recovery system, revealing a high efficiency of up to 76.7% in recovering low- and medium-temperature thermal energy [24].

The design of heat exchanger A is presented in Figure 1. The heat exchanger consists of two main areas called the evaporator and condenser, separated by a sealing flange. Fourteen heat pipes are inserted transversely inside the equipment with the role of extracting thermal energy from the primary agent inside the evaporator and transferring it to the secondary agent inside the condenser. Inside the condenser, there are three rings and two discs, creating a path for the secondary agent. The evaporator, condenser, and separation flange are made of stainless steel and the rings, discs, and heat pipes are made of copper. The working fluid of the heat pipes is distilled water with a fill ratio of 25%.

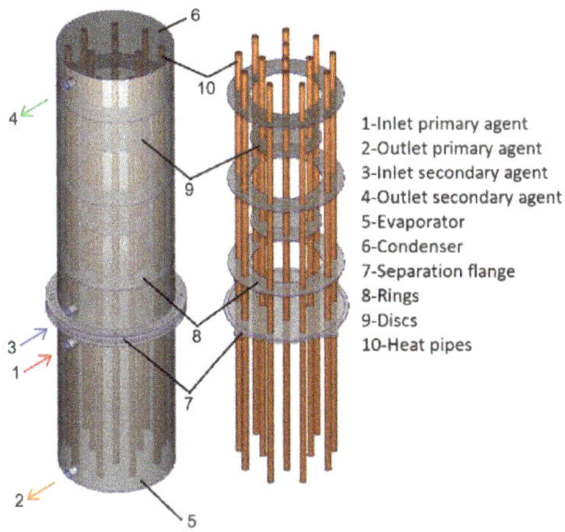

Figure 1. The design of the water–water heat exchanger (heat exchanger A).

The design of heat exchanger B is similar to the design of the heat exchanger, but the major difference is given by the fact that the secondary agent is directed through the condenser by a coil and the volume of the condenser is filled with a phase change material. The coil is made of copper, has a total length of 3.5 m, and has 7 turns with a diameter of 15 cm. The phase change material is coconut oil, and the necessary quantity to fill the condenser is 26.4 kg. The thermophysical properties of these phase change materials are presented in Table 1 [25].

Table 1. Thermophysical properties of coconut oil [24].

Mean specific heat of the solid	3.2 kJ/kg × K
Mean specific heat of the liquid	4.1 kJ/kg × K
Heat of fusion	249 kJ/kg
Melting temperature	35 °C

Figure 2 presents the design of heat exchanger B in 3D.

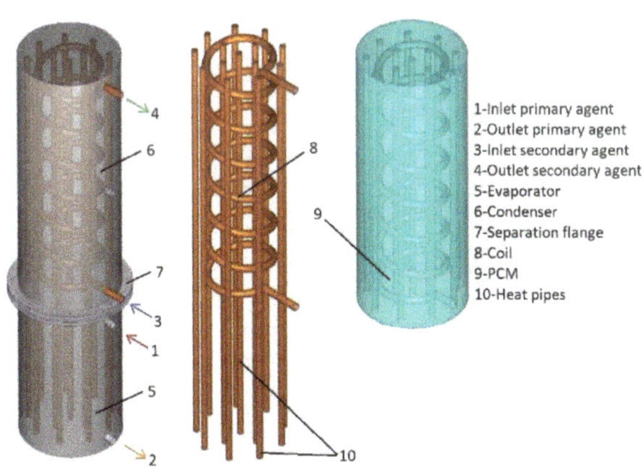

Figure 2. The design of the PCM heat exchanger (heat exchanger B).

The constructive details of the two heat exchangers can be viewed in Table 2.

Table 2. Constructive details of the components.

Component	Height/Length (m)	Diameter (m)
Evaporator	0.395	0.250
Condenser	0.640	0.250
Separation flange	0.005	0.300
Discs	0.005	0.150
Rings	0.005	0.246
Heat pipes	1.000	0.015
Coil	3.500	0.015

2.2. Experimental Setup

The thermal agent from which heat is recovered (primary agent) is water, heated by an 8 kW electric heater. The electric heater is connected to the heat recovery–storage system through flexible stainless-steel pipes. The circuit is equipped with a valve for filling/draining the evaporator and the primary circuit and a probe sheath for measuring the temperature of the primary agent.

The circuit of the secondary agent is directly connected to the water supply network of the laboratory. Two probe sheaths were installed at the inlet of condenser A and at the outlet of condenser B, connected to an electronic thermometer type LT BTM-4208SD which has a precision of ±0.4%. A thermal energy meter of type ELTERM CF 55 is also tracking the volume flow rate and the temperature of the secondary agent at the outlet of condenser B. Figure 3 presents a photo of the experimental stand from the laboratory.

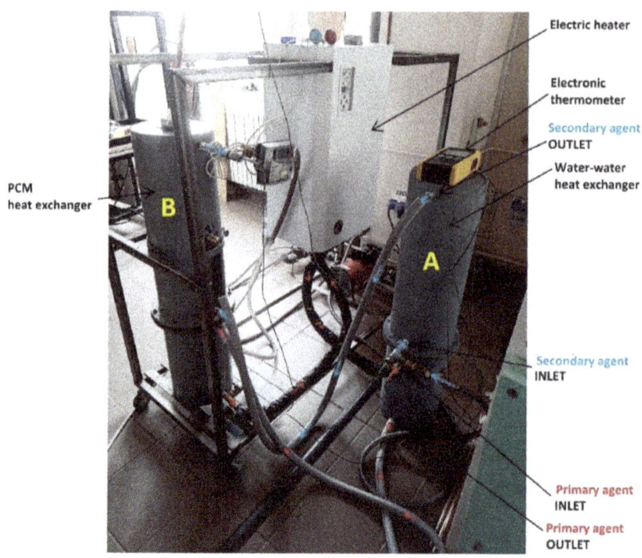

Figure 3. Photo of the experimental stand.

In this study, an electric heater is used to heat the primary agent, which flows through the evaporator of two heat exchangers, A and B. The heat pipes are used to extract thermal energy from the primary agent and transfer it to a secondary agent flowing through heat exchanger A and a phase change material (PCM) volume in heat exchanger B. The heated secondary agent, which exits condenser A, passes through the coil of condenser B, where it extracts additional temperature from the volume of coconut oil. Four sensors are used to measure the temperature of the primary and secondary agents, as well as the PCM. Sensor

CH1 tracks the temperature of the primary agent, while sensors CH2 and CH3 track the temperature of the secondary agent at the inlet and outlet of condenser, respectively. Sensor CH4 tracks the temperature of the PCM.

Figure 4 presents a schematic diagram of the experimental two-stage heat recovery–storage system.

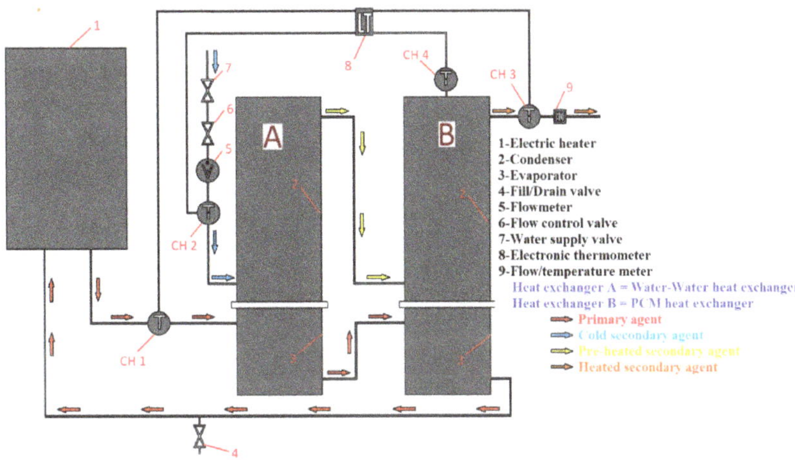

Figure 4. Schematic diagram of the 2-stage heat recovery–storage system.

2.3. *Tests Performed*

To test the efficiency of the equipment, a series of tests were performed with variations in the temperature of the primary agent and the volume flow rate of the secondary agent. The parameters are presented in Table 3. The volume flow rate of the primary agent was kept constant for all cases by the pump of the electric heater, and the temperature of the secondary agent was kept at 16 °C.

Table 3. Parameters of the experimental tests.

	Primary Agent		Secondary Agent	
	Temperature (°C)	Volume Flow Rate (L/min)	Temperature (°C)	Volume Flow Rate (L/min)
Test 1	60		16	1
Test 2	60		16	2
Test 3	60		16	3
Test 4	65		16	1
Test 5	65	24	16	2
Test 6	65		16	3
Test 7	70		16	1
Test 8	70		16	2
Test 9	70		16	3

Each test lasted 3 h, being staged according to Figure 5. The staging followed the behavior of the heat recovery–storage system and the ability to store the thermal energy of the coconut oil. In the first hour, the electric heater operated constantly to melt the coconut oil and transfer heat to the volume of the secondary agent in unit A. After one hour, the valve of the secondary agent was opened at the volume flow required by the test, thus starting its circulation through the heat exchangers. After the first two hours of operation, the electric heater was turned off, starting the cooling stage where the secondary agent absorbed the heat stored in the volume of phase-changing material.

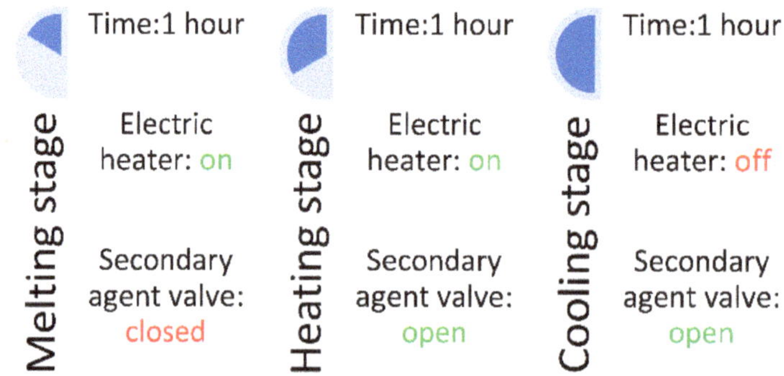

Figure 5. The stages of the tests.

This staging was aimed at the equipment's ability to recover heat in 2 scenarios: if the primary agent has an uninterrupted supply and in the case of an intermittent supply with the primary agent.

2.4. Numerical Simulations

For results with improved accuracy, the heat recovery system was subjected to numerical simulations and the results were compared to the experimental findings. The equipment was created in 3D with the Autodesk Inventor Professional 2018 software, and imported into the Autodesk Simulation CFD 2018 environment.

The first step is to assign a material to each component. The evaporator, the condenser, and the separation flanges are made of stainless steel, while the heat pipes and the coil are made of copper and the primary and secondary agents are water. Coconut oil does not exist in the program database, so a new material with its thermal properties was created.

The boundary conditions used are the same as in the experimental tests. The volume flow rate of the primary agent was set at 24 L/min and the temperature was set specifically for each test. The temperature of the secondary agent was set at 10 °C and the volume flow rates were set according to each scenario (Table 3).

To select the proper flow regime, the Reynolds number was calculated for each case. The results are centralized in Table 4. When the volume flow rate of the secondary agent is 1 L/min or 2 L/min, the Reynolds number is smaller than 2300, resulting in a laminar flow. When the volume flow rate is 3 L/min, the Reynolds number has a value of approximately 2460 which means that there is a transition regime.

Table 4. Reynolds number.

	Secondary Agent			
	Temperature (°C)	**Volume Flow Rate (L/min)**	**Reynolds**	**Regime**
Test 1	16	1	819.8	Laminar
Test 2	16	2	1640.4	Laminar
Test 3	16	3	2461.3	Transition
Test 4	16	1	820.0	Laminar
Test 5	16	2	1640.9	Laminar
Test 6	16	3	2460.1	Transition
Test 7	16	1	820.5	Laminar
Test 8	16	2	1640.6	Laminar
Test 9	16	3	2460.9	Transition

After applying the boundary conditions, the 3D model was discretized into a network of nodes. For the first simulation, the geometry was discretized into a network that consists of 1 million nodes. Then, the number of nodes was increased by 1 million for each simulation until the mesh reached 7 million nodes. After 5 million nodes, the results were very similar. The higher the number of discretization elements, the greater the processing power and more time was needed, so the number of elements for the calculation of the solution was optimized according to the criteria and the tests performed. The resulting mesh had 5.2 million nodes (Figure 6).

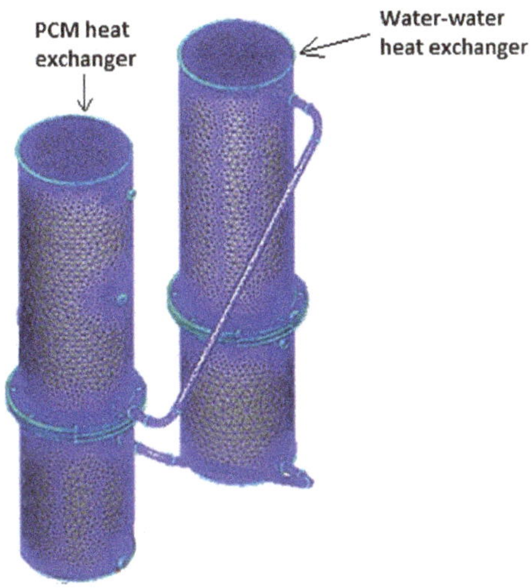

Figure 6. Meshing—2-stage heat recovery–storage system.

Experimental testing was conducted in the laboratory in order to evaluate the performance of the 2-stage heat recovery–storage system. The equipment was tested using several temperatures for the primary agent to check whether the heat pipes are capable of efficiently recovering heat in different scenarios, and different volume flow rates for the primary agent to check the quantity of heat recovered in different instances. Additionally, to check the efficiency of the heat storage, the measurements of temperatures were made in three different stages: a melting stage, a heating stage, and a cooling stage. A detailed description of these stages is given in Section 2.3.

The last step was the calculation of the maximum quantity of heat contained by the primary agent and the quantity of heat recovered by the secondary agent. By comparing these, the efficiency of the equipment can be established for each test.

3. Results and Discussions

The results of the experimental tests are centralized in Figure 7. T max represents the maximum temperature reached by the two-stage recovery system and T med represents the average temperature of the secondary agent at the outlet of the PCM heat exchanger during the heating stage.

In test 1, during the first 27 min of the heating stage, the difference between the temperature of the coconut oil and the temperature of the secondary agent varies between 0 and 2 °C. During this period, this difference increased to 3–5 °C, which is maintained until the end of the test. For test 2, the difference of 1–2 °C was only in the first 10 min of the heating stage, and later, this difference increased to 5–7 °C and remains constant. In test 3, the difference of 1–2 °C was shortened to 8 min but still reached 5–7 °C by the end of the test, as shown in Figure 8.

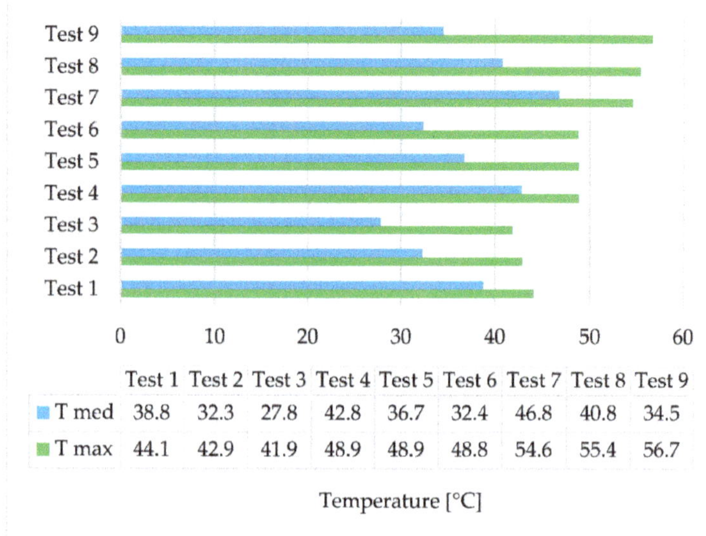

	Test 1	Test 2	Test 3	Test 4	Test 5	Test 6	Test 7	Test 8	Test 9
T med	38.8	32.3	27.8	42.8	36.7	32.4	46.8	40.8	34.5
T max	44.1	42.9	41.9	48.9	48.9	48.8	54.6	55.4	56.7

Figure 7. Average and maximum temperatures obtained.

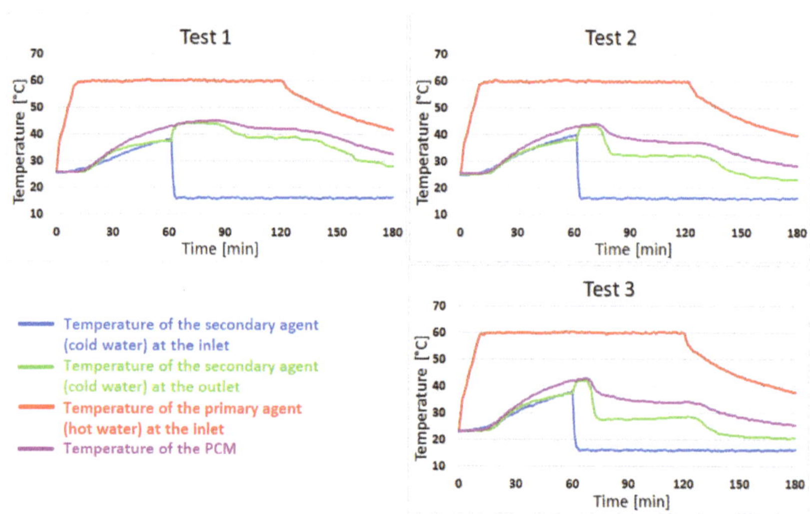

Figure 8. The results for tests 1, 2, and 3.

Test 4 shows almost no temperature difference between coconut oil and the secondary agent for the first 23 min of the heating stage, but the difference increases to 2–3 °C for the next 52 min and towards the second half of the cooling stage, this difference becomes 4–6 °C. The similarity of the temperature of coconut oil to that of the secondary agent lasts 13 min in test 5 with a difference of 0–2 °C. The difference increases quite suddenly at 5–8 °C and is maintained until the end of the test. For test 6, the duration of the 0–2 °C variation is shortened to only 7 min, and the temperature difference between the PCM and the secondary agent sharply increases to 5–9 °C, as can be seen in Figure 9.

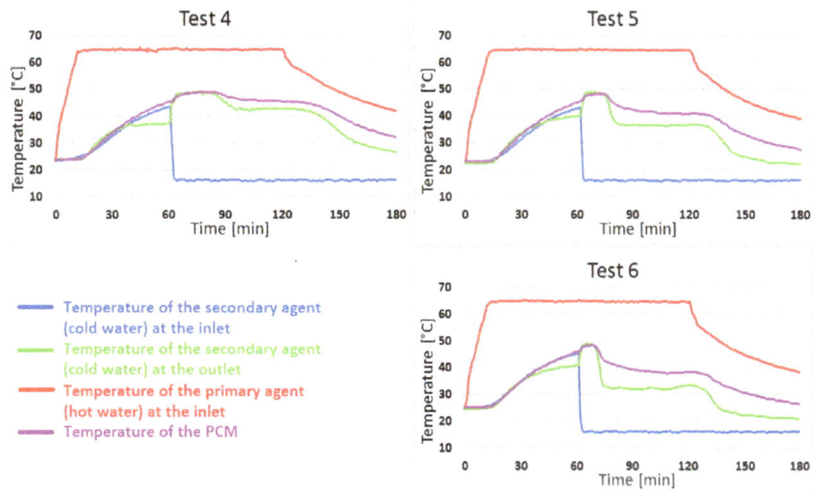

Figure 9. The results for tests 4, 5, and 6.

Test 7 shows the best results of the two-stage recovery system, with an average temperature of the secondary agent during the heating stage of 46.8 °C. The temperature of the secondary agent is higher than the temperature of the coconut oil in the first 20 min of the heating stage. In the last 40 min of this stage, the temperature at the outlet drops below the temperature of the phase change material by 3 °C. The difference then increases to 4–7 °C in the cooling stage and is maintained until the end of the test. Additionally, in tests 8 and 9, the temperature of the secondary agent is higher than that of coconut oil for a short time, but then decreases at the same rate as it, as can be seen in Figure 10.

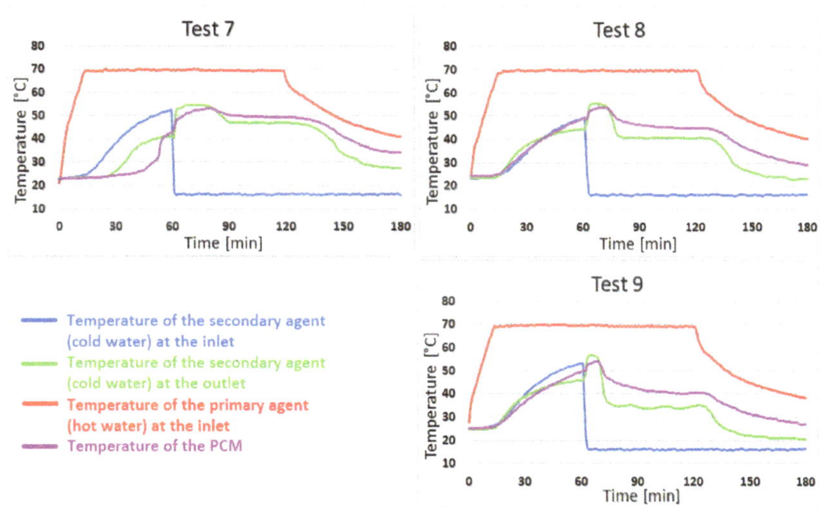

Figure 10. The results for tests 7, 8, and 9.

The temperatures of the secondary agent recorded at the outlet of the system were compared to the average temperatures from the experimental results and they are presented in Figure 11.

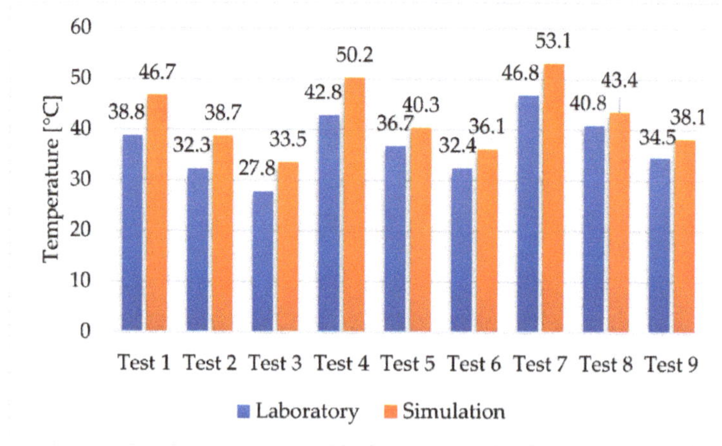

Figure 11. Comparison between the experimental and numerical results.

Since the simulations do not take into account the convective heat losses from the laboratory experiment, the temperature in the experimental results is lower than in the numerical simulations, with values between 3 °C and 8 °C. Since the temperature discrepancies between the experimental results and the simulation results are negligible, it can be concluded that the experimental results are reliable.

In Figure 12, some 3D temperature contours extracted from the results of the numerical simulations are presented for tests 7, 8, and 9 at the end of the cooling stage. The temperature of the secondary agent increases between 3 °C and 6 °C from the outlet of the first condenser to the outlet of the system but the main advantage of the two-stage heat recovery system is the storage of thermal energy, which cannot be tracked in the numerical simulations due to software limitations.

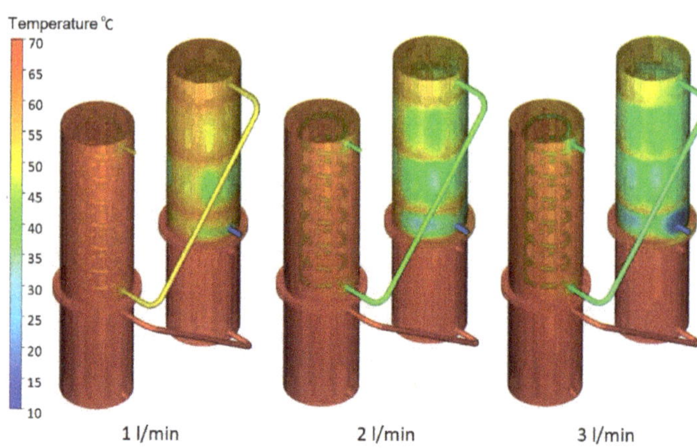

Figure 12. Three-dimensional temperature contours for tests 7, 8, and 9.

The heat quantity in the primary agent was calculated as the product of the mass flow rate, the specific heat capacity of water, and the temperature difference between evaporator A's inlet and evaporator B's outlet. The heat quantity of the secondary agent was determined in a comparable manner, using the secondary agent's mass flow rate, the specific heat capacity of water, and the temperature difference between T med (Figure 7) for each scenario and the cold secondary agent's temperature of 16 °C (Table 4). According to Equation (1), the efficiency of the two-stage heat recovery–storage system was calculated as a fraction of the heat recovered by the secondary agent and the maximum amount of heat produced by the primary agent. The results are shown in Figure 13.

$$\epsilon = Q_2/Q_1 \tag{1}$$

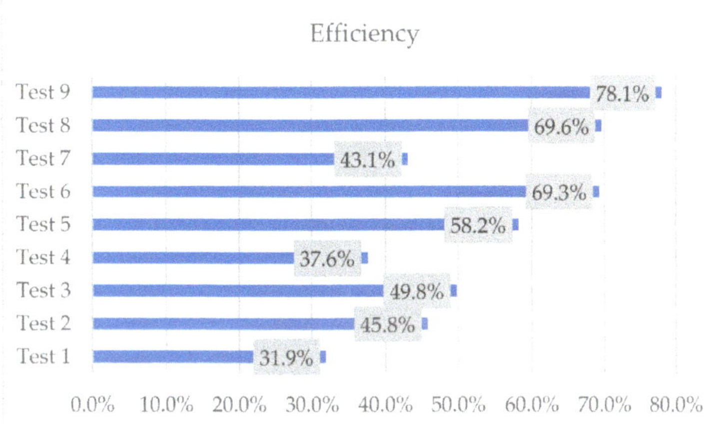

Figure 13. The efficiency of the 2-stage heat recovery–storage system.

The highest efficiency was obtained in test 9 when the volume flow rate of the secondary agent is 3 L/min and the temperature of the primary agent is 70 °C. Although the highest temperatures of the secondary agent were obtained with the smallest volume flow rate, the quantity of heat recovered is smaller.

4. Conclusions

An experimental investigation followed by numerical simulations was carried out on a two-stage heat recovery–storage system. From the experimental results, the highest temperature recorded was when the volume flow rate of the secondary agent was the lowest and the temperature of the primary agent was the highest, at 70 °C, with an average temperature of the secondary agent of 46.8 °C. However, the highest efficiency of the equipment regarding the quantity of the recovered heat was in test 9, with an efficiency of 78.1%.

The ability of the equipment to store heat was observed in all nine tests conducted. During the cooling stage, when the heat source is shut down, the secondary agent still keeps a higher temperature at the outlet of the equipment compared to the one at the inlet by 11.5 °C–4.1 °C, depending on its volume flow rate.

Compared to a common heat recovery system, the two-stage heat recovery–storage system comes with a heat storage capability, which makes the heat recovery process more efficient, especially when the heat source is intermittent. The organic phase change material used for heat storage is a plant-based product, which makes it environmentally friendly.

Another important advantage is the ability to recover heat passively through the gravitational heat pipes, which means that there is no additional pumping system required. The two-stage heat recovery–storage system also has a low cost of production and maintenance compared to common heat exchangers. The two thermal agents are completely separated since the mean of transporting the heat are the heat pipes, which means there are very small chances of accidental mixing. Additionally, if one of the heat pipes malfunctions, the equipment can still function at a slower pace until the heat pipe is replaced, and thus, the maintenance is simple.

Author Contributions: Conceptualization, R.S.V. and A.B.; Methodology, R.S.V., A.B., C.A., M.C.B. and S.E.V.; Software, R.S.V., S.E.V., M.B. and N.E.K.; Validation, A.B., C.A. and M.C.B.; Formal analysis, C.A. and M.B.; Investigation, R.S.V., A.B., M.C.B., S.E.V. and M.B.; Resources, A.B., M.C.B. and N.E.K.; Data curation, R.S.V., A.B., C.A., S.E.V. and M.B.; Writing—original draft, R.S.V., S.E.V. and M.B.; Writing—review & editing, A.B., C.A. and M.C.B.; Visualization, R.S.V., M.C.B. and S.E.V.; Supervision, A.B., C.A. and N.E.K.; Project administration, R.S.V., A.B., C.A., M.C.B. and N.E.K.; Funding acquisition, A.B. and N.E.K.. All authors have read and agreed to the published version of the manuscript.

Funding: This work was supported by a grant from the Romanian Ministry of Education and Research, CCCDI-UEFISCDI, project number PN-III-P2-2.1-CI-2017-0774, within PNCDI III.

Institutional Review Board Statement: Not applicable.

Informed Consent Statement: Not applicable.

Data Availability Statement: Not applicable.

Conflicts of Interest: The authors declare no conflict of interest.

Nomenclature

PCM	Phase change material
NEPCM	Nano-enhanced phase change material
HRSS	Heat recovery–storage system
LHTES	Latent heat thermal energy storage
T max	Maximum temperature
T med	Average temperature
ϵ	Efficiency of the equipment
Q_1	Quantity of heat of the primary agent
Q_2	Quantity of heat of the secondary agent

References

1. IEA (International Energy Agency). *Perspectives on the Energy Transition*; IEA: Paris, France, 2017.
2. Jouhara, H.; Khordehgah, N.; Almahmoud, S.; Delpech, B. Waste heat recovery technologies and applications. *Therm. Sci. Eng. Prog.* **2018**, *6*, 268–289. [CrossRef]
3. Jouhara, H.; Almahmoud, S.; Chauhan, A.; Delpech, B.; Bianchi, G.; Tassou, S.; Llera, R.; Lago, F.; Arribas, J.J. Experimental and theoretical investigation of a flat heat pipe heat exchanger for waste heat recovery in the steel industry. *Energy* **2017**, *141*, 1928–1939. [CrossRef]
4. Delpech, B.; Axcell, B.; Jouhara, H. Experimental investigation of a radiative heat pipe for waste heat recovery in a ceramics kiln. *Energy* **2019**, *170*, 636–651. [CrossRef]
5. Tian, E.; He, Y.-L.; Tao, W.-Q. Research on a new type waste heat recovery gravity heat pipe exchanger. *Appl. Energy* **2017**, *188*, 586–594. [CrossRef]
6. Mahajan, G.; Thompson, S.M.; Cho, H. Experimental characterization of an n-pentane oscillating heat pipe for waste heat recovery in ventilation systems. *Heliyon* **2018**, *4*, e00922. [CrossRef]
7. Mahajan, G.; Thompson, S.M.; Cho, H. Energy and cost savings potential of oscillating heat pipes for waste heat recovery ventilation. *Energy Rep.* **2017**, *3*, 46–53. [CrossRef]
8. Jouhara, H.; Anastasov, V.; Khamis, I. Potential of heat pipe technology in nuclear seawater desalination. *Desalination* **2009**, *249*, 1055–1061. [CrossRef]
9. Kuravi, S.; Trahan, J.; Goswami, D.Y.; Rahman, M.M.; Stefanakos, E.K. Thermal energy storage technologies and systems for concentrating solar power plants. *Prog. Energy Combust. Sci.* **2013**, *39*, 285–319. [CrossRef]
10. Mostafavi, A.; Parhizi, M.; Jain, A. Semi-analytical thermal modeling of transverse and longitudinal fins in a cylindrical phase change energy storage system. *Int. J. Therm. Sci.* **2020**, *153*, 106352. [CrossRef]
11. Ahmed, N.; Elfeky, K.E.; Lu, L.; Wang, Q.W. Thermal performance analysis of thermocline combined sensible-latent heat storage system using cascaded-layered PCM designs for medium temperature applications. *Renew. Energ.* **2020**, *152*, 684–697. [CrossRef]
12. Zhang, C.; Meng, Y.; Fan, Y.; Zhang, X.; Zhao, Y.; Qiu, L. Numerical study on heat transfer enhancement of PCM using three combined methods based on heat pipe. *Energy* **2020**, *195*, 116809. [CrossRef]
13. Yazdani, M.R.; Laitinen, A.; Helaakoski, V.; Farnas, L.K.; Kukko, K.; Saari, K.; Vuorinen, V. Efficient storage and recovery of waste heat by phase change material embedded within additively manufactured grid heat exchangers. *Int. J. Heat Mass Transf.* **2021**, *181*, 121846. [CrossRef]
14. Song, H.; Zhang, W.; Li, Y.; Yang, Z.; Ming, A. Exergy analysis and parameter optimization of heat pipe receiver with integrated latent heat thermal energy storage for space station in charging process. *Appl. Therm. Eng.* **2017**, *119*, 304–311. [CrossRef]

15. Hu, B.; Wang, Q.; Liu, Z. Fundamental research on the gravity assisted heat pipe thermal storage unit (GAHP-TSU) with porous phase change materials (PCMs) for medium temperature applications. *Energy Convers. Manag.* **2015**, *89*, 376–386. [CrossRef]
16. Amini, A.; Miller, J.; Jouhara, H. An investigation into the use of heat pipe technology in thermal energy storage heat exchangers. *Energy* **2017**, *136*, 163–172. [CrossRef]
17. Fan, Y.; Zhang, C.; Jiang, L.; Zhang, X.; Qiu, L. Exploration on two-stage latent thermal energy storage for heat recovery in cryogenic air separation purification system. *Energy* **2022**, *239*, 122111. [CrossRef]
18. Tiari, S.; Qiu, S. Three-dimensional simulation of high temperature latent heat thermal energy storage system assisted by finned heat pipes. *Energy Convers. Manag.* **2015**, *105*, 260–271. [CrossRef]
19. Khedher, N.B.; Rashad, A.B.; Kolsi, L.; Omri, M. Performance investigation of a vertically configured LHTES via the combination of nano-enhanced PCM and fins: Experimental and numerical approaches. *Int. J. Heat Mass Transf.* **2022**, *137*, 106246. [CrossRef]
20. Nie, C.; Deng, S.; Liu, J. Numerical investigation of PCM in a thermal energy storage unit with fins: Consecutive charging and discharging. *J. Energy Storage* **2020**, *29*, 101319. [CrossRef]
21. Morales-Ruiz, S.; Rigola, J.; Oliet, C.; Oliva, A. Analysis and design of a drain water heat recovery storage unit based on PCM plates. *Appl. Energy* **2016**, *179*, 1006–1019. [CrossRef]
22. Ladekar, C.L.; Chaudhary, S.K.; Khandare, S.S. Experimental Investigate for Optimization of Heat Pipe Performance in Latent Heat Thermal Energy Storage. *Mater. Today Proc.* **2017**, *4*, 8149–8157. [CrossRef]
23. Li, Z.; Lu, Y.; Huang, R.; Chang, J.; Yu, X.; Jiang, R.; Yu, X.; Roskilly, A.P. Applications and technological challenges for heat recovery, storage and utilisation with latent thermal energy storage. *Appl. Energy* **2021**, *283*, 116277. [CrossRef]
24. Vizitiu, R.S.; Burlacu, A.; Abid, C.; Verdes, M.; Balan, M.C.; Branoaea, M. Experimental and Numerical Study of Thermal Performance of an Innovative Waste Heat Recovery System. *Appl. Sci.* **2021**, *11*, 11542. [CrossRef]
25. Putri, W.A.; Fahmi, Z.; Sutjahja, I.M.; Kurnia, D.; Wonorahardjo, S. Thermophysical parameters of coconut oil and its potential application as the thermal energy storage system in Indonesia. *J. Phys. Conf. Ser.* **2016**, *739*, 012065. [CrossRef]

materials

MDPI

Article

Physical, Thermal, and Chemical Properties of Fly Ash Cenospheres Obtained from Different Sources

Andrei Shishkin [1,*], Vitalijs Abramovskis [1], Ilmars Zalite [2], Ashish Kumar Singh [1], Gundars Mezinskis [2], Vladimir Popov [3] and Jurijs Ozolins [1]

[1] Rudolfs Cimdins Riga Biomaterials Innovations and Development Centre of RTU, Institute of General Chemical Engineering, Faculty of Materials Science and Applied Chemistry, Riga Technical University, Pulka 3, K-3, LV-1007 Riga, Latvia

[2] Institute of Materials and Surface Technologies of the Riga Technical University, P. Valdena iela 7, LV-1048 Riga, Latvia

[3] Department of Materials and Engineering, Tel Aviv University, Tel Aviv 6997801, Israel

* Correspondence: andrejs.siskins@rtu.lv

Abstract: Cenospheres are hollow particles in fly ash, a by-product of coal burning, and are widely used as a reinforcement when developing low-density composites called syntactic foams. This study has investigated the physical, chemical, and thermal properties of cenospheres obtained from three different sources, designated as CS1, CS2, and CS3, for the development of syntactic foams. Cenospheres with particle sizes ranging from 40 to 500 μm were studied. Different particle distribution by size was observed, and the most uniform distribution of CS particles was in the case of CS2: above 74% with dimensions from 100 to 150 μm. The CS bulk had a similar density for all samples and amounted to around 0.4 $g \cdot cm^{-3}$, with a particle shell material density of 2.1 $g \cdot cm^{-3}$. Post-heat-treatment samples showed the development of a SiO_2 phase in the cenospheres, which was not present in the as-received product. CS3 had the highest quantity of Si compared to the other two, showing the difference in source quality. Energy-dispersive X-ray spectrometry and a chemical analysis of the CS revealed that the main components of the studied CS were SiO_2 and Al_2O_3. In the case of CS1 and CS2, the sum of these components was on average from 93 to 95%. In the case of CS3, the sum of SiO_2 and Al_2O_3 did not exceed 86%, and Fe_2O_3 and K_2O were present in appreciable quantities in CS3. Cenospheres CS1 and CS2 did not sinter during heat treatment up to 1200 °C, while sample CS3 was already subjected to sintering at 1100 °C because of the presence of a quartz phase, Fe_2O_3 and K_2O. For the application of a metallic layer and subsequent consolidation via spark plasma sintering, CS2 can be deemed the most physically, thermally, and chemically suitable.

Keywords: hollow microballoons; cenospheres; ceramics phase composition; chemical composition

check for updates

Citation: Shishkin, A.; Abramovskis, V.; Zalite, I.; Singh, A.K.; Mezinskis, G.; Popov, V.; Ozolins, J. Physical, Thermal, and Chemical Properties of Fly Ash Cenospheres Obtained from Different Sources. *Materials* **2023**, *16*, 2035. https://doi.org/10.3390/ma16052035

Academic Editor: Miguel Ángel Sanjuán

Received: 29 December 2022
Revised: 11 February 2023
Accepted: 17 February 2023
Published: 1 March 2023

1. Introduction

In recent years, syntactic foam (SF) composites have increasingly found use in various industries where superior mechanical strength at a low weight is desired. SF is made of polymer or metal in between hollow microspheres [1–3]. Syntactic foam with specific properties can be obtained by reinforcing hollow particles in a polymer, metallic, or ceramic matrix. Syntactic foams are increasingly being used for their excellent mechanical strength and stiffness-to-weight ratios [4,5] Such materials are used in aviation, space technology, and ship construction to manufacture individual parts [6,7].

A promising material for obtaining syntactic foam is cenospheres (CS), hollow aluminosilicate microballoons. CS is a waste product in the composition of fly-ash burning coal in thermal power plants (TPC), as a result of complex thermochemical processes [8]. Recently, coal thermal power plants have produced about 41% of global electricity, which is expected to increase to about 44% by 2030 [8]. This combustion of coal results in the production of large quantities of waste called fly ash. It is estimated that approximately

750 million tonnes of fly ash are produced annually worldwide, and this is increasing at a high rate. A large quantity of fly ash is disposed of in landfills and ash lagoons. Due to the heavy metal leaching and high cost of landfills, many applications have been suggested to convert this waste into value-added products. To increase the re-use of coal fly ash, several separation methods have been developed to segregate value-added components, such as aluminosilicates, magnetite, and CS from solid particles and unburned carbon [8]. CS can be separated from the ash by the method of flotation; in lagoons, they float to the surface of the water, from where they can be collected for further processing. The CS world market is worth up to USD 600 million annually, equivalent to 700–800 tonnes [9,10]. The main suppliers of CS in the world are China, India, Russia, Kazakhstan, and Ukraine. As a waste product of TPC operations, CS are available at a relatively low cost.

Cenospheres account for an average of 0.01 to 4.80% by mass of fly ash. They are characterised by a low bulk density (0.4–0.72 $g \cdot cm^{-3}$), very low thermal conductivity (about 0.065 $W \cdot m^{-1} \cdot K^{-1}$), and excellent stability in an alkaline solution and up to high temperatures. The particle size of CS ranges from about 5 to 500 µm [8,11]. By chemical composition, these materials are multi-component systems with a SiO_2–Al_2O_3–Fe_2O_3 content of approximately 90 wt% [12–14]. The utilisation of fly ash is determined by its properties, such as fineness, specific surface area, particle shape, hardness and freeze–thaw resistance [15,16].

The properties of CS and their significant accumulations can be recycled long-term and used to create new functional materials with a high added value. CS are very widely used in the manufacture of stuffed building materials with a reduced density and improved thermal insulation properties [9,10] and are also used as a filler and reinforcement in metal–matrix composites [17–19] ceramics–matrix composites [20–22], polymer–matrix composites [23,24] and various hybrid matrix composites [21,25]. Cenospheres can be employed with polymers and hybrid matrices intended for low-temperature applications [26–29] as well as for the development of geopolymer composites [20,30,31]. In both cases of material production, the working temperatures do not exceed 250 °C; therefore, these studies have not paid significant attention to the effect of temperature on the physical properties or CS stability. High-temperature cases for the CS have also been studied. Materials intended for applications such as fire protection [30,32] and thermal barriers [32] have employed CS, and typically experience temperatures above 900 °C. Other high-temperature conditions exist for CS, such as the fabrication of metal–matrix composites [33–36], where relatively high (up to 900–1200 °C) temperatures can be seen. However, these studies too lack the much-desired understanding of the changes in CS properties during the thermal treatment.

A promising direction for their use may be to manufacture lightweight and mechanically resistant multi-layered composite materials with specific properties. Previous work shows a proof-of-the-concept of obtaining novel (matrix-less) SF [37,38] by using CS from burnt coal from the Donetsk and Ekibastuz coal basins' TPC. As described in works [37,38], matrix-less SFs were obtained by the spark plasma sintering (SPS) of metallised CS (Me@CS) at a temperature range from 800 °C (for Cu) up to 1150 °C in the case of Ti and 316 steel coatings. A micron and a sub-micron layer of metal (Cu, Ti, and 316 steel) [37,38] were deposited via physical vapour deposition (PVD).

The development of SF from metallised CS has been studied previously and by other researchers. However, neither the effect of the heat treatment and processing on the properties of CS, nor the difference between the behaviour of CS obtained from different sources is well understood. It is especially important to understand this, given the prevailing geopolitical situation, which is making it challenging to source raw materials with the same ease and availability. Since the properties of CS depend on the type of coal and the conditions of a TPC, CS have a wide range of properties [8,11]. The authors believe that it is necessary to understand the influence of temperature on the structure, physical, and chemical properties in the heat treatment of CS available from different thermal power plants.

The purpose of this work is to understand the quality of the raw material, the effect of heat treatment on the CS, and their characterisation, which could be suitable for the preparation of novel matrix-less SF precursors. In this research, the phase change that occurs in the CS during the pre-treatment will be studied. The physical and chemical properties of raw CS and metallised CS are studied to better inform and guide subsequent processes, i.e., the development of SF and its mechanical properties.

2. Materials and Methods

The cenospheres used in the current study were a by-product of coal-fired plants from the Donetsk and Ekibastuz coal basins, burning them in various TPC. The origin, designation, and particle size characteristics of CS are shown in Table 1.

Table 1. Source designation and granulometric composition of CS.

Sample Designation	Coal Field	Grading Composition * (μm)
CS1	Donetsk	40–500
CS2	Donetsk	40–200
CS3	Ekibastuz	50–300

* according to the supplier data.

A determination of the particle size distribution (PSD) was carried out with the help of the vibratory sieve shaker Analysette 3 PRO (FRITSCH GmbH, Idar-Oberstein, Germany). Particles were fractioned using sieves with opening sizes of 0.05, 0.10, 0.15, 0.20, 0.25, and 0.30 mm. The bulk density was determined by using a bulk density tester (Scott volumeter, according to ASTM B 329-98, Copley, Nottingham, UK). To obtain the true particle density (TPD), a Quantachrome Ultrapyc 1200e (Anton Paar GmbH, Graz, Austria) automatic gas pycnometer was used.

Specific thermal conductivity was determined using LaserComp (ASV) Fox 600 measuring equipment, and the sample was freely poured into a layer with a thickness of 30 mm.

2.1. Determination of the Interparticle Void Fraction

Before producing the syntactic foams, it is essential to know the maximum achievable packing density or its opposite value—interparticle void fraction ε. This value gives the maximum volume fraction (VF) of CS that can be dispersed in the matrix. The maximum possible packing density for spherical particles is known to be 0.741 [39,40]. In the case of practically monodispersed particles, with possible dense packing (random close packing), this value can reach 0.64 [41–44], while in the case of real loose packing (random loose packing), it usually does not exceed 0.60 [41]. This means that an actual interparticle void fraction is not less than $\varepsilon = 1 - 0.60 = 0.40$. The determination of the interparticle void or volume filling coefficient ε was calculated using Equation (1):

$$\varepsilon = \frac{V - \left(V_{p+fl} - V_{fl}\right)}{V} \quad (1)$$

where:

V—the volume of the particles in bulk condition, prepared with Scott volumeter,
V_{p+fl}—the volume of the mixture (particles + fluid),
V_{fl}—the volume of the fluid.
A—method for determining the CS interparticle voids was developed.

The essence of this is as follows. The volume of the particles in the bulk condition (V) is the sum of the volume occupied by the particles and the interparticle void volume, which can be determined by filling the voids with a certain volume of fluid V_{fl}. Since the bulk density of CS is usually in the range of 0.38 to 0.43 $g \cdot cm^{-3}$, it is impossible to perform the experiment correctly using water. The forces acting on a dispersed particle in a viscous medium are described by Stokes's law, which states that the sedimentation rate of a particle under other independent conditions is proportional to the difference between the densities of the particle and the medium, and inversely proportional to the viscosity of the medium. The kinetic persistence of dispersed systems (or, more simply, the sedimentation speed of particles or the particle rise speed) is described by Stokes's law, Equation (2):

$$V_S = \frac{d^2 g\left(\rho_p - \rho_{fl}\right)}{18\,\mu} \quad (2)$$

where V_S is the sedimentation speed of particles in m·s^{-1} (if $\rho_p > \rho_{fl}$) or the particle rise speed to the surface of a liquid (if $\rho_p < \rho_{fl}$).

d—diameter of the particle (m),
g—acceleration of gravity (m·s^{-2}),
ρ_p—single particle apparent density (kg·m^{-3}),
ρ_{fl}—density of the fluid (kg·m^{-3}),
μ—dynamic viscosity of the fluid (Pa·s).

Based on Stokes's Formula (2), the sedimentation (stratification) speed is directly proportional to the particle diameter, the phase difference, and the square of the density of the environment, as well as inversely proportional to the viscosity of the environment. Therefore, to reduce the sedimentation rate, i.e., to increase the resistance to sedimentation, different methods can be used, such as increasing the viscosity of the environment using dispersants with a density close to the density of the substance.

The simplest way is to increase the viscosity of the environment dispersion. To control the viscosity, a solution of potato starch in distilled water (10 g·L^{-1}), with 0.01% (−)-Ethyl L-lactate (purists. p.a., cleaning grade, ≥98.0% (sum of enantiomers, GC), supplier—Fluka (now Sigma-Aldrich)) as surfactant, was used. The exact volume of CS (25 cm^3) under investigation was obtained with a Labulk-0302 Scott Volumeter (Version ISO 3923-2) to ensure the actual bulk density, and then 25 cm^3 of CS was mixed with a precise (100.0 cm^3) volume of starch solution. The system was then degassed, and a precise volume measurement was taken.

2.2. Determination of the Ratio between Intact and Defective Cenospheres

To obtain the highest quality syntactic foam from metallised CS, separating the intact CS from the damaged ones is necessary. The damaged CS can be: broken, porous and/or cracked from handling or processing. For this, the flotation method was chosen and implemented as follows. A solution consisting of 800 cm^3 of distilled water and 50 cm^3 of ethanol (to reduce surface tension) was prepared, and 100 g of CA was added and stirred with a glass rod for 60 s. The mixture was sifted for 5 min, and then the floating fraction (with the aid of a spatula and decantation) was carried to the next beaker with fresh washing liquid, and the procedure was repeated. After repeated settling of the upper layer, it was transferred to a Büchner funnel and washed with ethanol. Then, it was dried in an oven at 105 °C for 12 h with a layer of no more than 20 mm. To determine the ratio of intact and destroyed CS, the liquid collected below settlings was decanted carefully and the sediments were transferred to a Büchner funnel. It was then rinsed with distilled water, and then by ethanol, and finally dried in an oven at 105 °C for 12 h.

2.3. Morphology, Chemical and Phase Composition

A Scanning Electron Microscope (SEM) Zeiss EVO MA-15 EDS (used at 5 kV) equipped with INCA Energy 350 energy-dispersive X-ray spectroscopy (EDS) at 15 kV was used to evaluate the morphology and element composition of the CS. For the elemental composition via EDS, 10 points were used—from the surface and from the inside (fractured CS) of the cenospheres. For the phase-composition determination, powder X-ray diffraction (XRD) on a BRUKER D8 Advance diffractometer with a Cu anode, accelerating voltage of 40 kV, and beam current of 40 mA, with 0.02° per step, was applied. For quantitative analysis, the Rietveld method with internal standard mineral corundum was used.

A high-temperature optical microscope (HTM) (EM201 HT163, Hesse instruments, Germany) was used for 2D-dilatometry, which is an optical dilatometry—taking cross-section measurements of the specimen's area. A cylindrical shape (d = 5 and h = 7 mm) was used for the test. The heating mode was set as follows: up to 800 °C, the heating rate was 20 °C·min^{-1}, since up to this temperature, deformations of the CS are unlikely to occur; however, heating above 800 °C was carried out at a speed of 5 °C·min^{-1}.

Fourier transform infrared (FTIR) spectra were recorded using a Thermo Scientific Nicolet™ iSTM50 (Thermo Fisher) spectrometer in the Attenuated Total Reflectance (ATR) mode. Spectra were obtained over a range of wavenumbers from 400 cm^{-1} to 4000 cm^{-1}, co-adding 64 scans at a 4 cm^{-1} resolution. Before every measurement, a background spectrum was taken and deducted from the sample spectrum.

Loss of ignition (LOI) was determined at temperatures of 400 and 1000 °C, according to GOST 2642.2-86. The chemical analysis was performed according to GOST 32362-2013.

2.4. Material Preparation and Study

It is necessary to understand the thermal–structural relationship for CS in order to obtain advanced SF, since the method of foam synthesis is SPS. The sintering of metallised CS involves the melting and fusion of the metal coating that can reach temperatures in the range of 1100–1200 °C [37,38]. The composition of CS is refractory ceramics that can withstand high temperatures, but it is necessary to investigate the effects of this thermal treatment on the CS.

The process flowchart for the study is shown in Figure 1, which describes the treatment and characterization steps taken for cenospheres CS1, CS2 and CS3. The bulk density (ρ_{bk}), pycnometric density(ρ_{py}), interparticle void fraction (ε), particle size distribution (PSD), LOI, SEM, phase composition and 2D-dilatometry (HTM analysis) were determined. To ensure the high quality of the metal–CS foam by sintering of the metallised CS, it is necessary to separate the intact and damaged CS. For this purpose, the flotation method was used with a liquid medium—in this case, water. After the separation of damaged CS and unburned coal particles, the influence of 1100 and 1200 °C thermal treatment on the CS was studied. The thermal treatment was performed in a Nabertherm Laboratory Furnace L9/13 with a P330 controller (Nabertherm, Lilienthal, Germany), using an open alumina crucible (V = 300 cm^3), filled to 2/3 of its total volume, in an air atmosphere, with a heating rate of 5 °C·min^{-1} and dwell time of 30 min with subsequent natural cooling.

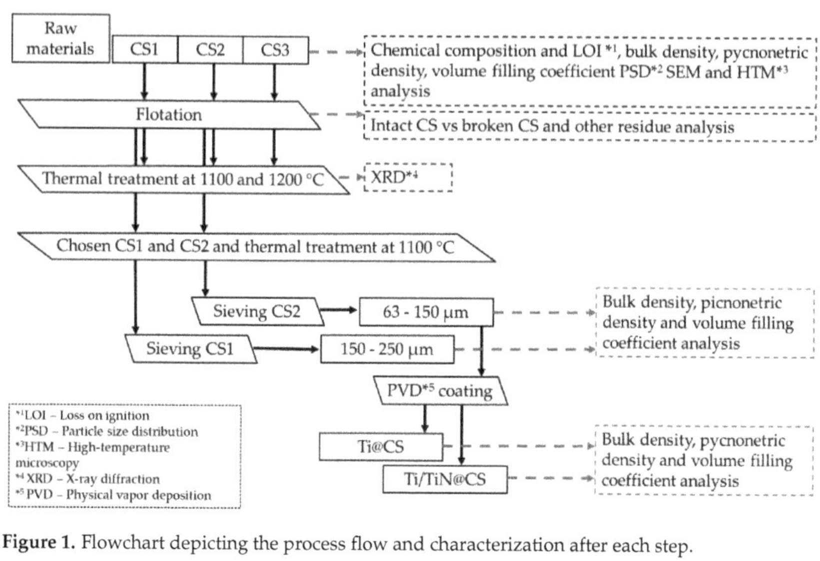

Figure 1. Flowchart depicting the process flow and characterization after each step.

3. Results

3.1. Granulometry Composition and Density of CS

For all types of CS, the particle size distribution was determined. As can be seen from Figure 2, the CS used in the experiments are predominantly (more than 40%) in the 100 to 150 μm size range. In the case of CS1, up to 14% are particles of <100 μm, and about 27% are particles of >150 μm. In the case of CS2, particles with average sizes of 100–150 μm are mainly present, while in CS3, particles (up to 42%) with sizes of 100–150 μm and particles of >200 μm (up to 15%) are observed. From a granulometric point of view, a more even distribution of CS particles is seen in the case of CS2: above 74% with dimensions from 100 to 150 μm.

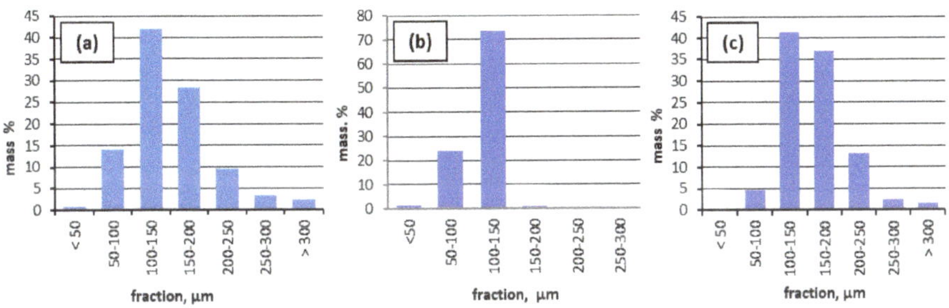

Figure 2. Particle size distribution (PSD) of the three types of cenospheres: (**a**) CS1, (**b**) CS2, and (**c**) CS3.

Important indicators are the bulk density of the material, the interparticle void fraction, as well as the density of the material of the globular wall. The experimental data are summarised in Table 2. The lowest void fraction (38.0 %) corresponds to the most polydispersed specimen—CS3. Additionally, the highest void fraction (46.0 %) corresponds to CS1 after sieving and flotation (specimen CS1, 150–250 μm, treated at 1100 °C). The lowest bulk density is observed in the case of CS2, which may be due to a more even particle size distribution, as was seen in Figure 2. Dispersions with different particle sizes during pouring are arranged more densely, which leads to an increase in bulk density and a decrease in the interparticle void volume. The pycnometric density after heat treatment at 1100 °C increased from 2.153 to 2.178 g·cm^{-3} in the case of the CS1 raw material. In the case of CS1, the pycnometric density mainly corresponds to the CS wall material, namely ceramics, and the density could change due to a phase composition change. In the case of CS2, the pycnometric density decreased from 2.272 to 2.185 g·cm^{-3}. This fact could be explained by the burning out of residual coal particles (C content up to 11.5 at.% and LOI = 0.4) during the thermal treatment (Tables 3 and 4), and the significant amount of broken CS (11.6%) content in the CS2 sample, which is separated later by the flotation method to prepare the CS 63–150 μm fraction. Ti- and Ti-TiN-coated CS have a greater pycnometric density, which is obvious, due to the deposited Ti and Ti-TiN.

Table 2. Bulk density of CS, the density of the material, and interparticle void fraction.

Material	Bulk Density, g·cm^{-3}	Pycnometric Density g·cm^{-3}	Voids, %
Raw CS1	0.415 ± 0.004	2.153 ± 0.001	40.0
Raw CS2	0.380 ± 0.002	2.272 ± 0.001	43.0
Raw CS3	0.411 ± 0.004	2.301 ± 0.001	38.0
CS2 63–150 μm treated at 1100 °C	0.39 ± 0.004	2.185 ± 0.001	44.0
CS1 150–250 μm treated at 1100 °C	0.410 ± 0.004	2.178 ± 0.001	46.0
Ti@CS *	0.400 ± 0.004	2.661 ± 0.001	44.0
Ti-TiN@CS *	0.420 ± 0.004	2.531 ± 0.001	43.0

* CS2 63–150 μm treated at 1100 °C and then PVD coated.

Table 3. Average CS1, CS2, and CS3 element analysis in at. %.

Sample	C	O	Al	Si	Fe	Ca	Ti	Na	K	Mg	Si/Al*
CS1	10.04	55.48	14.92	17.37	0.31	0.25	0.29	0.23	-	0.02	1.16
CS2	8.94	57.16	15.25	16.76	0.24	0.04	0.33	-	-	-	1.10
CS3	11.49	53.00	8.50	21.00	2.02	0.05	0.40	-	2.0	0.74	2.47

Si/Al*—Si/Al at. ratio.

Table 4. Chemical composition of cenospheres in wt.%.

Sample	SiO_2	Al_2O_3	Fe_2O_3	CaO	MgO	Na_2O	K_2O	LOI * 400 °C,%	LOI * 1000 °C,%
CS1	56.5	36.9	1.4	2.4	1.2	1.1	0.5	0.5	0.1
CS2	53.8	40.7	1.0	1.4	0.6	0.5	0.4	0.6	0.4
CS3	61.4	24.4	3.4	1.9	1.6	0.5	2.2	3.9	0.2

* LOI—Loss on ignition.

3.2. Chemical Composition

In order to more fully characterise the microspheres studied, it is important to find out and compare their chemical composition, which can differ significantly when burning different coals at different temperatures. Energy-dispersive X-ray spectroscopy data from the surface of the raw microspheres show that the main elements present in their shell are Al, Si, Fe, Ca, Mg, K, Na, O, and in significant quantities, C. The average compositions of the microsphere material elements from EDS analysis data are shown in Table 3.

It should be noted that the composition of the material is rather uneven for all samples. Samples CS1 and CS2 have the same elemental composition and Si/Al ratio, but sample CS3 is significantly different. The content of Al has practically been increased twofold, and Fe and K appear in the composition of the material in noticeable quantities. For all samples, the elemental analysis shows significant amounts of carbon, which could be explained by the carbon that was not fully burned.

The chemical analysis was performed by converting the analysed elements to oxides. For the LOI determination, the studied samples were thermally treated at temperatures of 400 and 1000 °C (see Table 4).

The results of the chemical analysis confirm the obtained EDS data—CS1 and CS2 are close enough in composition, but CS3 is significantly different. Additionally, the main differences are in the proportions of SiO_2 and Al_2O_3 and in the content of Fe_2O_3. It should also be mentioned that the mass loss during heating in the case of sample CS3 is on average seven times higher than in the other samples.

3.3. The Ratio between Intact and Defective Cenospheres

When analysing the samples CS1, CS2, and CS3, it was found that after the settling (segregation in floating and non-floating fractions) of samples CS2 and CS3 (Figure 3), the liquid turned black. After filtering, black deposits remained on the filter. Sediment analysis was not carried out, but these are believed to have been formed by particles of coal that had not burned out. In the sample, the CS1 liquid became greyish yellow.

The results of the floating and non-floating microsphere fractions, as well as the calculations of the percentage of the fraction are: 1.1 ± 0.05% defective microspheres in sample CS1, in CS2—11.6 ± 0.09%, in CS3—3.2 ± 0.08%. In the contents of samples CS2 and CS3, it is possible that coal particles that had not burned out were present.

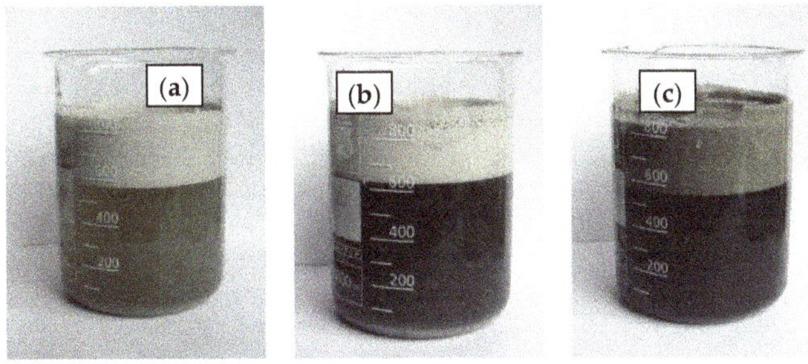

Figure 3. Images of beakers with water and CS1 (**a**), CS2 (**b**), and CS3 (**c**), showing the setup used to separate the broken and intact particles.

3.4. Determination of Crystalline Phases in Cenosphere Composition

To determine the phase composition of the CS and how it is affected by heat treatment, X-ray diffraction (XRD) phase analysis was performed on as-received and heat-treated CS. The heat treatment of the CS was performed at 1100 and 1200 °C in accordance with the process flow shown in Figure 1. Cenospheres form a small fraction of the fly ash that is produced as a by-product of the mineral component of coal when they burn at high temperatures, with the formation of crystallite mixtures and a glass phase. The diffractograms for CS1, CS2, and CS3 are shown in Figure 4. As shown by the X-ray studies, the diffractograms of samples CS1 and CS2 are very lax. In the output samples, mainly mullite $Al_6Si_2O_{13}$ and the amorphous phase, which can be attributed to the glass phase, can be observed. Mullite is the main crystalline phase, which occurs as a result of thermochemical transformations of the aluminosilicates contained in coal. When the samples are heated at 1100 °C and 1200 °C, the phase of cristobalite SiO_2 is shown, and its intensity increases with the increasing heat-treatment temperature (sample diffractograms are shown in Figure 4). The cristobalite (SiO_2) phase is not seen in the as-received samples from all sources, but upon heat treatment, it is seen that the amount of the cristobalite phase increases, showing that heat treatment leads to the formation of cristobalite. Along with that, the net amount of the mullite phase also increases, indicated by the increase in peak height.

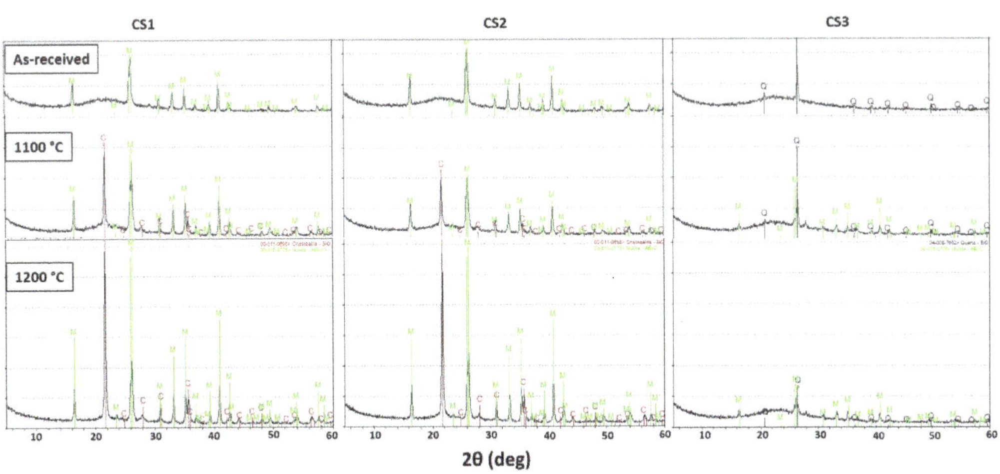

Figure 4. X-ray diffractograms of the CS at as-received and heat-treated conditions. Phase designations: C—cristobalite SiO_2; Q—quartz SiO_2; M—mullite $Al_6Si_2O_{13}$.

Sample CS3 differs from the rest in terms of phase composition, showing that it is less crystalline than CS1 and CS2. Additionally, SiO_2 is already present in the polymorphic modification of quartz in the untreated sample, and mullite forms after heat processing. Moreover, in sample CS3, the amount of the amorphous phase after heat treatment is much higher than in the other two samples (the characteristic plateau on the X-ray diffraction pattern is indicative of this), but in samples CS1 and CS2, it turned into a crystalline phase.

The diffractograms of sample CS3 are sharply different from CS1 and CS2. In the untreated material, it was possible to distinguish the quartz crystallographic phase only, but other phases in an amorphous state (also Al_2O_3). In the output sample, only the crystalline phase of quartz at 26.2 degrees at 2θand the X-ray phase are detected. With a decrease in the concentration of Al_2O_3, that is, with an increase in the SiO_2/Al_2O_3 ratio, the quartz phase prevails. The heat treatment of the samples causes the quartz phase to decrease and the formation of the mullite phase.

Figure 5 shows that the visible FT-IR spectra of the microspheres studied show a wide absorption band in the interval of 1058–1062 cm^{-1}, which can be attributed to the valence and deformation oscillations of Si-O-Si(Al) in the crystal lattice and v = 820 cm^{-1} to valence oscillations in the case of Si-O in CS1 and CS2. In the case of CS3, a pronounced band at 1008 cm^{-1} appears, which is attributed to valence oscillations in the bridge links in the Si-O-Si(Al) crystal lattice. A small maximum also indicates the presence of quartz in the case of CS3 at 778 cm^{-1} [45].

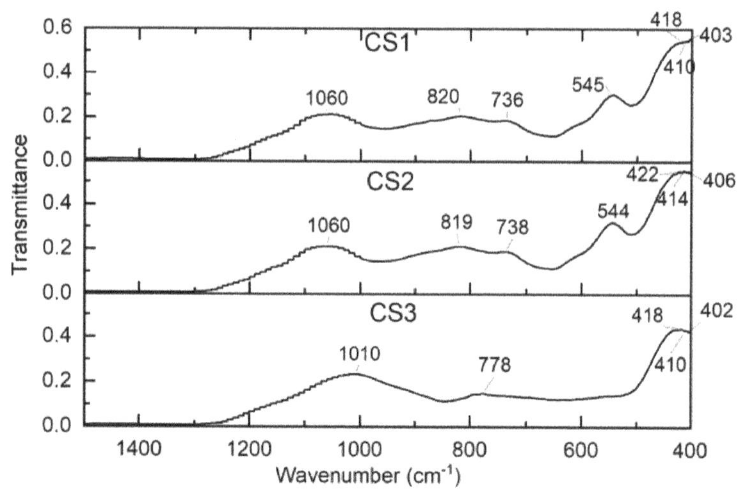

Figure 5. The FT-IR spectra of the cenospheres CS1, CS2, and CS3.

The various valence oscillations of Si-O-Al, Al-O, Si-OH, and Si-O can be attributed to the many absorption bands observed in the interval of 800–420 cm^{-1}. The positions of the absorption bands and their intensity in the IR spectra are significantly influenced by the ratio in the structure of Si/Al microspheres. Moreover, this is not about pure materials, but about a technical product with a sufficiently heterogeneous composition.

3.5. The Morphology of the CS

The chemical and phase composition of CS largely determines the morphology of their surface. Figure 6 shows SEM photomicrographs of the three cenospheres, with all three showing a different surface morphology.

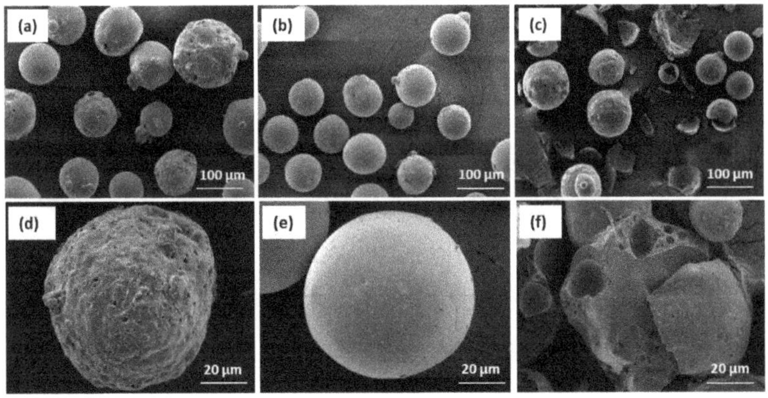

Figure 6. SEM micrographs of the cenospheres: CS1 (**a**,**d**); CS2 (**b**,**e**); CS3 (**c**,**f**).

The cenospheres in sample CS1 (Figure 6a) can be seen to have different globular features on the surface, and most of the surface is embossed and perforated. In the case of CS2 (Figure 6b), the particles are of uniform size with high sphericity and a smooth surface. The CS3 cenospheres (Figure 6c) have a wide variety of morphological forms and a significant amount of them are damaged. The walls of the particles have a distinct porosity. The variety of shapes and surface conditions of CS is explained by the very complex processes of transformation of oxide formation and crystallisation after coal burning during the separate stages of their combustion at high temperatures.

In order to be able to judge the thickness of the wall of the CS shells and their structures, resin-fixed CS grindings were made (Figure 7). In most cases, when the CS have a pronounced spherical shape, the shell-wall thickness is from 4 to 10 μm on average. As can be seen, the walls have closely spaced pores, with average pore sizes from 0.5 to 5.0 μm. Additionally, it is clearly seen that CS3 walls have the greatest and closest porosity among all the studied CS.

Figure 8 shows the Ti-deposited layer (122.8 nm) on the CS2 (Figure 8a), and Ti-TiN-deposited layers (279.1 nm) on the CS2 (Figure 8b). However, multiple measurements noted that the layer thickness of Ti (for Ti@CS) is between 100 and 350 nm; and in the case of Ti-TiN@CS, the layer of TiN nitride is between 90 and 350 nm and Ti between 150 and 380 nm. The deposited Ti and Ti-TiN coatings in all the observed specimens are intact and have no cracks, pores, or delamination.

3.6. *The Behaviour of the CS at Elevated Temperature*

Using HTM, 2D-dilatometry curves were obtained that describe changes in the volume or linear dimensions of CS under the influence of temperature. As can be seen from the curves in Figure 9, in the patterns made of CS1 and CS2, the deformation is negligible until the 1200–1325 °C temperature interval, but after this temperature, a slight shrinkage can be observed, which could be due to the deformation of the CS and their mutual compaction (softening begins).

As the samples reach 1350 °C, CS1 demonstrates greater (up to 5%) shrinkage than CS2. At the CS3 temperature above 1050 °C, a rapid shrinkage of the sample occurs, which is associated with the transition of the sample to a semi-pyroplastic state. This effect could be caused by the presence of K_2O and Fe_2O_3, resulting in a lower softening point (Tg) of the sample. Additionally, as can be seen in Figure 4, the sample CS3 contains a quartz phase, which may react with alumina to form mullite at 1050–1200 °C [46–48]: the amount of the quartz phase decreases and mullite appears. Starting from 1190–1200 °C, a sharp increase in volume begins, which is associated with the release of gases.

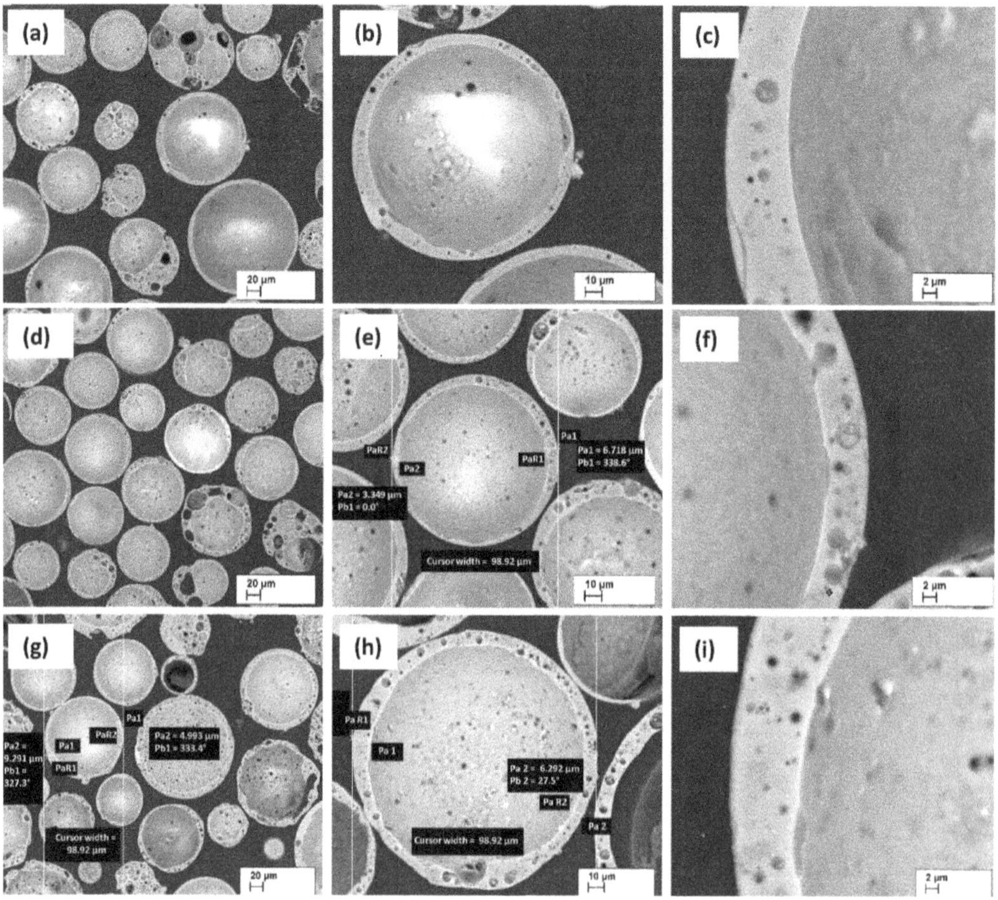

Figure 7. SEM micrographs of CS1 (**a**–**c**), CS2 (**d**–**f**), and CS3 (**g**–**i**) at 200, 500, and 2000 × times magnification, respectively.

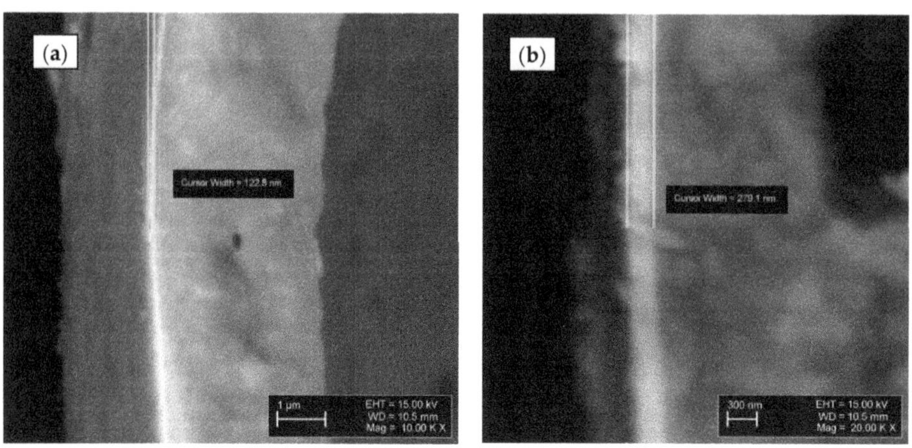

Figure 8. SEM micrographs of the Ti@CS (**a**) and Ti-TiN@CS (**b**).

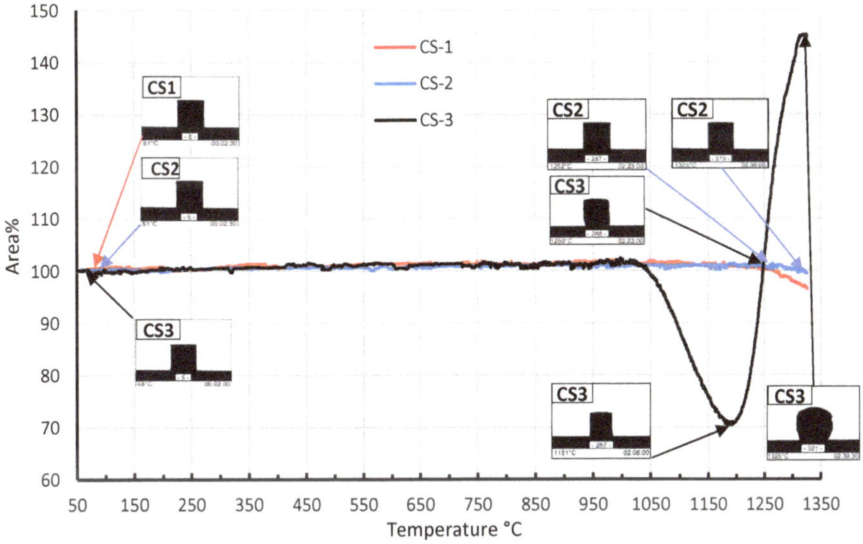

Figure 9. CS1, CS2 and CS3 2D-dilatometry curves in the 50–1325 °C temperature interval.

As can be seen in Figure 10, heating to 1200 °C does not cause visually detectable changes for cenospheres CS1 and CS2, but a slight change in the colour of CS1 is observed, which can be explained by the burnout of coal particles and the oxidation of the Fe present in small quantities, which is black, to Fe_2O_3, which has a reddish-brown colour. In the case of sample CS3, a significant discolouration and compaction of the sample is observed at 1100 °C, and is even more pronounced at 1200 °C. The visually observed phenomena are consistent with the HTM experiment data.

Figure 10. Images of cenospheres at various stages of heat treatment: (**a**) as-received, (**b**) 1100 °C and (**c**) 1200 °C.

3.7. Specific Thermal Conductivity Coefficient

Specific thermal conductivity is an important physical parameter for predicting the properties of CS-made composites. The thermal conductivities obtained from experiments are 0.109 ± 0.0005, 0.100 ± 0.0005, and 0.096 ± 0.0005 $W \cdot m^{-1} \cdot K^{-1}$ for CS1, CS2, and CS3. The data obtained on the thermal conductivity of CS allow us to conclude that if these particles are used as a filler in any matrix, a very low thermal conductivity can be achieved. This is primarily due to the fact that the freely-poured particles contain a lot of air, both internally (voids inside and pores in the walls) and between the particles.

The indicators that were determined during the research (chemical composition, thermal conductivity, bulk density) for samples CS1 and CS2 are rather similar, but for CS3, they differ somewhat. Sample CS3 differs from CS1 and CS2 considerably, with its thermal durability—these cenospheres sinter and shrink considerably. This may be due to the relatively greater number of particles with d of <63 μm, and a higher content of volatile and combustible carbon, at around 3.9 % (mass). It was also observed that a greater amount of the amorphous phase was present after heat treatment, with greater contents of SiO_2.

4. Conclusions

The physical and chemical properties of microballoons of aluminosilicates, called cenospheres, which are produced as a by-product (a component of fly ash) of burning coal, were studied. Several sources of fly ash from coal basins in thermal power plants were selected. Cenospheres with particle sizes ranging from 40 to 500 μm were studied. Different particle distribution by size was observed, and the most uniform distribution of CS particles was in the case of CS2: above 74% with dimensions from 100 to 150 μm. The CS bulk had a similar density for all samples and amounted to around 0.4 $g \cdot cm^{-3}$, with a particle shell material density of 2.1 $g \cdot cm^{-3}$.

Energy-dispersive X-ray spectrometry (EDS) and chemical analysis of the CS revealed that the main components of the studied CS were SiO_2 and Al_2O_3. In the case of CS1 and CS2, the sum of these components was on average from 93 to 95%. In the case of CS3, the sum of SiO_2 and Al_2O_3 did not exceed 86%, and Fe_2O_3 and K_2O were present in appreciable quantities in CS3.

The presence of SiO_2 and Al_2O_3 in the composition of the microballoons was also confirmed by spectroscopic FTIR studies. As shown by X-ray studies, the diffractograms of the samples CS1 and CS2 were very similar; in the output samples, one can mainly observe the mullite $Al_6Si_2O_{13}$ and the amorphous phase. In the output samples of CS3, mainly quartz and amorphous phases can be observed; the heat treatment caused a decrease in the quartz phase, and the crystalline phase of mullite appeared.

The thermal conductivity of all the CS samples varied from 0.096 to 0.109 $W \cdot m^{-1} \cdot K^{-1}$. The chemical composition, size distribution, and structure of the cenospheres affect their thermal stability. Cenospheres CS1 and CS2 did not sinter during heat treatment up to 1200 °C, while the sample CS3 was subjected to sintering already at 1100 °C, making it a bad choice for SF development. This behaviour in CS3 is attributed to the softening of the structure due to the presence of the quartz phase, Fe_2O_3 and K_2O.

Due to the properties discussed above, CS1, which has similar thermal characteristics to sample CS2, should be selected for further experiments of matrix-less SF composite preparation or future scale-up, having better thermal stability, a narrow grading composition, and more uniform particle size distribution (94.2% of particles have a size of 63–150 μm). The morphology is also uniform, with high sphericity and a very low presence of agglomerates. Although there was a significantly higher percentage of defective particles in the as-received CS, the post-sieved product was of high quality.

Author Contributions: Conceptualization, A.S. and J.O.; methodology, I.Z. and G.M.; validation, V.P. and A.K.S.; formal analysis, A.S.; investigation, V.A., I.Z. and A.S.; resources, A.S.; data curation, V.A. and A.S.; writing—original draft preparation, A.S., J.O. and G.M.; writing—review and editing, A.S.

and A.K.S.; visualisation, V.A., I.Z. and V.P.; supervision, J.O. and A.S.; project administration, A.S.; funding acquisition, A.S. All authors have read and agreed to the published version of the manuscript.

Funding: This research was funded by VPP AIPP project Nr. VPP-AIPP-2021/1-0015 "Combined lightweight, high-temperature resistant hybrid composite for combined protection of drones from Direct Energy Weapon". The article was published with the financial support from the Riga Technical University Research Support Fund.

Institutional Review Board Statement: Not applicable.

Informed Consent Statement: Not applicable.

Data Availability Statement: Not applicable.

Conflicts of Interest: The authors declare no conflict of interest.

References

1. Gupta, N.; Rohatgi, P.K. *Metal Matrix Syntactic Foams: Processing, Microstructure, Properties and Applications*; Gupta, N., Ed.; DEStech Publications, Inc.: Lancaster, UK, 2014; ISBN 978-1932078831.
2. John, B.; Nair, C. *Update on Syntactic Foams*; iSmithers: Shropshire, UK, 2010; ISBN 9781847351203.
3. Thiyagarajan, R.; Senthil Kumar, M. A Review on Closed Cell Metal Matrix Syntactic Foams: A Green Initiative towards Eco-Sustainability. *Mater. Manuf. Process.* **2021**, *36*, 1333–1351. [CrossRef]
4. Yousaf, Z.; Morrison, N.F.; Parnell, W.J. Tensile Properties of All-Polymeric Syntactic Foam Composites: Experimental Characterization and Mathematical Modelling. *Compos. Part B Eng.* **2022**, *231*, 109556. [CrossRef]
5. Singh, A.K.; Saltonstall, B.; Patil, B.; Hoffmann, N.; Doddamani, M.; Gupta, N. Additive Manufacturing of Syntactic Foams: Part 2: Specimen Printing and Mechanical Property Characterization. *JOM* **2018**, *70*, 310–314. [CrossRef]
6. Gupta, N.; Zeltmann, S.E.; Shunmugasamy, V.C.; Pinisetty, D. Applications of Polymer Matrix Syntactic Foams. *JOM* **2014**, *66*, 245–254. [CrossRef]
7. Rohatgi, P.K.; Gupta, N.; Schultz, B.F.; Luong, D.D. The Synthesis, Compressive Properties, and Applications of Metal Matrix Syntactic Foams. *JOM* **2011**, *63*, 36–42. [CrossRef]
8. Ranjbar, N.; Kuenzel, C. Cenospheres: A Review. *Fuel* **2017**, *207*, 1–12. [CrossRef]
9. Adesina, A. Sustainable Application of Cenospheres in Cementitious Materials—Overview of Performance. *Dev. Built Environ.* **2020**, *4*, 100029. [CrossRef]
10. Markets and Markets, *Cenosphere Market by Type (Gray Cenosphere, White Cenosphere) End-Use Industry (Refractory, Construction, Oil and Gas, Automotive, Paints and Coatings) and Region-Global Forecast*. Available online: https://www.marketsandmarkets.com/Market-Reports/cenosphere-market-99648171.html?gclid=CjwKCAiA9NGfBhBvEiwAq5vSy8PwvqJ7UI36UrrG12jtTfmDVawWQiuD5NWeksnBdwTvvYTklzxLoRoCat8QAvD_BwE (accessed on 21 February 2023).
11. Shao, Y.; Jia, D.; Liu, B. Characterization of Porous Silicon Nitride Ceramics by Pressureless Sintering Using Fly Ash Cenosphere as a Pore-Forming Agent. *J. Eur. Ceram Soc.* **2009**, *29*, 1529–1534. [CrossRef]
12. Rohatgi, P.K.; Matsunaga, T.; Gupta, N. Compressive and Ultrasonic Properties of Polyester/Fly Ash Composites. *J. Mater. Sci.* **2009**, *44*, 1485–1493. [CrossRef]
13. Anshits, N.N.; Vereshchagina, T.A.; Bayukov, O.A.; Salanov, A.N.; Anshits, A.G. The Nature of Nanoparticles of Crystalline Phases in Cenospheres and Morphology of Their Shells. *Glass Phys. Chem.* **2005**, *31*, 306–315. [CrossRef]
14. Bride, S.P.M.; Shukla, A. Processing and Characterization of a Lightweight Concrete Using Cenospheres. *J. Mater. Sci.* **2002**, *7*, 4217–4225.
15. Kulkarni, M.; Bambole, V.; Mahanwar, P. Effect of Particle Size of Fly Ash Cenospheres on the Properties of Acrylonitrile Butadiene Styrene-Filled Composites. *J. Thermoplast. Compos. Mater.* **2012**, *27*, 251–267. [CrossRef]
16. Chauhan, S.; Thakur, S. Effect of Micro Size Cenosphere Particles Reinforcement on Tribological Characteristics of Vinylester Composites under Dry Sliding Conditions. *J. Miner. Mater. Charact. Eng.* **2012**, *11*, 938. [CrossRef]
17. Rohatgi, P.K.; Kim, J.K.; Gupta, N.; Alaraj, S.; Daoud, A. Compressive Characteristics of A356/Fly Ash Cenosphere Composites Synthesized by Pressure Infiltration Technique. *Compos. Part A Appl. Sci. Manuf.* **2006**, *37*, 430–437. [CrossRef]
18. Dou, Z.Y.; Jiang, L.T.; Wu, G.H.; Zhang, Q.; Xiu, Z.Y.; Chen, G.Q. High Strain Rate Compression of Cenosphere-Pure Aluminum Syntactic Foams. *Scr. Mater.* **2007**, *57*, 945–948. [CrossRef]
19. Sankaranarayanan, S.; Nguyen, Q.B.; Shabadi, R.; Almajid, A.H.; Gupta, M. Powder Metallurgy Hollow Fly Ash Cenospheres' Particles Reinforced Magnesium Composites. *Powder Metall.* **2016**, *59*, 188–196. [CrossRef]
20. Wang, M.-R.; Jia, D.-C.; He, P.-G.; Zhou, Y. Microstructural and Mechanical Characterization of Fly Ash Cenosphere/Metakaolin-Based Geopolymeric Composites. *Ceram Int.* **2011**, *37*, 1661–1666. [CrossRef]
21. Shishkin, A.; Mironovs, V.; Zemchenkovs, V.; Hussainova, I. Metal Powder/Fly Ash Cenosphere/Modified Clay, Composite. In Proceedings of the Euro PM 2014 Congress and Exhibition, Salzburg, Austria, 21–24 September 2014.
22. Castellanos, A.G.; Mawson, H.; Burke, V.; Prabhakar, P. Fly-Ash Cenosphere/Clay Blended Composites for Impact Resistant Tiles. *Constr. Build Mater.* **2017**, *156*, 307–313. [CrossRef]

23. Bharath Kumar, B.R.; Zeltmann, S.E.; Doddamani, M.; Gupta, N.; Gurupadu, S.; Sailaja, R.R.N. Effect of Cenosphere Surface Treatment and Blending Method on the Tensile Properties of Thermoplastic Matrix Syntactic Foams. *J. Appl. Polym. Sci.* **2016**, *133*, 43881. [CrossRef]
24. Bharath Kumar, B.R.; Doddamani, M.; Zeltmann, S.E.; Gupta, N.; Gurupadu, S.; Sailaja, R.R.N. Effect of Particle Surface Treatment and Blending Method on Flexural Properties of Injection-Molded Cenosphere/HDPE Syntactic Foams. *J. Mater. Sci.* **2016**, *51*, 3793–3805. [CrossRef]
25. Irtiseva, K.; Lapkovskis, V.; Mironovs, V.; Ozolins, J.; Thakur, V.K.; Goel, G.; Baronins, J.; Shishkin, A. Towards Next-Generation Sustainable Composites Made of Recycled Rubber, Cenospheres, and Biobinder. *Polymers* **2021**, *13*, 574. [CrossRef] [PubMed]
26. Waddar, S.; Jeyaraj, P.; Doddamani, M. Influence of Axial Compressive Loads on Buckling and Free Vibration Response of Surface-Modified Fly Ash Cenosphere/Epoxy Syntactic Foams. *J. Compos. Mater.* **2018**, *52*, 2621–2630. [CrossRef]
27. Bharath, H.S.; Bonthu, D.; Prabhakar, P.; Doddamani, M. Three-Dimensional Printed Lightweight Composite Foams. *ACS Omega* **2020**, *5*, 22536. [CrossRef]
28. Doddamani, M.; Shunmugasamy, V.C.; Gupta, N.; Vijayakumar, H.B. Compressive and Flexural Properties of Functionally Graded Fly Ash Cenosphere-Epoxy Resin Syntactic Foams. *Polym. Compos.* **2015**, *36*, 685–693. [CrossRef]
29. Bharath Kumar, B.R.; Doddamani, M.; Zeltmann, S.E.; Gupta, N.; Ramesh, M.R.; Ramakrishna, S. Processing of Cenosphere/HDPE Syntactic Foams Using an Industrial Scale Polymer Injection Molding Machine. *Mater. Des.* **2016**, *92*, 414–423. [CrossRef]
30. Peng, X.; Li, H.; Hu, Y. Preparation of Metakaolin-Fly Ash Cenosphere Based Geopolymer Matrices for Passive Fire Protection. *J. Mater. Res. Technol.* **2023**, *23*, 604–610. [CrossRef]
31. Cheng, Z.; An, J.P.; Li, F.; Lu, Y.; Li, S. Effect of Fly Ash Cenospheres on Properties of Multi-Walled Carbon Nanotubes and Polyvinyl Alcohol Fibers Reinforced Geopolymer Composites. *Ceram. Int.* **2022**, *48*, 18956–18971. [CrossRef]
32. Brooks, A.L.; Shen, Z.; Zhou, H. Development of a High-Temperature Inorganic Synthetic Foam with Recycled Fly-Ash Cenospheres for Thermal Insulation Brick Manufacturing. *J. Clean Prod.* **2020**, *246*, 118748. [CrossRef]
33. Jha, N.; Mondal, D.P.; Goel, M.D.; Majumdar, J.D.; Das, S.; Modi, O.P. Titanium Cenosphere Syntactic Foam with Coarser Cenosphere Fabricated by Powder Metallurgy at Lower Compaction Load. *Trans. Nonferrous Met. Soc. China* **2014**, *24*, 89–99. [CrossRef]
34. Luong, D.; Lehmhus, D.; Gupta, N.; Weise, J.; Bayoumi, M. Structure and Compressive Properties of Invar-Cenosphere Syntactic Foams. *Materials* **2016**, *9*, 115. [CrossRef]
35. Peroni, L.; Scapin, M.; Fichera, C.; Lehmhus, D.; Weise, J.; Baumeister, J.; Avalle, M. Investigation of the Mechanical Behaviour of AISI 316L Stainless Steel Syntactic Foams at Different Strain-Rates. *Compos. B Eng.* **2014**, *66*, 430–442. [CrossRef]
36. Luong, D.D.; Shunmugasamy, V.C.; Gupta, N.; Lehmhus, D.; Weise, J.; Baumeister, J. Quasi-Static and High Strain Rates Compressive Response of Iron and Invar Matrix Syntactic Foams. *Mater. Des.* **2015**, *66*, 516–531. [CrossRef]
37. Shishkin, A.; Hussainova, I.; Kozlov, V.; Lisnanskis, M.; Leroy, P.; Lehmhus, D. Metal-Coated Cenospheres Obtained via Magnetron Sputter Coating: A New Precursor for Syntactic Foams. *JOM* **2018**, *70*, 1319–1325. [CrossRef]
38. Shishkin, A.; Drozdova, M.; Kozlov, V.; Hussainova, I.; Lehmhus, D. Vibration-Assisted Sputter Coating of Cenospheres: A New Approach for Realizing Cu-Based Metal Matrix Syntactic Foams. *Metals* **2017**, *7*, 16. [CrossRef]
39. Hales, T. A Proof of the Kepler Conjecture. *Ann. Math.* **2005**, *162*, 1065–1185. [CrossRef]
40. Hales, T.C. Historical Overview of the Kepler Conjecture. *Discrete. Comput. Geom.* **2006**, *36*, 5–20. [CrossRef]
41. SCOTT, G.D. Packing of Spheres: Packing of Equal Spheres. *Nature* **1960**, *188*, 908–909. [CrossRef]
42. Scott, G.D.; Kilgour, D.M. The Density of Random Close Packing of Spheres. *J. Phys. D Appl. Phys.* **1969**, *2*, 311. [CrossRef]
43. Bargiel, M.; Tory, E.M. Packing Fraction and Measures of Disorder of Ultradense Irregular Packings of Equal Spheres. II. Transition from Dense Random Packing. *Adv. Powder Technol.* **2001**, *12*, 533–557. [CrossRef]
44. Kapfer, S.C.; Mickel, W.; Mecke, K.; Schröder-Turk, G.E. Jammed Spheres: Minkowski Tensors Reveal Onset of Local Crystallinity. *Phys. Rev. E* **2012**, *85*, 030301. [CrossRef]
45. Reig, F.B.; Adelantado, J.V.G.; Moya Moreno, M.C.M. FTIR Quantitative Analysis of Calcium Carbonate (Calcite) and Silica (Quartz) Mixtures Using the Constant Ratio Method. Application to Geological Samples. *Talanta* **2002**, *58*, 811–821. [CrossRef] [PubMed]
46. Rouabhia, F.; Nemamcha, A.; Moumeni, H. Elaboration and Characterization of Mullite-Anorthite-Albite Porous Ceramics Prepared from Algerian Kaolin. *Ceramica* **2018**, *64*, 126–132. [CrossRef]
47. Nair, S.G.; Sreejith, K.J.; Srinivas, C.; Prabhakaran, K.; Devasia, R. Low Temperature Mullite Forming Pre-Ceramic Resins of High Ceramic Yield for Oxide Matrix Composites. *Ceram. Int.* **2022**, *48*, 18441–18451. [CrossRef]
48. Singh, A.K.; Sarkar, R. Nano Mullite Bonded Refractory Castable Composition for High Temperature Applications. *Ceram. Int.* **2016**, *42*, 12937–12945. [CrossRef]

materials

MDPI

Article

TiSiCN as Coatings Resistant to Corrosion and Neutron Activation

Matlab N. Mirzayev [1,2,3,4], Anca C. Parau [5], Lyubomir Slavov [6], Mihaela Dinu [5], Dimitar Neov [1,7], Zdravka Slavkova [7], Evgeni P. Popov [1,7,8], Maria Belova [1], Kanan Hasanov [2], Fuad A. Aliyev [1,9] and Alina Vladescu (Dragomir) [5,10,*]

1 Joint Institute for Nuclear Research, 141980 Dubna, Russia
2 Institute of Radiation Problems, Azerbaijan National Academy of Sciences, AZ1143 Baku, Azerbaijan
3 Scientific-Research Institute Geotecnological Problems of Oil, Gas and Chemistry, Azerbaijan State Oil and Industry University, AZ1010 Baku, Azerbaijan
4 Innovation & Research Center, Western Caspian University, AZ1001 Baku, Azerbaijan
5 National Institute of Research and Development for Optoelectronics INOE 2000, 409 Atomistilor St., 77125 Magurele, Romania
6 Institute of Electronics, Bulgarian Academy of Sciences, 1784 Sofia, Bulgaria
7 Institute of Solid-State Physics, Bulgarian Academy of Sciences, 1784 Sofia, Bulgaria
8 Institute for Nuclear Research and Nuclear Energy, Bulgarian Academy of Sciences, 1784 Sofia, Bulgaria
9 Institute of Geology and Geophysics of Azerbaijan, National Academy of Sciences, AZ1143 Baku, Azerbaijan
10 Physical Materials Science and Composite Materials Centre, Research School of Chemistry & Applied Biomedical Sciences, National Research Tomsk Polytechnic University, 30 Lenina Avenue, 634050 Tomsk, Russia
* Correspondence: alinava@inoe.ro; Tel.: +40-21-457-57-59

Abstract: The aim of the present paper was to evaluate the effect of neutron activation on TiSiCN carbonitrides coatings prepared at different C/N ratios (0.4 for under stoichiometric and 1.6 for over stoichiometric). The coatings were prepared by cathodic arc deposition using one cathode constructed of Ti88 at.%-Si12 at.% (99.99% purity). The coatings were comparatively examined for elemental and phase composition, morphology, and anticorrosive properties in 3.5% NaCl solution. All the coatings exhibited f.c.c. solid solution structures and had a (111) preferred orientation. Under stoichiometric structure, they proved to be resistant to corrosive attack in 3.5% NaCl and of these coatings the TiSiCN was found to have the best corrosion resistance. From all tested coatings, TiSiCN have proven to be the most suitable candidates for operation under severe conditions that are present in nuclear applications (high temperature, corrosion, etc.).

Keywords: carbonitrides; corrosion resistance; surface morphology; neutron activation analysis; crystal structure; potentiodynamic polarization

check for updates

Citation: Mirzayev, M.N.; Parau, A.C.; Slavov, L.; Dinu, M.; Neov, D.; Slavkova, Z.; Popov, E.P.; Belova, M.; Hasanov, K.; Aliyev, F.A.; et al. TiSiCN as Coatings Resistant to Corrosion and Neutron Activation. *Materials* **2023**, *16*, 1835. https://doi.org/10.3390/ma16051835

Academic Editor: Chunxu Pan

Received: 26 January 2023
Revised: 15 February 2023
Accepted: 21 February 2023
Published: 23 February 2023

1. Introduction

In the last few years, ceramics gained more popularity in nuclear fusion applications due to their high insulation properties, heat, and radiation resistance when compared to other commonly used isolators (i.e., plastics). Ceramic materials can be used for the elements of plasma facing (i.e., in diverters and as insulation coating of coils) or for the construction components of tokamaks. In the first case, the materials are permanently exposed to different types of ionizing radiation generated inside the fusion chambers and they should possess certain properties; on the other hand, in the second case, the irradiation will be applied just in case of emergency and the properties should be different than those necessary for constant exposure.

In the present paper, our attention is focused on materials subjected to continuous irradiation. This dangerous phenomenon is associated with the ion component of the internal tokamak plasma due to the deceleration length of ions inside the plasma-facing solid materials, which is short in comparison with that of the other disturbing radiations (i.e., X-ray photons, electrons, and neutrons). In the tokamak plasmas, two magnitudes of ion energies are reachable: plasma temperature (about 10 keV) and the beams of fast ions produced during neutral-beam plasma heating (about 100 keV). The diffusion through an insulating magnetic field of plasma inside a tokamak chamber is transformed into the plasma temperature of about <1 keV, while the energy of fast ions bombarding the plasma-facing elements becomes very wide; thus, based on what we presented above, we believe that the testing of the radiation effect on various types of ceramics is essential.

By comparing the coatings prepared by physical vapor deposition (PVD) methods, it can be seen that carbonitrides are widely used for applications where mechanical, tribological, and oxidation resistance are needed. Carbonitride coatings consist of "perfectly" mixed C and N in a f.c.c structure and combine the best characteristics of both parent components (carbide and nitride). For example, TiCN coatings provide superior mechanical resistance and thermal stability as compared to independent TiC and TiN coatings. That is due to the combination of the high ductility and the high melting point of TiC, and to the superior adhesion strength and the low internal stress of TiN [1]. Moreover, there are several configurations accessible for the carbonitrides, such as monolayer, multilayer, or graded structure, respectively, offering the possibility to alter the structure and composition in a precise way leading to adjustable properties [2]. Among all carbonitrides, those of ternary coatings containing one transition metal, such as TiCN, ZrCN, NbCN, or CrCN, are most widely used for industrial applications [3–6]. Even though TiCN and CrCN have already become regular commercial products, extensive research is still needed. In the last years, a new generation of carbonitrides was proposed by the addition of various metallic and/or non-metallic elements into the basic ternary matrix. For example, Constantin et al. reported that inserting small amounts of Zr, Nb, or Si into TiZrCN, TiNbCN, and TiSiCN systems leads to stress decrease and adhesion increase; also, the corrosion processes of high-speed steel in NaCl aggressive environment were significantly improved [7]. Up to this date, some complex Ti carbonitrides coatings prepared by PVD were studied (Cr and/or Si were used as alloying elements in structures such as TiCrCN [8], TiSiCN [9–11], TiCrNbCN [12], TiAlSiCN [13–15], TiCrSiCN [15–17], and TiNbCN [18–20]), revealing their superior properties for different applications.

The TiSiCN coatings were selected due to their known properties, such as high hardness, low friction, and wear performance [21]. Up to this date, TiSiCN coatings were prepared by magnetron sputtering [22] or CVD methods [9,11,23]. It was reported that their structure and mechanical characteristics were influenced by the Si and C contents in the composition [24]. For the TiSiCN coatings prepared by magnetron sputtering, many researchers reported that their strength is due to "nanocrystallite (nc) and amorphous (a) phases" interface [9,22,23,25–29]. This model was proposed by Veprek in 1995 for the "nc-TiN/a-Si_3N_4" coatings to describe the strengthening mechanism of the TiSiN nanocomposites [30]. Nevertheless, the topic related to the interface of "nc-nanocrystallite/a-amorphous" system is still under heated discussion due to the experimental confirmation deficiency for the interfaces state—whether it is amorphous or crystalline. Regarding the TiSiCN coatings prepared by CVD, it was reported that the performance, such as oxidation resistance, improved by increasing the Si content [31]; however, very limited information related to the corrosion resistance of TiSiCN coatings and the preparation by cathodic arc technique could be found in the literature [32–34].

To our knowledge, none of the mentioned carbonitrides have been used for nuclear fusion applications and there is no available information about the irradiation behavior of these materials. ZrC, TiC, AlN, ZrN, SiN, and TiN have been considered as inert coatings suitable for fusion reactors [35], but the existing information is rather about the effect of irradiation on these materials.

In the present paper, the corrosion behavior represents the subject of analysis, because of the high corrosion attack appearing in nuclear fusion and, therefore, coatings with high corrosion barriers are needed. For example, Konys et al. demonstrated that severe corrosion dissolved steel components by forming precipitates which blocked the system [36]. Moreover, the corrosion leads to a loss of mechanical integrity of materials. Thus, knowledge related to corrosion behavior of materials used in nuclear fusion is very important. For example, Cr based alloys are commonly used for molten salt reactors and information about the effect of Cr amount variation in the alloy and temperature on the corrosion of materials in the chloride salts is not clear. Further systematic studies should be well conducted for the understanding of the chloride salt corrosion mechanism. The challenges about the electrochemical processes in the molten salt reactor are considered to be relatively innovative and few papers on this subject are published; thus, the corrosion of materials is a more challenging topic in the case of molten salt nuclear systems compared to traditional water reactors due to the formation of a passive oxide layer on top of the surface, which is thermodynamically unfavorable in molten salts, leading to a restriction of materials is this area.

The TiSiCN coatings deposited on Ti6Al4V substrate were not so much discussed in the literature, despite their excellent wear and corrosion resistance suitable for many industrial applications, such as navigation, aerospace, engine turbine, and ocean exploration. Wang et al. reported that the C/N ratio in TiSiCN coatings obtained by arc ion plating has a great influence on the tribocorrosion resistance in artificial seawater (standard ASTM D1141-98) [24].

The aim of the present paper is to study the effect of neutron activation on TiSiCN carbonitride coatings prepared by the cathodic arc evaporation method in a reactive gas mixture of C_2H_2 and N_2. The coatings were obtained at different C/N ratio (0.4 and 1.6) to study the effect of C/N ratio variation on the neutron activation. The obtained coatings are investigated in terms of elemental and phase composition, crystalline structure, morphology, and corrosion resistance in aggressive saline solution (3.5% NaCl), in order to be considered suitable candidates for operation under severe conditions present in nuclear applications.

2. Materials and Methods

2.1. Deposition of the Coatings

The cathodic arc deposition system was used for the preparation of TiSiCN on Ti6Al4V alloy. The unit was equipped with one cathode constructed of Ti 88 at.%-Si 12 at.% (99.99% purity). The base pressure prior to coating was of 6×10^{-4} Pa; the samples were biased at -1000 V for 10 min in an Ar atmosphere at a pressure of 0.2 Pa. For each deposition the Si content was carefully adjusted up to 6 at. % such as the final (C + N)/(metal + Si) ratio to range between 0.6 and 0.8. Depending on the C/N ratio, under stoichiometric coating (0.4) is labeled TiSiCN-1 and over stoichiometric (1.6) as TiSiCN-2. The gas mass flow rates for TiSiCN-1 coating were kept at 25 sccm for CH_4 and at 65 sccm for N_2; for TiSiCN-2 the gas flow rates were inversed. Arc current was at 110 A and the substrate bias was -100 V. All parameters were kept constant for maximum 40 min, leading to a deposition temperature of 320 °C and thickness of the coatings of ~3 µm. More preparation details could be found in [7,34].

2.2. Corrosion Testing and Characterization of the Coatings

The corrosion resistance has been evaluated by the polarization technique in 3.5% NaCl (purchased from Sigma-Aldrich, Hamburg, Germany), using a potentiostat/galvanostat (VersaStat, Princeton Applied Research, Oak Ridge, TN, USA). A typical three-electrode cell was used: sample as working electrode (WE), platinum foil used as counter electrode (CE), and saturated Ag/AgCl-electrode as reference electrode (RE). The open circuit potential (OCP) was monitored until equilibrium occurred. The potentiodynamic curves were plotted from -1 V to +2 V vs. OCP. Corrosion potential (E_{corr}) and corrosion current density (i_{corr}) parameters were extracted from the Tafel extrapolations in the range of $\pm$50 mV. Polarization resistance (R_p) was calculated according to the procedure described in ASTM G59-97 standard (reapproved 2014). The corrosion rate (CR) has been calculated by the following formula, assuming that uniform corrosion has taken place over the whole immersed surface:

$$CR\,\frac{mm}{year} = \text{weight loss, g} \times \frac{cm^3}{\text{material density, g}} \times \frac{1}{\text{exposed area, cm}^2} \times \frac{10\ mm}{1\ cm} \times \frac{1}{\text{exposed time, h}} \times \frac{8760\ h}{1\ year} \quad (1)$$

The anti-corrosive properties of the coatings were estimated by calculating the protective efficiency (P_e) [37]:

$$P_e = \left(1 - \frac{i_{corr,\ coating}}{i_{corr,\ substrate}}\right) \times 100 \quad (2)$$

where $i_{corr,coating}$ and $i_{corr,substrate}$ are the corrosion current densities of coating and substrate, respectively.

The total porosity (P) of a coating was assessed using Elsener's empirical equation [37]:

$$P = \left(\frac{R_{p,substrate}}{R_{p,coating}}\right) \times 10^{\frac{-|\Delta E_{corr}|}{\beta_a}} \quad (3)$$

where $R_{p,substrate}$ and $R_{p,coating}$ are the polarization resistances of the substrate and coating, respectively, ΔE_{corr} represents the difference between the corrosion potentials of the coatings and substrate, and β_a is the anodic slope of the Tafel extrapolation of the substrate.

The EIS measurements were performed over $0.5 \div 10^3$ Hz frequency range by applying a sinusoidal signal of 10 mV RMS vs. EOC. Data were recorded by VersaStudio software and the fitting procedure was performed using ZView software.

The morphology of the surface before and after corrosion tests was investigated using a scanning electron microscope (Hitachi 3030PLUS, Tokyo, Japan). The surface roughness before and after corrosion tests was evaluated using a surface profilometer (Dektak 150, Bruker, Tucson, Arizona) with a stylus diameter of 2.5 μm over a length of 10 mm on two replicates in 10 randomly selected different areas in the centre of each sample. R_a (arithmetic average) and S_k (profile symmetry relative to the mean line) parameters were used to estimate the roughness of the investigated surfaces.

The elemental composition of both TiSiCN coatings was investigated using an equipped energy dispersive X-ray detector (EDS, Bruker) and a scanning electron microscope (Hitachi 3030PLUS). The crystalline structure and the phase composition were obtained by means of X-ray diffraction (XRD) using a Rigaku SmartLab diffractometer, with Cu Kα radiation (1.5405 nm) set up in $\theta/2\theta$ geometry range of 30–80° with a step of 0.02°/min. Additionally, grazing incidence XRD (GRXRD) was performed by fixing the incident angle at 3°. The phase identification was aspired by Rietveld profile fitting using the FullProf program and the crystallite sizes were determined from the XRD peak widths using the Scherrer formula.

2.3. *Fast Neutrons Irradiation and the Neutron Activation Analysis (NAA)*

The neutron irradiation of the prepared samples was performed at the IBR-2M pulsed reactor at the Frank Laboratory of Neutron Physics (JINR, Dubna, Russia). The conditions during irradiation were the nominal power of 1450 kW (average power of the reactor during the period cycle) and the cycle time of 288 h. The neutron activation analysis was used for the determination of fast neutrons flux density (by nickel) and fluence on the specimens. The investigated isotope (line in the spectrum) was ^{58}Co (810.7 keV) and the accepted effective section was 0.092 barn. Laboratory gamma-spectrometer Canberra GC10021 and Lynx multichannel analyzer were used to conduct NAA with high accuracy.

3. Results and Discussions

3.1. *Elemental Composition Surface Morphology (EDS)*

In Table 1, results of the EDS elemental composition analysis of the deposited coatings are shown. Synthesized under different conditions (C/N ranged from 0.4 to 1.6 stoichiometric and C + N(/(Metal + Si) ranged 0.6 to 0.8), the base elements (Ti, Si, C, and N) are predominant in research samples.

Table 1. Elemental composition determined by EDS for TiSiCN-1 and TiSiCN-2 coatings.

	Elemental Composition (at.%)								
Coatings	**Ti**	**Si**	**C**	**N**	**O**	**V**	**Al**	**(C + N)/Metal + Si**	**C/N**
TiSiCN-1	42.30	5.17	15.91	34.36	2.12	0	0.14	0.61	0.46
TiSiCN-2	44.12	4.57	29.16	19.19	2.90	0	0.10	0.71	1.52

3.2. X-ray Diffraction Analysis of TiSiCN

For confidence, in order to distinguish reflections belonging to the substrate and the coatings, the substrate without coating was measured alone (Figure 1a). The main substrate peaks were indexed as (100), (002), (101), (102), (110), (103), (112), and (201). Titanium alloy reflection (200) was also present, but with a lower intensity. Most of the reflections were indexed in space group: P63/mmc (α-Ti, hexagonal close packed). The lattice parameters proved to be a = b = 2.93706 Å and c = 4.69061 Å. According to R. Pederson [38], some residual amount of β phase (SG Im$\bar{3}$m) should remain and the peak at 2θ = 39.67° could be indexed as (110) of β-Ti. One can notice that the texture of the substrate is extreme, which is not surprising since Ti alloys tend to exhibit a very strong texture after the hot-rolling manufacturing process. Usually, it is described as {11-20}(0001). In our case, we find (10-10) normal component to be the strongest. The texture was considered as fiber described by FullProf incorporated exponential function $P_h = G_2 + (1-G_2)\exp(G_2\alpha_2)$ with texture parameter $G_2 > 7$ (from FullProf). The texture was very strong with main component (10-10), which diverges from (0001) mentioned in the literature. This, however, is perfectly possible, since we are unaware of the direction of cutting from the ingot.

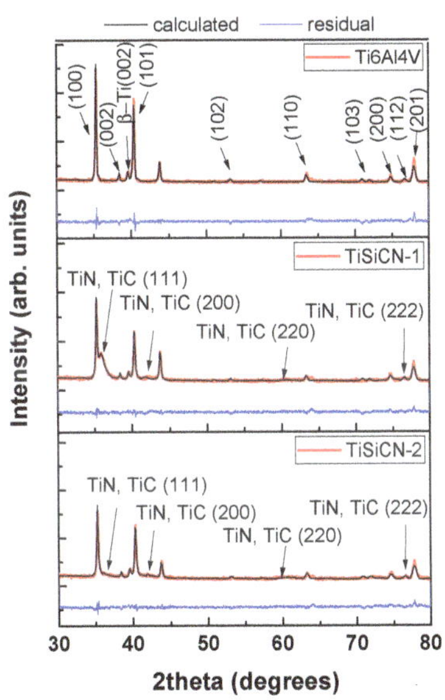

Figure 1. Rietveld fit of Ti6Al4V substrate (main phases—α-Ti [97.84 (2.16) wt.%] and β-Ti [2.16 (0.19) wt.%], TiSiCN-1, and TiSiCN-2.

The scattering density of the substrate, Ti6Al4V, is considerably higher than that of the coating; therefore, the superposition of substrate reflections was expected. Upon inspection of data from the TiSiC-1 sample (Figure 1b), it can be seen that the spectrum displayed complexity with superposition of individual phases of different features. The spectrum was characterized by much sharper peaks presumably belonging to Ti6Al4V substrate. Clearly, several substrate peaks emerged, at 2θ = 42.00°, 60.4°, and 72°; also, a broadened peak (2θ = 36°) superimposing (100) Ti6Al4V was observed. These positions coincide roughly with the peaks of the TiN structure found in the literature. The space group was Fm$\bar{3}$m (cubic) and the lattice parameter a = 4.31263 Å; however, inside the coating, the lattice appears distorted, with lattice parameter slightly above the literature value of 4.28 Å. One could assume that since TiC possesses the same space group, but higher a TiC phase is present in the coating, but the value for TiC falls out of the range of possible a. Therefore, more probable is the situation where the composition of the coating is not fixed, but C/N ratio has some continuous distribution and, hence, results in continuous dispersion of the cell dimensions (spanning). This novel result is reported for the first time. All these peaks are much broadened, and an attempt was made to estimate the size of coherent scattering blocks using Scherrer equation from (111) peak width. It yields upper limit value for d = 10 nm, which is the evidence for highly dispersed nanostructured coating. The specimen featured very sharp axial texture with overwhelming component (111) in diffraction spectrum; therefore, direction [111] is normal to the surface, with FullProf texture exponential parameter equals 9. The strength of the texture is a very strong texture component.

In the case of C/N ratio 1.6 (TiSiCN-2), it is expected that TiC will prevail in the coating. TiC crystallizes in the same group Fm$\bar{3}$m. XRD of the coating with predominant carbon content was shown in Figure 1c. An effect was observed, which is the slight shift of the coatings peaks towards lowest angles that indicated a content of TiC with the same crystal symmetry, but smaller unit cell parameters than TiN and correspondingly higher diffraction angles. The second difference is a drastic intensity decrease in (111) the reflection in comparison with TiN case and, hence, the decrease in the sharp (111) texture to a negligible one (FullProf texture parameter 0.12470). Lattice parameter was found a = 4.32 Å.

The comparison of XRD spectra before and after corrosion procedure in saline solution did not show any significant changes; however, for TiSiCN-2, we observe considerable decrease in intensities of the diffraction peaks. It could be that some, although little, X-ray intensity is taken away by the oxide layer formed on top of the surface. XRD spectra of grazing geometry showed an intriguing phenomenon for the investigated coatings (see Figure 2). In this case, the diffraction power decrease is much stronger than in the Bragg–Brentano geometry case and this time it mainly concerns diffraction from the coatings. The peak erosion is stronger in TiSiCN-2 than in TiSiCN-1, meaning under stoichiometric coating, which is more corrosion-resistant than the over stoichiometric, i.e., nitrogen has a positive influence on the chemical protection of the coating. All peaks are eroded after corrosion, but the effect manifests itself better on the peak at 2θ = 39.6°. The same is observed on the TiSiCN-2. The same peak exhibits the strongest decrease. We suspect that this reflection, although on the place of β-Ti (110), does not belong to it but to TiO_2-rutile phase (200) synthesized during the high-temperature coating deposition. Rutile, although not soluble in saline solution, apparently has worse adhesion with the rest of the coating layer and is washed away during the corrosion treatment.

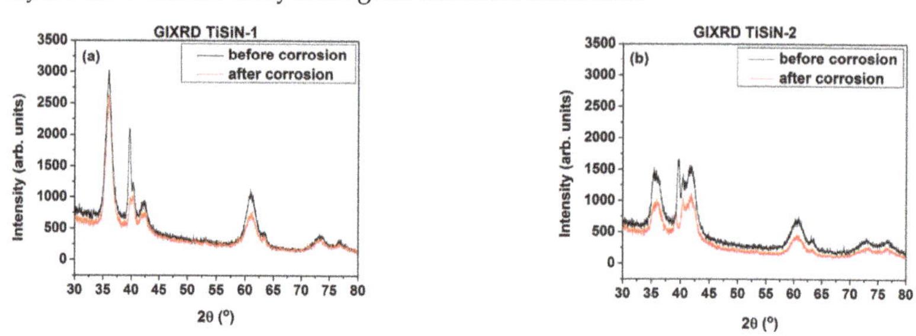

Figure 2. Comparison of GIXRD spectra before and after corrosion for: (**a**) TiSiCN-1 and (**b**) TiSiCN-2.

3.3. Surface Morphology

The surface morphology images (SEM), alongside the elemental mapping profiles of the TiSiCN coatings on Ti6Al4V alloy, before and after corrosion testing in 3.5% NaCl are presented in Figure 3.

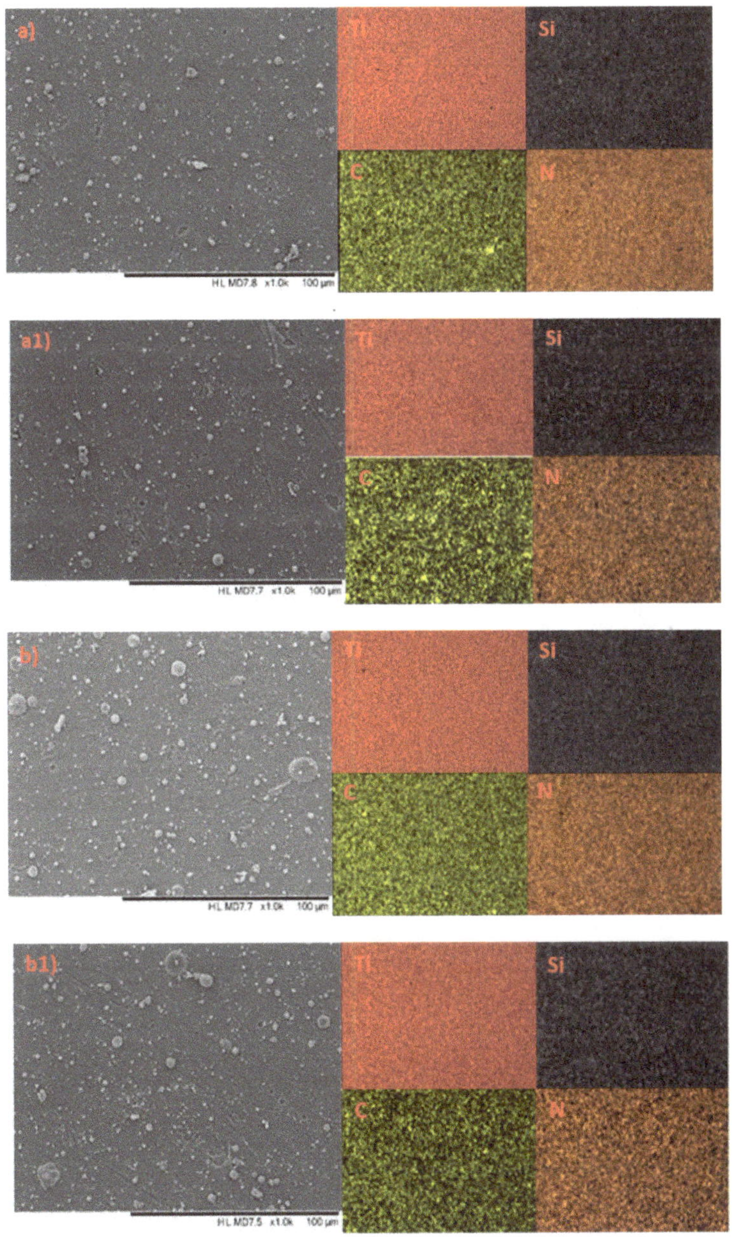

Figure 3. SEM images of TiSiCN coatings on Ti6Al4V alloy with elemental mapping profiles: (**a**) TiSiCN-1 before corrosion testing; (**a1**) TiSiCN-1 after corrosion; (**b**) TiSiCN-2 before corrosion testing; and (**b1**) TiSiCN-2 after corrosion.

Many micro-droplets with different diameters could be seen on the surfaces of the deposited pre-corrosion coatings (Figure 3a,b). The appearance of such micro-sized morphology features is typical not only for the cathodic arc technique but is observed as well for the same coatings that are fabricated using PEMS, CVD, or arc ion plating methods [24,39–42]. They are unavoidable and generated during the deposition process and tend to be more pronounced in thicker coatings [40]. From the figures, it could be seen that the micro-droplets differ in size and surface density. On the surface of TiSiCN-2, more and larger droplets are observed, with some reaching up to 20 μm in size. That could be in connection with the difference in the gas mass flow rates (CH_4 and N_2) in both samples during deposition, since intensity of defects and growth-induced stress are strongly dependent on the process parameters [43,44].

An assumption could be made regarding the formation process of such microdroplets in the samples under study here. In many papers, it is reported that TiSiCN coating consists predominantly of TiC_xN_y nanosized crystals embedded in SiC_xN_y)/a-C amorphous matrix [11,40,41]. It is possible that the lesser amount of nitrogen in the TiSiCN-2 system compared to TiSiCN-1 hinders the amorphous matrix formation, which is reported to play a major role in the nanocrystal grain growth, since it is an interface phase [9,45]. That could lead to more nucleation sites, larger particles to evolve (microdroplets), and also a greater likelihood of occurrence of inter-particle interactions; so, a nitrogen deficiency could lead to more vacant Ti sites available for oxidation and formation of a TiO2 layer on top, as illustrated in [46]. The EDS data for TiSiCN-2 (Table 1) somewhat reinforce such an assumption—lower nitrogen content coincides with increased percentage values for both Ti and O.

It should be noted that similar large surface formations have been shown to occur also on coatings without carbon (TiSiN) [46,47]. Studies on such formations reveal that they consist predominantly of Ti, N, and O [46,47]; thus, it seems that the different carbon content in the investigated TiSiCN samples should not play a role in the difference in both number and density of the microdroplets. On the other hand, it is reported that higher carbon content (as in TiSiCN-2) enhances the amorphous phase amount and favors reducing the size of crystallites [9,44]. On the carbon elemental profile map, some areas with increased concentration could be seen on both coatings (Figure 3a,b). In the titanium and nitrogen mapping profiles, the same areas lack in color, thus unveiling the absence of both elements. Since there is no titanium in these areas, the presence of carbide structures should be excluded and only the accumulation of amorphous carbon should be regarded in these locations. Again, it seems that the difference in carbon content in both coatings does not play a role in the formation of such carbon piles, since they are observed in both TiSiCN coatings.

Despite the presence of microdroplets, both coatings were uniformly deposited, without any major defects, such as pinholes, voids, or cracks. The different C/N ratio in the coatings is easily recognizable from the difference in density on C and N elemental profiles but does not have a major effect on the titanium homogeneity in the coatings. The smoother surface morphology in TiSiCN-1 (SEM images in Figure 3), could be associated with the bigger quantity of nitrogen, since it is essential for the formation of both TiN and Si_3N_4 amorphous matrix components.

Despite the aggressive corrosion medium, there was no change in the surface morphology. A decrease in the number of micro and macro particles of different sizes was observed only on the surface of the sample. The surface morphology of TiSiCN coatings can be due to carbon content, as previously reported [9]. Usually, these defects constitute preferential diffusion trails of aggressive species and can result in an accelerated coating failure [34]. In addition, Li investigated the surface morphology of TiSiCN coatings at the C/Si 4:0, 3:1, 2:2, 1:3, and 0:4 ratios and concluded the ideal surface morphology is formed in the ratio C:Si = 2:2 when no surface degradation occurs [21].

3.4. Surface Roughness

Table 2 contains the roughness parameters values before and after corrosion testing performed in 3.5% NaCl. R_a is the average arithmetic deviation from the mean line linked with the peaks profile height/depth and valleys present on the surface [48]. Both coated surfaces exhibited much higher R_a values than the uncoated substrate before and after corrosion investigation, indicating rougher surfaces. Such results were expected because of the observed morphology patterns in SEM images, including microdroplets and pores visible on the surface of both coatings. Again, in good agreement with SEM observations, TiSiCN-2 coating exhibits rougher surfaces than TiSiCN-1 coating before corrosion. It seems that the lower C/N ratio leads to smoother surface of the coating. Although the uncoated substrate showed its smoothest surface before corrosion (R_a~37 nm), after the corrosion attack R_a decreased to 67.9% of its original value, indicating a severe corrosive process. On the other hand, the deviation in the R_a values for both coated surfaces of 2.4% and 0.2% for TiSiCN-1 and TiSiCN-2, respectively, indicate that the coatings were almost intact after corrosion testing, signifying a less pronounced corrosion of these surfaces. The difference in the R_a values of the two coatings is statistically insignificant; thus, it could be concluded that both are not prone to significant changes upon corrosion attack. It could also be noted that the TiSiCN-2 rougher surface seems to be less susceptible to corrosion, compared to the smoother surface of the TiSiCN-1 sample.

Table 2. Roughness parameters of the investigated surfaces before and after corrosion tests (R_a—arithmetic average deviation from the mean line; S_k—skewness factor).

Substrate and Samples	Before Corrosion		After Corrosion	
	R_a (nm)	S_k	R_a (nm)	S_k
Ti6Al4V	36.69 ± 9.7	−0.11 ± 0.1	24.94 ± 2.7	−0.15 ± 0.1
TiSiCN-1	473.92 ± 19.0	1.99 ± 0.3	462.85 ± 35.3	1.71 ± 0.3
TiSiCN-2	545.55 ± 33.5	1.50 ± 0.3	544.59 ± 52.4	2.16 ± 0.6

Considering the skewness value, it is recognized that a positive value indicates better corrosion resistance [49]. By comparing the S_k values before corrosion, one may see that the uncoated substrate has negative value, unlike the two film specimens, indicating that the surface has more valleys than peaks, probably due to the polishing process. After corrosion testing, the S_k value is more pronounced, displaying the depression depth increase, meaning that the corrosive solution dug deep in the initial cavities found on the Ti6Al4V surface. The corrosion experiment has a different effect on the two coatings. While for the TiSiCN-1 the S_k value lowers, meaning some kind of smoothing, with the filling of available pits and the peaks becoming less pronounced, in the case of the TiSiCN-2 coating an opposite process unfolds (Table 2). This suggests a possible difference in the chemical composition of the constituents of the surfaces of the two coatings. The S_k values were close before and after corrosion of TiSiCN-1; thus, it could be concluded that the corrosive solution did not affect the coating surface to a big extent. On the contrary, in the case of TiSiCN-2, a significant increase in S_k value after corrosion could be observed, meaning that a severe corrosive process has taken place on the surface. It is possible that a lower C/N ratio in TiSiCN-1 to make the surface more prone to passivation process and quick development of an oxide layer, consequently increasing its corrosion resistance. Films with higher C/N ratio are more abundant in microdroplets and they are larger. It results in a rougher surface, which leads to a larger interfacial area with the corrosive environment and, therefore, an enhanced rate of the corrosion process is expected [49].

3.5. Corrosion Behaviour

The EIS data obtained for TiSiCN-coated TiAlV alloy are presented in Figure 4 as both Nyquist and Bode amplitude plots. As observed in Nyquist impedance plots, slightly higher semicircles were obtained for TiSiCN-1 coating, indicating a higher charge transfer resistance. The main frequency semicircle is modeled as a combination of parallel resistance and capacitance elements in series with electrolyte resistance. The electrical equivalent circuit (EEC) used for fitting the impedance data is also presented as an insert in Figure 4a. Two interfaces were considered: the porous outer oxide layer/coating-electrolyte interface (characterized by CPE_{coat} and R_{pore} parameters) and the inner layer formed at the electrolyte/dense film interface (characterized by CPEdl and Rct parameters). The same EEC was used also for the Ti6Al4V substrate, the model being an indicated circuit used for coated surfaces. According to Pan et al., a bilayer structure of oxide film is formed on the surface of titanium during the immersion in a saline environment: TiO_2 dense inner layer and porous Ti6Al4V alloy outer layer [50]. According to literature reports, the thickness and characteristics of the oxide film depends on the testing solution [50], but this is mainly formed of TiO_2, Ti_3O_5, Ti_2O_3, and TiO oxides [51–53]. I. Milošev et al. used the same model to fit the impedance data taking into consideration the formation of the mentioned oxides and suboxides formed on the Ti-based alloys immersed in different media, as indicated by XPS analysis [51,54].

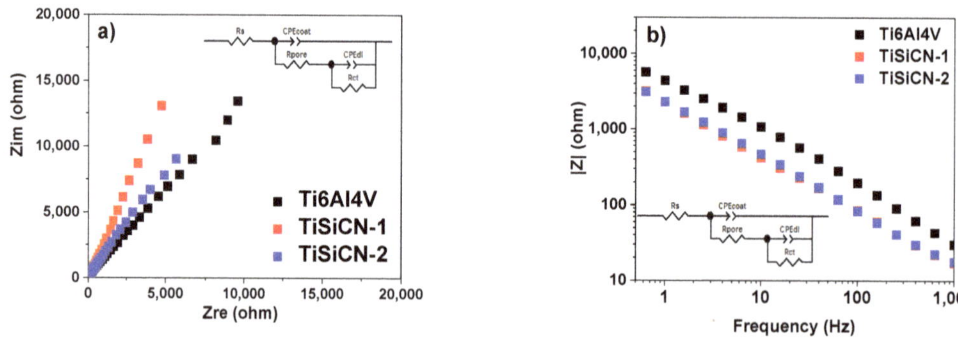

Figure 4. (**a**) Nyquist and (**b**) Bode amplitude plots for the investigated samples after 1 h immersion in 3.5% NaCl.

Table 3 presents the fitted parameters of impedance data obtained, along with χ^2 parameter as an indication of the goodness of fit. For Ti6Al4V alloy, notice the high value of the Q_{dl} (resistance associated to the current flow through the pores), which shows that the formed oxide layer allows the electrolyte to ingress through the pores. Indeed, the pore resistance indicated the lowest value as compared with the coated surfaces (R_{pore} = 4 kΩ cm^2). Instead, the highest value was showed by TiSiCN-1 coating (R_{pore} = 3626 kΩ cm^2), indicating a more compact structure with beneficial effects on electrochemical behavior when immersed in 3.5% NaCl solution. This coating also showed the highest α_{coat} parameter (0.77), followed by TiSiCN-2 (0.63), which are values related to non-uniform current distribution along the surface [55]. At the interface with the substrate, there is an almost defect-free surface, since α_{dl} showed high values near 1; thus, the constant phase element (CPE) used in this case could be seen as an ideal capacitor. Instead, at electrolyte-coating interface α_{coat} showed low values and a CPE is needed to take into consideration possible deviations from the ideal dielectric behavior [56]. As pointed out in literature, there are multiple factors associated with these deviations, which include surface disorder or inhomogeneity, geometric irregularities, working electrode porosity, or ether surface roughness [56]; thus, the surface properties of the electrode under investigation can have a possible contribution in electrochemical results. Considering the values presented in Table 2, one can note that the roughness measured for TiSiCN-1 before corrosion examination proved a smother surface since the Ra parameter showed a value of ~473 nm, whereas for TiSiCN-2 R_a, it was ~545 nm.

The high-frequency parameters Q_{coat} and R_{pore} represent the properties of the reactions at the electrolyte-coating interface. The results show that coating capacitance exhibits higher values for TiSiCN-coated Ti6Al4V, indicating a more active exchange with the NaCl solution, especially in the case of TiSiCN-2. Since there are insufficient data in the low-frequency region in the measured frequency range, it was not possible to determine the R_{ct} parameter. The χ^2 parameter was also presented and the low values have indicated a validation of the fitting procedure.

Table 3. The fitting results of EIS data for the investigated samples after 1 h immersion in 3.5% NaCl.

Sample	Rs (Ω cm^2)	Qcoat (μF s$^{(\alpha-1)}$ cm^{-2})	α_{coat}	R_{pore} (kΩ cm^2)	Q_{dl} (μF s$^{(\alpha-1)}$ cm^{-2})	α_{dl}	R_{ct} (kΩ cm^2)	χ^2
Ti6Al4V	9	76.17	0.53	4	5004	1.0	-	4.0×10^{-4}
TiSiCN-1	8	89.65	0.77	3626	12.21	1.0	-	2.0×10^{-4}
TiSiCN-2	5	120.71	0.63	17	10.65	0.98	-	4.0×10^{-4}

The evolution of the open circuit potential (E_{oc}) during the 1 h immersion is given in Figure 5a. Both coatings exhibited a constant evolution of E_{oc}, indicating a good coating stability in 3.5% NaCl solution during 1 h immersion tests. Under stoichiometric coatings (TiSiCN-1) exhibited more electropositive E_{oc} values compared to the over stoichiometric coatings (TiSiCN-2). The uncoated alloy displayed an unstable surface due to the fluctuations, despite the fact that it exhibited electropositive value of E_{oc}.

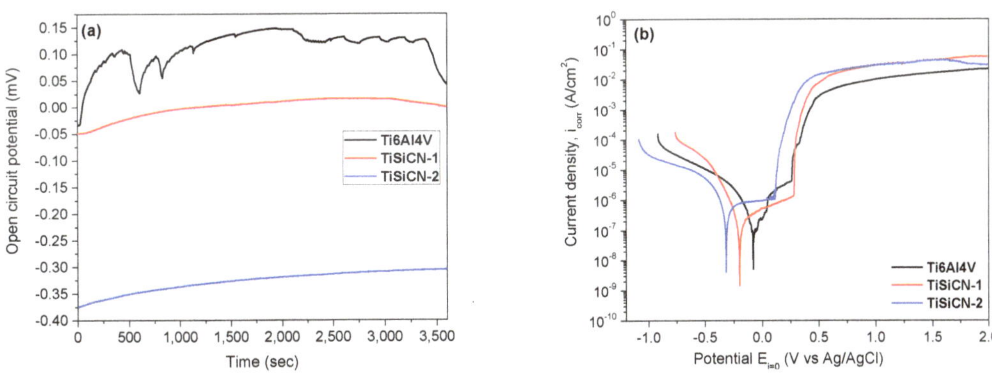

Figure 5. (**a**) Open circuit curves and (**b**) potentiodynamic polarization curves of the investigated surfaces after 1 h immersion in 3.5% NaCl.

The potentiodynamic polarization curves of the investigated surfaces are shown in Figure 5b, while the electrochemical parameters are presented in Table 4. The resistance to corrosion of the investigated samples could be estimated according to the following principles:

(1) Electropositivity signifies good resistance to corrosion: more electropositive corrosion potential value (E_{corr}) means that the material is nobler in the used electrolyte. Figure 5b reveals that the most noble corrosion potential was measured with uncoated Ti6Al4V alloy. TiSiCN-2 coatings exhibit the more electronegative value, indicating poor corrosion resistance;

(2) Surfaces with a low i_{corr} value demonstrated a good corrosion resistance: all coatings presented lower i_{corr} values compared to the uncoated Ti6Al4V alloy. TiSiCN-1 coating showed a lower i_{corr} value compared to TiSiCN-2 coating, including the uncoated substrate;

(3) Surfaces having higher anticorrosive properties demonstrate high polarization resistance (R_p) values. Considering the R_p values in Table 4, note that both coatings exhibit higher R_p value than the uncoated substrate;

(4) Porosity of the coatings was also considered: surfaces with low porosity have good anticorrosive properties. TiSiCN-1 coating has low porosity, indicating that the increase in C/N ratio leads to an increase in porosity, meaning the loss of corrosion resistance;

(5) The protective efficiency (P_e) was also considered: the highest P_e value was obtained for the TiSiCN-1 coating.

To summarize the electrochemical investigations, the increase in C/N ratio leads to a loss of the anticorrosive properties of the TiSiCN coatings in 3.5% NaCl solution.

Table 4. Electrochemical parameters of the investigated samples after 1 h immersion in 3.5% NaCl: E_{corr}: corrosion potential; i_{corr}: corrosion current density; R_p: polarization resistance; P: porosity; P_e: protective efficiency.

Substrate and Samples	E_{corr} (mV)	i_{corr} (nA)	R_p ($\Omega \times 10^{-3}$)	P_e (%)	P
Ti6Al4V	−78	222.123	0.195	-	-
TiSiCN-1	−199	25.281	8.795	88.6	0.016
TiSiCN-2	−316	94.347	1.816	57.5	0.058

3.6. Neutron Activation Analysis for Determination of Fast Neutrons Flux Density

The neutron activation analysis (NAA) principle lies in the interplay between neutrons and the nuclei from the investigated material. It generates gamma radiation, and its intensity is being determined. In the present research, we measured the neutrons fluency through the samples, as well as the density of the neutron flux in a certain energetical diapason. The neutron spectrum was taken down via neutron activation satellite sample. Nickel (Ni) was activated to ^{58}Co and its gamma activity was used for the estimation of the fast neutron flux density. What makes Ni appropriate is its initial capture cross-section from 1 MeV for neutrons. The reactions are:

$$^{60}\mathrm{Ni} + fast\ neutron \rightarrow triton + {}^{58}\mathrm{Co} \tag{4}$$

$$^{58}\mathrm{Ni} + fast\ neutron \rightarrow protons + {}^{58}\mathrm{Co} \tag{5}$$

In Figure 6 shows the energy cross-sections relations of the reactions. Judging by the upper right graph, only reaction (5) takes place in the energy range that is of importance for us. The reaction cross-section on the plateau part of the diagram equals tenths of a barn. Reaction (4) starts at > 16 MeV with a cross-section of tenths of a millibarn, as illustrated in the lower right graph (Figure 6).

Despite the general rule that the energy should not surpass 3 MeV in laboratory conditions, registered X-ray and X-ray radiation energies vary between 40 keV to 10 MeV. The Canberra GC10021 spectrometer has integral nonlinearity of 0.025% and that distinctly points to an energy-channel adequacy; also, the FWHM is 1.1 keV for the 122 keV line and 1.8 keV for the 1332 keV line.

For the neutron flux densities estimation, an operative cross-section of σ_{eff} = 92 mb was employed, as well as considering the ^{60}Ni nuclei to ^{58}Ni nuclei ratio in the measurement wire and cross-sections integral convolutions of the fulfilled reactions. Figure 7 presents the NAA satellite gamma-spectrum after illumination with specimens.

The location of the total absorption peak at E_y = 810.7 keV matches ^{58}Co (Figure 7). Its activity could be used in the fast neutrons flux density evaluation for each sample. Calculations determined the fast neutrons flux density and the fluency of both samples to be of 1.65×10^{7} n cm^{-2} s^{-1} and of 1.71×10^{13} n cm^{-2}, respectively. The gamma dose rate was estimated via dosimeter after the specimen removal from the irradiation unit. Prompt gamma decay, rather than delayed beta decay, would indicate sample activation. The dosimeter background level indications of 0.1–0.2 μSv\h revealed no sample activation. The absence of nucleus with significant cross-section to interact with the fast neutrons in the studied sample predicts this aspect. Bearing in mind the vast fast neutron flux and the fluencies reached, the samples non-activation could indicate their potential use under reactor analogous conditions. That would be so since it is well known that the most severe defect formation process is caused by fast neutrons. Future studies regarding the resulting number of defects in the crystal structure of such are necessary.

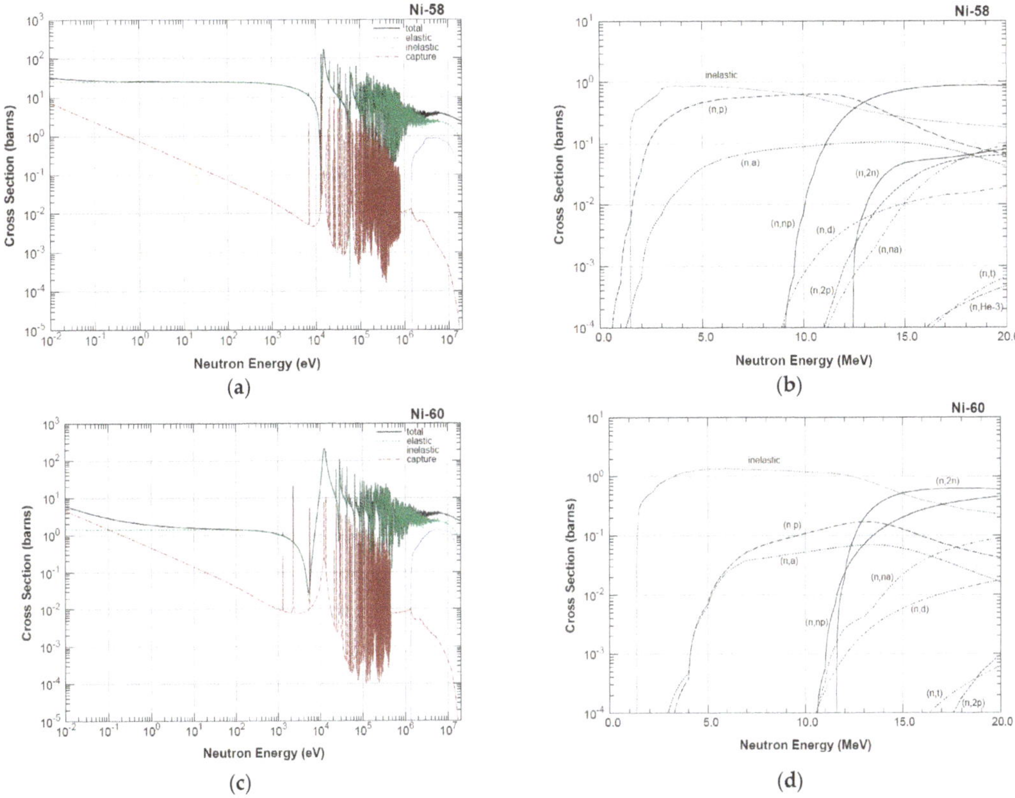

Figure 6. Energy dependences: (**a**,**c**)—effective nuclear cross-section of neutron interaction reactions with nuclei ^{58}Ni and ^{60}Ni; (**b**,**d**)—cross-sections of neutron capture reactions for ^{58}Ni and ^{60}Ni.

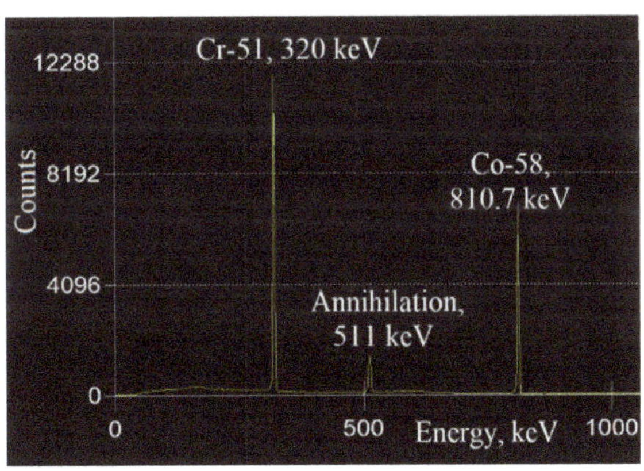

Figure 7. The measured gamma-spectrum of the NAA satellite after irradiation.

4. Conclusions

TiSiCN carbonitride coatings prepared at different C/N ratios of 0.4 and 1.6, respectively, were developed by cathodic arc evaporation. The SEM investigation showed the presence of micro-droplets of different sizes present on the surfaces, especially for the high carbon coating. Some shallow oxides might form during the immersion, as indicated by XRD, since there was a decrease in the diffraction lines intensities after the corrosion assessment; however, there was no visible change in the SEM surface morphology. Modification of surface roughness as compared with the pre-tested coatings was observed mainly in the case of TiSiCN obtained at 1.6 C/N ratio (TiSiCN-2). According to EIS-fitted parameters, the highest pore resistance value was shown by TiSiCN-1 coating, indicating a more compact structure with beneficial effects on electrochemical behavior; also, a more uniform current distribution along the surface was observed in this case, which could be linked with a smoother surface. During samples immersion, more electropositive E_{oc} values, lower i_{corr}, and a higher polarization resistance were exhibited by the under stoichiometric coatings (TiSiCN-1). On the other hand, the increase in C/N ratio leads to an increase in porosity and, consequently, to surfaces with lower anticorrosive properties. In the case of irradiation with 1.65×10^7 n/cm^2/sec fast neutron flux, short and long-lived isotopes were not observed in the samples, not even after 288 h. These results proved that both TiSiCN coatings can be successfully used in the severe conditions, which are characteristic to nuclear plants.

Author Contributions: Conceptualization and supervision, A.V. and M.N.M.; methodology, A.V., D.N., L.S. and E.P.P.; software, M.N.M. and F.A.A.; investigation, M.N.M., A.C.P., A.V., F.A.A., M.B., M.D., Z.S. and K.H.; resources, A.V. and M.N.M.; writing—original draft preparation, A.V.; writing—review and editing, A.V., A.C.P., D.N., L.S., M.B., M.D., Z.S., F.A.A., K.H. and E.P.P.; visualization, D.N.; Z.S., M.D. and E.P.P.; project administration, A.V. All authors have read and agreed to the published version of the manuscript.

Funding: This research was funded by a grant of the Ministry of Research, Innovation and Digitization, CNCS/CCCDI-UEFISCDI, project number COFUND-M-ERANET-3-HardCoat-1 (no. 311/2022), within PNCDI III.

Institutional Review Board Statement: Not applicable.

Informed Consent Statement: Not applicable.

Data Availability Statement: Not applicable.

Acknowledgments: The INOE team is grateful to the Core Program within the National Research Development and Innovation Plan 2022–2027, carried out with the support of MCID, project no. PN 23 05 (id: PN11N-03-01-2023) and through Program 1—Development of the national research-development system, Subprogram 1.2—Institutional performance-projects to finance the excellent RDI, contract no. 18PFE/30.12.2021. A.V.D. is grateful to the Tomsk Polytechnic University within the framework of the Tomsk Polytechnic University Competitiveness Enhancement Program grant.

Conflicts of Interest: The authors declare no conflict of interest. The funders had no role in the design of the study; in the collection, analyses, or interpretation of data; in the writing of the manuscript; or in the decision to publish the results.

References

1. Tang, J.-F.; Huang, S.-Y.; Lin, J.-H.; Yang, F.-C.; Chang, C.-L. Mechanical Properties of TiN Deposited in Synchronous Bias Mode through High-Power Impulse Magnetron Sputtering. *Surf. Coat. Technol.* **2022**, *434*, 128201. [CrossRef]
2. Braic, M.; Braic, V.; Balaceanu, M.; Zoita, C.N.; Kiss, A.; Vladescu, A.; Popescu, A.; Ripeanu, R. Structure and Properties of Zr/ZrCN Coatings Deposited by Cathodic Arc Method. *Mater. Chem. Phys.* **2011**, *126*, 818–825. [CrossRef]
3. Bull, S.J.; Bhat, D.G.; Staia, M.H. Properties and Performance of Commercial TiCN Coatings. Part 1: Coating Architecture and Hardness Modelling. *Surf. Coat. Technol.* **2003**, *163–164*, 499–506. [CrossRef]
4. Silva, E.; Rebelo de Figueiredo, M.; Franz, R.; Escobar Galindo, R.; Palacio, C.; Espinosa, A.; Calderon, V.S.; Mitterer, C.; Carvalho, S. Structure–Property Relations in ZrCN Coatings for Tribological Applications. *Surf. Coat. Technol.* **2010**, *205*, 2134–2141. [CrossRef]
5. Mesquita, R.A.; Schuh, C.A. Tool Steel Coatings Based on Niobium Carbide and Carbonitride Compounds. *Surf. Coat. Technol.* **2012**, *207*, 472–479. [CrossRef]
6. Warcholinski, B.; Gilewicz, A.; Kuklinski, Z.; Myslinski, P. Hard CrCN/CrN Multilayer Coatings for Tribological Applications. *Surf. Coat. Technol.* **2010**, *204*, 2289–2293. [CrossRef]
7. Constantin, L.; Braic, M.; Dinu, M.; Balaceanu, M.; Braic, V.; Farcau, C.; Vladescu, A. Effects of Zr, Nb, or Si Addition on the Microstructural, Mechanical, and Corrosion Resistance of TiCN Hard Coatings. *Mater. Corros.* **2016**, *67*, 929–938. [CrossRef]

8. He, J.; Nan, Z.; Mi, P.; Chen, K.; Qin, Y. In Situ Nanostructured (TiCr)CN Coating by Reactive Plasma Spraying. *Ceram. Int.* **2018**, *44*, 712–717. [CrossRef]
9. Ma, S.L.; Ma, D.Y.; Guo, Y.; Xu, B.; Wu, G.Z.; Xu, K.W.; Chu, P.K. Synthesis and Characterization of Super Hard, Self-Lubricating Ti–Si–C–N Nanocomposite Coatings. *Acta Mater.* **2007**, *55*, 6350–6355. [CrossRef]
10. Xie, X.; Li, J.; Dong, M.; Zhang, H.; Wang, L. Structure and Properties of TiSiCN Coatings with Different Bias Voltages by Arc Ion Plating. *Surf. Topogr.* **2018**, *6*, 014003. [CrossRef]
11. Wang, H.; Ou, Y.; Zhang, X.; Liao, B.; Ou, X.; Luo, J.; Pang, P.; Chen, L.; Hua, Q.; Bao, M. Tribocorrosion Behaviors of TiSiCN Nanocomposite Coatings Deposited by High Power Impulse Magnetron Sputtering. *Mater. Res. Express* **2020**, *7*, 076407. [CrossRef]
12. Bondar, O.V.; Postolnyi, B.O.; Kravchenko, Y.A.; Shypylenko, A.P.; Sobol, O.V.; Beresnev, V.M.; Kuzmenko, A.P.; Zukowski, P. Fabrication and Research of Superhard (Zr-Ti-Cr-Nb)N Coatings. *Acta Phys. Pol. A* **2015**, *128*, 867–871. [CrossRef]
13. Zhang, X.; Li, J.; Xiao, J.; Pi, J.; He, G.; Chen, L.; Zeng, Y.; Jiang, J. Effects of Si Addition on Structure and Mechanical Properties of TiAlSiCN Coatings. *Surf. Coat. Technol.* **2019**, *362*, 21–26. [CrossRef]
14. Golizadeh, M.; Kuptsov, K.A.; Shvyndina, N.V.; Shtansky, D.V. Multilayer SiBCN/TiAlSiCN and AlOx/TiAlSiCN Coatings with High Thermal Stability and Oxidation Resistance. *Surf. Coat. Technol.* **2017**, *319*, 277–285. [CrossRef]
15. Shtansky, D.V.; Kuptsov, K.A.; Kiryukhantsev-Korneev, P.V.; Sheveiko, A.N.; Fernandez, A.; Petrzhik, M.I. Comparative Investigation of Al- and Cr-Doped TiSiCN Coatings. *Surf. Coat. Technol.* **2011**, *205*, 4640–4648. [CrossRef]
16. Li, Q.; Jiang, F.; Leng, Y.; Wei, R.; Huang, N. Microstructure and Tribological Properties of Ti(Cr)SiCN Coating Deposited by Plasma Enhanced Magnetron Sputtering. *Vacuum* **2013**, *89*, 168–173. [CrossRef]
17. Xin, X.; Tengfei, Z.; Yu, W.; Yan, Y.; Yongxiang, L.; Dong, X.; Nan, H. Microstructure and Mechanical Properties of Ti(Cr)SiC(O)N Coatings Deposited by Plasma Enhanced Magnetron Sputtering. *Rare Met. Mater. Eng.* **2017**, *46*, 1762–1767. [CrossRef]
18. Chaparro, W.A.; Martin, C.R.; López, E.V. Synergy between Erosion-Corrosion of Steel AISI 4140 Covered by a Multilayer TiCN /TiNbCN, at an Impact Angle of 90°. *Dyna* **2013**, *80*, 101–108.
19. Caicedo, J.C.; Guerrero, A.; Aperador, W. Determination of Multilayer Effect Evidence on Metal Carbon-Nitride System. *J. Alloys Compd.* **2019**, *785*, 178–190. [CrossRef]
20. Bondarev, A.V.; Kiryukhantsev-Korneev, P.V.; Levashov, E.A.; Shtansky, D.V. Tribological Behavior and Self-Healing Functionality of TiNbCN-Ag Coatings in Wide Temperature Range. *Appl. Surf. Sci.* **2017**, *396*, 110–120. [CrossRef]
21. Li, W.; Liu, P.; Xue, Z.; Ma, F.; Zhang, K.; Chen, X.; Feng, R.; Liaw, P.K. Microstructures, Mechanical Behavior and Strengthening Mechanism of TiSiCN Nanocomposite Films. *Sci. Rep.* **2017**, *7*, 2140. [CrossRef] [PubMed]
22. El-Rahman, A.M.A.; Wei, R. A Comparative Study of Conventional Magnetron Sputter Deposited and Plasma Enhanced Magnetron Sputter Deposited Ti–Si–C–N Nanocomposite Coatings. *Surf. Coat. Technol.* **2014**, *241*, 74–79. [CrossRef]
23. Abraham, S.; Choi, E.Y.; Kang, N.; Kim, K.H. Microstructure and Mechanical Properties of Ti-Si-C-N Films Synthesized by Plasma-Enhanced Chemical Vapor Deposition. *Surf. Coat. Technol.* **2007**, *202*, 915–919. [CrossRef]
24. Wang, Y.; Li, J.; Dang, C.; Wang, Y.; Zhu, Y. Influence of Carbon Contents on the Structure and Tribocorrosion Properties of TiSiCN Coatings on Ti6Al4V. *Tribol. Int.* **2017**, *109*, 285–296. [CrossRef]
25. Wei, R. Plasma Enhanced Magnetron Sputter Deposition of Ti–Si–C–N Based Nanocomposite Coatings. *Surf. Coat. Technol.* **2008**, *203*, 538–544. [CrossRef]
26. Guo, Y.; Ma, S.; Xu, K. Effects of Carbon Content and Annealing Temperature on the Microstructure and Hardness of Super Hard Ti–Si–C–N Nanocomposite Coatings Prepared by Pulsed d.c. PCVD. *Surf. Coat. Technol.* **2007**, *201*, 5240–5243. [CrossRef]
27. Jeon, J.-H.; Choi, S.R.; Chung, W.S.; Kim, K.H. Synthesis and Characterization of Quaternary Ti–Si–C–N Coatings Prepared by a Hybrid Deposition Technique. *Surf. Coat. Technol.* **2004**, *188–189*, 415–419. [CrossRef]
28. Lin, H.-M.; Duh, J.-G.; Wei, R.; Rincon, C.; Lee, J.-W. The Effect of Microstructure and Composition on Mechanical Properties in Thick-Layered Nanocomposite Ti–Si–C–N Coatings. *Surf. Coat. Technol.* **2010**, *205*, 1460–1464. [CrossRef]
29. Onoprienko, A.A.; Ivashchenko, V.I.; Dub, S.N.; Khyzhun, O.Y.; Timofeeva, I.I. Microstructure and Mechanical Properties of Hard Ti–Si–C–N Films Deposited by Dc Magnetron Sputtering of Multicomponent Ti/C/Si Target. *Surf. Coat. Technol.* **2011**, *205*, 5068–5072. [CrossRef]
30. Vepřek, S.; Reiprich, S.; Shizhi, L. Superhard Nanocrystalline Composite Materials: The TiN/Si 3 N 4 System. *Appl. Phys. Lett.* **1995**, *66*, 2640–2642. [CrossRef]
31. Endler, I.; Höhn, M.; Schmidt, J.; Scholz, S.; Herrmann, M.; Knaut, M. Ternary and Quarternary TiSiN and TiSiCN Nanocomposite Coatings Obtained by Chemical Vapor Deposition. *Surf. Coat. Technol.* **2013**, *215*, 133–140. [CrossRef]
32. Marchin, N.; Ashrafizadeh, F. Surface Structural Characterization of TiSiN and TiSiCN Coatings Produced by Cathodic Arc Physical Vapor Deposition. *Iran. J. Surf. Sci. Eng.* **2019**, *39*, 47–58.
33. Eriksson, A. Cathodic Arc Synthesis of Ti-Si-C-N Thin Films–Plasma Analysis and Microstructure Formation. Ph.D. Thesis, Linköping University Sweden, Linköping, Sweden, 2012.
34. Pruncu, C.I.; Braic, M.; Dearn, K.D.; Farcau, C.; Watson, R.; Constantin, L.R.; Balaceanu, M.; Braic, V.; Vladescu, A. Corrosion and Tribological Performance of Quasi-Stoichiometric Titanium Containing Carbo-Nitride Coatings. *Arab. J. Chem.* **2017**, *10*, 1015–1028. [CrossRef]
35. Meyer, M.K. Composite Fuel (Cermet, Cercer). In *Comprehensive Nuclear Materials*; Elsevier: Amsterdam, The Netherlands, 2012; pp. 257–273. ISBN 9780081028667.

36. Konys, J.; Krauss, W.; Holstein, N. Aluminum-Based Barrier Development for Nuclear Fusion Applications. *Corrosion* **2011**, *67*, 026002-1–026002-6. [CrossRef]
37. Elsener, B.; Rota, A.; Böhni, H. Impedance Study on the Corrosion of PVD and CVD Titanium Nitride Coatings. *Mater. Sci. Forum* **1991**, *44–45*, 29–38. [CrossRef]
38. Pederson, R. Microstructure and Phase Transformation of Ti-6Al-4V. Ph.D. Thesis, Luleå University of Technology, Luleå, Sweden, 2002.
39. Hatem, A.; Lin, J.; Wei, R.; Torres, R.D.; Laurindo, C.; de Souza, G.B.; Soares, P. Tribocorrosion Behavior of Low Friction TiSiCN Nanocomposite Coatings Deposited on Titanium Alloy for Biomedical Applications. *Surf. Coat. Technol.* **2018**, *347*, 1–12. [CrossRef]
40. Qin, C.P.; Zheng, Y.G.; Wei, R. Cavitation Erosion Behavior of Nanocomposite Ti–Si–C–N and Ti/Ti–Si–C–N Coatings Deposited on 2Cr13 Stainless Steel Using a Plasma Enhanced Magnetron Sputtering Process. *Surf. Coat. Technol.* **2010**, *204*, 3530–3538. [CrossRef]
41. Qiu, L.; Du, Y.; Wu, L.; Wang, S.; Zhu, J.; Cheng, W.; Tan, Z.; Yin, L.; Liu, Z.; Layyous, A. Microstructure, Mechanical Properties and Cutting Performances of TiSiCN Super-Hard Nanocomposite Coatings Deposited Using CVD Method under the Guidance of Thermodynamic Calculations. *Surf. Coat. Technol.* **2019**, *378*, 124956. [CrossRef]
42. Lin, J.; Wei, R.; Bitsis, D.C.; Lee, P.M. Development and Evaluation of Low Friction TiSiCN Nanocomposite Coatings for Piston Ring Applications. *Surf. Coat. Technol.* **2016**, *298*, 121–131. [CrossRef]
43. Abd El-Rahman, A.M.; Wei, R.; Raaif, M.; El-Hossary, F.M.; Hammad Fawey, M.; Abo El-kassem, M. Effect of N2/TMS Gas Ratio on Mechanical and Erosion Performances of Ti-Si-C-N Nanocomposite Coatings. *J. Mater. Eng. Perform.* **2020**, *29*, 3324–3333. [CrossRef]
44. Marchin, N.; Ashrafizadeh, F. Effect of Carbon Addition on Tribological Performance of TiSiN Coatings Produced by Cathodic Arc Physical Vapour Deposition. *Surf. Coat. Technol.* **2021**, *407*, 126781. [CrossRef]
45. Du, H.; Liu, P.; Li, W.; Zhang, K.; Ma, F.; Chen, X.; Yang, J.; Feng, R.; Liaw, P.K. Effects of C/Si Ratios on Structures and Behaviors of NbSiCN Nanocomposite Films Synthesized by Reactive Magnetron Sputtering. *Mater. Charact.* **2020**, *167*, 110466. [CrossRef]
46. Wan, Q.; Wu, Z.Y.; Liu, Y.; Yang, B.; Liu, H.D.; Ren, F.; Wang, P.; Xiao, Y.Y.; Zhang, J.; Zhang, G.D. Lead-Bismuth Eutectic (LBE) Corrosion Mechanism of Nano-Amorphous Composite TiSiN Coatings Synthesized by Cathodic Arc Ion Plating. *Corros. Sci.* **2021**, *183*, 109264. [CrossRef]
47. Wan, Q.; Chen, Y.M.; Liu, H.D.; Yang, B. Investigation on Oxidation Behaviors of Ti-Si-N Coating at High Temperature. *Anti-Corros. Methods Mater.* **2018**, *65*, 125–130. [CrossRef]
48. Finšgar, M.; Petovar, B.; Xhanari, K.; Maver, U. The Corrosion Inhibition of Certain Azoles on Steel in Chloride Media: Electrochemistry and Surface Analysis. *Corros. Sci.* **2016**, *111*, 370–381. [CrossRef]
49. Evgeny, B.; Hughes, T.; Eskin, D. Effect of Surface Roughness on Corrosion Behaviour of Low Carbon Steel in Inhibited 4 M Hydrochloric Acid under Laminar and Turbulent Flow Conditions. *Corros. Sci.* **2016**, *103*, 196–205. [CrossRef]
50. Pan, J.; Thierry, D.; Leygraf, C. Electrochemical Impedance Spectroscopy Study of the Passive Oxide Film on Titanium for Implant Application. *Electrochim. Acta* **1996**, *41*, 1143–1153. [CrossRef]
51. Milošev, I.; Metikoš-Huković, M.; Strehblow, H.-H. Passive Film on Orthopaedic TiAlV Alloy Formed in Physiological Solution Investigated by X-Ray Photoelectron Spectroscopy. *Biomaterials* **2000**, *21*, 2103–2113. [CrossRef]
52. Qin, P.; Chen, L.Y.; Liu, Y.J.; Jia, Z.; Liang, S.X.; Zhao, C.H.; Sun, H.; Zhang, L.C. Corrosion and Passivation Behavior of Laser Powder Bed Fusion Produced Ti-6Al-4V in Static/Dynamic NaCl Solutions with Different Concentrations. *Corros. Sci.* **2021**, *191*, 109728. [CrossRef]
53. Bignon, Q.; Martin, F.; Auzoux, Q.; Miserque, F.; Tabarant, M.; Latu-Romain, L.; Wouters, Y. Oxide Formation on Titanium Alloys in Primary Water of Nuclear Pressurised Water Reactor. *Corros. Sci.* **2019**, *150*, 32–41. [CrossRef]
54. Milošev, I.; Kosec, T.; Strehblow, H.-H. XPS and EIS Study of the Passive Film Formed on Orthopaedic Ti–6Al–7Nb Alloy in Hank's Physiological Solution. *Electrochim. Acta* **2008**, *53*, 3547–3558. [CrossRef]
55. Hirschorn, B.; Orazem, M.E.; Tribollet, B.; Vivier, V.; Frateur, I.; Musiani, M. Determination of Effective Capacitance and Film Thickness from Constant-Phase-Element Parameters. *Electrochim. Acta* **2010**, *55*, 6218–6227. [CrossRef]
56. Córdoba-Torres, P. Relationship between Constant-Phase Element (CPE) Parameters and Physical Properties of Films with a Distributed Resistivity. *Electrochim. Acta* **2017**, *225*, 592–604. [CrossRef]

materials

MDPI

Article

The Influence of Annealing at 500 and 900 °C on the Structure and Mechanical Properties of $Al_xCoCrFeNi$ Alloys

Marzena Tokarewicz [1,*], Małgorzata Grądzka-Dahlke [1], Katarzyna Rećko [2] and Magdalena Łępicka [1]

[1] Department of Materials and Production Engineering, Faculty of Mechanical Engineering, Bialystok University of Technology, ul. Wiejska 45 C, 15-351 Białystok, Poland
[2] Faculty of Physics, University of Bialystok, K. Ciołkowskiego 1L, 15-245 Białystok, Poland
* Correspondence: m.tokarewicz@sd.pb.edu.pl

Abstract: The AlCoCrFeNi high-entropy alloy is sensitive to heat treatment. The aim of the present study was to test a similar correlation for $Al_xCoCrFeNi$ alloys with less than equimolar aluminum content. This paper presents a study of the annealing effect on the structure and mechanical properties of selected alloys. $Al_xCoCrFeNi$ alloys (x = 0, 0.5, 0.7) were fabricated by the induction melting method. The obtained specimens were annealed at 500 °C and 900 °C. A detailed study of the changes in crystalline structure due to annealing was conducted. Three-point bending and hardness tests were carried out for the as-cast and annealed specimens to determine selected mechanical properties. The study confirmed that increasing the aluminum content in the $Al_xCoCrFeNi$ alloy improves mechanical properties. For the alloy with aluminum content x = 0.7, hardness increased by 187% and yield strength by 252% compared to the alloy without aluminum. A significant effect of annealing on the crystalline structure of the $Al_{0.7}CoCrFeNi$ alloy was found, but this was not followed by changes in mechanical properties.

Keywords: high-entropy alloys; $Al_xCoCrFeNi$; AlCoCrFeNi; mechanical properties; crystal structure

check for updates

Citation: Tokarewicz, M.; Grądzka-Dahlke, M.; Rećko, K.; Łępicka, M. The Influence of Annealing at 500 and 900 °C on the Structure and Mechanical Properties of $Al_xCoCrFeNi$ Alloys. *Materials* **2023**, *16*, 1245. https://doi.org/10.3390/ma16031245

Academic Editors: Corneliu Munteanu, Ioan Doroftei and Liyuan Sheng

Received: 20 December 2022
Revised: 29 January 2023
Accepted: 30 January 2023
Published: 1 February 2023

1. Introduction

Traditional alloys consist of one or two main components and contain alloying additives to improve their properties [1]. High-entropy alloys (HEAs) were first described in 2004 [2], so it is a relatively new generation of materials. They contain 5 to 13 elements in amounts ranging from 5 to 35 at.% [3]. HEAs exhibit much better properties than conventional alloys, which will help to overcome many barriers and challenges that are generated by the rapid development of technology. This new group of materials may have high corrosion resistance, yield strength, or hardness, as well as many other useful properties in construction [4,5]. Due to their unique properties, high-entropy alloys can be used in many areas such as the energy and transportation industries, which require many low-density and high-strength components [6]. Since HEAs were developed, many high-entropy alloy systems have been characterized in the literature, e.g., TiNbTaZrMo [7], FeNiMnCr [8], CuCoNiCrAlFeTiV [9], and AlCoCrCuFeNiSi [10].

AlCoCrFeNi was one of the first studied high-entropy alloys and is now one of the most popular. Researchers often study the mechanical properties of the AlCoCrFeNi alloy and the influence of the varying amounts of elements on its properties [11–13]. Among all alloy components, aluminum shows the largest influence on the structure and properties. According to the research, depending on the aluminum content, the phases occurring in the alloy change from fcc into an fcc and bcc mixture to bcc [14,15]. The concentration of aluminum in $Al_xCoCrFeNi$ also significantly affects its mechanical properties, e.g., increasing the aluminum content in the $Al_xCoCrFeNi$ alloy improves its strength while decreasing its ductility [16] and positively affects its wear properties [17]. In addition, studies show that the microstructure of $Al_xCoCrFeNi$ alloy changes with the aluminum content [18].

The properties of high-entropy alloys are also determined by the preparation method, selected manufacturing parameters, and heat treatment. HEAs are most commonly obtained by casting and powder metallurgy. Currently, additive methods are also used. The manufacturing method has an impact on the microstructure and the number of defects present in the resulting sample [19]. The effect of annealing on the high-entropy AlCoCrFeNi alloy was extensively investigated. Liang et al. [20] noticed that changes in the crystal structure of AlCoCrFeNi alloy are present after annealing at temperatures higher than 600 °C. Above 600 °C, precipitation of the fcc phase begins. At 800 °C, a new sigma phase of $Fe_{0.5}Cr_{0.5}$ begins to form. The results of Cheng et al. [21] also confirm the changes occurring in the AlCoCrFeNi alloy after annealing. The authors of [22] carried out a detailed study of the equimolar AlCoCrFeNi alloy after annealing at different temperatures. They clearly showed that a sigma phase appears in the 900 °C range, which was not identified below 700 °C or above 1000 °C.

The aim of the present study was to test similar relationships of the effect of characteristic temperatures on changes in structure and mechanical properties for $Al_xCoCrFeNi$ alloys with aluminum contents lower than equimolar. $Al_xCoCrFeNi$ (where: x = 0; 0.5; 0.7) alloys were obtained by induction melting in an argon atmosphere. The properties of the obtained samples were investigated in the as-cast state and after annealing at 500 °C and 900 °C. A three-point bending test was performed to determine the mechanical properties. The hardness, microstructure, and the occurring phases in the alloys were investigated.

2. Materials and Methods

High-entropy alloys $Al_xCoCrFeNi$ (x = 0, 0.5, 0.7) were obtained by induction melting in a protective argon atmosphere. Pure metals were melted in a crucible placed in an induction coil. The purity of the raw materials was above 99.9%. The current induced in the charge caused the metallic components to melt and mix. The samples were remelted three times to achieve chemical homogeneity. The liquid alloy was then poured into a copper mold with a diameter of Ø = 32 mm and a height of h = 10 mm and then cooled to ambient temperature. The samples were cut on a wire cutter into the shape of beams with dimensions 24 × 3.5 × 1.5 mm for strength tests and cylinders with a diameter of Ø = 12 mm and a height of h = 5 mm for structural analyses. Next, part of them was annealed at 500 °C or at 900 °C in a tube furnace under a protective argon atmosphere (details in Table 1).

Table 1. List of $Al_xCoCrFeNi$ alloy specimens used for testing.

No Series	Al Content	Alloy	Annealing
1.0			as-cast
1.1	x = 0	CoCrFeNi	500 °C
1.2			900 °C
2.0			as-cast
2.1	x = 0.5	$Al_{0.5}CoCrFeNi$	500 °C
2.2			900 °C
3.0			as-cast
3.1	x = 0.7	$Al_{0.7}CoCrFeNi$	500 °C
3.2			900 °C

Crystal structures of the samples were analyzed by means of X-ray diffraction (XRD) using an Empyrean Panalytical powder diffractometer (Malvern Panalytical, Malvern, UK), with 40 kV and 30 mA and Cu Kα radiation, λ = 1.540598 Å in the Bragg–Brentano geometry. A 2θ range from 20° to 135° was covered using a step of 0.026261° and a count time of 400 s per point. The standard used during X-ray refinements was LaB_6 C660 crystallizing in a cubic system Pm-3m (space group no. 221). The phase analysis was carried out based on the Inorganic Crystal Structure Database (ICSD) using the HighScore program [23]. During the analysis of the annealing effect on the microstructure for simplicity, we concentrated on the strongest reflections, i.e., (111) FCC and (110) BCC structures. The crystallites (grain) size (D) and microstrain effect (ε) were estimated according to the linear Williamson–Hall method (Table 2).

Table 2. Phase analyses carried out by the use of the ICSD database with respect to referenced cubic space groups number, the unit cell parameters, and the percentage of phase volume contribution. Vol % obtained from XRD data.

Sample		Crystal Structure Data	Vol. %	Graphic Symbol	Crystallite Size (nm)
CoCrFeNi	as-cast	3.6524(3) Å/Fm-3m no. 225	19	ordered fcc/F_1	3009.7
		3.5644(1) Å/Fm-3m no. 225	81	ordered fcc/F_2	
	500 °C	3.5768(2) Å/Fm-3m no. 225	100	ordered fcc/F_2	41.9
	900 °C	3.6022(2) Å/Fm-3m no. 225	18	ordered fcc/F_1	86.3
		3.5712(1) Å/Pm-3m no. 221	82	disordered fcc/F_{2d}	
$Al_{0.5}CoCrFeNi$	as-cast	2.8609(3) Å/Im-3m no. 229	9	ordered bcc/I	29.8
		3.5915(1) Å/Fm-3m no. 225	91	ordered fcc/F_2	37.3
	500 °C	2.8637(1) Å/Im-3m no. 229	3	ordered bcc/I	
		3.5796(1) Å/Pm-3m no. 221	97	disordered fcc/F_{2d}	34.2
	900 °C	2.8642(1) Å/Im-3m no. 229	11	ordered bcc/I	41.2
		3.5707(2) Å/Pm-3m no. 221	89	disordered fcc/F_{2d}	37.2
$Al_{0.7}CoCrFeNi$	as-cast	2.8584(5) Å/Im-3m no. 229	6	ordered bcc/I	31.3
		3.5948(1) Å/Fm-3m no. 225	94	ordered fcc/F_2	51.1
	500 °C	2.8612(2) Å/Im-3m no. 229	54	ordered bcc/I	33.4
		3.5926(3) Å/Fm-3m no. 225	46	ordered fcc/F_2	58.5
	900 °C	3.6578(4) Å/Fm-3m no. 225	1	ordered fcc/F_1	59.8
		3.5750(1) Å/Pm-3m no. 221	89	disordered fcc/F_{2d}	
	disordered I/I_d	5.7335(8) Å/Fm-3m no. 225	3	ordered $L2_1$/L	
		3.6054(3) Å/Pm-3m no. 221	7	disordered fcc/F_{3d}	

For microstructure observation, the samples were ground and polished to a mirror finish, then they were chemically etched in aqua regia. The chemical composition and microstructures were analyzed by a high-resolution scanning electron microscope (SEM-FIB DualBeam Scios 2, Thermo Scientific, Waltham, MA, USA) equipped with an energy-dispersive spectrometer (EDS). Hardness was measured by the Vickers method under a load of 98 N using an INNOVATEST hardness tester (Innovatest Europe BV, Maastricht, The Netherlands). Three-point bending tests were performed on an MTS Insight testing machine (MTS Insight, MTS System Corporation, Eden Prairie, MN, USA). The span length during the examination was 15 mm. To avoid shear load and cutting, 5 mm diameter rollers were used on the press head and for support. A scheme of the three-point bending is illustrated in Figure 1.

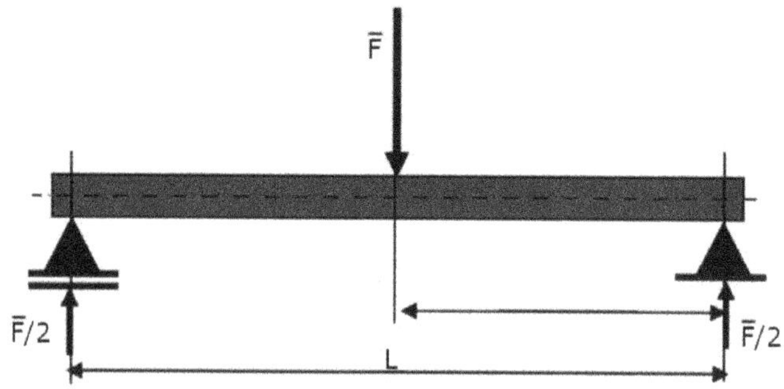

Figure 1. Scheme of the three-point bend test.

The specimens were bent at room temperature with a constant speed of 0.005 mm/s. A load–displacement curve was determined from the obtained data. To determine the conventional bending yield strength $R_{B0.2}$, it was assumed that the extreme fibers of the specimen elongate by ε = 0.2%. For this assumption, the deflection arrow $f_{B0.2}$, which correlates to such elongation, was calculated using the formula:

$$f_{B0.2} = \frac{\varepsilon L^2}{6h}, \quad (1)$$

where ε = 0.2%, L is the span length during the measurement, and h is the height of the sample. For the calculated deflection arrow, the load value P was read off from the obtained load–displacement diagrams. The value of $R_{B0.2}$ was then calculated using the formula:

$$R_{B0.2} = \frac{PL}{4W_g}, \quad (2)$$

where P is the load value read from the load–displacement diagram for the calculated value of the deflection arrow $f_{B0.2}$; W_g is the moment of inertia of the section area about the neutral axis calculated from:

$$W_g = \frac{bh^2}{6}, \quad (3)$$

where b is the width of the sample.

To calculate the flexural modulus, the following formula was used:

$$E_{flex} = \frac{F * L^3}{d * 48 * I}, \quad (4)$$

where F is the applied load, d is displacement, and I is the moment of inertia calculated from:

$$I = \frac{b * h^3}{12}. \quad (5)$$

3. Results and Discussion

3.1. Homogeneity and Crystal Ordering (Phase Analysis)

Room temperature XRD measurements disclosed several different types of cubic crystal structures. Moreover, a few of them only exhibit a well-ordered state (see Table 2). For better readability, disordered phases are marked with a lower d index. In the case of a complex composition of the sample, the predominant phase is indexed as the "1", while the residual ones are indexed according to decreasing volume percentage contributions. According to X-ray diagrams, the series of as-cast HEA alloys and those annealed in different temperatures reveal the coexistence of fcc and bcc phases.

Figure 2 shows the bi-phase nature of the as-cast CoCrFeNi (red diagram). It is noteworthy that both phases disclose the ordered crystal structure of the face-centered cubic (F) type. The annealing procedure carried out at 500 °C leads to a single-phase product (blue diagram in Figure 2), while the higher annealing temperature results again in a double phase in the sample and a disordered predominant F_2 structure (green diagram). Neither effect is desired. Moreover, mechanical textures of the [020] type in the double-phased samples and [111] as typical for single-phase sample were discovered.

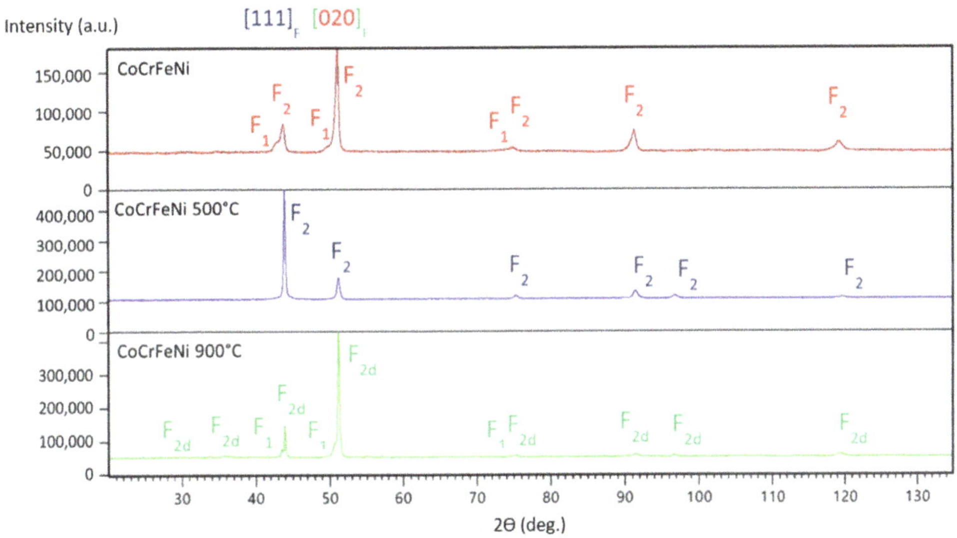

Figure 2. X-ray diagrams of as-cast CoCrFeNi samples and those annealed at different temperatures. The identified phases have been marked according to the graphic symbols described in Table 2. The directions of the mechanical textures are shown at the top of the diagram set.

In both figures below, there are no homogeneous systems in the samples with different aluminum admixtures, either in non-annealed dies or annealed systems. Systems in which face-centered cubic F and body-centered cubic (I) phases coexist are systematically observed.

The set of X-ray diagrams shown in Figure 3 confirms that all samples are at least double-phase. Moreover, in this series, the mechanical texture is not a problem. It is noteworthy that the $Al_{0.5}$CoCrFeNi sample annealed at 500 °C revealed only traces of a body-centered cubic structure (I).

Regardless of the history of heat treatment, all samples of the $Al_{0.7}$CoCrFeNi series exhibit the coexistence of fcc and bcc phases (see Figure 4). Undoubtedly, the increase in the annealing temperature of the system leads to an increasingly complex phase distribution. The green diagram in Figure 4 illustrates phase analysis results obtained for $Al_{0.7}$CoCrFeNi annealed at 900 °C. The sample is a multiphase system for which four phases were identified. Several extra peaks remained unrecognized in the range of lowest scattering angles, up to 40 degrees, which can be defined as peaks from the σ-phase, as is often reported for the equimolar alloy AlCoCrFeNi [22]. Moreover, with a limited number of detected intensities, the $L2_1$ ordered phase (marked in Table 2 as L) and I_d, i.e., the disordered body-centered cubic one, become indistinguishable.

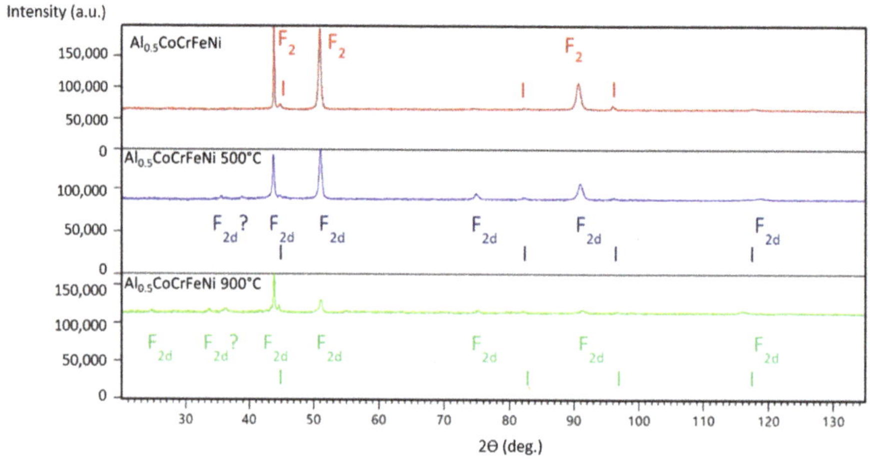

Figure 3. X-ray diagrams of as-cast $Al_{0.5}CoCrFeNi$ samples and those annealed at different temperatures. The identified phases have been marked according to the graphic symbols described in Table 2.

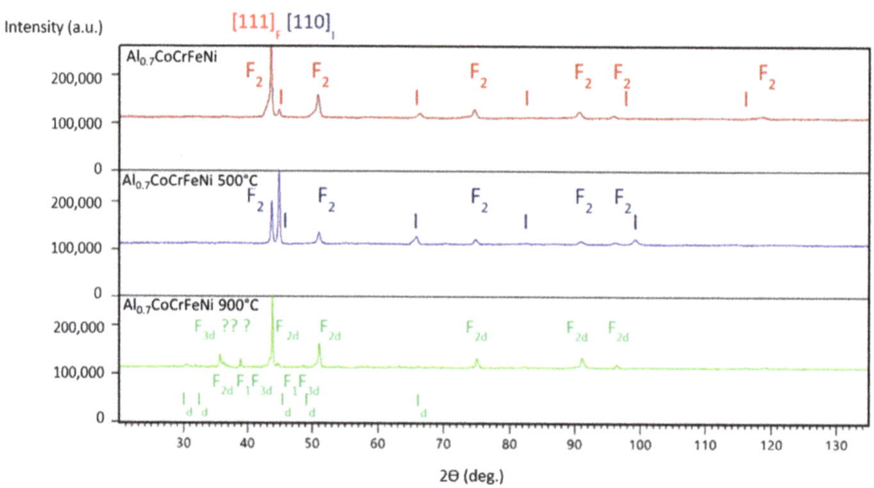

Figure 4. X-ray diagrams of as-cast $Al_{0.7}CoCrFeNi$ samples and those annealed at different temperatures. The identified phases have been marked according to the graphic symbols described in Table 2. The directions of the mechanical textures are shown at the top of the diagram set.

In summary, XRD studies show that annealing at 500 °C and 900 °C did not significantly affect the crystalline structure of the x = 0 alloy. These results correspond well with the data in [24]. For the alloy with low aluminum content, $Al_{0.2}CoCrFeNi$, they identified only the presence of the fcc phase, also after annealing at 800 °C and 1000 °C. However, in the case of higher aluminum content, especially for the $Al_{0.7}CoCrFeNi$ alloy, additional heat treatment caused significant structural changes. Annealing at 500 °C had a stabilizing effect on the bcc phase compared to the as-cast sample. The percentages of the fcc and bcc phases changed. The data in the literature confirm that an increase in aluminum content causes an increase in the proportion of the bcc phase [25]. However, annealing at 900 °C resulted in a permanent transformation to the fcc structure, which corresponds well with the data obtained by the authors of [26]. The appearance of additional peaks for alloys at aluminum contents x = 0.5 and x = 0.7, which can be defined as the σ-phase, should be noted. Many authors investigating $Al_xCoCrFeNi$ alloys with higher aluminum content highlight the presence of this phase at temperatures around 800–900 °C, which may affect the mechanical properties of the materials [22,27,28]. However, a detailed analysis of the obtained diffractograms does not indicate the clear presence of the σ-phase. The explanation may be that the aluminum content in the investigated alloys is too low to initiate the transformation with the formation of this phase, which is also confirmed by the results presented in the work [27] for alloys with aluminum contents x = 0.75 and x = 1.25. Similar results were obtained by the authors of [28,29]. It should be pointed out that the authors of [30] also did not observe the σ-phase during annealing of $Al_xCoCrFeNi$ alloys (x = 0; 0.1; 0.3; 0.5).

3.2. Microstructure and Chemical Composition

Figure 5 shows the microstructure of the as-cast CoCrFeNi alloy (Figure 5a) and those annealed at 500 °C and 900 °C (Figure 5b,c, respectively). All images show a homogeneous structure without any inclusions, which corresponds well to the XRD analysis results. In the case of this alloy, annealing did not change the microstructure.

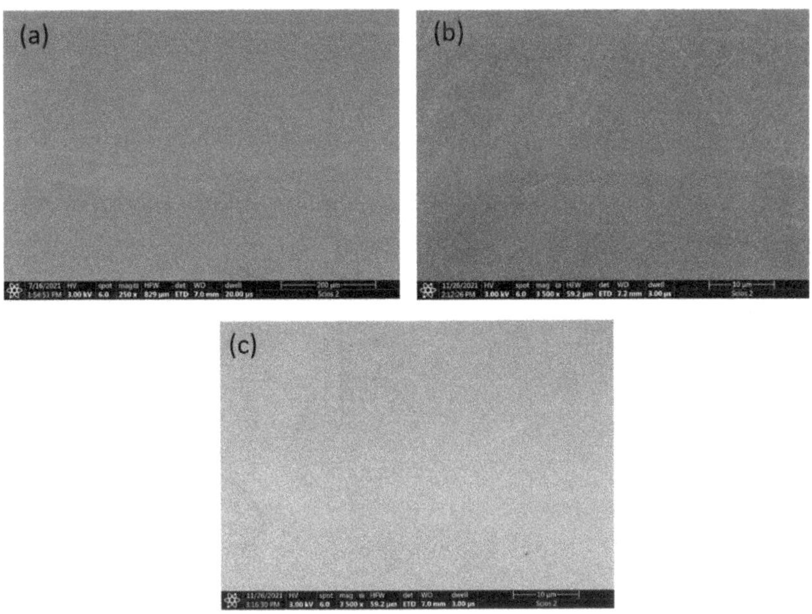

Figure 5. SEM micrographs of CoCrFeNi alloy: (**a**) as-cast, (**b**) annealed at 500 °C, and (**c**) annealed at 900 °C.

The microstructure of the $Al_{0.5}CoCrFeNi$ alloy is biphasic. At lower magnification, there are clearly visible dendrites of the phase that is a solid solution of the constituent elements (Figure 6a), while a mixture of phases can be observed in the interdendritic spaces (Figure 6b). The dendritic areas are poorer in aluminum, while the interdendritic spaces have a significantly higher aluminum content. Annealing did not significantly affect the quantitative ratio of dendritic areas and interdendritic spaces (Figure 6c,d). Only in the case of sample 2.2 did the image of the interdendritic mixture change after annealing at 900 °C—there was a coagulation of the light phase precipitates, which is visible as spherical precipitates (Figure 6d).

The images in Figure 7 show the microstructures of the series 3 alloys before and after heat treatment. In this case, a two-phase structure is also visible. The arrangement of the areas of the individual phases is not dendritic, as in the case of series 2 alloys; it is more like a lamellar Widmanstatten structure (Figure 7a–d). It should be noted that the arrangement and shape of the two phases' precipitates are similar for all samples in this series.

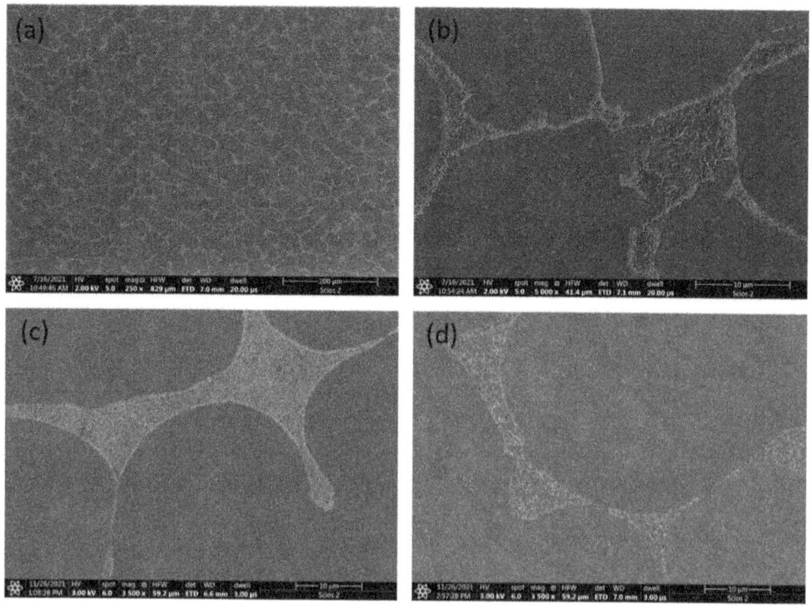

Figure 6. SEM micrographs of $Al_{0.5}$CoCrFeNi alloy: (**a**,**b**) as-cast, (**c**) annealed at 500 °C, and (**d**) annealed at 900 °C.

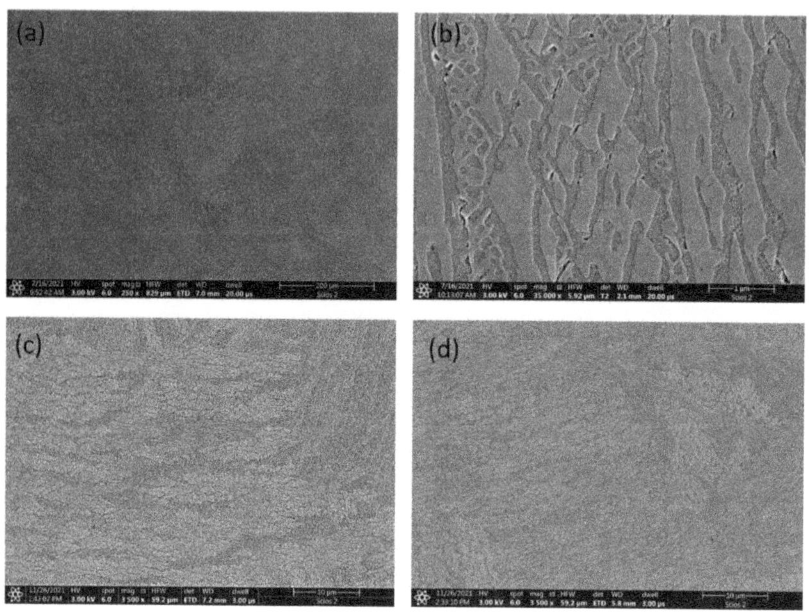

Figure 7. SEM micrographs of $Al_{0.7}$CoCrFeNi alloy: (**a**,**b**) as-cast, (**c**) annealed at 500 °C, and (**d**) annealed at 900 °C.

Table 3 shows the results of the chemical composition analysis of the tested as-cast alloys. The average values obtained from the larger measurement plots correspond approximately to the assumed values of the constituent elements. In the case of alloys with aluminum contents of x = 0.5 and x = 0.7, due to the two-phase structure, the chemical composition of the main phase (dendrite or side plate) and the spaces between the separations of this phase were also analyzed. It can be seen that there are convergences between these results. The dendrite (x = 0.5) and inter-sideplate (x = 0.7) phases are richer in aluminum compared to the averaged values for these materials. In contrast, the interdendrite and side-plate phases are poorer in aluminum. Note that the chemical composition of the separated phases is similar for both alloys (x = 0.5 and x = 0.7), only the quantitative contribution of these phases is different (Figures 6 and 7).

Table 3. Chemical composition of obtained alloys, EDX analysis.

Alloys	Area	Element (at. %)				
		Al	Co	Cr	Fe	Ni
CoCrFeNi	global	-	24.82	24.50	25.59	25.09
$Al_{0.5}CoCrFeNi$	global	10.96	22.72	21.27	23.01	22.04
	dendrite	9.18	23.81	21.66	24.83	20.52
	interdendrite	16.71	19.03	22.96	19.12	22.19
$Al_{0.7}CoCrFeNi$	global	13.82	21.86	21.40	22.13	20.79
	side plate	10.45	22.89	22.11	23.85	20.71
	inter-sideplate	16.25	19.75	23.28	21.17	19.55

3.3. Mechanical Properties

The effect of heat treatment on the selected mechanical properties of the obtained alloys was also analyzed. The results of the alloy hardness measurements are shown in Figure 8. With increasing aluminum content, the hardness of the alloy becomes greater. The received results are consistent with other studies [15,31]. The reason for such a large increase in hardness may be the phases occurring in the alloy. An increase in the aluminum concentration in the alloy leads to the change of the fcc phase into a mixture of the fcc and bcc, and then into the bcc.

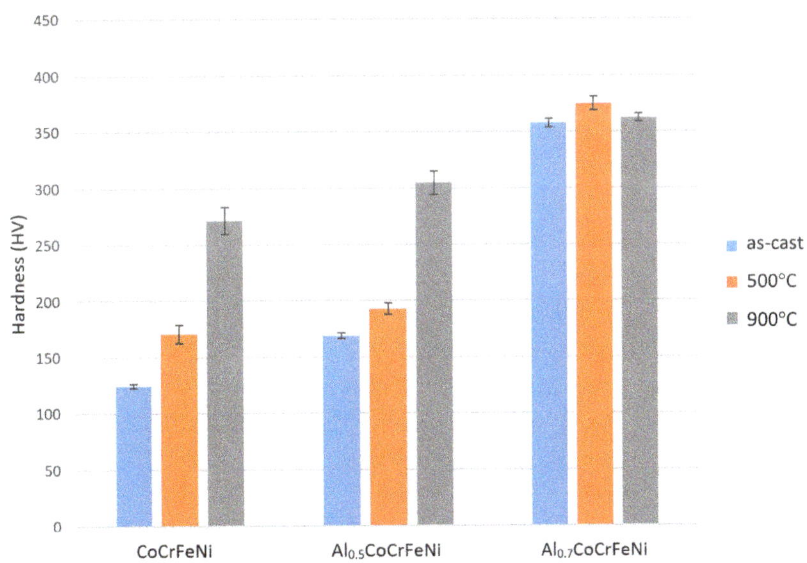

Figure 8. The average hardness of obtained alloys.

In the case of CoCrFeNi and $Al_{0.5}CoCrFeNi$ alloys, the heat treatment had a bigger influence on the hardness than in the $Al_{0.7}CoCrFeNi$ alloy. Annealing at 900 °C increased the hardness of the CoCrFeNi alloy by 218% and the $Al_{0.5}CoCrFeNi$ alloy by 181% compared to the initial alloys. In the case of these two materials (series 1 and 2), annealing at 500 °C resulted in a significantly lower increase in hardness than annealing at 900 °C. In the series 3 alloy, $Al_{0.7}CoCrFeNi$, a slightly different result was observed. The heat treatment had a smaller influence on properties. Annealing at 500 °C resulted in an increase in hardness of only 10%, while at 900 °C the increase is minimal. Similar results for the $Al_{0.75}CoCrFeNi$ alloy were obtained by the authors of [27].

Figure 9 shows the averaged load–displacement curves determined from the results received during the three-point bending test.

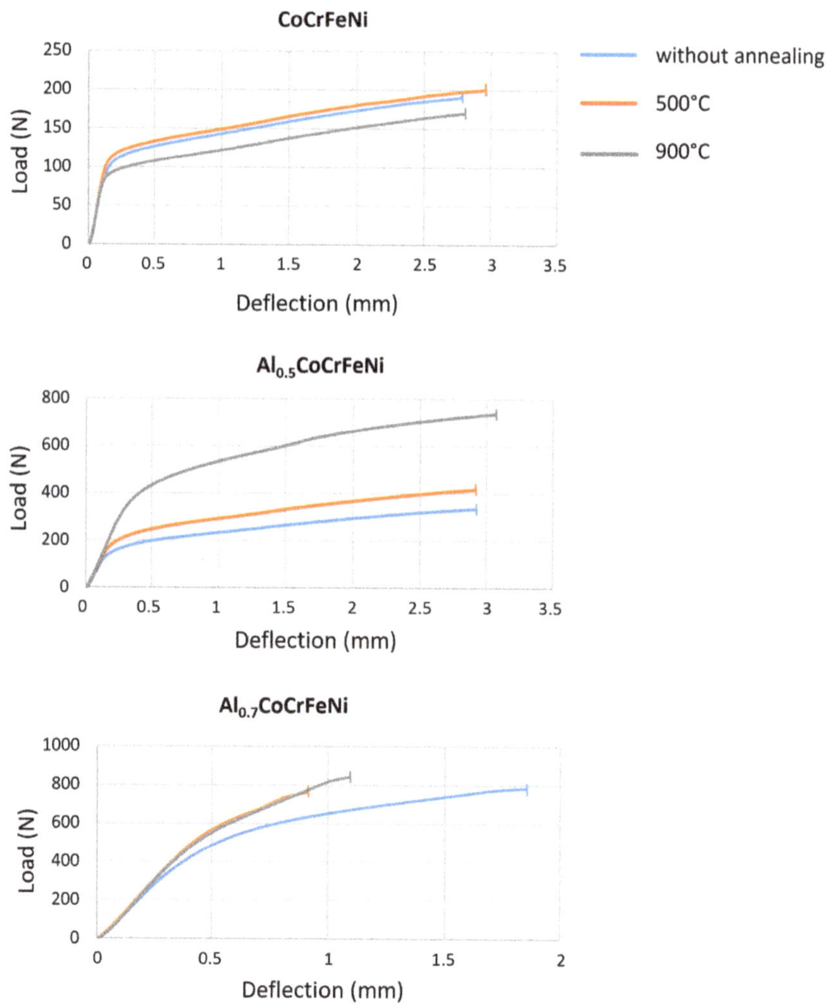

Figure 9. Averaged load–displacement curves.

Selected alloy properties determined from the three-point bending test are shown in Figures 10 and 11.

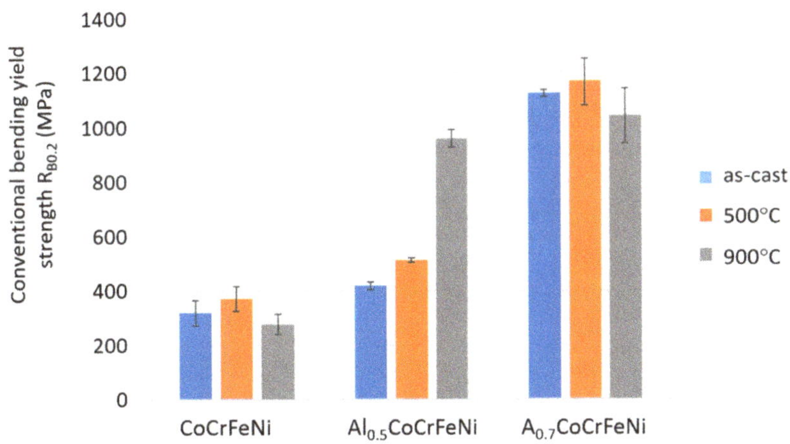

Figure 10. Determined conventional yield strength $R_{0.2}$.

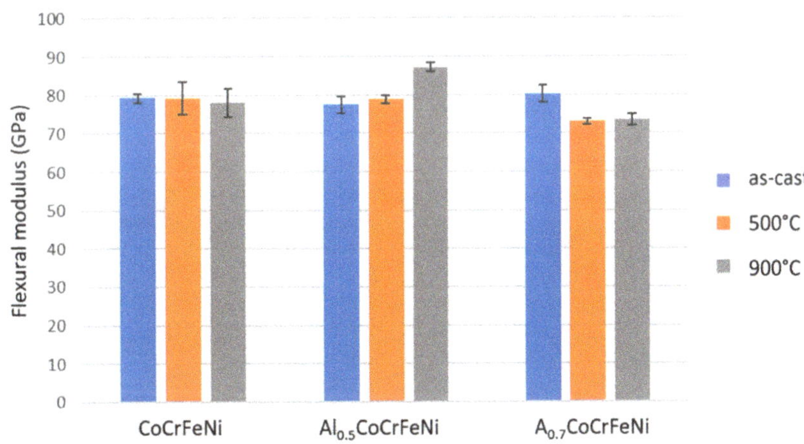

Figure 11. Determined flexural modulus.

Figure 10 shows that annealing at 500 °C increased the conventional bending yield strength of all samples. In this case, the flexural modulus after heat treatment at 500 °C did not change for the CoCrFeNi sample, but increased for $Al_{0.5}$CoCrFeNi and decreased significantly for $Al_{0.7}$CoCrFeNi (Figure 5). Heat treatment at 900 °C resulted in a large increase in the conventional bending yield strength of the $Al_{0.5}$CoCrFeNi sample, while for the other samples the obtained values were lower than before annealing. The same is observed for the determined flexural modulus.

3.4. Correlation between Structure and Mechanical Properties

The results of the study clearly showed the significant influence of the aluminum content in the Al_xCoCrFeNi alloy and the annealing process at different temperatures on the mechanical properties. Which strengthening mechanism plays the greatest role here, solution strengthening, related to the difference in atomic radii of aluminum compared to the other constituent elements, phase transformations of the fcc and bcc structure with different mechanical properties, or interactions at the microstructural level?

Solution strengthening should play an important role, as confirmed by the authors of [32]. It could be expected that an increase in aluminum content from x = 0 to x = 0.5 would result in greater strengthening than between alloys x = 0.5 and x = 0.7. Comparing the data shown in Figures 8 and 10, it can be seen that the properties of the $Al_{0.5}CoCrFeNi$ alloy are slightly higher than those of CoCrFeNi. In contrast, at an aluminum content of x = 0.7, both hardness and yield strength increase by more than 100% compared to the alloy without aluminum.

Many authors believe that the change in the crystal structure is mainly responsible for the strengthening of $Al_xCoCrFeNi$ alloys. Usually, the mixing enthalpy (H_{mix}) and the valence electron concentration (VEC) are employed to predict the stability of fcc and bcc solid solutions in these alloys. According to calculations [32,33] at low aluminum content there is a homogeneous fcc phase, over 10% of the aluminum atomic content a bcc phase appears, and its proportion increases with the aluminum concentration. A single-phase bcc structure occurs for x = 1. This is confirmed by the XRD observations of many researchers [34]. Certainly, the phase structure affects the mechanical properties of the alloys. According to the results of the authors of [34], the hardness of the $Al_xCoCrFeNi$ alloy improves from x = 0 to x = 0.9, while a further increase in aluminum content to x = 1.5 does not affect the hardness growth. In addition, heat treatment of materials also causes phase transformations, e.g., for x = 0.5 above 800 °C a bcc $\rightarrow$ fcc transformation occurs [35]. The authors of [24] observe that annealing at 600 °C causes an increase in the contribution of the bcc phase, resulting in improved hardness. However, heat treatment above 800 °C causes a decrease in hardness, which they justify by the separation of Al and Ni from the fcc matrix. Many authors point out the strong influence of the σ phase on the strengthening of the AlCoCrFeNi alloy [22,28]. This phase appears during annealing at 800 °C, and it disappears at temperatures above 1000 °C. However, it only occurs at higher aluminum contents (above x = 0.9) [25,34]. Based on the data obtained in the present work, it is difficult to justify the changes in strength properties on the basis of phase transformations. In the case of the CoCrFeNi alloy (x = 0), practically no changes were observed in the crystalline structure and microstructure due to annealing (single-phase fcc structure). There was also no change in the bending yield strength. However, this does not correlate with the change in hardness—after annealing at 900 °C, the hardness increased more than twice compared to the initial material. Thermal stresses were unlikely to affect the properties—the alloy was cooled slowly with the furnace. For an alloy with an aluminum content of x = 0.5, a good correlation of the hardness and yield strength results was obtained: after annealing at 500 °C, these parameters increased by about 14% and 22%, respectively, while a significant increase of 81% and 129% was obtained after annealing in 900 °C. In the case of this alloy, it could be justified by the presence of a hard bcc phase, which appears with increasing aluminum content, but quantitative XRD studies do not confirm the increased proportion of the bcc phase after heat treatment at 900 °C compared to the initial material.

It appears that the most possible connection can be made between the changes in mechanical properties and the microstructure of the investigated alloys obtained by induction melting. In the case of the series 1 alloy (without aluminum), we are considering a single-phase structure. Annealing had almost no effect on the change in microstructure and yield strength values. It is interesting that the hardness of this alloy increased with the annealing temperature, which does not correspond to the parameter $R_{B0.2}$. In contrast, for the series 2 alloy (x = 0.5) the results of hardness and $R_{B0.2}$ measurements are analogous—there was a significant increase in mechanical properties after annealing at 900 °C. This can be connected with changes in the microstructure of the alloy (Figure 6). The microstructure of the investigated $Al_{0.5}CoCrFeNi$ alloy consists of dendrites of the fcc phase which are deficient in aluminum, and a small proportion of interdendritic spaces with a mixture of phases with increased aluminum content. It is possible to observe slight changes in the microstructure in the interdendrite areas after annealing at 900 °C (Figure 6d). This may be caused by the migration of atoms at elevated temperatures and the appearance of precipitates at phase boundaries in interdendritic spaces. Explaining this aspect requires more detailed research. The authors of [35] suggest an effect of coherent nanoparticles on the strengthening of the AlCrFeNiV alloy. In contrast, the $Al_{0.7}CoCrFeNi$ alloy shows the highest mechanical properties regardless of the additional annealing, even though in this case the heat treatment had a significant effect on the changes in the crystal structure. In the microstructure of this alloy, the phases that are richer and poorer in aluminum are evenly distributed over the entire surface of the sample, which was hardly affected by annealing.

The mechanism of strengthening in $Al_xCoCrFeNi$ alloys with increasing Al content during annealing is complex and requires further detailed analyses with complex consideration of phase and microstructural transformations. Nevertheless, the effect is very significant, which raises the importance of such studies.

It should be mentioned that the results presented by other researchers investigating similar alloys are also often inconclusive, particularly regarding the mechanical properties and their changes under annealing. This fact is well illustrated by a summary prepared by the authors [22] for the equimolar alloy AlCoCrFeNi. In addition, significant differences are observed depending on the way the material is prepared (mechanical alloying, SPS [36], induction melting, which the authors used, laser deposition [32], or plasma sprayed in the case of layer deposition [24]).

4. Conclusions

The research presented in this article analyzes the effect of annealing at 500 °C and 900 °C of selected $Al_xCoCrFeNi$ alloys obtained by induction melting. The following conclusions can be drawn from the achieved results:

- No significant changes in the microstructure of the studied alloys were observed as a result of the heat treatment. The alloy without aluminum was characterized by a single-phase structure. However, in the microstructure of alloys with an aluminum content of $x = 0.5$ and $x = 0.7$, the presence of two phases differing in aluminum content can be observed.
- Changes in the crystalline structure of the alloys are related to changes in the aluminum content. The addition of aluminum enhances the appearance of the bcc phase. In the case of CoCrFeNi and $Al_{0.5}CoCrFeNi$ alloys, annealing at 500 °C and 900 °C did not significantly affect the changes in the crystalline structure. On the other hand, a significant effect of annealing on the crystalline structure of the $Al_{0.7}CoCrFeNi$ alloy was noted. By annealing at 900 °C, a permanent transformation to fcc occurred.
- These transformations are not reflected in changes in the mechanical properties of the tested materials. In the case of CoCrFeNi and $Al_{0.5}CoCrFeNi$ alloys, annealing at 900 °C resulted in a strengthening of the materials during the three-point bending test. In contrast, no significant changes in mechanical properties were observed for the $Al_{0.7}CoCrFeNi$ alloy as a result of annealing.
- As the aluminum content increases, the mechanical properties of the $Al_xCoCrFeNi$ alloys improve, particularly at an aluminum content of $x = 0.7$.

Author Contributions: Conceptualization, M.G.-D. and M.T.; methodology, M.G.-D. and M.T.; validation, M.G.-D., M.T., K.R. and M.Ł.; formal analysis, M.T.; investigation, M.T., K.R. and M.Ł.; resources, M.G.-D.; data curation, M.T. and K.R.; writing—original draft preparation, M.T., K.R. and M.G.-D.; writing—review and editing, M.G.-D. and M.Ł.; visualization, M.T.; supervision, M.G.-D.; project administration, M.G.-D.; funding acquisition, M.G.-D. All authors have read and agreed to the published version of the manuscript.

Funding: This research was funded by the program of the Minister of Science and Higher Education of Poland named "Regional Initiative of Excellence" in 2019–2022 project number 011/RID/2018/19 amount of financing 12,000,000 PLN".

Institutional Review Board Statement: Not applicable.

Informed Consent Statement: Not applicable.

Data Availability Statement: The data presented in this study are available on request from the corresponding author.

Acknowledgments: This work was partially supported by the National Science Centre under grant OPUS no. 2018/31/B/ST3/00279.

Conflicts of Interest: The authors declare no conflict of interest.

References

1. Evans, D.; Chen, J.; Bokas, G.; Chen, W.; Hautier, G.; Sun, W. Visualizing Temperature-Dependent Phase Stability in High Entropy Alloys. *Npj Comput. Mater.* **2021**, *7*, 151. [CrossRef]
2. Yeh, J.-W.; Chen, S.-K.; Lin, S.-J.; Gan, J.-Y.; Chin, T.-S.; Shun, T.-T.; Tsau, C.-H.; Chang, S.-Y. Nanostructured High-Entropy Alloys with Multiple Principal Elements: Novel Alloy Design Concepts and Outcomes. *Adv. Eng. Mater.* **2004**, *6*, 299–303. [CrossRef]
3. Yeh, J.W.; Chen, Y.L.; Lin, S.J.; Chen, S.K. High-Entropy Alloys—A New Era of Exploitation. *Mater. Sci. Forum* **2007**, *560*, 1–9. [CrossRef]
4. Wu, W.-H.; Yang, C.-C.; Yeh, J.-W. Industrial Development of High-Entropy Alloys. *Eur. J. Control* **2006**, *31*, 737–747. [CrossRef]
5. Gao, M.C.; Miracle, D.B.; Maurice, D.; Yan, X.; Zhang, Y.; Hawk, J.A. High-Entropy Functional Materials. *J. Mater. Res.* **2018**, *33*, 3138–3155. [CrossRef]
6. Yeh, J.-W.; Lin, S.-J. Breakthrough Applications of High-Entropy Materials. *J. Mater. Res.* **2018**, *33*, 3129–3137. [CrossRef]
7. Todai, M.; Nagase, T.; Hori, T.; Matsugaki, A.; Sekita, A.; Nakano, T. Novel TiNbTaZrMo High-Entropy Alloys for Metallic Biomaterials. *Scr. Mater.* **2017**, *129*, 65–68. [CrossRef]
8. Kumar, N.A.P.K.; Li, C.; Leonard, K.J.; Bei, H.; Zinkle, S.J. Microstructural Stability and Mechanical Behavior of FeNiMnCr High Entropy Alloy under Ion Irradiation. *Acta Mater.* **2016**, *113*, 230–244. [CrossRef]
9. Yeh, J.-W.; Lin, S.-J.; Chin, T.-S.; Gan, J.-Y.; Chen, S.-K.; Shun, T.-T.; Tsau, C.-H.; Chou, S.-Y. Formation of Simple Crystal Structures in Cu-Co-Ni-Cr-Al-Fe-Ti-V Alloys with Multiprincipal Metallic Elements. *Metall. Mater. Trans. A* **2004**, *35*, 2533–2536. [CrossRef]
10. Yang, C.-C.; Hang Chau, J.L.; Weng, C.-J.; Chen, C.-S.; Chou, Y.-H. Preparation of High-Entropy AlCoCrCuFeNiSi Alloy Powders by Gas Atomization Process. *Mater. Chem. Phys.* **2017**, *202*, 151–158. [CrossRef]
11. López Ríos, M.; Socorro Perdomo, P.P.; Voiculescu, I.; Geanta, V.; Crăciun, V.; Boerasu, I.; Mirza Rosca, J.C. Effects of Nickel Content on the Microstructure, Microhardness and Corrosion Behavior of High-Entropy AlCoCrFeNix Alloys. *Sci. Rep.* **2020**, *10*, 21119. [CrossRef] [PubMed]
12. Shun, T.-T.; Hung, W.-J. Effects of Cr Content on Microstructure and Mechanical Properties of AlCoCrxFeNi High-Entropy Alloy. *Adv. Mater. Sci. Eng.* **2018**, *2018*, 5826467. [CrossRef]
13. Guo, L.; Xiao, D.; Wu, W.; Ni, S.; Song, M. Effect of Fe on Microstructure, Phase Evolution and Mechanical Properties of (AlCoCrFeNi)100-XFex High Entropy Alloys Processed by Spark Plasma Sintering. *Intermetallics* **2018**, *103*, 1–11. [CrossRef]
14. Tong, C.-J.; Chen, Y.-L.; Yeh, J.-W.; Lin, S.-J.; Chen, S.-K.; Shun, T.-T.; Tsau, C.-H.; Chang, S.-Y. Microstructure Characterization of AlxCoCrCuFeNi High-Entropy Alloy System with Multiprincipal Elements. *Metall. Mater. Trans. A* **2005**, *36*, 881–893. [CrossRef]
15. Kao, Y.-F.; Chen, T.-J.; Chen, S.-K.; Yeh, J.-W. Microstructure and Mechanical Property of As-Cast, -Homogenized, and -Deformed AlxCoCrFeNi ($0 \leq x \leq 2$) High-Entropy Alloys. *J. Alloys Compd.* **2009**, *488*, 57–64. [CrossRef]
16. Joseph, J.; Jarvis, T.; Wu, X.; Stanford, N.; Hodgson, P.; Fabijanic, D.M. Comparative Study of the Microstructures and Mechanical Properties of Direct Laser Fabricated and Arc-Melted AlxCoCrFeNi High Entropy Alloys. *Mater. Sci. Eng. A* **2015**, *633*, 184–193. [CrossRef]
17. Joseph, J.; Haghdadi, N.; Shamlaye, K.; Hodgson, P.; Barnett, M.; Fabijanic, D. The Sliding Wear Behaviour of CoCrFeMnNi and AlxCoCrFeNi High Entropy Alloys at Elevated Temperatures. *Wear* **2019**, *428–429*, 32–44. [CrossRef]
18. Geanta, V. Chemical Composition Influence on Microhardness, Microstructure and Phase Morphology of AlxCrFeCoNi High Entropy Alloys. *Rev. Chim.* **2018**, *69*, 798–801. [CrossRef]
19. Alshataif, Y.A.; Sivasankaran, S.; Al-Mufadi, F.A.; Alaboodi, A.S.; Ammar, H.R. Manufacturing Methods, Microstructural and Mechanical Properties Evolutions of High-Entropy Alloys: A Review. *Met. Mater. Int.* **2020**, *26*, 1099–1133. [CrossRef]
20. Liang, J.-T.; Cheng, K.-C.; Chen, S.-H. Effect of Heat Treatment on the Phase Evolution and Mechanical Properties of Atomized AlCoCrFeNi High-Entropy Alloy Powders. *J. Alloys Compd.* **2019**, *803*, 484–490. [CrossRef]
21. Cheng, K.-C.; Chen, J.-H.; Stadler, S.; Chen, S.-H. Properties of Atomized AlCoCrFeNi High-Entropy Alloy Powders and Their Phase-Adjustable Coatings Prepared via Plasma Spray Process. *Appl. Surf. Sci.* **2019**, *478*, 478–486. [CrossRef]
22. Zhang, X.; Liu, L.; Yao, K.; Duan, K.; Wu, F.; Zhao, R.; Zhang, Y.; Shang, J.; Chen, M. The Phase Composition Characteristics of AlCoCrFeNi High Entropy Alloy Heat-Treated by Simple Normalizing Treatment and Its Effects on Mechanical Properties. *J. Alloys Compd.* **2022**, *926*, 166896. [CrossRef]
23. Degen, T.; Sadki, M.; Bron, E.; König, U.; Nénert, G. The HighScore suite. *Powder Diffr.* **2014**, *29*, S13–S18. [CrossRef]
24. Zhang, F.; Wang, L.; Yan, S.; Yu, G.; Chen, J.; He, J.; Yin, F. Effect of Al Content on Microstructure and Mechanical Properties of Atmosphere Plasma Sprayed AlxCoCrFeNi High-Entropy Alloy Coatings under Post-Annealing. *Surf. Coat. Technol.* **2022**, *446*, 128804. [CrossRef]
25. Liu, M.; Zhang, S.; Li, F.; Wang, L.; Wang, Z.; Wang, Z.; Sha, Y.; Shen, Q. Effect of Annealing Treatment on the Microstructure and Hardness of AlxCoCrFeNi (x = 0.75, 1.25) High Entropy Alloys. *J. Mater. Eng. Perform.* **2022**, *31*, 1444–1455. [CrossRef]
26. Chou, H.-P.; Chang, Y.-S.; Chen, S.-K.; Yeh, J.-W. Microstructure, Thermophysical and Electrical Properties in AlxCoCrFeNi ($0 \leq x \leq 2$) High-Entropy Alloys. *Mater. Sci. Eng. B* **2009**, *163*, 184–189. [CrossRef]
27. Rao, J.C.; Diao, H.Y.; Ocelík, V.; Vainchtein, D.; Zhang, C.; Kuo, C.; Tang, Z.; Guo, W.; Poplawsky, J.D.; Zhou, Y.; et al. Secondary Phases in AlxCoCrFeNi High-Entropy Alloys: An in-Situ TEM Heating Study and Thermodynamic Appraisal. *Acta Mater.* **2017**, *131*, 206–220. [CrossRef]

28. Meshi, L.; Linden, Y.; Munitz, A.; Salhov, S.; Pinkas, M. Retardation of the σ Phase Formation in the AlCoCrFeNi Multi-Component Alloy. *Mater. Charact.* **2019**, *148*, 171–177. [CrossRef]
29. Wang, Q.; Wang, H.; Gao, T.; Qian, L.; Zeng, L.; Chen, Q.; Liu, X. Microstructure Evolution and Mechanical Properties of an Annealed Dual-Phase CrFeCoNiAl0.7 High-Entropy Alloy. *Intermetallics* **2022**, *149*, 107657. [CrossRef]
30. Anber, E.A.; Smith, N.C.; Liaw, P.K.; Wolverton, C.M.; Taheri, M.L. Role of Al Additions in Secondary Phase Formation in CoCrFeNi High Entropy Alloys. *APL Mater.* **2022**, *10*, 101108. [CrossRef]
31. Wang, W.-R.; Wang, W.-L.; Wang, S.-C.; Tsai, Y.-C.; Lai, C.-H.; Yeh, J.-W. Effects of Al Addition on the Microstructure and Mechanical Property of AlxCoCrFeNi High-Entropy Alloys. *Intermetallics* **2012**, *26*, 44–51. [CrossRef]
32. Huang, L.; Sun, Y.; Amar, A.; Wu, C.; Liu, X.; Le, G.; Wang, X.; Wu, J.; Li, K.; Jiang, C.; et al. Microstructure Evolution and Mechanical Properties of AlxCoCrFeNi High-Entropy Alloys by Laser Melting Deposition. *Vacuum* **2021**, *183*, 109875. [CrossRef]
33. Zhang, C.; Zhang, F.; Chen, S.; Cao, W. Computational Thermodynamics Aided High-Entropy Alloy Design. *JOM* **2012**, *64*, 839–845. [CrossRef]
34. Wang, W.-R.; Wang, W.-L.; Yeh, J.-W. Phases, Microstructure and Mechanical Properties of AlxCoCrFeNi High-Entropy Alloys at Elevated Temperatures. *J. Alloys Compd.* **2014**, *589*, 143–152. [CrossRef]
35. Wang, L.; Wang, L.; Zhou, S.; Xiao, Q.; Xiao, Y.; Wang, X.; Cao, T.; Ren, Y.; Liang, Y.-J.; Wang, L.; et al. Precipitation and Micromechanical Behavior of the Coherent Ordered Nanoprecipitation Strengthened Al-Cr-Fe-Ni-V High Entropy Alloy. *Acta Mater.* **2021**, *216*, 117121. [CrossRef]
36. Bhattacharya, R.; Annasamy, M.; Cizek, P.; Kamaraj, M.; Muralikrishna, G.M.; Hodgson, P.; Fabijanic, D.; Murty, B.S. Evolution of Phase Constitution with Mechanical Alloying and Spark Plasma Sintering of Nanocrystalline AlxCoCrFeNi (x = 0, 0.3, 0.6, 1 Mol) High-Entropy Alloys. *J. Mater. Res.* **2022**, *37*, 959–975. [CrossRef]

materials

MDPI

Article

Innovative Passive and Environmentally Friendly System for Improving the Energy Performance of Buildings

Andrei Burlacu [1,*], Gavril Sosoi [1], Chérifa Abid [2], Marinela Barbuta [1], Marina Verdes [1], Robert Stefan Vizitiu [1,*] and Marius Branoaea [1]

1 Faculty of Civil Engineering and Building Services, "Gheorghe Asachi" Technical University of Iasi, 700050 Iasi, Romania
2 Aix-Marseille Université, CNRS, IUSTI, 13453 Marseille, France
* Correspondence: andrei.burlacu@academic.tuiasi.ro (A.B.); robert-stefan.vizitiu@academic.tuiasi.ro (R.S.V.)

Abstract: The aim of the study is to develop a system for converting, accumulating, and delivering solar energy that is based on the development of an innovative solar panel with heat pipes and a heat storage wall, for the construction of passive structures. The novel aspect of this experiment is the utilization of concrete walls that have different recyclable materials added to their structure in various proportions. The solar energy from the sunny façades is transformed by this system into thermal energy, which is then transferred by integrated heat pipes in a massive element with high thermal inertia. Using insulated shutters, thermal energy can be stored during the day and released at night to keep the room at a comfortable temperature. In order to integrate the modules into the solar recovery system, four concrete samples were cast with a blend of standard and waste aggregates. Four heat fluxes of 100 W/m^2, 150 W/m^2, 200 W/m^2, and 250 W/m^2 were applied to each global system. Thermal imaging data and numerical simulations both supported the findings of temperature sensors. The most effective mixture, fly ash and chopped PET, delivered temperatures that were, on average, 3.3% higher at the end of the charging cycle than those measured for the control sample. The discharging cycle of the concrete block with fly ash and sawdust was the most effective, with an average temperature loss of 5.0 °C as compared to 5.5 °C for the control sample, on average.

Keywords: energy efficiency; passive system; solar heat flux; heat pipes

Citation: Burlacu, A.; Sosoi, G.; Abid, C.; Barbuta, M.; Verdes, M.; Vizitiu, R.S.; Branoaea, M. Innovative Passive and Environmentally Friendly System for Improving the Energy Performance of Buildings. *Materials* **2022**, *15*, 7224. https://doi.org/10.3390/ma15207224

Academic Editor: Antonio Caggiano

Received: 16 September 2022
Accepted: 12 October 2022
Published: 17 October 2022

1. Introduction

Energy efficiency in buildings is a constant challenge for researchers, who must come up with new solutions to reduce energy consumption. New technologies proposed and developed for energy performance in buildings must take into account two criteria: low cost of production and environmentally-friendly manufacturing.

Energy performance in buildings became a priority in recent decades, and the use of passive systems represents a great alternative due to their capability to convert or to store heat received from solar radiation. The use of green walls is one of the passive methods for lowering energy use in buildings. Their effects have been researched in both warm and cold climates, including oceanic [1], temperate [2], and subtropical [3]. These passive energy-saving solutions have also been categorized by Perez et al. [4]. Louis et al. [5] examined how green walls affect the thermal characteristics of the building envelope and how solar radiation absorption, reflection, and transmissivity are impacted.

Trombe walls are also used to make buildings more energy efficient. They have also been researched in a variety of climates, including the desert regions of Iran [6], the cold regions of Turkey [7], and the city of Athens, Greece [8]. Various studies have examined the possibility of improving Trombe walls by analyzing the conditions of ventilation openings and occlusion device operation [9], or by creating innovative designs [10].

Buildings can also save energy by using walls that include phase-change materials. Researchers have looked into a number of ways to introduce PCM into the wall structure.

In order to preserve similar comfort levels throughout the year, Faraji et al. [11] quantitatively analyzed the feasibility of replacing the thick and heavy external walls utilized in Mediterranean nations with thin and light thermal mass walls. The findings demonstrate that internal temperature variations in the wall using PCM are significantly reduced. Omari et al. [12] examined how the addition of micro-dispersed PCM-composite boards alters a wall's thermal behavior throughout the year.

In addition, Meng et al. [13] and Xie et al. [14] examined the behavior of rooms with PCM-composite walls throughout the year. Chen et al. [15] examined another usage of PCM walls by putting them in solar greenhouses. The suggested approach improved the wall's ability to store and release heat, raising the average soil temperature, indoor air temperature, and daily effective accumulative temperature, on average.

The use of double-skin façades is another passive energy-saving technology. Studies like those by Dama et al. [16] and Souza et al. [17] use experimental and numerical analysis to examine how introducing a double-skin façade to a building influences heat gains inside the building. Zhang et al. [18] attempted to improve the DFS by figuring out the ideal vertical air channel thickness. They discovered that the thickness should be less than 0.6 m for air ducts intended to boost ventilation and less than 0.2 m for those intended to increase warm air supply. Amorphous silicon photovoltaic blinds may be used to generate power on-site while also reducing heat absorption or loss through glazing, according to research by Luo et al. [19]. To lessen the additional rise in cooling loads in buildings, Li et al. [20] inserted a PCM into a double-skin façade.

Another solution for improving the performance of passive systems is to integrate heat pipes in the wall structure. A study by Zhang et al. [21] proposed this type of passive solar energy utilization technology, where the heat transfer and energy-saving characteristics of the wall were analyzed theoretically and experimentally. In a similar manner, Tan et al. [22] investigated the heat transfer and energy-saving capabilities of wall-implanted heat pipes during the heating season.

The heat transfer performance of the proposed systems was validated by conducting the average equivalent heat transfer coefficient experiment. Heat pipe evaporation section length and diameter were also optimized to calculate the energy-saving potential. They reported that the proposed system wall, implanted with heat pipes, has remarkable energy-saving potential. Liu et al. [23] analyzed the possibility of improving the structure of heat pipes implanted in walls, for a better heat transfer performance.

Li et al. [24] and Zhang et al. [25] investigated the dynamic heat transfer performance of the wall implanted with heat pipes during heating season. According to their report, the suggested method can save 21.61% of energy, and the heat flow ratio to the inside of the wall is 4:1. In the same manner, the wall with heat pipes implanted was analyzed during the winter season by Sun et al. [26].

The performance of a proposed heat pipe-assisted solar wall was investigated by Michael et al. [27] through simulation and experiments. The heat pipe model was calculated using electrical resistances, and to simulate hourly performance, MatLab codes were used. The investigation of the heat pipe-assisted solar wall was continued by Brian et al. [28] who, based on the numerical and experimental results, proposed a full-scale model to analyze the performance during the heating season. Continuing the research on the subject, Brian et al. [29] proposed the use of the solar heat pipe system to reduce thermal gains in the cold season.

In this study, we evaluated a passive heating system that harnesses solar energy and converts it into thermal energy through the use of heat pipes inserted in walls. The novel aspect of this experiment is the utilization of concrete walls that have different recyclable materials added to their structure in various proportions.

2. Materials and Methods

2.1. Description of the Global System

The newly-designed passive system was built in four modules (Figure 1). Module one consists of a solar radiation simulator built using six lamps that are arranged in two columns with 3 lamps in each column. The second module consists of a wood shell with a height of 0.8 m, a width of 0.6 m, and a thickness of 0.15 m, called the air gap. A layer of high-density polystyrene is applied inside the shell. The air gap has a glass layer in front and a black painted steel plate at the back. The glass layer is 0.05 m thick and was placed to allow the solar radiation to pass through and reach the absorber plate.

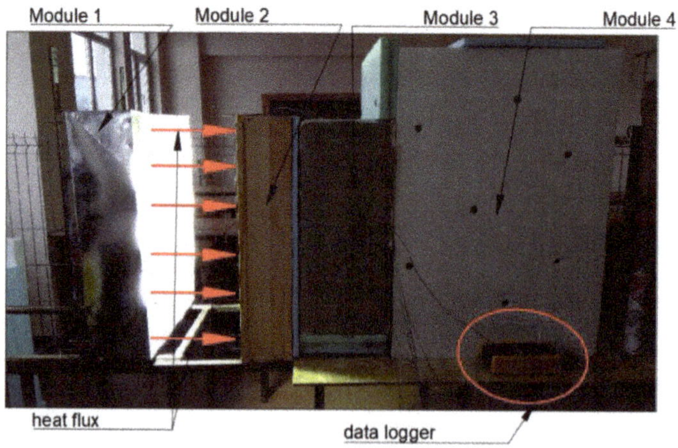

Figure 1. The experimental setup of the global system.

The third module consists of an insulated concrete wall equipped with heat pipes. The concrete block has a length of 0.6 m, a width of 0.4 m, and a thickness of 0.2 m. Four concrete samples were cast with standard aggregates and waste aggregates in the mixture: CW01—corresponds to the concrete control sample; CW02—corresponds to concrete with 10% fly ash replacement for cement and 20% chopped PET replacement for the sand mass; CW03—corresponds to concrete with 10% fly ash replacement for cement and 20% sawdust replacement for the sand mass; CW04—corresponds to concrete with 10% fly ash replacement for cement, 10% granular polystyrene with dimensions between 1 and 4 mm replacement for the sand mass and 20% granular polystyrene with dimensions between 4 and 8 mm replacement for the aggregate sort. The 4th module consists of the insulated room built from wood plates. This module is insulated using a polystyrene layer of 0.1 m at the exterior to prevent heat loss. The concrete blocks were built according to previous research [30], which analyzed the density, compressive strength, and thermal conductivity of several concrete samples mixed with various percentages of fly ash, chopped PET, sawdust, and granular polystyrene.

2.2. Experimental Approach to the Global System

The designed model was subjected to experimental investigations in the laboratory. The air temperature inside the laboratory measured during the experimental investigations was 15 °C. The solar radiation simulator was used to replicate the heat flux emitted by solar radiation. The heat fluxes proposed for the experimental investigations of the heat pipe-equipped concrete walls are 100 W/m^2, 150 W/m^2, 200 W/m^2, and 250 W/m^2.

The discharge of the heat captured by the heat pipe was measured on the external surface of the wall inside module 4 and in the center of module 4. For both systems, constant heat flux was applied to the glass side of the solar collector for 10 h, which was considered the charging time. At the end of the charging cycle, the solar module was turned off and the discharging cycle began, continuing for the next 14 h. Temperature measurements were made with type K thermocouples and recorded by a data logger on a memory card. There were 8 sensors on the surface of the wall, and 3 sensors in the middle of module 4, placed at different heights. The data logger recorded the temperatures every 60 s during the charging and discharging cycle. The placement of the sensors are presented in Figure 2.

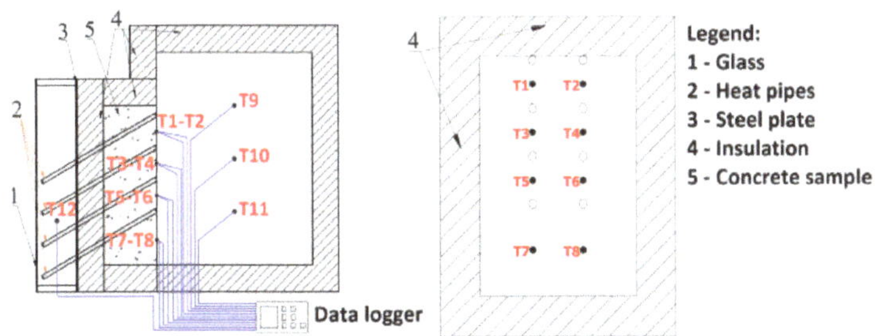

Figure 2. Experimental setup and thermocouples position.

2.3. Numerical Approach to the Global System

To choose an optimum design for the experimental stand, the global system was designed in 3D using Autodesk Inventor software and subjected to numerical simulations using Autodesk CFD software. The geometry of the global system presented in Figure 3 was reproduced at a real scale for the modules built in the experimental model.

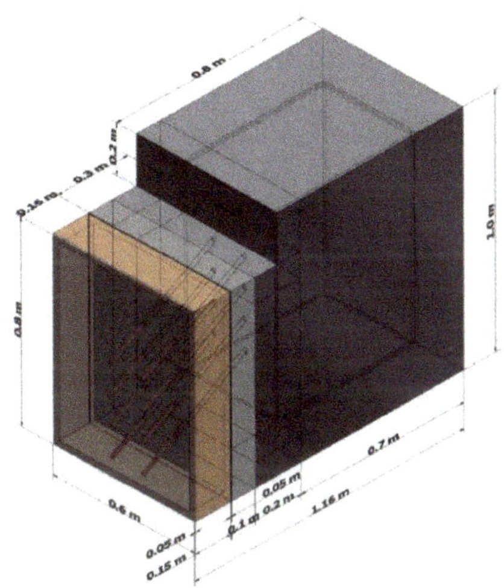

Figure 3. The 3D design of the global system.

The geometry was imported in Autodesk CFD, suppressing the components with no use in the global system to reduce the computational time required to reach the convergence. The simulations ran based on the governing equations of fluid dynamics or Navier-Stokes equations: the continuity, momentum, and energy equation coupled to the energy conservation equation. The software resolved the conjugate heat transfer where conduction, convection, and radiation heat transfer are coupled with fluid flow.

A constant temperature of 15 °C was applied to the exterior air volume, which reproduced the initial conditions of laboratory indoor air. Adiabatic conditions were imposed for the insulated components, leading to a heat flux of $\varphi = 0$ W/ m^2 applied on the polystyrene surfaces. Various boundary conditions of heat fluxes, from 100 to 250 W/m^2, were applied on the absorber plate and heat pipe evaporator surfaces. Due to the software's limitation with regard to simulating the heat pipe multiphase change when the evaporation-condensation process occurs, a new material with a very high and constant thermal conductivity was created and assigned to the heat pipe inner volume. From the software database, materials were assigned: glass for the glass; wood for the wood shell of the solar heat absorber and for the room; and copper for the heat pipes. The geometry was discretized in 300,000 elements using the mesh generator of the software. Based on automatic local curvature, the meshing of the solid and fluid zones was generated at a medium size while the contact between them was generated at a smaller size for better capture of the heat transfer. Assuming a laminar incompressible flow at hydrostatic pressure, the heat transfer mechanisms by radiation, convection, and conduction were proposed to be resolved during the simulation.

3. Results

3.1. Experimental Results

3.1.1. Charging Cycle

All 4 concrete samples were subjected to a 10 h charging cycle and a 14 h discharging cycle. The data logger recorded the temperatures for all 11 sensors every 60 s. Tables 1–4 present the average temperature at the surface of the concrete blocks recorded by the sensors at specific moments, after one hour, five hours, and at the end of the charging cycle (10 h).

Table 1. The average temperature on the surface of the wall for CW01.

Heat Flux	After 1 h	After 5 h	After 10 h
100 W/m^2	16.0 °C	20.1 °C	23.7 °C
150 W/m^2	16.1 °C	20.7 °C	24.5 °C
200 W/m^2	16.2 °C	21.2 °C	24.9 °C
250 W/m^2	16.2 °C	22.3 °C	27.2 °C

Table 2. The average temperature on the surface of the wall for CW02.

Heat Flux	After 1 h	After 5 h	After 10 h
100 W/m^2	15.9 °C	20.0 °C	23.3 °C
150 W/m^2	16.1 °C	21.3 °C	25.1 °C
200 W/m^2	16.7 °C	22.8 °C	26.9 °C
250 W/m^2	17.4 °C	23.8 °C	28.2 °C

Table 3. The average temperature on the surface of the wall for CW03.

Heat Flux	After 1 h	After 5 h	After 10 h
100 W/m^2	15.9 °C	19.9 °C	23.3 °C
150 W/m^2	16.0 °C	20.7 °C	24.0 °C
200 W/m^2	16.2 °C	21.5 °C	25.1 °C
250 W/m^2	16.3 °C	22.3 °C	26.1 °C

Table 4. The average temperature on the surface of the wall for CW04.

Heat Flux	After 1 h	After 5 h	After 10 h
100 W/m^2	16.2 °C	20.3 °C	23.3 °C
150 W/m^2	16.2 °C	20.5 °C	23.6 °C
200 W/m^2	16.2 °C	21.4 °C	24.3 °C
250 W/m^2	16.4 °C	22.1 °C	26.4°C

3.1.2. Discharging Cycle

The discharging cycle begins once the solar simulator is turned off, after 10 h. The cycle lasts for 14 h, thus completing a 24 h cycle. The following figures present the temperature variation of all 8 sensors placed on the external face of the concrete wall, for all of the samples tested. Figure 4 presents the variation of temperature recorded by the sensors for the control sample.

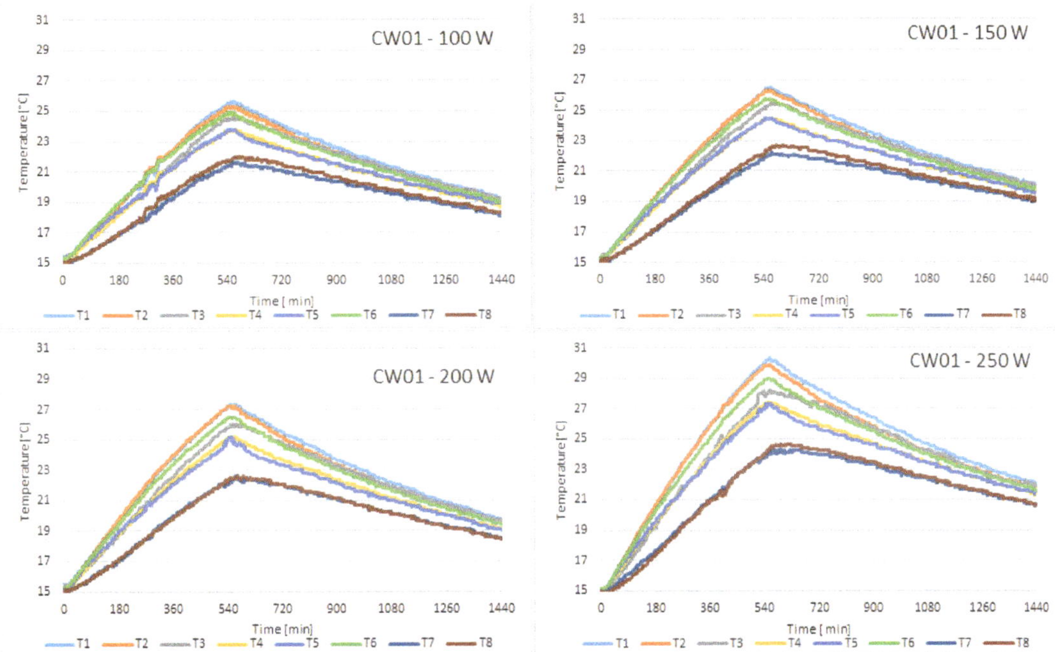

Figure 4. Temperature variation recorded by the sensors for concrete sample CW01.

In the first hour, the heat pipes increased the average temperature of concrete sample CW01 by 1 °C in all 4 tests. We can observe that when the heat flux is 250 W/m^2, the temperature rise is the fastest. The temperature peak inside the room is 24.7 °C at the end of the charging cycle.

Compared to CW01, the temperature on the face of CW02 sample is closer to uniformity on the 8 sensors. It can be observed in Figure 5 that the temperature difference between the T2 sensor which is placed in the upper zone and the T7 sensor which is placed in the lower zone of the concrete block is smaller for CW02 than for CW01. This means the thermal conductivity of this concrete with fly ash and chopped PET is superior to simple concrete.

Probe CW03 revealed the most uniform temperature distribution, but the peak temperatures were smaller than the temperatures recorded for CW01. With the highest heat flux, the temperature of sensor T1 barely reached 29 °C at the end of the charging cycle, as can be observed in Figure 6.

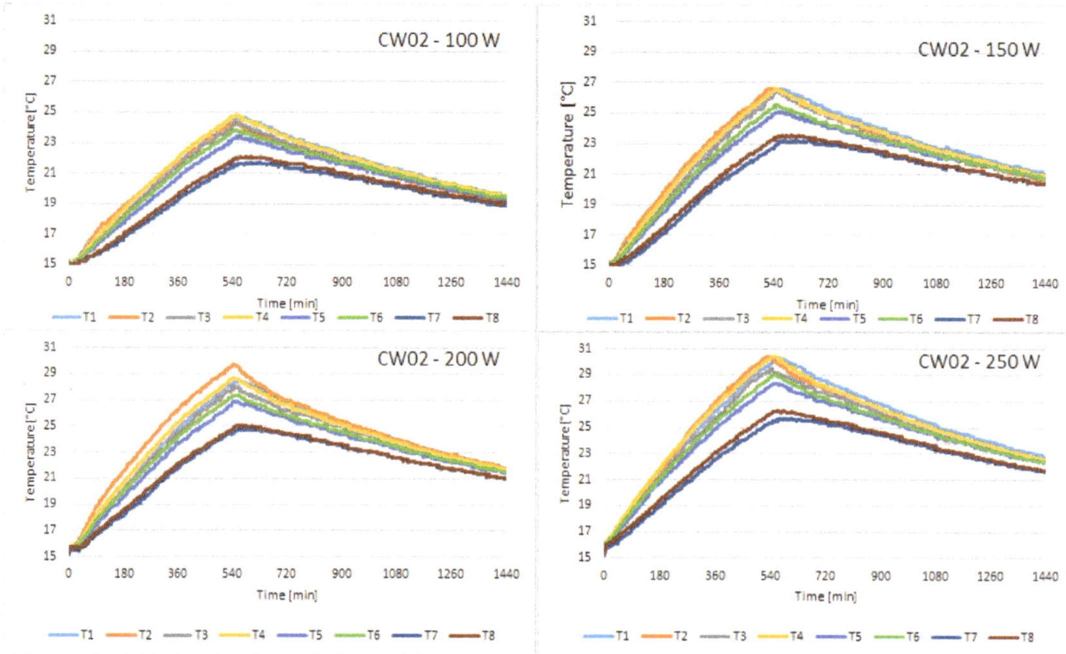

Figure 5. Temperature variation recorded by the sensors for concrete sample CW02.

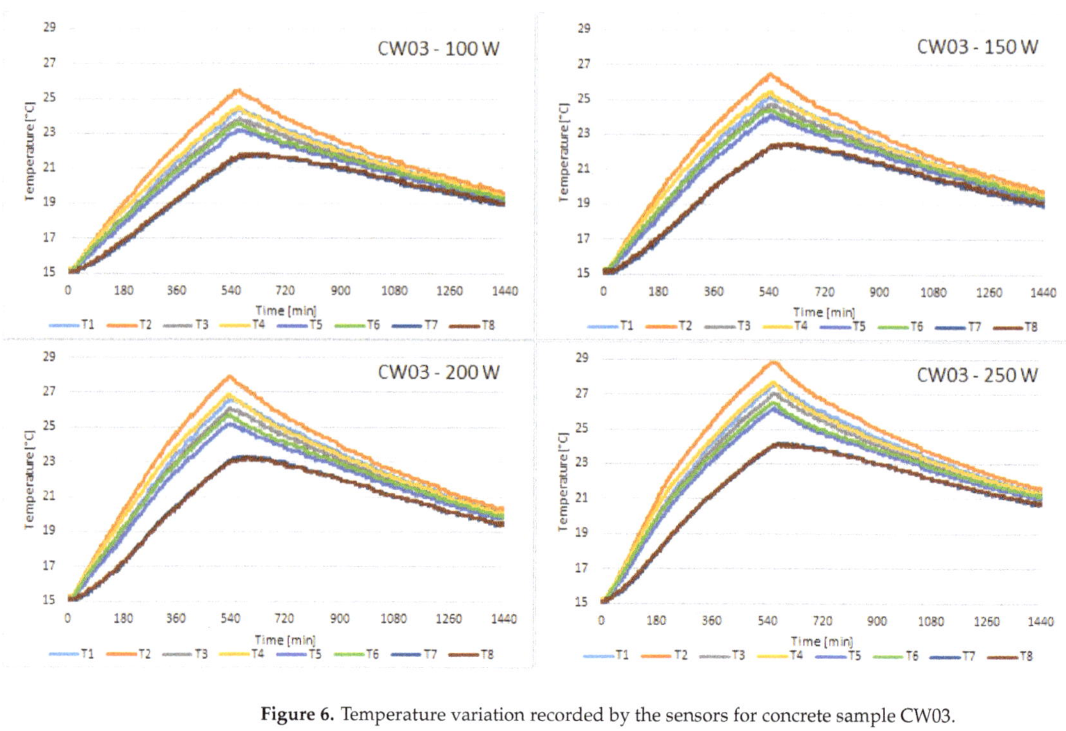

Figure 6. Temperature variation recorded by the sensors for concrete sample CW03.

The temperatures recorded for probe CW04 were very similar to those recorded for probe CW03. Figure 7 presents the variation of temperature recorded by the sensors.

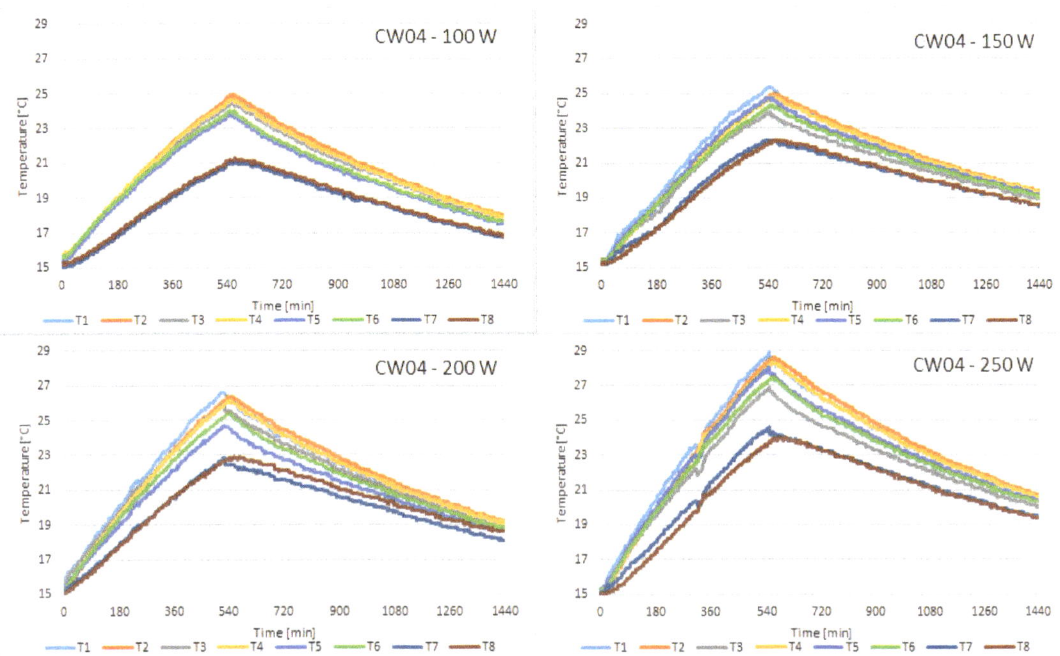

Figure 7. Temperature variation recorded by the sensors for concrete sample CW04.

With the temperature values recorded by the sensor, the conductive heat transfer was calculated at the end of the charging cycle and at the end of the discharging cycle using Equation (1), constituting the heat gain of the wall.

$$Q = \frac{k \times A \times \Delta T}{d} \, [W] \quad (1)$$

When the imposed heat flux is at 100 W/m^2, all 4 concrete samples behave similarly; however, when the heat flux is increased we can observe that CW02 has the highest heat gain (Figure 8).

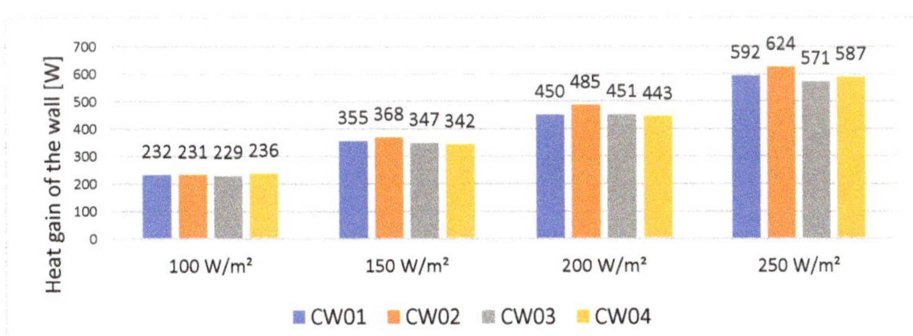

Figure 8. Heat gain of the wall.

Considering that the area of the wall is 0.24 m^2, the heat flux through the wall is presented in Table 5.

Table 5. Heat flux through the wall.

Heat Flux	CW01	CW02	CW03	CW04
100 W/m^2	1174.2 W/m^2	1143.9 W/m^2	1139.8 W/m^2	1149.6 W/m^2
150 W/m^2	1774.1 W/m^2	1818.2 W/m^2	1736.5 W/m^2	1715.1 W/m^2
200 W/m^2	2233.3 W/m^2	2405.3 W/m^2	2250.0 W/m^2	2200.0 W/m^2
250 W/m^2	2971.8 W/m^2	3106.1 W/m^2	2856.3 W/m^2	2925.0 W/m^2

3.1.3. Infrared Camera Measurements

At the end of the charging cycle, each concrete block was photographed using a FLIR thermal camera. The sensors of the camera were set on the temperature thermocouples to compare the values recorded by the sensors to the values recorded by the infrared camera. Photos of the CW02 concrete block were chosen as an example since they represent the highest temperatures reached by a concrete block during this experiment. The results can be observed in Figure 9.

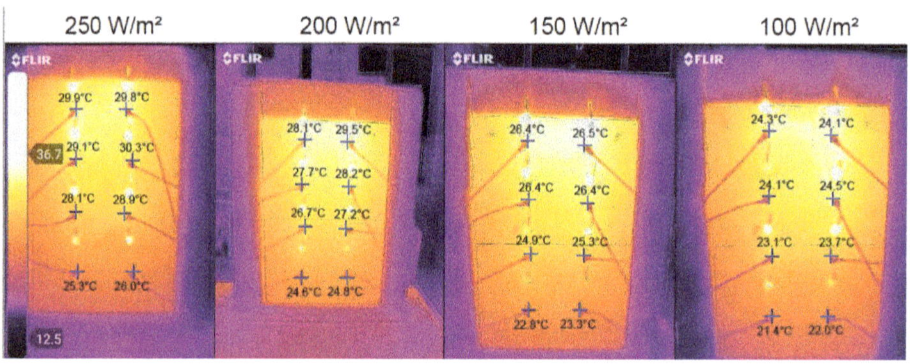

Figure 9. Infrared camera photos for CW02.

Table 6 presents the comparison between the temperatures recorded by the sensors and the temperatures recorded by the infrared camera. Since the accuracy of the camera is ±5 °C, the results were slightly smaller than the experimental results, by 0.1 to 0.3 °C, which are within the admissible measurement error range. The rest of the experiments had similar results.

Table 6. Temperature comparison between experimental and infrared camera results for CW02.

		T1	T2	T3	T4	T5	T6	T7	T8
Sensors	250 W/m^2	30.2 °C	30.0 °C	29.2 °C	30.4 °C	28.3 °C	29.0 °C	25.4 °C	26.2 °C
IR		29.9 °C	29.8 °C	29.1 °C	30.3 °C	28.1 °C	28.9 °C	25.3 °C	26.0 °C
Sensors	200 W/m^2	28.4 °C	29.6 °C	27.9 °C	28.5 °C	26.8 °C	27.3 °C	24.6 °C	24.9 °C
IR		28.1 °C	29.5 °C	27.7 °C	28.2 °C	26.7 °C	27.2 °C	24.6 °C	24.8 °C
Sensors	150 W/m^2	26.6 °C	26.5 °C	26.5 °C	26.6 °C	25.0 °C	25.5 °C	23.0 °C	23.4 °C
IR		26.4 °C	26.5 °C	26.4 °C	26.4 °C	24.9 °C	25.3 °C	22.8 °C	23.3 °C
Sensors	100 W/m^2	24.6 °C	24.2 °C	24.3 °C	24.8 °C	23.3 °C	23.8 °C	21.5 °C	22.0 °C
IR		24.3 °C	24.1 °C	24.1 °C	24.5 °C	23.1 °C	23.7 °C	21.4 °C	22.0 °C

3.2. Results of the Numerical Simulations

The numerical simulations were focused on the charging cycle. The global system geometry was designed in Autodesk Inventor, reproducing the components built in the experimental model at a real scale. In order to speed up the convergence process, the geometry was loaded into Autodesk CFD while the thermal transfer-related components were suppressed. Around the global system, an air volume of 4.9 m^3 was generated to reproduce the laboratory conditions. Using the automatic mesh generator, the geometry was discretized, resulting in 300,000 elements. Figure 10a,b present the geometry imported in the CFD Simulation environment, and the meshing applied to the 3D model, respectively. Based on automatic local curvature, the meshing of the solid and fluid zones was generated at a medium size while the contact between them was generated at a smaller size to better capture the heat transfer. Assuming a laminar incompressible flow at hydrostatic pressure, the heat transfer mechanisms by radiation, convection, and conduction were proposed to be resolved during the simulation.

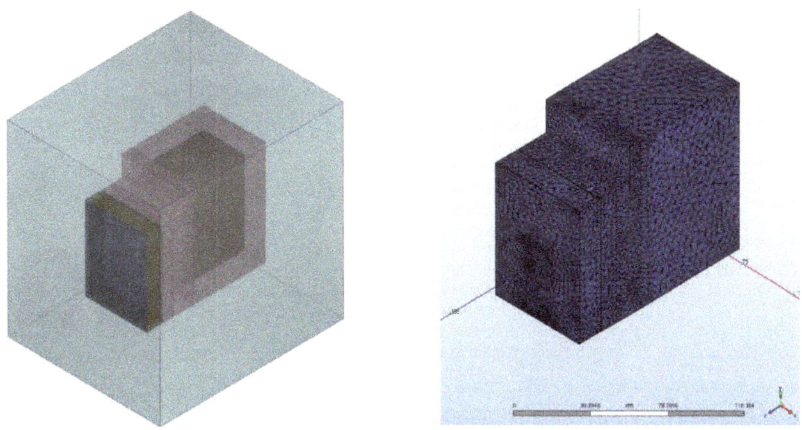

(a) (b)

Figure 10. The global system imported in the simulation environment: (**a**) the global system inside the air volume; (**b**) meshing of the global system.

All four concrete blocks were created in the CFD simulation environment and were assigned a custom material with their specific thermal properties. To see the influence of heat pipes inserted inside the concrete block, the first 3D model was created without heat pipes inserted and was subjected to a heat flux of 250 W/m^2. It was named CW00. The comparison is presented in Table 7. The influence of heat pipes can be observed for every block.

Table 7. The average temperatures resulted from numerical simulations.

	CW00	CW01	CW02	CW03	CW04
Wall temp.	21.3 °C	26.8 °C	30.3 °C	24.9 °C	29.7 °C
Room temp.	19.6 °C	27.2 °C	30.6 °C	25.3 °C	29.9 °C

All 4 concrete blocks were subjected to numerical simulations. Figure 11 presents a section of the global system using the CW02 concrete module, where how heat is transferred from the frontal part of the system to the room using the heat pipes inserted in the wall can be observed. The right image presents the face of the wall inside the room at the end of the charging cycle. A photograph taken with the infrared camera for CW03 with an applied heat flux of 250 W/m^2 was compared to an image from the simulations, and the results were very similar. The comparison is presented in Figure 12.

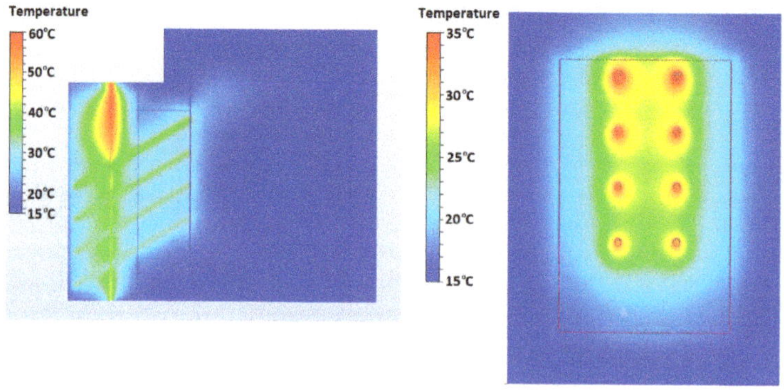

Figure 11. Numerical simulation for CW02—100 W/m^2.

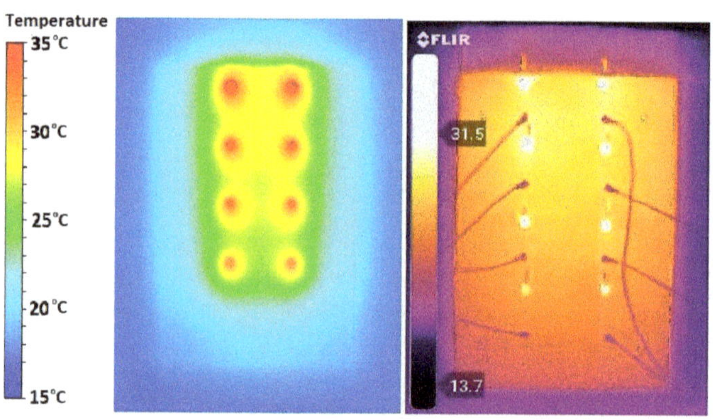

Figure 12. Comparison between simulation and infrared camera for CW03—250 W/m^2.

The same boundary conditions used in the tests were applied to numerical simulations of each global system containing the four different types of concrete block. The average temperature of the external face of the wall was determined. The results are presented in Table 8.

Table 8. Results of the numerical simulations.

	100 W/m^2	150 W/m^2	200 W/m^2	250 W/m^2
CW01	24.4 °C	25.9 °C	26.4 °C	28.2 °C
CW02	24.4 °C	26.8 °C	29.1 °C	30.3 °C
CW03	24.2 °C	25.6 °C	26.6 °C	27.8 °C
CW04	24.5 °C	25.3 °C	25.2 °C	27.7 °C

4. Discussion

All 4 concrete samples showed promising results. Sample CW02, a mixture of concrete, fly ash, and chopped PET, registered the highest temperature in the experiments at the end of the charging cycle, at 28.2 °C. The numerical simulations confirmed this temperature peak.

Compared to the control sample CW01, sample CW03 behaved very similarly during the charging cycle. For all of the heat fluxes studied in the experiments, the temperature difference between the time when the lamps were switched off and the time when the discharging cycle ended was marginally smaller in CW03. That means the concrete sample with fly ash and sawdust is better at storing heat than the control sample.

Temperature differences between the concrete sample CW04 with fly ash and granular polystyrene, and CW03, were from 0.2 to 0.6 °C. However, their behavior was different during the discharging phase. CW04 had the smallest heat storage capacity, with an average temperature change of 6 °C between hours 10 and 24. A more detailed view of the temperature drop during the discharging phase is presented in Table 9. These numbers were calculated using the difference in average temperature between minute 600 and minute 1440 from the experiments.

Table 9. The temperature drop during the discharging phase.

	100 W/m^2	150 W/m^2	200 W/m^2	250 W/m^2
CW01	5.0 °C	5.0 °C	6.1 °C	6.0 °C
CW02	4.3 °C	4.7 °C	5.8 °C	6.4 °C
CW03	4.1 °C	5.0 °C	5.6 °C	5.2 °C
CW04	6.1 °C	5.0 °C	6.1 °C	6.8 °C

The results recorded by the temperature sensors were confirmed by thermal imaging photos and also by numerical simulations. The temperatures resulting from the numerical simulations were slightly higher than those of the experiments, with values between 1 and 3 °C. This can be explained by the fact that the software does not take into account convective heat losses.

5. Conclusions

The main goal of the research was to convert solar energy from bright building façades in an efficient way, putting forth a conversion-accumulation-solar energy delivery system based on the construction of an inventive solar panel with heat pipes and a heat storage wall for creating passive buildings. This system converts the solar energy from the sunny façades into thermal energy, which is then transferred by the integrated heat pipes in a large element with high thermal inertia. The thermal energy is saved during the day, and is released at night by using insulated shutters to maintain a comfortable temperature in the room.

Four concrete samples were cast with standard aggregates and waste aggregates in the mixture: CW01—corresponds to the concrete control sample; CW02—corresponds to concrete with fly ash and chopped PET; CW03—corresponds to concrete with fly ash and sawdust; CW04—corresponds to concrete with fly ash and granular polystyrene. These were integrated into a solar recovery system. Each global system was subjected to four heat fluxes of 100 W/m^2, 150 W/m^2, 200 W/m^2, and 250 W/m^2. The results were recorded by temperature sensors, and were confirmed by thermal imaging photos and also by numerical simulations.

CW02 had the most efficient mixture, resulting in an increased efficiency of 3.3% higher on average at the end of the charging cycle than the efficiency of CW01, the control sample. CW03 was the most efficient during the discharging cycle, losing on average 5.0 compared with 5.5 °C, the average temperature drop of CW01.

The original idea of the passive, eco-friendly system used to increase the energy efficiency of buildings may be quickly and cheaply installed in modules. The system can be used in climate zones where moderate solar heat flux is present throughout the entire year since it can be applied to all buildings, old and new, meeting heat demand even with low heat fluxes applied. The heat pipe system, in addition to improving a building's energy efficiency, has a major economic impact by lowering overall heating costs and a large environmental impact by lowering CO_2 emissions. Also, recycling materials are included in the concrete mixture, which benefits the environment.

The final conclusion of this study is that the suggested global system is a practical and effective way to recover solar energy from sunny building façades in order to create passive structures, even in situations with low solar radiation intensities of 100 to 250 W/m^2.

Author Contributions: Conceptualization, A.B., G.S., C.A. and M.B. (Marinela Barbuta); methodology, A.B., G.S., C.A., M.B. (Marinela Barbuta) and R.S.V.; software, G.S., R.S.V. and M.B. (Marius Branoaea); validation, A.B., C.A., M.B. (Marinela Barbuta), M.V. and R.S.V.; formal analysis, A.B., G.S., C.A., M.B. (Marinela Barbuta) and R.S.V.; investigation, A.B., G.S., C.A., M.B. (Marinela Barbuta), R.S.V. and M.B. (Marius Branoaea); resources, A.B., C.A., M.B. (Marinela Barbuta) and M.V.; data curation, A.B., G.S., C.A., M.B. (Marinela Barbuta), R.S.V. and M.B. (Marius Branoaea); writing—original draft preparation, G.S., R.S.V. and M.B. (Marius Branoaea); writing—review and editing, A.B., C.A., M.B. (Marinela Barbuta) and M.V.; visualization, A.B., C.A., M.B. (Marinela Barbuta), M.V. and R.S.V.; supervision, A.B., C.A., M.B. (Marinela Barbuta) and M.V.; project administration, A.B., G.S., C.A., M.B. (Marinela Barbuta) and M.V.; funding acquisition, A.B. and M.V. All authors have read and agreed to the published version of the manuscript.

Funding: This work was supported by a grant from the Romanian Ministry of Education and Research, CCCDI-UEFISCDI, project number PN-III-P2-2.1-PED-2019-3112, within PNCDI III.

Institutional Review Board Statement: Not applicable.

Informed Consent Statement: Not applicable.

Data Availability Statement: Not applicable.

Acknowledgments: This work was supported by a grant from the Romanian Ministry of Education and Research, CCCDI-UEFISCDI, project number PN-III-P2-2.1-PED-2019-3112, within PNCDI III. Our thanks to the team involved in the research investigations in the PhD thesis "Contribution to the study of a heat pipe system to improve the energy performance of buildings" by Gavril Sosoi during doctoral program at "Gheorghe Asachi" Technical University of Isai, Doctoral School of Civil Engineering and Building Services and Université d'Aix-Marseille, Ecole Doctorale Science Pour l'Ingénieur: Mécanique, Physique, Micro et Nanoélectronique, Institut Universitaire des Systèmes Thermique Industriels/UMR 7343.

Conflicts of Interest: The authors declare no conflict of interest.

Nomenclature

CW01	Concrete control sample
CW02	Concrete with fly ash and chopped PET
CW03	Concrete with fly ash and sawdust
CW04	Concrete with fly ash and granular polystyrene
φ	Heat flux [W/m^2]
T_x	Thermocouple sensors
Q	Heat gain of the wall [W]
k	Thermal conductivity [W/mK]
A	Area [m^2]
ΔT	Temperature difference [°C]
d	Width [m]

References

1. Djedjig, R.; Belarbib, R.; Bozonnetba, E. Experimental study of green walls impacts on buildings in summer and winter under an oceanic climate. *Energy Build.* **2017**, *150*, 403–411. [CrossRef]
2. Mazzali, U.; Peron, F.; Romagnoni, P.; Pulselli, R.M.; Bastianoni, S. Experimental investigation on the energy performance of Living Walls in a temperate climate. *Build. Environ.* **2013**, *64*, 57–66. [CrossRef]
3. Wong, I.; Baldwin, A.N. Investigating the potential of applying vertical green walls to high-rise residential buildings for energy-saving in sub-tropical region. *Build. Environ.* **2016**, *97*, 34–39. [CrossRef]
4. Pérez, G.; Rincón, L.; Vila, A.; González, J.M.; Cabeza, L.F. Green vertical systems for buildings as passive systems for energy savings. *Appl. Energy* **2011**, *88*, 4854–4859. [CrossRef]
5. Louis, S.; Lee, H.; Jim, C.Y. Energy benefits of green-wall shading based on novel-accurate apportionment of short-wave radiation components. *Appl. Energy* **2019**, *238*, 1506–1518.

6. Özbalta, T.G.; Kartal, S. Heat gain through Trombe wall using solar energy in a cold region of Turkey. *Sci. Res. Essays* **2010**, *5*, 2768–2778.
7. Bellos, E.; Tzivanidis, C.; Zisopoulou, E.; Mitsopoulos, G.; Antonopoulos, K.A. An innovative Trombe wall as a passive heating system for a building in Athens—A comparison with the conventional Trombe wall and the insulated wall. *Energy Build.* **2016**, *133*, 754–769. [CrossRef]
8. Rabani, M.; Kalantar, V.; Dehghan, A.A.; Faghih, A.K. Experimental study of the heating performance of a Trombe wall with a new design. *Sol. Energy* **2015**, *118*, 359–374. [CrossRef]
9. Briga-Sá, A.; Boaventura-Cunhab, J.; Lanzinhaa, J.-C.; Paivaa, A. Experimental and analytical approach on the Trombe wall thermal performance parameters characterization. *Energy Build.* **2017**, *150*, 262–280. [CrossRef]
10. Dong, J.; Chen, Z.; Zhang, L.; Cheng, Y.; Sun, S.; Jie, J. Experimental investigation on the heating performance of a novel designed trombe wall. *Energy* **2019**, *168*, 728–736. [CrossRef]
11. Faraji, M. Numerical study of the thermal behavior of a novel Composite PCM/concrete wall. *Energy Procedia* **2017**, *139*, 105–110. [CrossRef]
12. Kamal, E.O.; Yves, L.G.; Pascal, B. Analysis of micro-dispersed PCM-composite boards behavior in a building's wall for different seasons. *J. Build. Eng.* **2016**, *7*, 361–371.
13. Meng, E.; Hang, Y.; Bo, Z. Study of the thermal behavior of the composite phase change material (PCM) room in summer and winter. *Appl. Therm. Eng.* **2017**, *126*, 212–225. [CrossRef]
14. Xie, J.; Wang, W.; Liu, J.; Pan, S. Thermal performance analysis of PCM wallboards for building application based on numerical simulation. *Sol. Energy* **2018**, *162*, 533–540. [CrossRef]
15. Chen, C.; Ling, H.; Zhai, Z.J.; Li, Y.; Yang, F.; Han, F.; Wei, S. Thermal performance of an active-passive ventilation wall with phase change material in solar greenhouses. *Appl. Energy* **2018**, *216*, 602–612. [CrossRef]
16. Alessandro, D.; Diego, A.; Olena, K.L. Naturally ventilated double-skin façade in modeling and experiments. *Energy Build.* **2017**, *144*, 17–29.
17. Souza, L.C.O.; Souza, H.A.; Rodrigues, E.F. Experimental and numerical analysis of a naturally ventilated double-skin façade. *Energy Build.* **2018**, *165*, 328–339. [CrossRef]
18. Zhang, T.; Yang, H. Determination of the optimal thickness of vertical air channels in double-skin solar façades. *Energy Procedia* **2019**, *158*, 1255–1260. [CrossRef]
19. Luo, Y.; Zhang, L.; Liu, Z.; Xie, L.; Wang, X.; Wu, J. Experimental study and performance evaluation of a PV-blind embedded double skin façade in winter season. *Energy* **2018**, *165*, 326–342. [CrossRef]
20. Li, Y.; Darkwa, J.; Su, W. Investigation on Thermal Performances of an Integrated Phase Change Material Blind System for Double Skin Façade Buildings. *Energy Procedia* **2019**, *158*, 5116–5123. [CrossRef]
21. Zhang, Z.; Sun, Z.; Duan, C. A new type of passive solar energy utilization technology—The wall implanted with heat pipes. *Energy Build.* **2014**, *84*, 111–116. [CrossRef]
22. Tan, R.; Zhang, Z. Heat pipe structure on heat transfer and energy saving performance of the wall implanted with heat pipes during the heating season. *Appl. Therm. Eng.* **2016**, *102*, 633–640. [CrossRef]
23. Liu, C.; Zhang, Z.; Shi, Y.; Ding, Y. Optimisation of a wall implanted with heat pipes and applicability analysis in areas without district heating. *Appl. Therm. Eng.* **2019**, *151*, 486–494. [CrossRef]
24. Li, Z.; Zhang, Z. Dynamic heat transfer characteristics of wall implanted with heat pipes in summer. *Energy Build.* **2018**, *170*, 40–46. [CrossRef]
25. Zhang, Z.; Ding, Y.; Guo, C. Dynamic heat transfer performance of wall implanted with heat pipes and its energy saving characteristics during the heating season. *Energy Procedia* **2019**, *158*, 1155–1160. [CrossRef]
26. Sun, Z.; Zhang, Z.; Duan, C. The applicability of the wall implanted with heat pipesin winter of China. *Energy Build.* **2015**, *104*, 36–46. [CrossRef]
27. Michael, V.A.; Brian, S.R.; Ellen, G.B.; Sharp, M.K. Simulated and experimental performance of a heat pipe assisted solar wall. *Sol. Energy* **2012**, *86*, 1552–1562.
28. Brian, S.R.; Nicholas, E.C.; Andrea, K.-K.; Ellen, G.B.; Sharp, M.K. Heating season performance of a full-scale heat pipe assisted solar wall. *Sol. Energy* **2013**, *87*, 76–83.
29. Robinson, B.S.; Sharp, M.K. Reducing unwanted thermal gains during the cooling season for a solar heat pipe system. *Sol. Energy* **2015**, *115*, 16–32. [CrossRef]
30. Sosoi, G.; Abid, C.; Barbuta, M.; Burlacu, A.; Balan, M.C.; Branoaea, M.; Vizitiu, R.S.; Rigollet, F. Experimental Investigation on Mechanical and Thermal Properties of Concrete Using Waste Materials as an Aggregate Substitution. *Materials* **2022**, *15*, 1728. [CrossRef]

materials

MDPI

Article

Evaluation of the Fatigue Behaviour and Failure Mechanisms of 52100 Steel Coated with WIP-C1 (Ni/CrC) by Cold Spray

Viorel Goanta [1], Corneliu Munteanu [1,2,*], Sinan Müftü [3,*], Bogdan Istrate [1,*], Patricia Schwartz [4], Samuel Boese [3,4], Gehn Ferguson [5] and Ciprian Ionut Morăraș [1]

1 Mechanical Engineering, Mechatronics and Robotics Department, Mechanical Engineering Faculty, "Gheorghe Asachi" Technical University of Iasi, 700050 Iasi, Romania; viorel.goanta@academic.tuiasi.ro (V.G.); ciprian-ionut.moraras@academic.tuiasi.ro (C.I.M.)
2 Technical Sciences Academy of Romania, 26 Dacia Blvd., 030167 Bucharest, Romania
3 Department of Mechanical and Industrial Engineering, Northeastern University, Boston, MA 02115, USA; s.boese@northeastern.edu
4 Kostas Research Institute, Northeastern University, Burlington, MA 01803, USA; t.schwartz@northeastern.edu
5 Army Research Laboratories, Aberdeen Proving Ground, Maryland, MD 21005, USA; gehn.ferguson@jhuapl.edu
* Correspondence: corneliu.munteanu@academic.tuiasi.ro (C.M.); s.muftu@northeastern.edu (S.M.); bogdan.istrate@academic.tuiasi.ro (B.I.)

Abstract: Cold spray technique has been major improved in the last decades, for studying new properties for metals and alloys of aluminum, copper, nickel, and titanium, as well as steels, stainless steel and other types of alloys. Cold sprayed Ni/CrC coatings have the potential to provide a barrier as well as improved protection to steels. Fatigue characteristics of 52100 steel coated with Ni/Chrome-Carbide (Ni/CrC) powder mixture by using cold gas dynamic spray are investigated. Fatigue samples were subjected to symmetrically alternating, axially applied cyclic fatigue loading until failure. The test was stopped if a sample survived more than 5×10^6 cycles at the applied stress. Fracture surfaces for each sample were examined to investigate the behaviour of the coating both at high stress levels and at a high number of stress cycles. Scanning electron microscopy was used to assess the damage in the interface of the two materials. Good fatigue behaviour of the coating material was observed, especially at low stresses and a high number of cycles. Details of the crack initiation region, the stable crack propagation region and the sudden crack expansion region are identified for each sample. In most of the samples, the initiation of the crack occurred on the surface of the base material and propagated into the coating material. The possible effects of coatings on the initial deterioration of the base material and the reduction of the lifespan of the coated system were also investigated. The aim of the paper was to study the interface between the base material and the coating material at the fatigue analysis for different stresses, highlighting the appearance of cracks and the number of breaking cycles required for each sample.

Keywords: coating; fatigue; microscopic observations; fissure; crack initiation; fatigue limit

check for updates

Citation: Goanta, V.; Munteanu, C.; Müftü, S.; Istrate, B.; Schwartz, P.; Boese, S.; Ferguson, G.; Morăraș, C.I. Evaluation of the Fatigue Behaviour and Failure Mechanisms of 52100 Steel Coated with WIP-C1 (Ni/CrC) by Cold Spray. *Materials* **2022**, *15*, 3609. https://doi.org/10.3390/ma15103609

Academic Editor: Yurii Sharkeev

Received: 22 April 2022
Accepted: 16 May 2022
Published: 18 May 2022

1. Introduction

Coatings are used to enhance surface properties or protect the base material from degradation caused by the interaction with the external environment such as corrosion, attack of various chemicals, erosion, effects of high temperature or other external aggressions such as the high-velocity impact of particles/components. The researchers must also consider the compatibility between the base material and coating material. In addition to many other properties that need to be considered, obtaining the mechanical properties of the coating itself must also be considered. Depending on the chemical composition and coating process used, the coating and base materials may differ from each other in terms of physico-mechanical properties [1]. Major differences between such properties of coating materials and base materials include thermal expansion, specific deformation, yield limit, fatigue behaviour and creep [2,3]. Important differences in properties can cause damage to the interface between the coating and the base material and failure of the structure of the coating material. If significant differences occur between the properties of the two materials, it can lead to a significant reduction in the lifespan of the assembly, basic material—coating material. As a result, the mechanical properties of the coating systems, as well as the prevailing mechanisms of damage will be determined by the critical property relevant to the entire assembly, basic material—the coating material. In case the thickness of the coating is much lower than that of the base material, the critical property is the difference between the specific deformations of the two materials for dynamic loading [4]. Surface temperature variations may result in changes in the mechanical and physical properties of the coating. In addition, micro-defects due to effects such as stress corrosion cracking can also occur. It is also possible for the coating process, depending on the method used, to damage the surface of the base material and/or to introduce local residual stress that is of the same sign as those arising from the mechanical loading [5–9]. This may cause premature destruction of a coated component. It is well known that surface quality plays an important role in the fatigue life of a component subjected to dynamic loading. A coating may have a beneficial role in terms of protection against environmental effects, but other operating conditions such as the fatigue life of the component must not be adversely affected. Any coating system that improves certain properties relative to the uncoated material, may shorten the fatigue life of a component due to cracks that begin in the coating and propagate in the substrate material [5,6]. In this context, research has focused on investigating the effect of coatings on the fatigue behavior of the resulting components [10–16]. This work showed different opinions regarding the mechanical characteristics, especially the fatigue analysis, following the coating process [17]. It is reasonable to expect that the change of mechanical properties depends on a number of factors that need to be investigated separately for each case.

Cold spray is a particle deposition technology in which micron sized particles are accelerated through a converging-diverging de Laval nozzle to velocities that can reach supersonic velocities [18]. Upon impact, the particles undergo severe plastic deformation and adhere to the substrate through a combination of metallic bonding and mechanical interlocking. The particles gain momentum and some heat from the hot gas flowing through the nozzle but remain below the melting temperature of the material. Thus, most of the material properties of the powder are preserved in the coating, and cold spray can be considered a solid-state material deposition technology. A wide range of ductile metal powders has been shown to work well with this technology when deposited on ductile metal surfaces. Ceramic powders, which tend to be good insulators and have good wear resistance, can be successfully deposited by using a ductile material as the matrix [19]. Cold spray metal matrix composite coatings have even been used as a coating to improve the fatigue life of additive manufactured components [20].

The present work aims to assess the effect of coating 52100 steel with a 0.5 mm thick Ni/CrC mixture by using cold spray technology on the fatigue life of the combined material system. To this end alternating—symmetrical axial stresses are applied on coated, cylindrical specimens. This subjects the sample to compression and tensile cycles. The tensile side of the cycle is useful to study the endurance of the coating. The cylindrical shape ensures that the stresses introduced into the sample are uniform throughout the cross-section of the sample. Thus, any defects that will lead to the onset of fatigue can be highlighted. The mechanical behaviour of the coated material system, the coating, the interface area, the crack initiation area and the regions of stable or unstable crack propagation are investigated. The paper also compared the fatigue resistance of 52100 steel given in reference [21] and the behaviour of the coating assembly presented in this paper, respectively, coating 52100 steel with a Ni/CrC mixture by using the cold spray technology.

2. Materials and Methods

The morphological aspect of the surfaces was determined using the SEM Quanta 200 3D (FEI, Brno, Czech Republic) electron microscope with LFD (Large Field Detector), a High Voltage of 20 kV, a spot size of 5 and a working distance of 15 mm. The X-ray diffractions (XRD, Panalytical, Almelo, The Netherlands) were performed using an Xpert PRO MPD 3060 facility from Panalytical (Almelo, The Netherlands), with a Cu X-ray tube ($K\alpha$ = 1.54051 Å), 2 theta: 10°–90°, step size: 0.13, time/step: 51 s, and a scan speed of 0.065651°/s.

Figure 1 presents the diffractograms for the base and coating material. Knowing the chemical composition of the powder [20], the predominant phase of NiCr (cubic structure -ICDD 96-901-2971) was obtained at a 2 Theta angle of approximately 44° and the secondary phase of Cr_3C_2 (cubic structure-ICDD 96-901-1599) at a 2 Theta angle of approximately 53° and 77°. In the case of the basic material, the predominant phase is Fe (cubic structure -ICDD 96-901-6602), around the 2 Theta angle of 45°, and the secondary phases contain Fe and Cr3C2 at the 2 Theta angles of 65° and 83°.

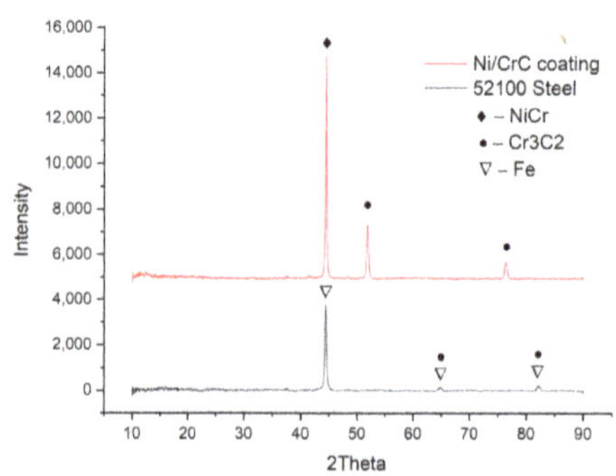

Figure 1. XRD pattern of coated sample and the base material.

The dimensions of the cylindrical specimens for testing the fatigue characteristics of the coated 52100-steel were determined according to the ASTM E4666-15 standard (Figure 2a). The specimens were coated with a mixture of Chrome-Carbide and Ni powder by using a VRC, Gen-III cold spray machine (VRC Metal Systems, LLC, SD, USA). The parameters used for spraying the powder mixture are given in Table 1. The powder mixture is commercially known as WIP-C1 [22]. Before the surface is coated with WIP-C1 powder a thin layer of bond coat that consists of WIP-BC1 powder was applied on the surface with a 60° nozzle orientation. This operation has multiple effects to shot-peen, clean and roughed the surface, in addition to creating a thin layer of WIP-BC1 coating. Additionally, the bond layer improves the adhesion of the WIP-C1 layer. The cylindrical specimens were mounted on a lathe and the WIP-C1 coating was applied by moving the nozzle axially, while holding it normal to the specimen axis. The spray conditions used to deposit the coatings are given in Table 1 and in references [20,23].

A standard tensile test (ASTM E8/E8M-16) was used to determine the material response of the coated system. This test was carried out on three coated samples, to ensure that the material system used in cyclic fatigue tests is well characterized. The specimen dimensions and a coated specimen are shown in Figure 2a,b. A typical stress-strain curve of the coated system is shown in Figure 3. The initial yield stress of the material before the application of cyclic loading is determined from these tests among other material characteristics. These include the stress level at which compatibility issues between the strains of the base and coating material become significant. To this end, the surface of the coated samples was monitored with a video camera during the test to look for the initiation of possible damage in the form of surface cracks.

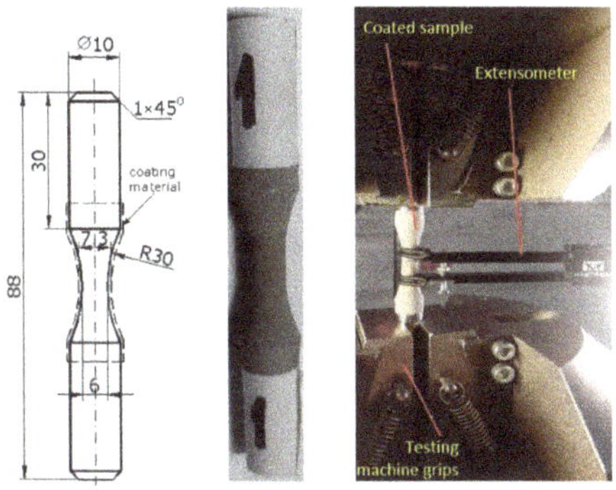

(**a**) (**b**) (**c**)

Figure 2. (**a**) The shape and dimensions of the sample (in mm). (**b**) Coated sample. (**c**) Attaching the sample to the testing machine.

Table 1. Cold spray deposition parameters used to prepare the fatigue specimens.

Parameter	Value
Gas	Nitrogen
Pressure	6.2 MPa (900 psi)
Temperature	675 °C
Nozzle ID	WC NNZL0060
Nozzle throat size	2 mm
Powder feeder speed	8 rpm
Powder feeder gas glow	105 slm
Standoff distance	25 mm
Spray angle	90°
Nozzle traverse speed	250 mm/s
Nozzle step distance	0.5 mm
Layer thickness	0.127 mm
Target coating thickness	0.508 mm
Powder	WIP-C1
Bond coat	WIP-BC1 and 60°

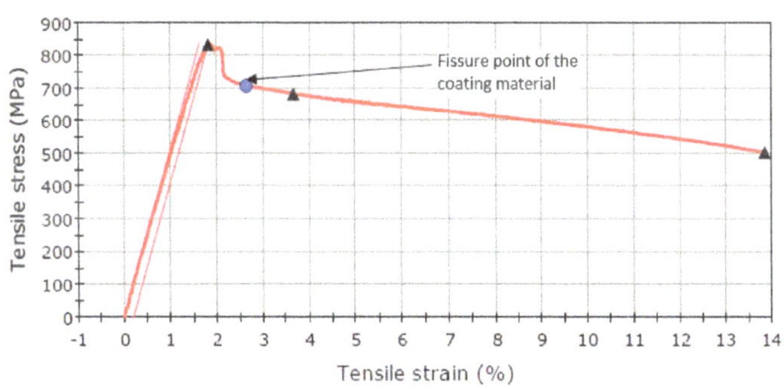

Figure 3. Characteristic stress-strain curve, specific to a coated sample.

The static loading was performed on a universal testing machine, Instron 8801 (Instron, Norwood, MA, USA) (Figure 2c). An extensometer was mounted in order to measure strain accurately (Figure 2c). Since the total deformation of the sample was not very large, it was possible to acquire data with the extensometer until the sample broke. A video of the calibration area was made for capturing the moment of the appearance of the crack in the coating material (Figure 4a). The point on the stress-strain curve indicating the moment of the appearance of the crack was found by overlapping the time signatures of the video and that of the extensometer data. A broken coated sample can be seen in Figure 4b.

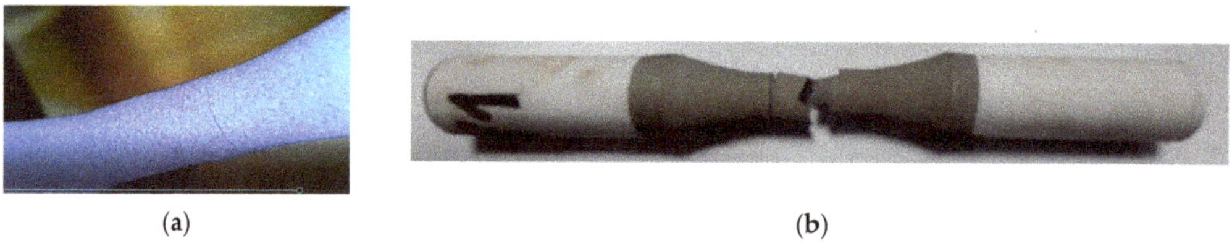

Figure 4. Sample subjected to static tensile load. (**a**) Image from the video that captures the moment of cracking of the coating material at the static tensile load; (**b**) broken sample after static tensile.

Both from the examination of the fractured sample (Figure 4b) and from the stress-strain curve given in Figure 3 the following observations can made: (i) the shape of the stress-strain curve is dominated by the behavior of the base material, 52100-steel); (ii) the elastic region is relatively wide; high values of the load are reached on a linear path and the yield point is close to the maximum stress; (iii) a material hardening region is not observed; after reaching the offset yield point, $Rp_{0.2}$ or $\sigma_{p0.2}$, the material undergoes a pronounced elongation, but without further increase of stress; (iv) on the stress-strain curve, the cracking point of the coating material was marked, which was determined by the study of the videos captured during the test, as explained above; this point corresponds to a strain of 2.64%; (v) cracking of the coating material does not occur in the elastic region of the stress-strain curve; (vi) the strain at which the cracking of the coating material takes place can be obtained locally, where a microcrack is initiated in the base material and where, under the conditions of existence of a stress concentrator (crack for example), the local deformations become large; (vii) given that the total strain is 14%, the base material undergoes significant plastic deformation, hence the contraction is observed in Figure 4b; (viii) the coating material suffers significant damage; (ix) it is found that the $\sigma p_{0.2}$ is 831 MPa, a value from which it starts to decrease in fatigue loading.

3. Results and Discussion

The ASTM E466-15 standard is used for fatigue testing. The shape and dimensions of the samples are shown in Figure 2a. The fatigue tests were conducted after a symmetrical, alternating cycle. For the fatigue testing, the initial load must be in the elastic region (Figure 3). The fatigue test is carried out on nine samples, starting with a load value corresponding with the stress close to the yield limit and continuing by decreasing the load. The first maximum stress was 76.5% of the yield limit value σ_{max} = 0.765, $\sigma_{p0.2}$ = 636 MPa. Table 2 shows the maximum stress values used in the tests.

Table 2. Maximum stress σ_{max} and the number of cycles to failure N obtained for Ni/CrC coated 52100-steel samples.

Sample No.	σ_{max} (MPa)	N
2	636	5300
3	530	34,660
4	477	878,965
7	459	2,287,070
5	449	3,456,873
9	446	4,424,120
8	438	5,451,948
1	420	5,519,600

The fracture surfaces were investigated at macro- and micro-scales by using optical and scanning electron microscopy. Observations of fracture of the coating material in different areas of the surfaces are presented. It is expected that the base and the coating materials will experience different strain levels where $\varepsilon_{base} > \varepsilon_{coating}$. Significant damage to the coating material could be expected to occur, given that the two materials deform differently, and large deformations occur at higher values of the stress cycles. In order to be able to observe this, images were taken from the fatigue test, from the central and calibration areas of the samples. From the SEM images (Figure 5) the grip of the coating material on the base material was a good one and there were no significant differences in the fatigue behaviour of the coating and base material.

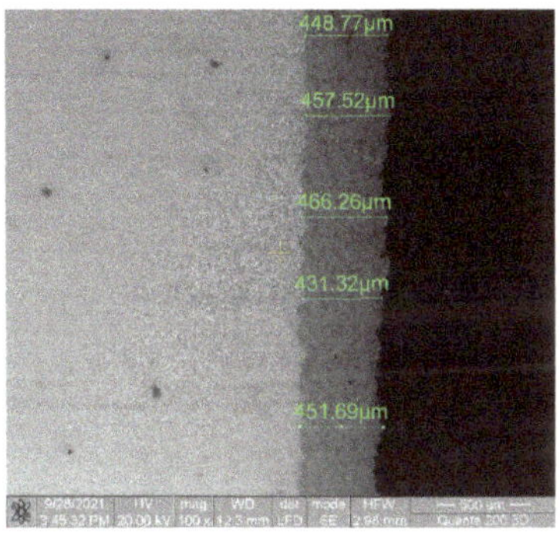

Figure 5. SEM image with cross-section at the interface between the base and the coated materials.

As can be seen in Figure 6, no damage to the coating material was observed even at more than five million cycles, nearly up until fracture. The macroscopic image analysis shows a very good behaviour of the coating material during the test where the material experienced both tensile and compressive strains.

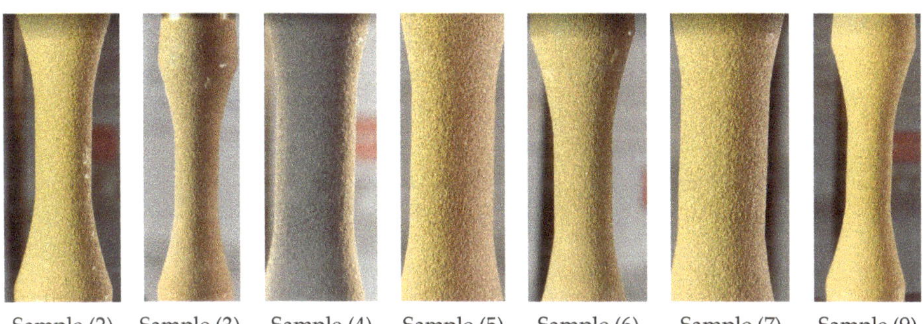

Figure 6. Outer surface during fatigue tests—different samples, near fracture point.

The moment of propagation of the fatigue crack is captured in Figure 7. It is known that, depending on the material, a relatively large number of fatigue cycles may be required to initiate a crack. After initiating the crack, depending also on the value of the tensile stress, the number of cycles for fatigue crack propagation may vary. In Figure 7, the visible crack (below the hand drawn line) is the one propagated by fatigue; a relatively small number of cycles were required until the final, sudden breakage of the sample soon after this crack appeared. From this figure, it can be observed a macroscopic propagation of the crack and the sample is close to the final breakage. No detachment of the coating material, near the front of the propagated crack, is seen. As a result, the propagation of the crack in the coating material follows the same path as in the base material, without any collateral damage to the coating material. From a macroscopic point of view, the coating behaves well during cyclic tests with alternating, symmetrical, axial loading.

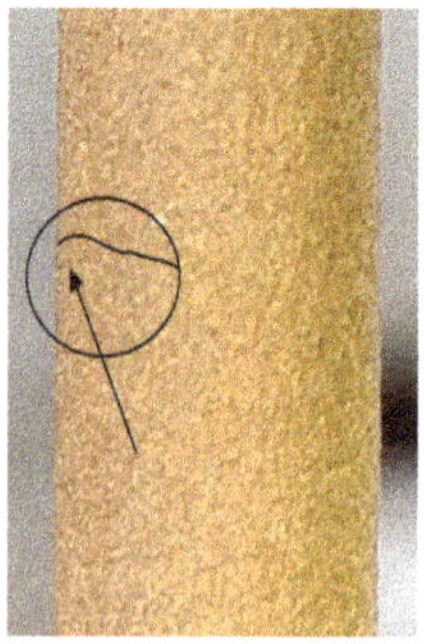

Figure 7. The appearance of fatigue cracks during fatigue loading.

As mentioned earlier, the tests were conducted from higher stress values (resulting in lower values of the number of stress cycles) to lower stress values. The stress cycle values of 5 million have been set as a limit. If the sample did not brake up to this number of stress cycles, the test was stopped, regardless of the stress values. Table 2 shows the number of cycles to failure for each maximum stress level.

For the fatigue analysis, 9 samples of 52100 steel were used as the base material, (Figure 8), coated using the method presented above. Samples 1 and 8 did not break even after 5 million cycles, as a result, the test was stopped, and the samples remained unbroken. The other samples fractured at different values of loads and number of cycles. In what follows observations about each sample are presented in detail.

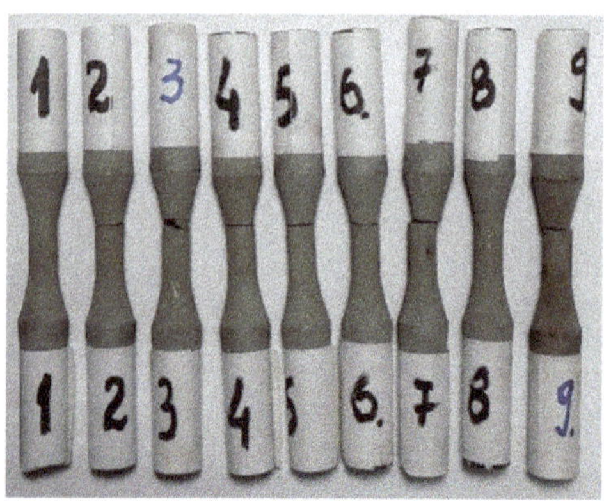

Figure 8. Samples after loading by alternate-symmetric fatigue tests.

3.1. Samples 1 and 8 (σ_{max} (MPa); N), (420; 5,519,600) and (438; 5,451,948)

Samples 1 and 8 did not break even after more than five million cycles of stress reversals. Examining the outer surface of the sample, no damage to the coating material was found (Figure 9). For Sample 1, the coefficient resulting from the ratio of the test loads to the yield limit is: $K_{\sigma F} = \sigma_{test}/\sigma_y = 438/831 = 0.53$. As a result, we can state that when the loading is at 53% of the yield limit, and if the sample doesn't fracture, the coating material withstands high values of the number of cycles, not causing significant damage to the surface. This statement must be taken with caution, since it is not known what happens at the interface between the two materials, in the area where the first plastic deformations of the base material occur. At low loads, in the elastic region, the deformations of the base material are not very different from the deformations of the coating material [19]. Under these conditions, the tension and compression of the coating materials are similar to the expansion and compression of the base material. Although the stress is applied over 5 million cycles, the loads being low, the coating material withstands very well within the stresses of cyclic fatigue, without experiencing significant damage.

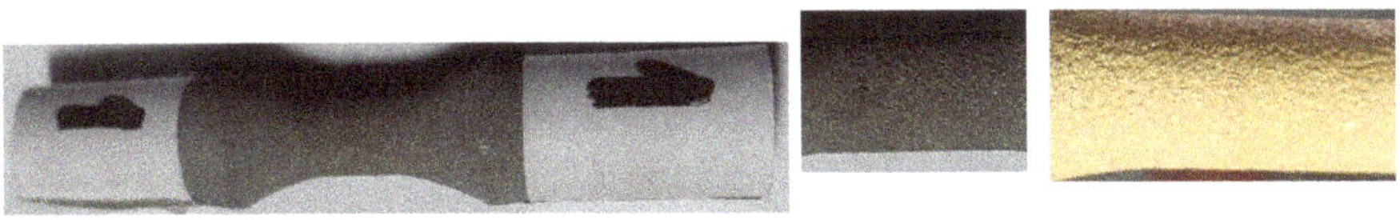

(a) (b) (c)

Figure 9. Macroscopic appearance of the outer surface after fatigue stress—Sample 1 σ_{max} = 420 MPa, N = 5,519,600 cycles. (**a**) the unbroken sample after more than 5 million cycles (8×). Macroscopic image of the surface for (**b**) 32× and (**c**) 45× magnification. Sample 8 had a similar appearance.

The macroscopic and microscopic observations were made on the fracture surfaces, of each failed sample separately. The crack initiation and the behaviour of the coating material are presented in descending order starting with the maximum load.

3.2. Sample 2 (σ_{max} (MPa), N), (636, 5300)

Macroscopic observations: Longitudinal cracks were observed in the coating material when the sample was subjected to high stresses, close to the flow limit (Figure 10a). In the area of the crack initiation, the surface of the coating material is flat, without obvious macroscopic cracks in the plane of fracture. Throughout the crack, the surface of the coating material remains flat, there being no detachment from the material (Figure 10b). At the boundary between the areas of the slowly propagating crack and suddenly propagating crack due to fatigue, significant damage to the coating material is seen (Figure 10c). Longitudinal cracks that are not seen in the other samples were found here. The longitudinal cracks appear only in the area of the suddenly propagating crack region. In the area of stable propagation of the crack front, no such cracks are found.

Microscopic observations: In the area of crack initiation, there is no significant damage to the coating (Figure 10d), like a macroscopic detachment from the material, cracks inside the coating material, and cracks in the particles that are in the coating material. In the zone of the crack initiation (Figure 10e), deterioration can be found in the area of the coating material with microscopic detachment from the material. In this area, no damage to the coating material is found on the outside of the sample and it can be concluded that the crack was initiated, first in the base material, after which it extended in both directions, both in the coating material, outwards and in the base material, inwards. In Figure 10e detachment of the coating material from the base material is observed. This occurred as a result of the differences between the Young modulus of the two materials, thus of the differences in specific deformation, emphasized earlier in the paper. As a result, in the area of crack initiation, the local strain of steel reached the value of 2.64%.

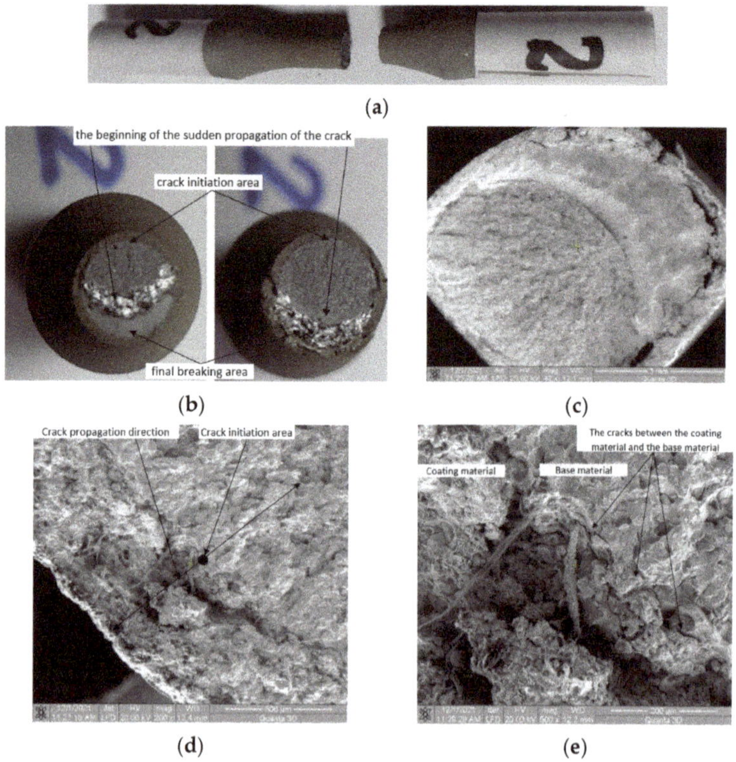

Figure 10. Macroscopic and microscopic appearance of broken surfaces for Sample 2, σ_{max} = 636 MPa, N = 5300 cycles. (**a**) Sample 2 broken by fatigue. (**b**) Fracture surfaces. (**c**) SEM image of the surface at 42× magnification. SEM image of crack initiation region with (**d**) 200× and (**e**) 500× magnification.

3.3. Samples 3 and 4, (σ_{max} (MPa), N), (530, 34,660) and (477, 878,965)

Macroscopic observations: Figure 11a–c show that at the end of the crack propagation period, the fracture surface changes direction, from being perpendicular to the force direction, to an orientation of approximately 45°. This can be explained as follows. While moving in the base material, the crack front encounters resistance to advancing in the initial direction; given that a large part of the surface has been broken, the outer force, which introduces high loads, acts unbalanced in relation to the remaining fracture surface (Figure 11f). Given the high loads in the remaining uncracked surface, the movement of the dislocations in the base material, under the action of tangential loads, is done in an accelerated manner. In the region where the front of the propagated crack changes direction, significant detachment of the coating material from the base material is found (Figure 11c).

Microscopic observations: In the final fractured surface, large damage to the coating material is noted (Figure 11c). For all samples, the large damage to the coating material in the final fracture surface is due to the large deformation differences between the base and coating material. The base material having a high deformation (in the plastic domain) in the final fracture surface, also trains the coating material. Thus, the coating, not being able to deform in the same way as the base material, deteriorates by the appearance of large cracks. In the area of crack initiation (Figure 11d,e), there is no significant damage to the coating material. Detachment of the coating material from the base material due to the deformation differences between the two materials is observed.

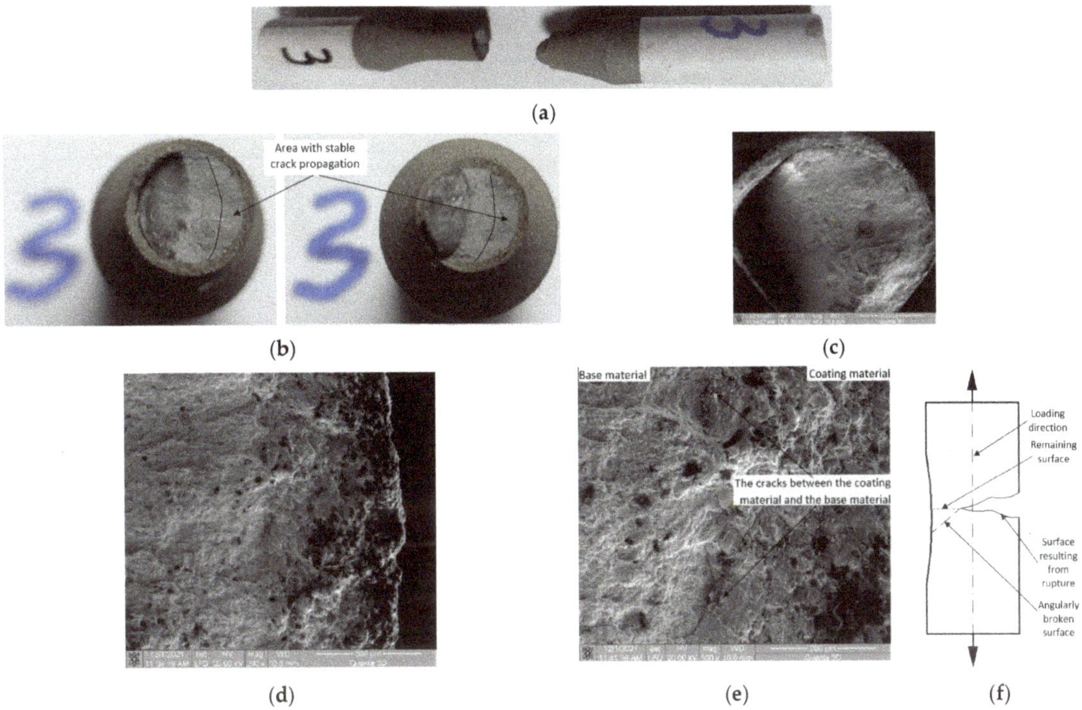

Figure 11. Macroscopic and microscopic appearance of broken surfaces for Sample 3, σ_{max} = 530 MPa, N = 34,660 cycles. (**a**) Sample 3 broken by fatigue. (**b**) Fracture surfaces. (**c**) SEM image of the surface at 45× magnification. SEM image of the area where the crack initiated with (**d**) 200× and (**e**) 500× magnification. (**f**) The remaining area before total failure.

3.4. Samples 4 and 7, (σ_{max} (MPa), N), (477, 878,965) and (459, 2,287,070)

Macroscopic observations: For Samples 4 and 7 the conclusions are similar. The loads used for Samples 4 and 7 are increasingly low. This results in an increasingly large number of fatigue cycles before a fracture occurs. The fracture occurs by stable propagation of the crack through fatigue and no deflection of the crack is observed (Figure 12a–c). However, in these samples, the load is higher than in the following samples and the area of stable crack propagation is smaller (see the corresponding figures in the following samples). No damage is observed due to the removal/detachment of particles or by the appearance of large cracks, in the final fracture surface (Figure 12c).

Microscopic observations: In the area of crack initiation by fatigue, no significant damage to the coating material is found (Figure 12c). At higher magnifications, in the area of crack initiation, a detachment is revealed by showing a radial crack through the coating material (Figure 12d,e).

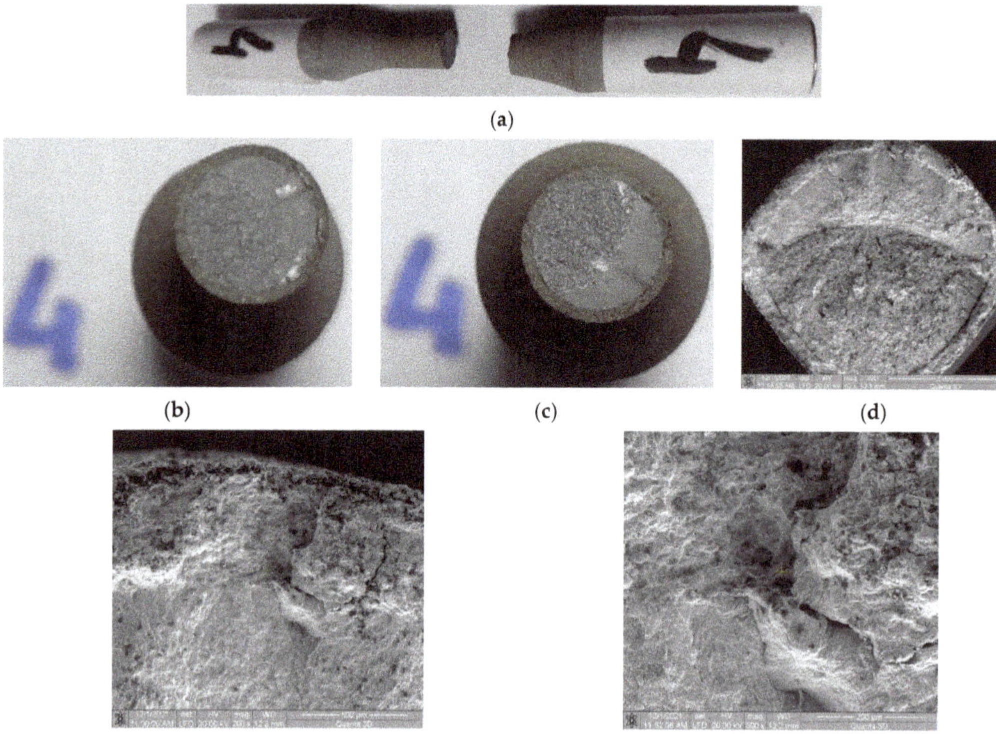

(a)

(b) (c) (d)

(e) (f)

Figure 12. Macroscopic and microscopic appearance of broken surfaces for Sample 4, σ_{max} = 477 MPa, N = 878,965 cycles. (**a**) Sample 4 broken by fatigue. (**b**,**c**) Fracture surfaces. (**d**) SEM image of the fracture surface at 42× magnification. SEM image of the area where a crack initiated with (**e**,**f**) 200×, and 500× magnification.

3.5. Samples 5 and 9, (σ_{max} (MPa), N), (449, 3,456,873) and (446, 4,424,120)

Macroscopic observations: For Samples 5 and 9 the conclusions are similar. These samples were tested at low loads, and more than 50% of the fracture surface is obtained by stable propagation of the crack. The coating uniformly broke throughout its circumference as shown in Figure 13a,b. A detachment of the coating material from the base material is found at the beginning of the sudden propagation of the crack front (Figure 13c).

Microscopic observations: In the area of crack initiation by fatigue, no significant damage to the coating material is observed (Figure 13d,e). A very fine crack between the base material and the coating material meaning a detachment between the two materials appears in the area of initiation of the fatigue crack. In this sample, inter- and intra-granular microcracks (Figure 13e), are observed more pronounced than in the previous samples, especially near the interface area between the two materials.

The data presented in Table 2 is plotted in a Wöhler diagram in Figure 14. Sample 8 (438 MPa) and Sample 1 (420 MPa) did not break after 5 million cycles. The appearance of the diagram is a conventional one, with a fatigue limit of around 446 MPa, given the imposed limit of 5 million cycles.

Tests performed by other researchers on the uncoated 52100 material have shown that no limit to well-defined fatigue can be determined [20]. On the other hand, the base material, in this case, 52100 steel significantly influences the appearance of the S-N durability curve.

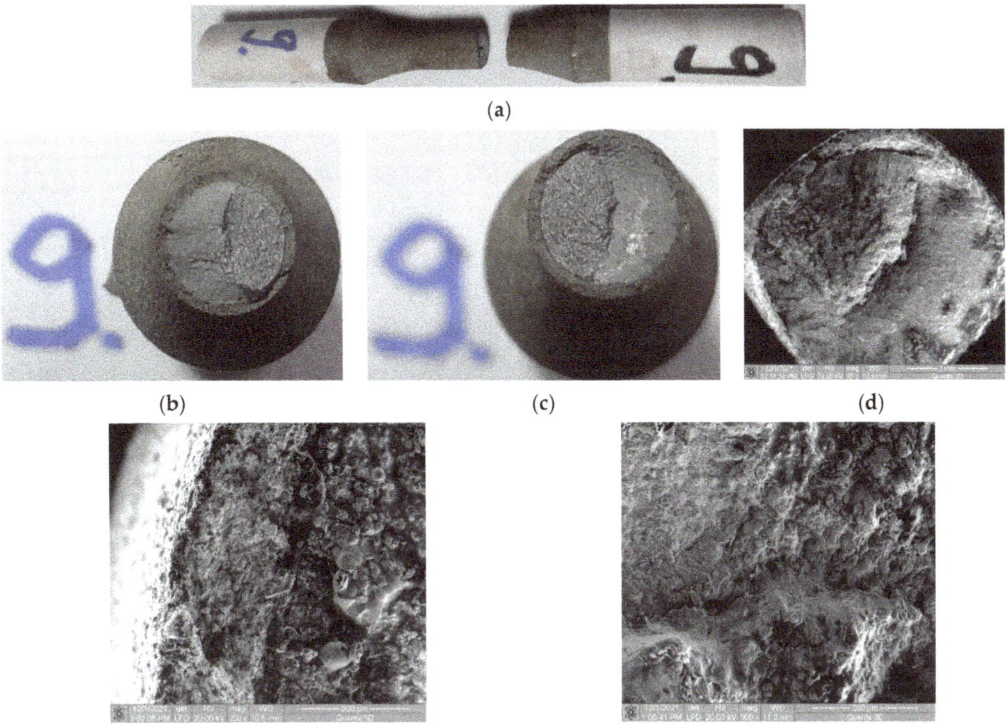

(**a**) (**b**) (**c**) (**d**) (**e**) (**f**)

Figure 13. Macroscopic and microscopic appearance of broken surfaces for Sample 9, σ_{max} = 446 MPa, N = 4,424,120 cycles. (**a**) Sample 9 broken by fatigue. (**b**,**c**) Image of fractured surfaces. (**d**) SEM image of the fracture surface at 42× magnification. (**e**,**f**) SEM images of the area where a crack initiated with 200×, and 500× magnification.

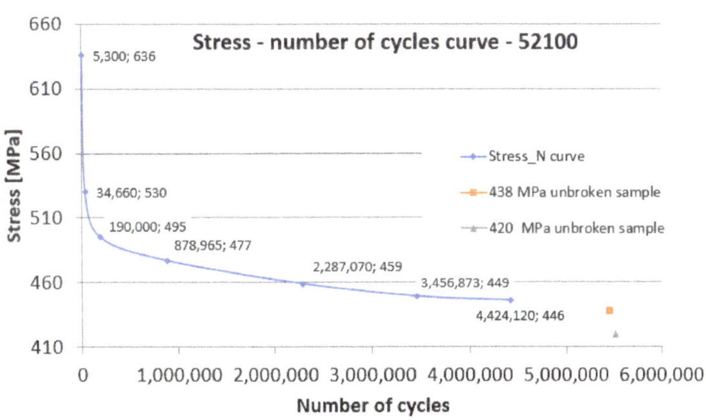

Figure 14. Wöhler curve for fatigue test of the 52100-steel coated with WIP-C1 powder.

4. Conclusions

The coated parts will have to work in the area of the high durability of Wöhler's curve meaning at relatively low stress levels. If parts are subjected to high stresses, significant damage to the coating material may occur, caused by differences in deformation, in the local plastic domain, between the base material and the coating material. The behaviour in the area of crack initiation is of particular interest, due to the greatest number of stress cycles used to initiate the crack and, by design, we do not have to reach the area of sudden propagation of the crack. An area of interest is also the stable propagation of the fatigue crack, where, if the crack is observed, early measures can be taken to replace/repair the cracked part.

The microscopic study of the surfaces resulting from the fatigue fracture shows the following:

- The two areas of the same surface are highlighted, an area characteristic of the propagation by fatigue in a long time and an area characteristic of the sudden crack propagation, until the final fracture;
- The area of stable fracture propagation by fatigue by finer grains; the area from the sudden breakage is characterized by coarser grains;
- In the crack initiation area, there are no pores or major cracks between the coating material and the base material;
- The initiation of the crack took place on the surface of the base material, then extended to the coating material and inside the base material;
- Crack initiation taking place on the surface of the base material suggests that applying the coating is unlikely to negatively impact the fatigue properties of the base material;
- The coating material did not suffer detachment at the beginning of fatigue cracking;
- In the whole area where the crack was propagated by stable fatigue (lower stresses), the coating material behaved very well, there being no detachment of the coating material from the base material;
- Lack of coating detachment at the crack initiation and stable crack growth indicate very good bonding between coating and substrate;
- Detachments of the coating material from the base material can be observed in the area of sudden crack propagation. There are pores or detachment, which is justified given the high values of deformations in this area and given the difference between the specific deformations (strains) of the two materials.

Author Contributions: Conceptualization, C.M., V.G., G.F., P.S. and S.M.; methodology, V.G.; software, B.I.; validation, C.M. and V.G.; formal analysis, B.I. and C.I.M.; investigation, C.M., B.I., P.S. and S.B.; resources, S.M.; data curation, V.G. and C.M.; writing—original draft preparation, V.G.; writing—review and editing, V.G., S.M. and C.M.; visualization, B.I. and G.F.; supervision, C.M. and S.M.; project administration, C.M. and S.M.; funding acquisition, S.M. All authors have read and agreed to the published version of the manuscript.

Funding: This work was sponsored in part by the Army Research Laboratories under grant number W911NF-20-2-0024. Any opinions, findings and conclusions or recommendations expressed in this material are those of the authors and do not necessarily reflect the views of the U.S. Government.

Institutional Review Board Statement: Not applicable.

Informed Consent Statement: Not applicable.

Data Availability Statement: Not applicable.

Conflicts of Interest: The authors declare no conflict of interest.

References

1. Baragetti, S.; Tordini, F. A Review of the Fatigue Behaviour of Components Coated with Thin Hard Corrosion-Resistant Coatings. *Open Corros. J.* **2011**, *4*, 9–17. [CrossRef]
2. Jafarlou, D.M.; Walde, C.; Champagne, V.K.; Krishnamurty, S.; Grosse, I.R. Influence of cold sprayed Cr3C2-Ni coating on fracture characteristics of additively manufactured 15Cr-5Ni stainless steel. *Mater. Des.* **2018**, *155*, 134–147. [CrossRef]
3. Ruffing, C.; Ivanisenko, Y.; Kerscher, E. A comparison of the fatigue and fracture behavior of high strength ultrafined grained medium carbon steel SAE 1045 with high strenght bearing steel SAE 52100. *Procedia Struct. Integr.* **2016**, *2*, 3240–3247. [CrossRef]
4. Grünling, H.W.; Schneider, K.; Singheiser, L. Effect of coating on the mechanical properties of the substrate, particularly in high and low-cycle fatigue and corrosion-fatigue. *Mater. Sci. Eng.* **1987**, *88*, 177–189. [CrossRef]

5. Baragetti, S.; La Vecchia, G.M. Variables affecting the fatigue resistance of PVD-coated components. *Int. J. Fatigue* **2005**, *27*, 1541–1550. [CrossRef]
6. Wang, Q.; Birbilis, N.; Huang, H.; Zhang, M.-X. Microstructure characterization and nanomechanics of cold-sprayed pure Al and Al-Al2O3 composite coatings. *Surf. Coat. Technol.* **2013**, *232*, 216–223. [CrossRef]
7. Cunha, L.; Andritschky, M. Residual strees, surface defects and corrosion resistance of CrN hard coatings. *Surf. Coat. Technol.* **1999**, *111*, 158–162. [CrossRef]
8. Karlssona, L.; Hultmanb, L.; Sundgren, J.-E. Influence of residual stresses on the mechanical properties of TiCxN1-x (x = 0, 0.15, 0.45) thin films deposited by arc evaporation. *Thin Solid Films* **2000**, *371*, 167–177. [CrossRef]
9. Lin, E.; Nault, I.; Ozdemir, O.C.; Champagne, V.K.; Nardi, A.; Muftu, S. Thermo-Mechanical Deformation History and the Residual Stress Distribution in Cold Spray. *J. Therm. Spray Technol.* **2020**, *29*, 1424–1436. [CrossRef]
10. Baragetti, S.; La Vecchia, G.M. Fatigue behavior and FEM modeling of thin-coated components. *Int. J. Fatigue* **2003**, *25*, 1229–1238. [CrossRef]
11. Ferreira, J.A.M.; Costa, J.D.M.; Lapa, V. Fatigue behaviour of 42CrMo4 steel with PVD coatings. *Int. J. Fatigue* **1997**, *19*, 293–299. [CrossRef]
12. Saruki, K.; Hotta, S.; Fujita, H.; Arai, T. Fatigue strength of steels with thin hard coating. *Thin Solid Films* **1989**, *181*, 383–395. [CrossRef]
13. Su, Y.L.; Yao, S.H.; Wei, C.S.; Wu, C.T. Evaluation on the tension and fatigue behavior of various PVD coated materials. *Thin Solid Films* **1998**, *322*, 218–224. [CrossRef]
14. Berríos, J.A.; Teerb, D.G.; Puchi-Cabrera, E.S. Fatigue properties of a 316L stainless steel coated with different TiNx deposits. *Surf. Coat. Technol.* **2001**, *148*, 179–190. [CrossRef]
15. Barbera-Sosa, J.G.; Teer, D.G.; Puchi-Cabrera, E.S. Fatigue properties of a 316L stainless steel coated with different ZrN deposits. *Surf. Coat. Technol.* **2004**, *179*, 145–157.
16. Ma, K.J.; Bloyce, A.; Bell, T. Examination of mechanical properties and failure mechanisms of TiN and Ti−TiN multilayer coatings. *Surf. Coat. Technol.* **1995**, *76–77*, 297–302. [CrossRef]
17. Kima, K.R.; Suh, C.M.; Murakami, R.I.; Chung, C.W. Effect of intrinsic properties of ceramic coatings on fatigue behavior of Cr–Mo–V steels. *Surf. Coat. Technol.* **2003**, *171*, 15–23. [CrossRef]
18. Champagne, V.K. The Cold Spray Materials Deposition Process. In *Fundamentals and Applications*; CRC Press: Boca Raton, FL, USA, 2007.
19. Winnicki, M. Advanced Functional Metal-Ceramic and Ceramic Coatings Deposited by Low-Pressure Cold Spraying: A Review. *Coatings* **2021**, *11*, 1044. [CrossRef]
20. Jafarlou, D.M.; Ferguson, G.; Tsaknopoulos, K.L.; Chuang, A.C.; Nardi, A.; Cote, D.; Champagne, V.; Grosse, I.R. Structural integrity of additively manufactured stainless steel with cold sprayed barrier coating under combined cyclic loading. *Addit. Manuf.* **2020**, *35*, 101338. [CrossRef]
21. Romanowicz, P.J.; Szybinski, B. Fatigue Life Assessment of Rolling Bearings Made from AISI 52100 Bearing Steel. *Materials* **2019**, *12*, 371. [CrossRef]
22. Young, B.; Heelan, J.; Langan, S.; Siopis, M.; Walde, C.; Birt, A. Novel Characterization Techniques for Additive Manufacturing Powder Feedstock. *Metals* **2021**, *11*, 720. [CrossRef]
23. Zhang, R.Y.; Muftu, S. Elastic impact of a spherical particle with a long, stationary, fixed Timoshenko beam. *J. Sound Vib.* **2021**, *495*, 115892. [CrossRef]

materials

Article

Experimental Investigation on Mechanical and Thermal Properties of Concrete Using Waste Materials as an Aggregate Substitution

Gavril Sosoi [1,2], Cherifa Abid [2], Marinela Barbuta [1], Andrei Burlacu [1,*], Marius Costel Balan [1,*], Marius Branoaea [1], Robert Stefan Vizitiu [1] and Fabrice Rigollet [2]

[1] Faculty of Civil Engineering and Building Services, "Gheorghe Asachi" Technical University of Iasi, 700050 Iasi, Romania; gavrilsosoi@gmail.com (G.S.); marinela.barbuta@academic.tuiasi.ro (M.B.); marius.branoaea@tuiasi.ro (M.B.); robert.vizitiu@tuiasi.ro (R.S.V.)

[2] IUSTI UMR 7343, Aix-Marseille Université, 13453 Marseille, France; cherifa.abid@univ-amu.fr (C.A.); fabrice.rigollet@univ-amu.fr (F.R.)

* Correspondence: andrei.burlacu@tuiasi.ro (A.B.); marius-costel.balan@academic.tuiasi.ro (M.C.B.)

Abstract: The continuous growth of the concrete industry requires an increased quantity of cement and natural aggregates year after year, and it is responsible for a major part of the global CO_2 emissions. These aspects led to rigorous research for suitable raw materials. Taking into account that these raw materials must have a sustainable character and also a low impact on environmental pollution, the replacement of the conventional components of concrete by residual waste can lead to these targets. This paper's aim is to analyze the density, compressive strength and the thermal conductivity of nine concrete compositions with various rates of waste: four mixes with 10%, 20%, 40% and 60% chopped PET bottles aggregates and 10% fly ash as cement partial substitution; a mix with 60% waste polystyrene of 4–8 mm and 10% fly ash; a mix with 20% waste polystyrene of 4-8 mm, 10% waste polystyrene of 0–4 mm and 10% fly ash; a mix with 50% waste polystyrene of 4–8 mm, 20% waste polystyrene of 0–4 mm and 20% fly ash two mixes with 10% fly ash and 10% and 40% waste sawdust, respectively. Using 60% PET aggregates, 60% polystyrene granules of 4–8 mm, or 20% polystyrene of 0–4 mm together with 50% polystyrene of 4–8 mm led to the obtainment of lightweight concrete, with a density lower than 2000 kg/m^3. These mixes also registered the best results from a thermal conductivity point of view, after the concrete mix with 40% saw dust. Regarding compressive strength, the mix with 10% PET obtained a result very close to the reference mix, while those with 20% PET, 40% PET, 30% polystyrene, and 10% saw dust, respectively, registered values between 22 MPa and 25 MPa, values appropriate for structural uses.

Keywords: cement based concrete; ecological concrete; waste aggregates; compressive strength; thermal conductivity

Citation: Sosoi, G.; Abid, C.; Barbuta, M.; Burlacu, A.; Balan, M.C.; Branoaea, M.; Vizitiu, R.S.; Rigollet, F. Experimental Investigation on Mechanical and Thermal Properties of Concrete Using Waste Materials as an Aggregate Substitution. *Materials* **2022**, *15*, 1728. https://doi.org/10.3390/ma15051728

Academic Editor: Luigi Coppola

Received: 17 January 2022
Accepted: 18 February 2022
Published: 25 February 2022

1. Introduction

Energy consumption in the world is increasing and, therefore, a need for taking into consideration the ongoing energy crisis and the impact it has on the environment is present. In terms of energy demand, the built environment is a net consumer of energy, demanding over 36% of the global energy, and upwards to 50% of raw materials worldwide. In terms of the environmental impact, buildings are responsible for over 39% of the global greenhouse gas emissions [1].

It is estimated that: 1 ton of Portland cement produces 0.96 tons of CO_2 and 1 ton of concrete produces 0.108 tons CO_2. According to experts, the concrete industry is responsible for 7 to 9% of the global greenhouse gas emissions [2,3].

The use of wastes as an aggregate substitution in construction materials represents a great alternative, especially in cement based concretes, due to large scale use in the construction industry. By using waste materials, such as chopped PET bottles, granular polystyrene, and waste wood, as aggregates in the concrete mixture, an ecological and performant concrete can be obtained, such as lightweight concrete, that can be used to reduce the weight of buildings. Another significant advantage of this type of concrete mixture is the valorification of waste materials that can produce a negative impact on the environment.

The purpose of this paper is to investigate the density, compressive strength and thermal conductivity of various concrete mixes using waste as aggregates and cement substitutions. The study implied the development of ten concrete mixes, which included a reference concrete mix made with conventional components only (river aggregates and cement), and nine concrete compositions with waste materials as aggregates and cement partial replacements, as follows: four mixes with 10%, 20%, 40% and 60% chopped PET bottles aggregates and 10% fly ash as cement partial substitution; a mix with 60% waste polystyrene of 4–8 mm and 10% fly ash; a mix with 20% waste polystyrene of 4–8 mm, 10% waste polystyrene of 0–4 mm and 10% fly ash; a mix with 50% waste polystyrene of 4–8 mm, 20% waste polystyrene of 0–4 mm and 20% fly ash two mixes with 10% fly ash and 10% and 40% waste sawdust, respectively. Using the waste materials mentioned can contribute significantly to reducing waste quantity and CO_2 emissions, by decreasing the produced quantity of cement, and by natural aggregates' use in a concrete mix.

If waste, such as chopped PET, granular polystyrene, and waste sawdust, are to be used as aggregates in the composition of concrete, then the pollution and impact of the construction sector on the environment will decrease, these facts represent a great advantage both in terms of recycling and reducing the consumption of natural ingredients, such as sand.

2. State of the Art

There are various recent studies on the use of waste chopped polyethylene terephthalate, granular polystyrene, and waste wood (sawdust) in the composition of the concrete mixture by replacing natural aggregates [4–11].

Furthermore, by investigating the mechanical properties of the concrete with waste chopped PET bottles, granular polystyrene, and waste wood as a substitution in the concrete mix, researchers obtained results that highlight the fact that the mechanical properties of fresh and hardened concrete are being influenced by the ratio of the replaced material in the concrete mix [12–21].

In terms of research trends, scientist have analyzed the incorporation of various materials into concrete in order to reduce the required raw materials, ranging from crushed recovered concrete and ceramics, vegetal byproducts and even plastics. Some examples of such studies are presented in Table 1.

2.1. Concrete with PET and Polystyrene

The use of PET and polystyrene in concrete mixtures, both in structural or lightweight concrete, was analyzed in a number of studies with the aim to reduce the waste in the world and reduce the quantity of required raw materials. The analyzed properties of these concretes include—but are not limited to—compressive strength, density, tensile strength, elasticity modulus, flexural strength, durability, thermal properties and fire resistance.

In their study on concrete with polystyrene with a ratio ranging from 20% to 80% of the total volume and a resin mixture of 0.5%, 1%, and 1.5% of the total weight, Kaya and Kar concluded that, as the added components percentage increase, the porosity of the concrete increases and the mechanical properties, such as density, tensile strength, compressive strength, and thermal conductivity, decrease [11].

Table 1. Summary of research papers on concrete with waste materials.

NR.	Contents of the Paper	Year	Ref.
1	The study analyzed the substitution of raw materials with fine—fMRA (25% and 50%) and coarse—cMRA (0%, 25% and 50%) mixed recycled aggregates. The study determined that the optimal percentage of an fMRA replacement is 25%.	2021	[22]
2	The paper highlighted a correlation and regression analysis of lightweight concretes with presoakedred ceramic waste aggregate, expanded clay coarse aggregate as waste aggregate and varying degrees of copper coated steel fibre used as reinforcement.	2022	[23]
3	This study analyzed the mechanical properties and durability of medium quality concrete with a replacement of the natural coarse aggregate (NCA) with various percentages of recycled coarse aggregate, RCA, (25%, 50%, 75% and 100%). The results reveal few differences between the concrete with RCA and the reference concrete in terms of the mechanical properties and durability related measurements.	2021	[24]
4	This research involved the experimental investigation of using fly ash admixture collected by a wet process as a replacement of the fine part of the aggregates in concrete. The experimental results show a favourable behaviour of the concretes based on fly ash to sulphate aggressive actions in the first part of life.	2020	[25]
5	The purpose of this paper was analyzing the performance of concrete with granules of corn cob and sunflower with a sodium silicate solution treatement (0%, 20% and 40%). The experimental analysis highlighted that the concrete with vegetal raw materials had a reduced density but, at the same time, a reduced compressive strength as vegetal aggregates have higher water absorption capacity. Sunflower concrete presents superior mechanical strengths in comparison to corn cob concrete.	2020	[26]
6	This study, through a statistical experimental analysis, evaluates the performance of concrete reinforced with recycled PET fibres with various fibre doses and aspect ratios. The results highlighted that the introduction of recycled PET fibres into concrete provides residual strength capacity to the concrete, with a reduced effect on the volumetric weight, ultimate flexural and compressive strength.	2021	[27]
7	The experimental research involved analysing the properties of concrete with sawdust as a replacement of sand containing 0%, 5%, 10% and 15% of sawdust. Through adding sawdust the results showed that the biological oxygen demand increased as the replacement percentage increased, but the compressive strength decreased.	2019	[28]
8	The study researched the properties of fly ash as a basic raw material used in the production of concrete. The inclusion of fly ash in concrete, referenced through multiple studies, provide increased mechanical and microstructure properties in comparison to the use of cement alone.	2020	[29]
9	The paper analyzed the incorporation of waste materials, such as sugarcane bagasse ash (SCBA), metakaolin (MK), and millet husk ash (MHA), and determines their effect on the fresh, hardened properties and embodied carbon of concrete. The study revealed that the use of SCBA, MK, and MHA up to 10–15% separate and combined as ternary cementitious material (TCM) in concrete provides ideal results for structural applications.	2022	[30]
10	The researchers reviewed the impact natural seawater has on the properties of concrete. The study points out that, in the long term, seawater is harmful to the concrete structures built in marine environments and that, through the use of supplementary cementitious materials (SCMs) such as rice husk ash, coal bottom ash and blast furnace slag in the concrete mix, the degradation is reduced. Another interesting fact is that, through adding seawater during the curing and mixing of concrete, the tensile properties of the concrete are increased.	2021	[31]

In their research, Saikia and Brito performed an experimental investigation on specimens for compressive strength, split tensile strength, and flexural strength, observing that an increase in the concentration of the PET ratio as aggregate in the mixture leads to a decrease in the compressive strength, split tensile strength and flexural strength of the concrete [12].

The study published by Rahmani et al. investigated the mechanical properties of cubic and cylindrical concrete specimens with 5% to 15% substitution of sand with particles of PET and different water to cement ratios. The experimental study revealed that the compressive and flexural strength initially tend to increase, however, they tend to decrease after a period of time. Moreover, the experiments highlighted that the samples containing PET particles have a lower density, splitting tensile strength, modulus of elasticity, and workability [13].

Jahidul et al. carried out an experimental investigation on the properties of PET aggregate concrete by replacing between 0 and 50% of the volume of coarse aggregates, observing that this type of concrete provided better workability than regular concrete while using a similar w/c ratio, the density of the PET concrete was reduced by 4 to 10%, thus reducing the self-weight of the structural element. The optimal substitution ratio was 20%, this percentage ensuring comparable compressive strength with natural aggregate concrete [15].

Maldonado-Bandala et al. observed similar results regarding mechanical parameters such as density and compressive strength—that these parameters decrease as the ratio of polystyrene increases—but the experiment revealed that the corrosion protection of the concrete with polystyrene is greatly superior to conventional concrete [16].

Studies performed by Sayadi et al. revealed that adding expanded polystyrene into the concrete mixture translates to a decrease in thermal parameters such as thermal conductivity and fire resistance [17].

2.2. Concrete with Sawdust

Sawdust, as a byproduct of the wood industry, has been researched and implemented in various forms since the 20th century in regions such as the United State of America, Asia (Malaysia and Singapore) and Europe (Germany and the UK). The possible applications of this material are significant, as highlighted by the fact that this material, in various forms and adaptations, has been used for building elements such as walls and floors [32].

Sales et al. analyzed the incorporation of untreated wood shavings into concrete to obtain lightweight concrete. The study highlighted that, through the introduction of sawdust into the concrete mixture, some key parameters were improved: the thermal conductivity of the concrete was improved and the mass density was reduced. At the same time, the sawdust presented a good adherence to the concrete matrix and the structure of the mixture was homogeneous [33].

Sofi et al. investigated the compressive strength of concrete samples with a volume of added sawdust ranging from 0% to 15% in the concrete mix at two curing periods, at 7 and 28 days. The results obtained after the curing periods show that the compressive strength is higher for concrete samples with a lower substitution percentage of sawdust [18].

The implementation of sawdust in concrete has not reached the level of widespread adoption due to the limitations of sawdust, which has low compressive strength. At the same time, the advantages of concrete with sawdust (reduction in structural weight and, as a consequence, reduced loads on the foundation, reduced damage phenomena and increased the lifespan of the structure, along with easier handling and reduced consumption of raw materials) justify the increasing interest in this type of material.

2.3. Concrete with Fly Ash

Fly ash is a byproduct of burning coal in power plants in the form of ultrafine solid residue. In numerous developed and developing countries, such as US, China and India, where coal is used, it is one of the largest solid wastes. Fly ash has been implemented as an additive in concrete since the mid-20th century as a replacement for some raw materials such as cement or as a mixture with a clinker [34].

The analysis performed by multiple researchers highlights the increase in interest towards ecological concrete and its potential to reduce the carbon footprint buildings impose on the environment. Even though some mechanical parameters are not as high with the substitution of natural aggregates, these types of walls can be applied in nonstructural elements.

3. Experimental Works

3.1. Materials

3.1.1. Cement and Fly Ash

The cement used in the concrete mixture is Portland cement type CEM II 45.2 R produced in Romania following the EN 197-1 standard, this fact is evidenced both from the producer's multiple declarations of conformity, performance certificates and through independent tests conducted on a cement sample. The cement has the following components: Portland clinker maximum dosage 94%, limestone 6–20%, and auxiliary components 0–5% [35,36].

The fly ash used as a replacement for the cement, which is released with the exhaust fuel gases, presented in Figure 1 is a byproduct from the Holboca Electric Power Plant in Iasi, with a bulk density ranging from 2400 to 2550 kg/m^3, specific surface area of 520 m^2/kg, particle size ranging from 0.01 to 400 μm and, as main chemical components, O_2 (43.3%), Si (30.8%), Al (19.2%) and Fe (3.05%) [37,38].

Figure 1. Fly ash from Holboca Electric Power Plant.

3.1.2. Aggregates

The aggregates used in the composition of the concrete can be classified into:

Natural aggregates such as sand with a specific gravity of 2.68, bulk density 1700 kg/m^3, and absorption of 1%, and two sorts of river stone (sort I 4–8 mm and sort II 8-16 mm) with a specific gravity of 2.62, bulk density 1400 kg/m^3, and absorption of 0.9%, presented in Figure 2.

Figure 2. Natural aggregates used in concrete mix.

Waste aggregates, evidenced by Figure 3, such as:

- Chopped PET from bottles (having dimensions between 0–4 mm), resulting from cutting discarded polyethylene terephthalate (PET) bottles. After the cutting of the material it was sieved in order to obtain the desired particle dimensions. This waste

material presents high stability and nonreactivity with substances. The unit weight of the chopped PET was determined experimentally and has a value of 433 kg/m^3 [39];
- Granular polystyrene (with dimensions 0–4 mm and 4–8 mm), which can be reused in the shape of granules, resulting from shredding, cutting or thermal treatment;
- Waste wood/sawdust (with dimensions 0–4 mm), is a byproduct resulting from the wood industry. The collected sawdust was air dried and afterwards sieved, in order to imitate the size of the natural aggregates. Sawdust in terms of chemical composition is comprised of carbon (60.8%), hydrogen (5.2%), oxygen (33.8%), and nitrogen (0.9%). The primary components of dry wood are cellulose, lignin, hemicelluloses, and small amounts (5–10%) of extraneous materials. The unit weight of the sawdust was determined experimentally and has a value of 168 kg/m^3 [39,40].

(a) (b) (c)

Figure 3. Waste aggregates used in the concrete mixture: (**a**) chopped PET bottle; (**b**) granular polystyrene; (**c**) sawdust/waste wood.

3.2. Concrete Mixture

The concrete mixture without any substitutions was made according to the following recipe: sand 803 kg/m^3, river gravel with 4–8 mm in diameter 384 kg/m^3, river gravel with 8–16 mm in diameter 559 kg/m^3, cement 360 kg/m^3, water 180 l/m^3 and a 1% superplasticizer additive of the cement mass.

In order to conduct the experiments and determine the mechanical parameters and the advantages and disadvantages of the partial substitution of the natural aggregate with either chopped PET, granular polystyrene, or waste wood (sawdust), nine types of different concrete mixtures were considered, with the mixing proportions and the quantities of the used materials presented in Table 2. The concrete was casted into 10 cubic shape specimens and 6 plate shape specimens for each concrete mixture.

The concrete mixture realization process involved adding the materials and water in the two-blade mixer, the mixing process was finished after 30–40 min.

Samples S1 to S4 also contain fly ash, which replaced 10% of the cement, and from 10% to 60% chopped PET replacing the same mass volume of cement.

In the concrete mix using waste granular polystyrene, three samples in which 10% of cement was replaced with fly ash, and the substitution of natural aggregates used were replaced by 60% of the sort I aggregate mass 4–8 mm were replaced with granular polystyrene 4–8 mm, resulting the concrete sample notated S5.

Following this, the concrete mixture was changed once again by replacing two natural aggregates in order to create sample S6, as follows: 90% of the sand was used, the remaining 10% being substituted by 10% of the sand mass with granular polystyrene 0-4 mm in volume, and, as a secondary substitution, only 80% of the original sort I, with a size of the granules of 4–8 mm, 20% from the sort I mass was replaced with granular polystyrene 4–8 mm in volume and a further substitution of 10% of the cement was made with fly ash.

Table 2. Concrete mixture with waste substitution.

Concrete Sample		Materials							
		Sand		Rocks (4–8 mm)		Rocks (8–16 mm)		Cement	
		Percent	Volumic Mass	Percent	Volumic Mass	Percent	Volumic Mass	Percent	Volumic Mass
S0	natural aggregate	100%	803.16	100%	384.12	100%	558.72	100%	360
	waste substitution	0% PET		0% PET		0% PET		0% fly ash	
S1	natural aggregate	90%	722.84	100%	384.12	100%	558.72	90%	324
	waste substitution	10% PET		0% PET		0% PET		10% fly ash	
S2	natural aggregate	80%	642.52	100%	384.12	100%	558.72	90%	324
	waste substitution	20% PET		0% PET		0% PET		10% fly ash	
S3	natural aggregate	60%	481.9	100%	384.12	100%	558.72	90%	324
	waste substitution	40% PET		0% PET		0% PET		10% fly ash	
S4	natural aggregate	40%	321.26	100%	384.12	100%	558.72	90%	324
	waste substitution	60% PET		0% PET		0% PET		10% fly ash	
S5	natural aggregate	100%	803.16	40%	153.64	100%	558.72	90%	324
	waste substitution	0% polystyrene		60% polystyrene 4–8 mm		0% polystyrene 8–16 mm		10% fly ash	
S6	natural aggregate	90%	722.84	80%	307.29	100%	558.72	90%	324
	waste substitution	10% polystyrene 0–4 mm		20% polystyrene 4–8 mm		0% polystyrene 8–16 mm		10% fly ash	
S7	natural aggregate	80%	642.52	50%	192.06	100%	558.72	80%	288
	waste substitution	20% polystyrene 0–4 mm		50% polystyrene 4–8 mm		0% polystyrene 8–16 mm		20% fly ash	
S8	natural aggregate	90%	722.84	100%	384.12	100%	558.72	90%	324
	waste substitution	10% saw dust		0% saw dust		0% saw dust		10% fly ash	
S9	natural aggregate	60%	481.9	100%	384.12	100%	558.72	90%	324
	waste substitution	40% saw dust		0% saw dust		0% saw dust		10% fly ash	

For the concrete sample S7, another modification in the concrete mixture was made, in which 20% of the cement was replaced with fly ash and two more natural aggregates were partially replaced with granular polystyrene, replacing 20% of the sand mass with granular polystyrene 0–4 mm in volume and 50% of the sort I mass 4–8 mm was replaced by 50% granular polystyrene 4–8 mm in volume.

The last two samples, S8 and S9, were realized by further modifying the mixture: 10% of the cement was replaced with fly ash in both concrete samples and a percentage of the sand was replaced with its corresponding volume of sawdust. In the case of sample S8 only 10% of the sand was replaced and in the case of sample S9, 40% of the sand mass was replaced with the same volume of sawdust.

3.3. *Preparation of Concrete Samples*

Following the mixing process, the concrete mixtures were poured into cubic shapes with the side length of 150 mm, Figure 4a, and rectangular prismatic shapes with the following dimensions 150 mm in length, 120 m width, and 20 mm depth, Figure 4b, in order to perform the experimental measurements.

The concrete samples in cubic shapes were made for the experiment of the compressive strength and density investigation; the rectangular prismatic concrete samples (plate concrete samples) were made for the experiment of the heat transfer coefficient investigation.

After the preparation, the concrete samples were left for curing for 28 days in laboratory conditions at a controlled temperature.

(**a**) (**b**)

Figure 4. View of concrete samples: (**a**) cubic shape concrete samples, (**b**) plate shape concrete samples.

4. Characterization of Concrete Properties

4.1. Density

Experimental investigation on the density of the concrete samples was carried out according to the EN12390-7 standard [35] on tree specimens for every type of concrete mix with waste substitution by determining the mass and the volume of the specimens. The volume was determined with the reference method by water displacement after ensuring that the specimens were in saturated conditions. The mass in water was measured by raising the water level from a tank until the stirrup without the specimens was touching the bottom of the tank. After this procedure, the water surface surplus was removed by wiping the specimens and the mass of the specimens in air conditions was measured. With the values obtained we calculated the volume of the specimens with the Formula (1):

$$V = \frac{m_a - [(m_{st} + m_w) - m_{st}]}{\rho_w} \quad (1)$$

The mass of the specimens was measured after the curing period, according to the EN12390-7 standard 35.

The density can be calculated with the formula:

$$\rho = \frac{m}{V} \quad (2)$$

4.2. Compressive Strength

The preparing and keeping of the concrete samples were performed respecting the indications of EN 12390-2 standard [41]. Following the standard instructions the concrete samples were cured for a period of 28 days. During the curing period the samples were kept in water for 7 days after that in dry conditions at constant temperature of 20 °C with the maximum tolerance ±2 °C.

The compressive strength was determined according to EN 12390-3 standard [42] on tree specimens for every type of concrete mix with waste substitution on the hardened cubic concrete samples with the edge of 150 mm. The samples were tested after the curing period with a press having a maximum load force of 2000 kN. The preparation of the specimen for the experimental investigation on compressive strength was realized by centering the specimen on the lower platen.

In the testing procedure presented in Figure 5, a first initial rate load force of 0.6 $N/mm^2 \cdot s$ constant at the beginning is applied on the area of the concrete sample without shock, after applying the first load force to the specimen, the load was increased continuously at a constant rate ±10% until the specimen reached the maximum load failure.

Figure 5. Experimental test of the compressive strength.

4.3. *Discussion and Analysis*

The experimental determination of the concrete density is presented in Figure 6. From the measurements, the fact that an increase in the concentration of added waste aggregates leads to a decrease in the concrete density. For the obtained values for the densities of the concrete samples with waste as aggregate substitution, the uncertainty on the results calculated for the density is from 3.6% to 36.88%.

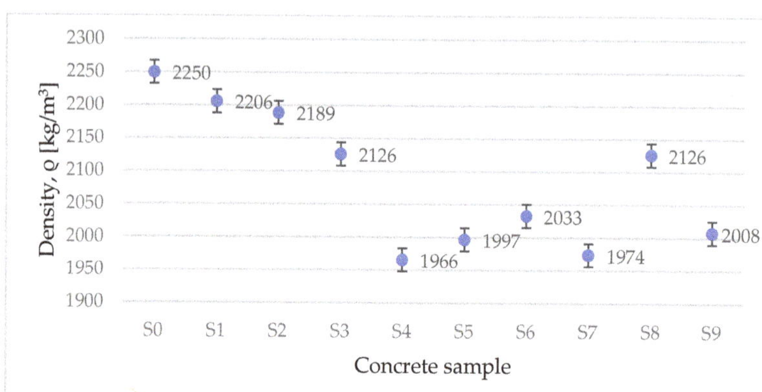

Figure 6. Variation of density for tested concrete samples.

In the testing procedure presented in Figure 6, an initial load force of 0.6 $N/mm^2 \cdot s$ constant force is applied at the beginning on the surface area of the concrete sample without shock. After applying the first load force to the specimen, the load was increased continuously at a constant rate of ±10% until the specimen reached the maximum load failure.

The uncertainty of the obtained results was calculated for the compressive strength on the studied specimens and it resulted in a minimum of 0.3% and a maximum value of 1.35%.

The variation of compressive strength presented in Figure 7 represents the average of the measured results on tree concrete specimens. As we expected, the compressive strength is decreasing by using even a small percentage of waste as a substitution.

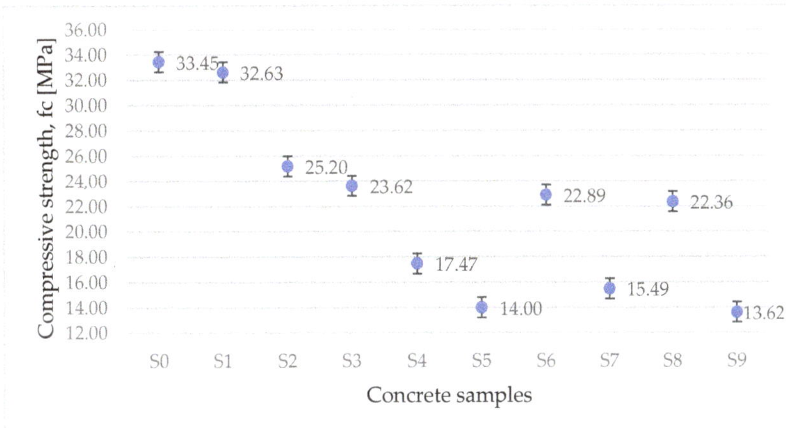

Figure 7. Variation of compressive strength for tested concrete samples.

In Figure 8a–d the variation of the density versus compressive strength of the concrete samples with waste substitution compared with the concrete sample with no waste replacement was analyzed.

A first analysis was performed by comparing the density versus compressive strength of concretes in which one natural aggregate was replaced with the same waste used in the concrete mixture, in this case, the natural aggregate sand was substituted with waste chopped PET.

By comparing the obtained results Figure 8a, in which only one natural aggregate was replaced, we observed that, up to a ratio of 20% substitution of natural aggregate with waste chopped PET, the results of the compressive strength show that the concrete can be used as a structural element and, due to it having a relatively high density, the concrete can be used as a heat-accumulating wall. Meanwhile, by increasing the waste substitution up to 60%, we observed that the compressive strength and density decrease, with the result that the concrete structure can be used as a self-sustained material or even as a heat-accumulating wall.

The second analysis was carried out on the concrete samples, substituting the natural aggregates with granular polystyrene. The results presented in Figure 8b highlight the fact that, in the case where a high percentage of aggregate sand, the compressive strength is greatly reduced and the density is slightly lower in comparison to the results of concrete samples with two natural aggregates as a substitution. The types of concrete with a high percentage of aggregates substitution can be used as a self-sustained structure and thermal insulating material, while replacing a small percentage of the natural aggregates with waste. Continuing the analyses of the mechanical properties for the concrete samples with waste wood substitution, Figure 8c, the results show that a higher percentage of substitution has the same effect as in the case of using waste chopped PET and granular polystyrene.

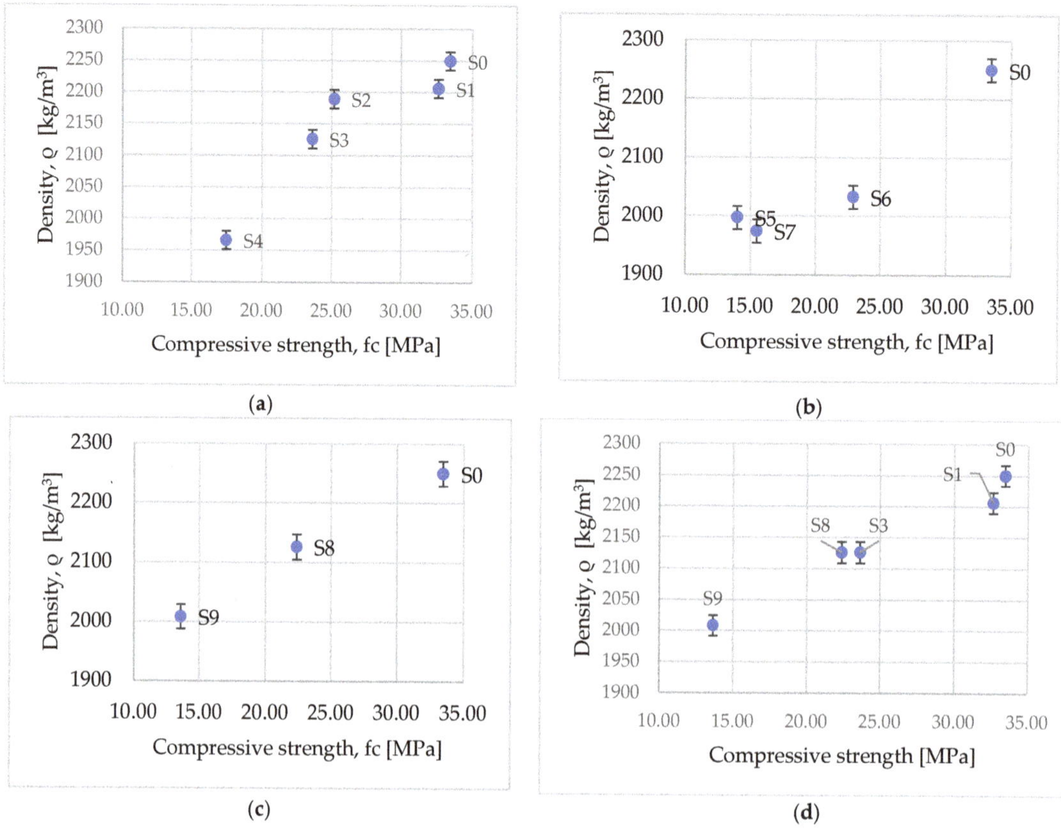

Figure 8. Variation of density versus compressive strength for concrete with waste: (**a**) concrete with waste chopped PET; (**b**) concrete with granular polystyrene; (**c**) concrete with sawdust; (**d**) concrete with waste chopped PET and sawdust.

A final analysis on the mechanical properties for the concrete with different waste substitution at the same percentage as the waste replacement is presented in Figure 8d, which highlights that the concrete has the same density at 40% substitution with waste chopped PET as at 10% of the substitution with sawdust, resulting that the waste wood is negatively affecting the mechanical properties even at a reduced percentage of substitution, the concrete could be used as self-sustained material or as heat accumulator wall.

Results from Figure 8 were obtained after testing three specimens for the same concrete mix with waste aggregates. The values presented in Figure 8 were obtained by calculating the average of the measured values obtained at the end of the test.

5. Characterization of Concrete Thermal Properties

5.1. Experimental Set Up and Process

The thermal properties were determined using an experimental device, developed at IUSTI Laboratory, which is based on the hot plane method with temperature measurements on both sides of the sample. The characterization is carried out at room temperature. A scheme of the experimental device is shown in Figure 9.

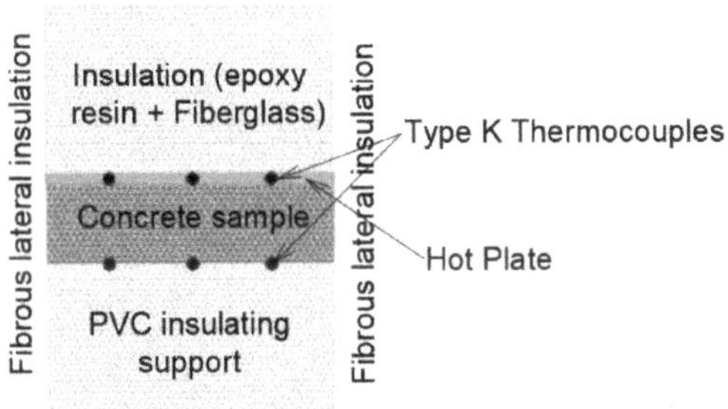

Figure 9. Experimental device.

The concrete sample is heated in the front face by a heating sole (hot plane) (Figure 9) consisting of a flat electrical resistance pasted to a copper plate. This part of the sample is grooved in order to allow the placement of three K thermocouples in contact with the sample. Three other thermocouples are placed in grooves of the sample holder that will be in contact with the 'rear face' of the sample (Figure 9). A thermal paste provides good contact between the faces of the sample, the hot plate, and the sample holder. Thanks to the uniformity of the supplied heat flux and the limitation of the lateral thermal losses, the heat transfer in the sample are assumed to be 1D. For each test, the experimental parameters to adjust are the heat flux supplied to the hot plane, the heating duration, and the test duration. Thus, for these various samples, the duration of the heating is 6 min for a heat flux of 16 W.

To achieve these measurements, the concrete sample was first positioned in the thermal device (Figure 9), and then heated for 6 min; then the heating was stopped and the cooling of the sample was awaited until it reached its steady state. The temperature signals were recorded; the signals corresponding to the 3 K thermocouples in the front face of the sample (Figure 9) were averaged to provide the thermogram of the front face; the same process was performed for the signals given by the 3 K thermocouples at the rear face of the sample (Figure 9).

A transient thermal model (thermal quadrupoles) was used to calculate the theoretical thermogram of the rear face, by convolution of the thermogram of the front face of the sample. This theoretical thermogram is a function of the following parameters:

- $t_d = e^2/a$
- $a = k/\rho Cp$
- $x = b/b_{PVC}$, corresponding to the ratio of the samples upon the sample's holder (PVC) effusivities.

The effusivity is given by:

- $b = \sqrt{k\rho c_p} = k/\sqrt{a} = \rho c_p \sqrt{a}$

These parameters were then identified by a least squares method, minimizing the gap between the theoretical and experimental thermograms, over range duration around the diffusive time.

The appropriate following parameters are then extracted.

- The thermal diffusivity a;
- The thermal effusivity b;
- The thermal conductivity $k = b\sqrt{a}$;
- The thermal capacity $c_p = b/\rho\sqrt{a}$.

The uncertainties calculation were carried out taking into account:

- The error of the thermocouples measurements (standard deviation: 0.005°C);
- The error due to the considered known parameters for determining a, b, k, ρ, and C_p, i.e.,: error on the thickness e: 1%; error on the PVC effusivity b_{PVC}: 10% and density error ρ: 4%.

First, a calibration was achieved to determine the sample holder (PVC) thermal diffusivity and effusivity. It was found that b_{PVC} is around (518 ± 10%) $Jm^{-2}K^{-1}s^{-0.5}$, which corresponds to values given by the manufacturer.

5.2. Results and Discussion

For each sample, the same process was applied in order to determine the various thermal parameters. In Figure 10, the thermal conductivity is plotted for the various considered concrete samples.

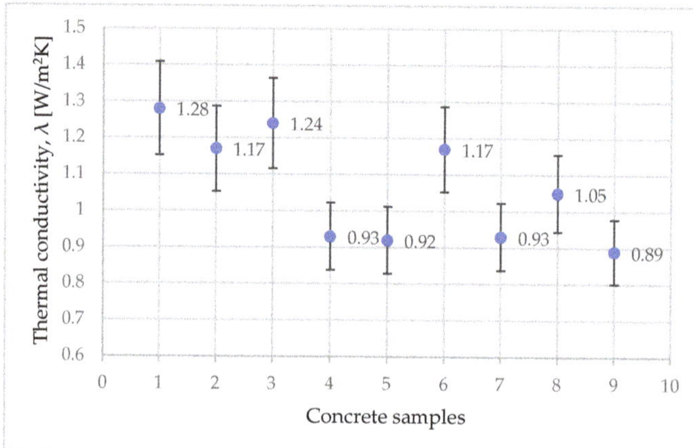

Figure 10. Variation of thermal conductivity.

From the thermal conductivity measurements, one can observe that the values of thermal conductivity are lower than the thermal conductivity of the standard concrete (S0). This result is consistent with the results found in the literature, since the thermal conductivity of the added waste is below the concrete one. Furthermore, reducing the thermal conductivity of concrete would provide better insulation for the building.

Following the measurements, Figure 11 was created highlighting the variation of thermal conductivity of the samples versus their density. As for compressive strength, the same behavior is highlighted; indeed, low density leads to low thermal conductivity.

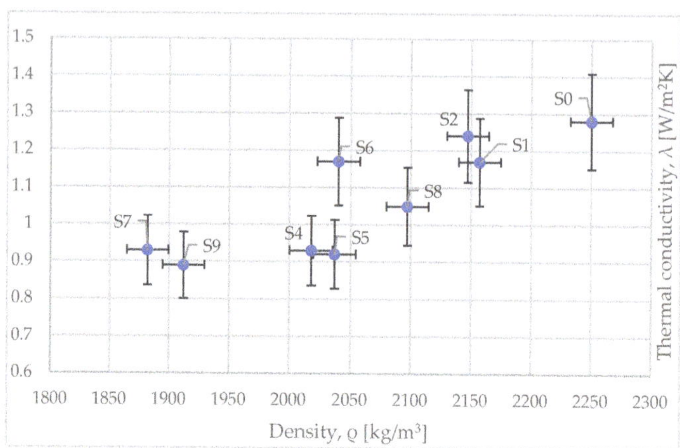

Figure 11. Variation of thermal conductivity versus density.

In order to illustrate the effect of the addition of waste on the global behavior of the concrete, another figure was created, Figure 12, which shows the compressive strength ratio versus the thermal conductivity.

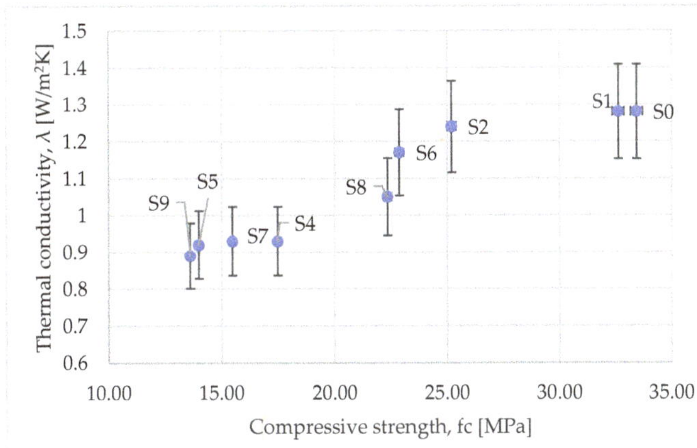

Figure 12. Variation of compressive strength versus density.

From Figure 12, one can notice the existence of three zones:

- The first one corresponds to concrete formed by the addition of a high concentration of waste, typically higher or equal to 50% (S4, S5, S7, and S9); the thermal conductivity was approximately reduced by almost 25%, while the compressive strength was reduced by 45–60%. One can consider that this zone is rather characterized by a constant thermal conductivity.
- The second zone gathers a moderate concentration of waste (S2, S3, S6, and S8), showing that the compressive strength is rather constant (a reduction of around 25%) whereas the thermal conductivity ratio is reduced by between approximately 5 and 20%. This zone will be considered as a zone with a constant compressive strength.
- Finally, the third zone, corresponding to the low concentration of waste, behaves mechanically as the standard concrete, the only noticeable difference being the lower thermal conductivity.

These results fall into accordance with the results found in the literature and infer the fact that, depending on the type of application, the appropriate concentration of waste aggregate can provide significant advantages in comparison to regular concrete. For example, through adding a low concentration of waste materials, the mechanical properties of the newly formed concrete are similar to those of the regular concrete, being good for the structural elements of buildings or, through adding an increased quantity of waste materials, the mixture provides a lower material conductivity, helping insulate the building and preventing heat loss [22–25,27–29].

6. Conclusions

The experimental study investigated the use of waste materials as a partial substitution of aggregates and cement in manufacturing concrete, and analyzed the mechanical properties via the compressive strength, density variations, and the thermal properties via the thermal conductivity parameter.

Three zones of behaviors were highlighted, the first one, corresponding to high waste concentration, leads approximately to a constant reduction in the thermal conductivity. The second, characterized by moderate waste concentration, behaves as a constant compressive strength. In addition, finally, the third one, defined by very low waste concentration, induces only a very small reduction of the thermal conductivity with the compressive strength approximately the same as standard concrete.

By analyzing the compressive strength of the concrete with waste aggregates substitution, the study highlighted that the concrete can be used as a structural material such as self-sustained walls for low waste concentration, the differences in mechanical properties being negligible. On the other hand, high concentrations of waste aggregate in the concrete lead to a decrease in thermal conductivity, providing good solution for nonstructural walls with an increased insulation capacity in comparison to regular concrete walls, preventing heat loss.

One of the most important key factors is that, through reusing the waste and not using conventional aggregates, the carbon footprint of the building is reduced, this being a significant factor in the development of the construction sector in the future.

The studies revealed that a 10% replacement of the cement with fly ash and a 10% replacement of the sand with chopped PET had minimal impact on the properties of concrete, but on an environmental scale, if concrete were to have a portion of the natural aggregates replaced, this would have a significant impact both in terms of the reduction in necessary materials and in terms of pollution, as the waste materials would be given a new purpose.

Author Contributions: Conceptualization, C.A., M.B. (Marinela Barbuta), A.B., M.B. (Marius Branoaea) and R.S.V.; data curation, F.R.; formal analysis, G.S.; funding acquisition, M.C.B.; investigation, G.S.; methodology, G.S.; project administration, C.A., M.B. (Marinela Barbuta) and A.B.; resources, M.C.B.; supervision, C.A., M.B. (Marinela Barbuta) and A.B.; visualization, G.S.; writing—original draft. All authors have read and agreed to the published version of the manuscript.

Funding: This work was supported by a publications grant of the "Gheorghe Asachi" Technical University from Iasi, Romania (TUIASI), project number GI/P28/2021.

Institutional Review Board Statement: Not applicable.

Informed Consent Statement: Not applicable.

Data Availability Statement: All data are contained within the article.

Acknowledgments: The authors are thankful to Nathalie EHRET and Fabrice RIGOLLET from IUSTI Laboratory, Aix-Marseille Université, CNRS, Marseille, France for the support provided to realize the experimental investigation for determining the thermal properties of the concrete samples.

Conflicts of Interest: The authors declare no conflict of interest.

Nomenclature

PET	Polyethylene terephthalate
V	The volume of the specimen [m^3]
m_a	The mass of the specimen in air [kg]
m_{st}	The apparent mass of the immersed stirrup [kg]
m_w	The apparent mass of the immersed specimen [kg]
ρ_w	The density of water, at 20 °C [kg/m^3]
ρ_c	The density of the specimen [kg/m^3]
m	The mass of the specimen [kg]
t_d	The diffusive time of the sample
e	The sample thickness [m]
a	Thermal diffusivity [m^2/s]
b	Thermal effusivity [$Jm^{-2}K^{-1}s^{-0.5}$]
k	Thermal conductivity [W/mK]
C_p	Thermal capacity [J/kgK]
x	The ratio of the sample

References

1. Santamouris, M.; Vasilakopoulou, K. Present and future energy consumption of buildings: Challenges and opportunities towards decarbonisation. *E-Prime Adv. Electr. Eng. Electron. Energy* **2021**, *1*, 100002. [CrossRef]
2. Andrew, M. Global CO_2 emissions from cement production, 1928–2018. *Earth Syst. Sci. Data* **2019**, *11*, 1675–1710. [CrossRef]
3. Guo, R.; Wang, J.; Bing, L.; Tong, D.; Ciais, P.; Davis, S.J.; Andrew, R.M.; Xi, F.; Liu, Z. Global CO2 uptake of cement in 1930–2019. *Earth Syst. Data* **2020**, *13*, 1791–1805. [CrossRef]
4. Mohammed, A.A. Modelling the mechanical properties of concrete containing PET waste aggregate. *Constr. Build. Mater.* **2017**, *150*, 595–605. [CrossRef]
5. Nursyamsi, W.S.B.Z. The Influence of Pet Plastic Waste Gradations as Coarse Aggregate Towards Compressive Strength of Light Concrete. *Procedia Eng.* **2017**, *171*, 614–619. [CrossRef]
6. Pereira, E.L.; de Oliveira Junior, A.L.; Fineza, A.G. Optimization of mechanical properties in concrete reinforced with fibers from solid urban wastes (PET bottles) for the production of ecological concrete. *Constr. Build. Mater.* **2017**, *149*, 837–848. [CrossRef]
7. Cadere, C.A.; Barbuta, M.; Rosca, B.; Serbanoiu, A.A.; Burlacu, A.; Oancea, I. Engineering properties of concrete with polystyrene granules. *Procedia Manuf.* **2018**, *22*, 288–293. [CrossRef]
8. Gregorova, V.; Ledererova, M.; Stefunkova, Z. Investigation of Influence of Recycled Plastics from Cable, Ethylene Vinyl Acetate and Polystyrene Waste on Lightweight Concrete Properties. *Procedia Eng.* **2017**, *195*, 127–133. [CrossRef]
9. Xu, Y.; Jiang, L.; Xu, J.; Li, Y. Mechanical properties of expanded polystyrene lightweight aggregate concrete and brick. *Constr. Build. Mater.* **2012**, *27*, 32–38. [CrossRef]
10. Duan, P.; Yan, C.; Zhou, W.; Luo, W. Fresh properties, mechanical strength and microstructure of fly ash geopolymer paste reinforced with sawdust. *Constr. Build. Mater.* **2016**, *111*, 600–610. [CrossRef]
11. Kaya, A.; Kar, F. Properties of concrete containing waste expanded polystyrene and natural resin. *Constr. Build. Mater.* **2016**, *105*, 572–578. [CrossRef]
12. Saikia, N.; de Brito, J. Mechanical properties and abrasion behaviour of concrete containing shredded PET bottle waste as a partial substitution of natural aggregate. *Constr. Build. Mater.* **2014**, *52*, 236–244. [CrossRef]
13. Rahmani, E.; Dehestani, M.; Beygi, M.H.A.; Allahyari, H.; Nikbin, I.M. On the mechanical properties of concrete containing waste PET particles. *Constr. Build. Mater.* **2013**, *47*, 1302–1308. [CrossRef]
14. Gesoglu, M.; Güneyisi, E.; Hansu, O.; Etli, S.; Alhassan, M. Mechanical and fracture characteristics of self-compacting concretes containing different percentage of plastic waste powder. *Constr. Build. Mater.* **2017**, *140*, 562–569. [CrossRef]
15. Islam, M.J.; Meherier, M.S.; Islam, A.K.M.R. Effects of waste PET as coarse aggregate on the fresh and harden properties of concrete. *Constr. Build. Mater.* **2016**, *125*, 946–951. [CrossRef]
16. Maldonado-Bandala, E.E.; Nieves-Mendoza, D.; Romero-López, R.; Tobias-Jaramillo, R.; Almeraya-Calderón, F.; Barrios-Durstewitz, C.P.; Núñez Jaquez, R.E. Electrochemical and Mechanical Properties of Lightweight Concrete Blocks with Expanded Polystyrene Foam. *Int. J. Electrochem. Sci* **2015**, *10*, 472–485.
17. Sayadi, A.A.; Tapia, V.; Neitzert, R.; Clifton, G. Effects of expanded polystyrene (EPS) particles on fire resistance, thermal conductivity and compressive strength of foamed concrete. *Constr. Build. Mater.* **2016**, *112*, 716–724. [CrossRef]
18. Sofi, A.; Saxena, A.; Agrawal, P.; Sharma, A.; Sharma, K. Strength Predictions of Saw Dust and Steel Fibres in Concrete. *Int. J. Innov. Res. Sci. Eng. Technol.* **2015**, *4*, 12473–12477.
19. Bajpai, P.; Brar, A.; Nara, A.; Dudeja, N. Effect on strength properties of concrete by using waste wood ash as partial replacement of cement. *Int. J. Adv. Technol. Eng. Sci.* **2017**, *5*, 545–549.
20. Zeggar, M.L.; Azline, N.; Safiee, N.A. Fly ash as supplementry material in concrete: A review. *IOP Conf. Ser. Earth Environ. Sci.* **2019**, *357*, 012025. [CrossRef]
21. Susanti, E.; Istiono, H.; Komara, I.; Pertiwi, D.; Septiarsilia, Y.; Syahputra, F.K. Effect of fly ash to water-cement ratio on the characterization of the concrete strength. *IOP Conf. Ser. Mater. Sci. Eng.* **2021**, *1010*, 012035. [CrossRef]
22. González, D.; Caballero, P.P.; Fernández, D.B.; Vidal, M.M.J.; Sáez del Bosque, I.F.; Martínez, C.M. The Design and Development of Recycled Concretes in a Circular Economy Using Mixed Construction and Demolition Waste. *Materials* **2021**, *14*, 4762. [CrossRef] [PubMed]
23. Horňáková, M.; Lehner, P. Analysis of Measured Parameters in Relation to the Amount of Fibre in Lightweight Red Ceramic Waste Aggregate Concrete. *Mathematics* **2022**, *10*, 229. [CrossRef]
24. Tran, D.V.P.; Allawi, A.; Albayati, A.; Cao, T.N.; El-Zohairy, A.; Nguyen, Y.T.H. Recycled Concrete Aggregate for Medium-Quality Structural Concrete. *Materials* **2021**, *14*, 4612. [CrossRef] [PubMed]
25. Diaconu, L.I.; Rujanu, M.; Diaconu, A.C.; Serbanoiu, A.A.; Babor, D.; Plian, D. Improvement of the concrete behaviour to sulphate corrosion using fly ash admixture collected by wet process. *IOP Conf. Ser. Mater. Sci. Eng.* **2020**, *789*, 012018. [CrossRef]
26. Grădinaru, C.M.; Muntean, R.; Serbanoiu, A.A.; Ciocan, V.; Burlacu, A. Sustainable Development of Human Society in Terms of Natural Depleting Resources Preservation Using Natural Renewable Raw Materials in a Novel Ecological Material Production. *Sustainability* **2020**, *12*, 2651. [CrossRef]
27. Meza, A.; Pujadas, P.; Meza, L.M.; Pardo-Bosch, F.; López-Carreño, R.D. Mechanical Optimization of Concrete with Recycled PET Fibres Based on a Statistical-Experimental Study. *Materials* **2021**, *14*, 240. [CrossRef] [PubMed]
28. Suliman, N.H.; Razak, A.A.A.; Mansor, H.; Alisibramulisi, A.; Amin, N.M. Concrete using sawdust as partial replacement of sand: Is it strong and does not endanger health? *MATEC Web Conf.* **2019**, *258*, 01015. [CrossRef]

29. Bouaissi, A.; Li, L.Y.; Abdullah, M.M.A.B.; Ahmad, R.; Razak, R.A.; Yahya, Z. Fly Ash as a Cementitious Material for Concrete. In *Zero-Energy Buildings—New Approaches and Technologies*; IntechOpen: London, UK, 2020.
30. Bheel, N.; Ali, M.O.; Khahro, S.H.; Keerio, M.A. Experimental study on fresh, mechanical properties and embodied carbon of concrete blended with sugarcane bagasse ash, metakaolin, and millet husk ash as ternary cementitious material. *Environ. Sci. Pollut. Res.* **2022**, *29*, 5224–5239. [CrossRef] [PubMed]
31. Mangi, S.A.; Makhija, A.; Raza, M.S.; Khahro, S.H.; Jhatial, A.A. Comprehensive Review on Effects of Seawater on Engineering Properties of Concrete. *Silicon* **2021**, *13*, 4519–4526. [CrossRef]
32. Paramasivam, P.; Loke, Y.O. Study of sawdust concrete. *Int. J. Cem. Compos. Lightweight Concr.* **1980**, *2*, 57–61. [CrossRef]
33. Sales, A.; Rodrigues de Souza, F.; Nunes dos Santos, W.; Zimer, A.M.; Almeida, F.d.C.R. Lightweight composite concrete produced with water treatment sludge and sawdust: Thermal properties and potential application. *Constr. Build. Mater.* **2010**, *24*, 2446–2453. [CrossRef]
34. Joshi, C.; Lohita, R.P. *Fly Ash in Concrete: Production, Properties and Uses*; Gordon & Breach Science Publishers: London, UK, 1997.
35. EN12390-7. *Testing Hardened Concrete—Part 7: Density of Hardened Concrete*; British Standards Institution: London, UK, 2019.
36. Holcim. Available online: https://www.holcim.ro/en/business/products/cement/cem-iia-ll-425r (accessed on 9 November 2021).
37. Buema, G.; Lisa, G.; Kotova, O.; Ciobanu, G.; Ivaniciuc, L.; Favier, L.; Harja, M. Application of thermal analysis to improve the preparation conditions of zeolitic materials from flying ash. *Environ. Eng. Manag. J.* **2021**, *20*, 377–388.
38. Bhatt, A.; Priyadarshini, S.; Mohanakrishnan, A.A.; Abri, A.; Sattler, M.; Techapaphawit, S. Physical, chemical, and geotechnical properties of coal fly ash: A global review. *Case Stud. Constr. Mater.* **2019**, *11*, e00263. [CrossRef]
39. Bărbuţă, M.; Ţăranu, N.; Harja, M. Wastes Used in Obtaining Polymer Composite. *Environ. Eng. Manag. J.* **2009**, *8*, 1145–1150.
40. Horisawa, S.; Sunagawa, M.; Tamai, Y.; Matsuoka, Y.; Miura, T.; Terazawa, M. Biodegradation of nonlignocellulosic substances II: Physical and chemical properties of sawdust before and after use as artificial soil. *J. Wood Sci.* **1999**, *45*, 492–497. [CrossRef]
41. EN 12390-2. *Testing Hardened Concrete—Making and Curing Specimens for Strength Tests;* British Standards Institution: London, UK, 2019.
42. EN 12390-3. *Testing Hardened Concrete—Part 3: Compressive Strength of Test Specimens; British Standards Institution: London, UK, 2019.*

materials

MDPI

Article

In Situ Compensation Method for Precise Integral SQUID Magnetometry of Miniscule Biological, Chemical, and Powder Specimens Requiring the Use of Capsules

Katarzyna Gas * and Maciej Sawicki *

Institute of Physics, Polish Academy of Sciences, Aleja Lotnikow 32/46, PL-02668 Warsaw, Poland
* Correspondence: kgas@ifpan.edu.pl (K.G.); mikes@ifpan.edu.pl (M.S.)

Abstract: Steadily growing interest in magnetic characterization of organic compounds for therapeutic purposes or of other irregularly shaped specimens calls for refinements of experimental methodology to satisfy experimental challenges. Encapsulation in capsules remains the method of choice, but its applicability in precise magnetometry is limited. This is particularly true for minute specimens in the single milligram range as they are outweighed by the capsules and are subject to large alignment errors. We present here a completely new experimental methodology that permits 30-fold in situ reduction of the signal of capsules by substantially restoring the symmetry of the sample holder that is otherwise broken by the presence of the capsule. In practical terms it means that the standard 30 mg capsule is seen by the magnetometer as approximately a 1 mg object, effectively opening the window for precise magnetometry of single milligram specimens. The method is shown to work down to 1.8 K and in the whole range of the magnetic fields. The method is demonstrated and validated using the reciprocal space option of MPMS-SQUID magnetometers; however, it can be easily incorporated in any magnetometer that can accommodate straw sample holders (i.e., the VSM-SQUID). Importantly, the improved sensitivity is accomplished relying only on the standard accessories and data reduction method provided by the SQUID manufacturer, eliminating the need for elaborate raw data manipulations.

Keywords: SQUID; integral magnetometry; in situ compensation; gelatin capsules; capsules mounting; amorphous specimens; straw sample holder; powder samples; SnTe; turmeric

check for updates

Citation: Gas, K.; Sawicki, M. In Situ Compensation Method for Precise Integral SQUID Magnetometry of Miniscule Biological, Chemical, and Powder Specimens Requiring the Use of Capsules. *Materials* **2022**, *15*, 495. https://doi.org/10.3390/ma15020495

Academic Editor: In-Sang Yang

Received: 18 December 2021
Accepted: 6 January 2022
Published: 10 January 2022

1. Introduction

Commercial general-purpose superconducting quantum interference devices (SQUID) have become ubiquitous experimental tools in laboratories where magnetic characterization and investigations of nanometer-scale objects is important and frequently indispensable in providing the vital insight into investigations. The range of objects include, in particular, ultrathin films [1,2], nanoparticles [3,4], and inclusions in magnetically dilute systems [5–7], 2D [8,9], and 1D [10,11] systems, or systems in which their surface matters [12], like topological insulators [13–15]. In the prevailing number of cases, the materials of interest are on a solid state substrate, which make the mounting of the sample relatively easy; a breadth of relevant experimental codes has been elaborated and methods for eliminating various pitfalls have been presented [16–20], which were reviewed by Pereira [21]. In this context, a new window of experimental opportunities was opened by the elaboration of methodology and full quantification of the in situ substrate compensational method [22], which allows for truly precise integral magnetometry of nanometer scale objects with absolute sensitivity down to 10^{-8} emu [23] or of thin antiferromagnetic layers [24].

However, there exists a wide spectrum of subjects in which SQUID magnetometry cannot be performed according to the rules elaborated for solid state objects. These are shapeless bodies, like irregular small bits or fragments, powdered materials, generally understood soft condensed materials [25], and all other chemical compounds and biological specimens, frequently of a strong spintronics [26] or therapeutic relevance [27–30]. These systems usually exhibit very weak magnetic properties, and they are predominantly diamagnets. Only a handful of them allow for the use of methods developed within the solid state approach to magnetometry—these that can be bound by weakly magnetic glues without any degradation effects. One of such very convenient bonding agents is the so-called GE varnish [31], which dissolves very easily in ethanol and works (bonds) down to the lowest temperatures. If the investigated material does not decompose or change its properties in ethanol, the investigated substance can be stabilized in the GE varnish and be affixed firmly to common sample holders, like rectangular pieces of semiconducting compound, and then mounted into plastic straws [32] or deposited on Si [25,33,34]. In all other cases, different, usually custom devised approaches had to be developed [35]. As far as the present authors can judge, none of them can be regarded as magnetically transparent or being able to provide truly quantitative results.

In this context, small capsules appear to be the solution of choice in magnetometry of small quantities of shapeless materials, which evade other known methods of sample mounting for the measurements [36,37]. There are many varieties of capsules available, but for the magnetometry the capsules should be first of all small, to minimize their own magnetic contribution to the research, and colorless (transparent), to minimize the amount of dyes that usually add a paramagnetic contribution to their generally diamagnetic response. These and other contaminants make their magnetic properties difficult to standardize. The second serious issue related to the capsule-based magnetometry is the lack of simple and magnetically clean means to firmly fix the specimen-containing capsule to the magnetometer sample holder. By far the most popular Quantum Design (QD) [38] lines of magnetometers—MPMS-XL and the MPMS-3 VSM SQUID—have provided means to use transparent drinking straws for sample housing. Indeed, most of the SQUID magnetometry research has been performed with an aid of these straws [20,39,40]; other, more clean but elaborate solutions are exercised less often [19,22].

It turns out the inner diameter of the straws provided by QD exceeds the outer diameter of capsules that can be obtained from the same source. Neither size 4 gelatin capsules (GC) nor #5 polycarbonate capsules (PCC) can be stabilized firmly within these straws, so various means of their stabilization have been exercised. Most of these ad hoc solutions, like indentations of the straw below and above the capsule [37], insertion of pieces of cotton wool, various plastic sticks, or short pieces of somehow deformed straws might have proven sufficient for the fixing problem, but each of them introduces their own magnetic contamination to the measurement process. Therefore, these solutions work very well only when the researched signals are strong, exceeding at least by an order magnitude the total magnetic signal of the capsule and its harness within the straw. However, frequently, the researched material is characterized by a weak magnetic response, say of a range typical for common diamagnetic substances, and/or for various technological reasons only miniscule amounts are available for the studies. Therefore, a new approach to precision integral magnetometry is needed to give a better experimental account in these situations where the investigated signals are far smaller than that of the capsule and its surroundings.

In this report we provide a complete solution to tackle the challenges related to precision integral magnetometry of miniscule amounts of shapeless substances. The method we put forward here combines both the capsules and modified plastic straws in a magnetically clean manner that allows the determination of the absolute magnitude of the magnetic moment of single-milligram-small diamagnetic samples with 1% accuracy. The method is validated for a crystalline body and exemplified for a powder sample. It has clear potential to considerably impact the research of these liquid solutions and wet biological specimens that will withstand the dry environment of the SQUID sample chamber without marked modifications to their properties. The latter line of research has not yet been exercised by the authors.

Our report is organized as follows. We first introduce the method of permanent plastic straw modification that produces straws with customized (reduced) internal diameter. This makes them suitable to accommodate capsules and provides sufficiently strong grip permitting 4 cm long scans in 1 Hz oscillating mode of the fast averaging mode, the so-called reciprocal space option (RSO) of the MPMS systems. We then introduce the method of in situ compensation of the signal of the specimen-housing capsule by substantially restoring the symmetry of the specimen holder that is otherwise broken by the presence of the capsule. We provide the complete record of the assembly process of the capsule compensational sample holder (C-CSH) and show that a 30-fold reduction of the apparent mass of the capsule, that is, the mass of the capsule that is sensed by the SQUID, is readily possible. We finally validate our method by measuring the complete temperature dependence of the magnetic susceptibility of a 2.1 mg small crystallite of diamagnetic SnTe in our C-CSH. Other measurements of a 2.9 mg sample of common turmeric powder exemplifies the applicability of our approach to investigations of equally small amounts of powders, which we quantify numerically. All of these accomplishments are done using only data reduction that is based on the standard output files of the MPMS system (*.dat files). It is shown that none of the excessive post-measurement data treatment is necessary. The method outlined here has been proven to be suitable for measurement from 2 to approximately 330 K and in the whole range of the magnetic fields available within the magnetometer and to be compatible with the RSO mode of measurements.

2. Materials and Methods

The main purpose of this report is to provide the reader with the complete method developed to in situ compensate for a large magnetic contribution brought about by common capsules, frequently employed to house soft matter or just shapeless specimens, which are difficult to be fixed to a reasonable sample holder for the intended magnetic studies. Therefore, in this section we stipulate only that the main building blocks of the self-compensating assembly are polycarbonate or gelatin capsules arranged in trains with (nearly) identical neighbors, which are all fixed into thermoplastically modified plastic drinking straws. All these elements are available directly from Quantum Design (San Diego, CA 92121, USA), and we advise to use this source for the sake of standardization. Gelatin capsules (GC) of size 4, or #4, are provided as item QDS-AGC1. The polycarbonate (PCC) ones of size 5 (smaller), or #5, can be ordered as item QDS-AGC3. Both are packed in bunches of 100. However, the inner diameter of the straws provided by QD (about 5.3 mm) exceeds the outer diameter of standard capsules, which are about 4.6 and 5 mm for PCC #5 and GC #4, respectively.

2.1. The Capsules

Apart from the mismatch of the external diameters of capsules with the inner diameter of straw, which determines the amount of effort required to fix them to a sample holder, there are other important factors that need to be discussed here. These are: (i) their magnetic susceptibility, which determines the magnitude of the magnetic response or the signal brought by the capsule to the total magnetic moment m registered during the measurement, (ii) their mass, which quantifies the weight of the sample that can be housed in the given capsule to produce an easy distinguishable magnetic response, and (iii) alignment (proper centering) issues of small specimens. The last factor is somehow the derivative of the former two. In these respects, there are virtually no problems or constraints for investigation of ferromagnetic materials of which even nanograms can dwarf the response of the capsule, particularly at a weak magnetic field. The method of compensation we put forward here will be certainly suitable for such samples, but in the report, we put our emphasis on weakly magnetic substances that are characterized by the weakest magnetic responses—the diamagnetic bodies. Our effort has been directed to minimize as much as possible the response of the capsule, so as to bring forward the weak or even very weak response of the specimen and to quantify its magnitude with a very high accuracy.

Both types of capsules have different sizes and masses. The #5 PCCs range between 22 and 35 mg and the #4 GCs between 30 and 44 mg, but it turns out that both types are characterized by a similar mass susceptibility χ of about $-6.3(2) \times 10^{-7}$ emu/g/Oe. Therefore, for the sake of simplicity and to relate the results to the similarly diamagnetic specimens of a very small mass, wherever possible, the magnetic results are converted to the mass of an idealized, yet quite typical, mass of polycarbonate capsules, i.e., to 30 mg. In that sense this conversion will make the conclusion of our findings independent of the magnitude of the magnetic field *H*, at which the measurements are taken, as both signals of the capsule(s) and the specimen scale similarly with *H*. Additionally, for the sake of argument, we define a standard reference diamagnetic material (SRDM) and assign to it magnetic susceptibility $\chi_{SRDM} = -5 \times 10^{-7}$ emu/g/Oe, that is, a kind of broadly understood average between room temperature magnitudes of χ of capsules and other commonly met materials as: Si ($\chi_{Si} \cong -1.2 \times 10^{-7}$ emu/g/Oe), GaAs ($\chi_{GaAs} \cong -2.25 \times 10^{-7}$ emu/g/Oe) [41], or sapphire ($\chi_{sapphire} \cong -3.5 \times 10^{-7}$ emu/g/Oe) [17].

2.1.1. Hygroscopicity of Gelatin Capsules—Time Domain Studies

We decided to standardize the magnetic results to the mass of typical polycarbonate capsules because they can be regarded as stable in ambient conditions, in stark contrast to the gelatin ones, which are highly hygroscopic. As the result, they are changing their mass in response to the surrounding humidity. The effect is particularly strong after insertion a typical GC to the SQUID chamber. In Figure 1, we compare time dependencies of *m* of both types of capsules. The effects seen for the GC are paramount, its time dependence of *m* is very substantial. There are some basic features which need to be highlighted:

1. Within 20 h spent in the dry residual helium atmosphere in the SQUID sample chamber, the capsule lost nearly 10% of its magnetic response. This ties precisely with the equivalent loss of about 3 mg of its weight—recorded accordingly at the key points of this investigation (the numbers are given in the caption). The equivalent mass of the capsule is indicated on the right Y-scale. This magnitude of change of mass is 50% greater than the mass of the sample that is used for the validation of our C-CSH presented further in the report;
2. Two-thirds of this change takes place within the first 3 h. In other words, the change of the signal during the typical time needed for the measurement of a standard *m(H)* hysteresis curve corresponds to the full mass (that is to the total signal) of about 2 mg of SRDM;
3. Upon removing the sample from the SQUID for about 2 h, the capsule regains approximately 0.8 mg of its weight, but when reinserted to the chamber, the process restarts again and continues with a similar pace. We assign these environmental changes to desorption and absorption of water vapor;
4. The same test executed for a polycarbonate capsule indicated only a very weak gain in *m* (an equivalent of less than 0.2 mg of its mass) seen within first 20 min of the test. This process saturates swiftly and no *m* dependence on time is subsequently detected, as expected for a polycarbonate material.

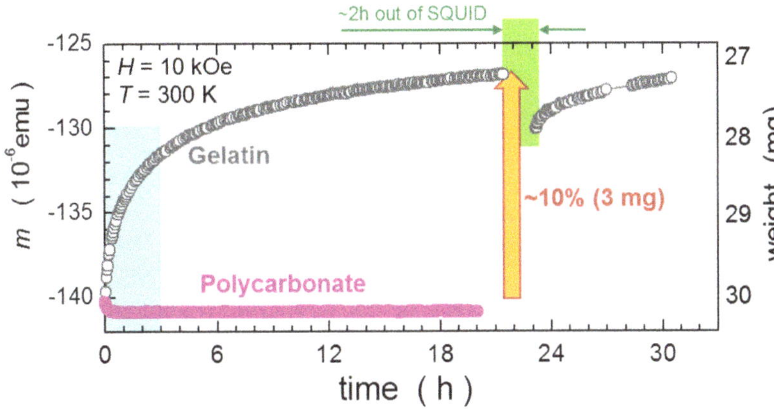

Figure 1. Comparison of time dependencies of the magnetic moment m of typical gelatin (grey symbols) and polycarbonate (magenta) capsules. All magnetic data are normalized to the standard 30 mg capsule and the equivalent scale of mass is given on the right Y-scale of the graph, established from weights of the capsules taken just before inserting to and just after removing from the SQUID sample chamber. The green band in the background indicates the time the gelatin capsule spent in open air during a break in the measurements. The light-blue band at the beginning of the graph marks the typical time needed for the execution of a standard hysteresis curve. The capsules are harness-free mounted in a diameter-reduced straw, as explained in Section 2.2.

2.1.2. Temperature Dependence

The temperature dependent studies of single capsules are summarized in Figure 2. Similar to the Figure 1, we scale m of GC to that of PCC and both relate to the standard mass of 30 mg. Importantly, for this measurement the GC has been kept in the pumped SQUID chamber for nearly 5 h to somehow saturate the dehydration of the gelatin. There are two major features of these results:

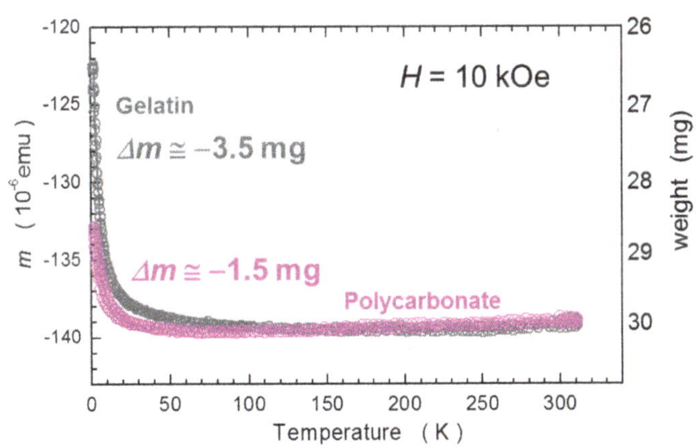

Figure 2. Temperature dependence of the magnetic moment m of single gelatin (grey) and polycarbonate (magenta) capsules. Magnetic data are normalized between themselves and a related standard mass of 30 mg. The magnitude of the paramagnetic shift of m at low temperatures is expressed as an equivalent "loss of mass" of the capsules, Δm. The capsules are harness-free mounted in a diameter-reduced straw, as explained in Section 2.2.

1. The temperature induced changes are relatively small down to about 50 K and relate to about 0.2 mg of SRDM. This is an encouraging result; however, the gelatin capsule had to be seasoned in the SQUID for about 5 h before the *T*-scan commenced;
2. Below about 50 K, strong paramagnetic contributions occur, which corrupt the otherwise ideal diamagnetism of the capsules. It is instructive to relate the reduction of the magnetic response of the capsules to an apparent change of mass of a sample corresponding to SRDM. We find that this change is as strong as above 3 mg for GCs, and is approximately half the magnitude in PCCs. This finding puts a strong constraint on precision magnetometry at low temperatures of miniscule samples requiring encapsulation. Importantly, this effect is practically impossible to compensate for due to unavoidable misalignment errors, discussed in the next section. The only positive aspect of this feature is that it is fully reversible with *T*.

In the main section of this report we show that all these detrimental properties to precision magnetometry of miniscule specimens can be substantially reduced—more than ten times—that is, it can be brought to the level corresponding to 0.2 mg of SRDM.

2.1.3. Capsules as Containers for SQUID Magnetometry (Centering Issues)

The factor that practically precludes precision integral magnetometry using capsules as housing carriers of miniscule samples of a very small m is the challenge of the adequate centering of the whole sample holder (alignment) and related inability of the extraction of the credible magnitudes of m of the scrutinized specimen. We remark first that there is no centering problem when the investigated material fills the whole volume of the capsule, as exemplified by the first four capsules in Figure 3. The magnetic moment of the sample is distributed evenly along the length of the sample and the MultiVu centering routine will direct the user to the proper alignment, marked as the black dashed line. There are also no big issues with establishing the accurate magnitude of m of such massive amounts of material. The reference measurement of the empty capsule will provide the correct magnitudes of the background signal. Alternatively, the studies can be carried out within the automatic background subtraction (ABS) method frame.

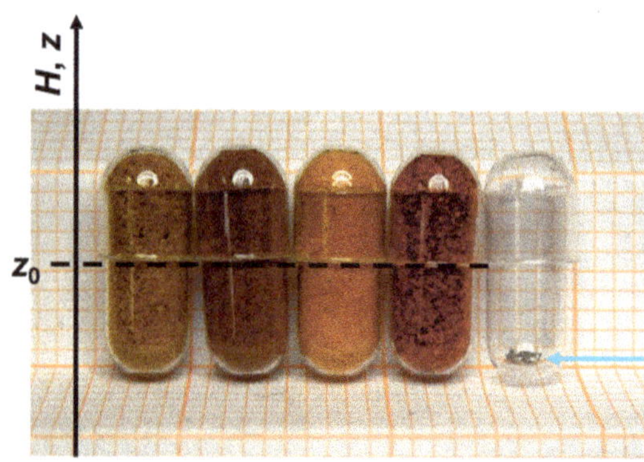

Figure 3. Some examples of the use of small capsules (#5 in this example) for measurements of substances that otherwise would be difficult to reliably and cleanly mount in the SQUID sample chamber. From the left, the first four are fully loaded, so no alignment problems are expected. The magnetic "center of gravity", marked by the dashed black line, coincides with that of the capsules. These capsules house common curry, cinnamon, turmeric, and red pepper powders. The fifth capsule houses a small, 2.1 mg, crystallite of SnTe, used later to validate the in situ compensation method put forward in this report. As indicated, the crystallite sitting at the bottom of the capsule, is shifted by about 5 mm from the center of the capsule, causing serious experimental errors due to alignment problems, discussed in this section.

The problems begin when a small specimen with a weak magnetic response occupies only the bottom end of the capsule, as exemplified by the fifth capsule in Figure 3. Such an object will cause only a small deviation to the dominant response of the capsule. Certainly, one can use a custom-made magnetic marker attached to the bottom of the capsule (Appendix A) to center the sample holder at this location. There is, however, a limited accuracy, or reproducibility in placing the marker, so the alignment errors are unavoidable. We evaluate the misalignment errors by multiple fitting of the SQUID response function (SRF, Appendix B) to the typical scan of a standard capsule, each time setting a different center position z_0 for SRF. The resulting dependence of m on z_0 is depicted in Figure 4 by pink bullets. For convenience, we express the magnitude of m by the corresponding weight. As can be seen, m drops sizably when we move away from the center of the capsule, z_0 = 2.0 cm. More informative is the derivative of this dependency, as it directly quantifies the magnitude of the misalignment errors. We read that it is as high as 10% of the mass of the capsule or simply about 3 mg of the SRDM per one millimeter of the misalignment, when one wants to extract the moment located at the vicinity of the bottom of the capsule. Clearly, only strong magnetic materials can be accurately investigated in such a manner.

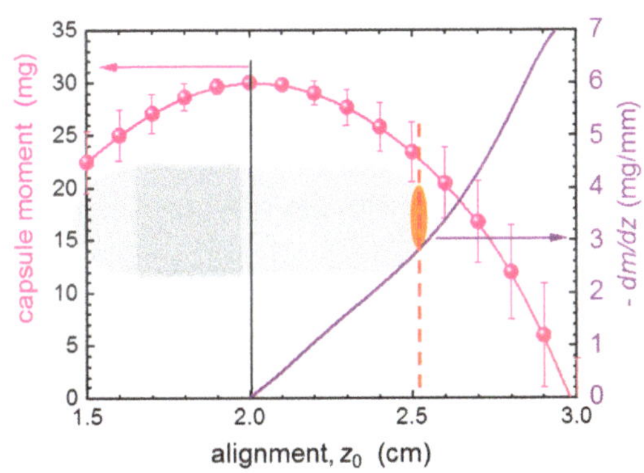

Figure 4. The position dependence of the magnetic moment calculated from a scan of a typical #5 capsule (30 mg of mass) expressed as the corresponding mass (pink bullets). The magnitude of this change defines the absolute value of the misalignment error with respect to the center position (a black vertical line at z_0 = 2.0 cm). The purple solid line yields the derivative of the former, that is the magnitude of the uncertainty due to the misalignment. The sketch of a #5 capsule at the background brings a sense of the real scale. The orange oval mimics a small object of investigation residing at the bottom of the capsule at the position marked by a dashed red line.

2.2. *Reduction of the Diameter of the Straws*

The clear drinking straws can be regarded as the main workhorses of the commercial integral magnetometers due to their "stealthy" properties. Interestingly, these properties are not related to their negligible weak magnetic properties. On the contrary, the magnetic susceptibility of the material of the straw is quite high for a diamagnetic substance: $\chi_{\text{straw}} \cong 8.8 \times 10^{-7}$ emu/g/Oe. At least, such a number has been established at room temperature for a 5 mm long piece of a straw coming from the same batch of straws used in this study. This is some 60–70% more than that of the capsules and of SRDM.

The straws earned their reasonably good reputation [20,32] as they are sufficiently homogeneous and long, so that during the scanning movement no appreciable changes of the magnetic flux are sensed by the detection circuit of the SQUID. The same straws constitute therefore the basic scaffolding within which the in situ compensation assembly is nested. However, the typical internal diameter of the straw is about 5.2 mm, so it exceeds the external diameter of the either type of capsules considered for the use in the magnetometric effort. The capsules move freely along the whole length of the unmodified straw. Since it is hard to design a magnetically stealthy harness for the stabilization of their position in the straw, we present a method for adequate reduction of the diameter of the straw that allows for precise, reproducible, and above all, harness-free firm fixing of capsules. Most conveniently, the design we put forward withstands even very long measurement runs performed in the oscillatory RSO mode without any noticeable change of its alignment with respect to the sensing coils, so no changes to the everyday experimental routines are necessary.

The key factor allowing the required modification of the diameter of the straws is their thermoplasticity. The straws become plastic at approximately 80–100 °C, and only if they are cooled in the adequate mold do they preserve the shape imposed by that mold, at least until reheated back to be plastic again. As the capsules have their external diameter smaller than the straws, we reduce the diameter of the straws by the following procedure:

- We select four straight straws that we clean thoroughly in ethanol. One of them is left aside;
- Next, we incise the wall of the remaining three straws along almost their whole length. That is, the cutting starts approximately 6–8 mm from one end and in one steady move is continued until the other end of the straw is reached. A sharp and clean scalpel blade should be used. The incision should be as smooth as possible, torn edges should be avoided—they are likely sources of additional magnetic signals, which makes such a straw useless as the sample holder;
- We now insert the first of these three cut straws into the uncut one. Obviously, despite being cut along their length, the cut straws preserve their original diameter. Therefore, to nest a cut straw into the uncut one, the inserted straw has to curl a bit to reduce its external diameter to the internal diameter of the uncut one. We press the straw as far as it goes. It stops where the incision begins;
- We continue the process with the second straw. This one has to curl more to nest into the reduced clearance of the already joined first two straws;
- The third inserted straw has to curl even more, so a noticeable resistance can be felt. One can skip this point if serious problems occur. Actually, it is the second straw that we aim for. The reduction of its diameter is sufficient to firmly hold the #5 PCC capsule. All four (three) combined straws should be now heated to approximately 80 °C. Any laboratory heater will do. At this temperature, the material softens and the straws are re-formed to their new diameter set by the outer straw. In this sense the external uncut straw serves as a mold for the whole stack. The whole assembly ready for baking is depicted in Figure 5;
- We typically allow the straws to stay there for about half an hour, after which we let them to naturally cool down to room temperature. After the cool down, the straws can be separated. Three (two) straws with new diameters has been formed. We mark them as S0 (the uncut one) and consecutively, S1, S2, and S3. We confirm upon multiple tests that neither the smooth incision nor the baking and shape modifications introduce any noticeable modification to the magnetic homogeneity of the straws.

The thickness of the wall of the straw is about 0.15 mm. At the first glance this may indicate that straw S1 has its internal diameter reduced by 0.3 mm, S2 by 0.6 mm, and so on. However, the clearance of straw S1 is smaller due to the overlap of the walls along the incision. Therefore, its clearance is reduced by three thicknesses of the wall, that is by 0.45 mm, to approximately 4.8 mm. Straw S2 has to curl even more in S0 and S1, so its clearance is further reduced to approximately 4.3 mm. We note that this is already less than the external diameter of the PCC #5, and straw S2 is found to be the best suitable to house these capsules.

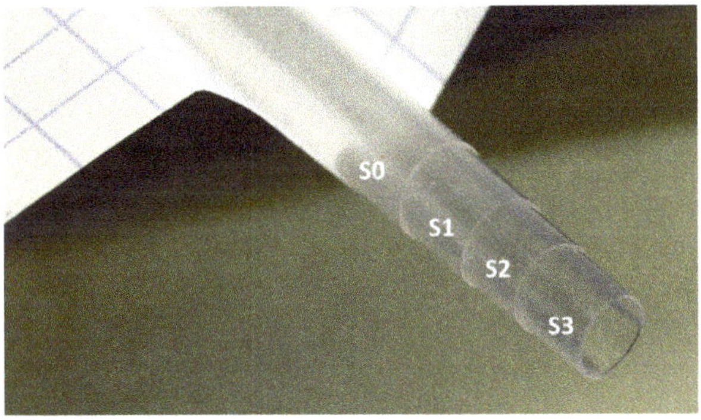

Figure 5. A complete set of three cut straws (S1, S2 and S3) inserted into the original straw S0. To nest in S0, the next straws have to progressively reduce their diameter by adequate curling.

The main usefulness of the modification described above stems from the existence of the incision along the nearly whole length of the straws. This permits the insertion of object(s) larger than the current clearance of the straw. The walls of the cut straw step apart in the vicinity of the object(s) to locally increase the clearance. Effectively a bulge is formed, as exemplified in Figure 6. Within the bulge, the walls of the straw press against the object, say a capsule (but it can be a rectangular sample of other crystalline material), and the friction between the walls and the objects holds the object firmly in place. The grip is so sufficient that the authors have never experienced any shift of the position of a #5 capsule in an S2 straw, even being loaded up to 180 mg. The measurements of the empty capsules presented in Figures 1 and 2 were performed having the investigated capsules clamped by the expanding walls of an S2 straw. The same straw accommodates #4 capsules, except the bulge gets larger.

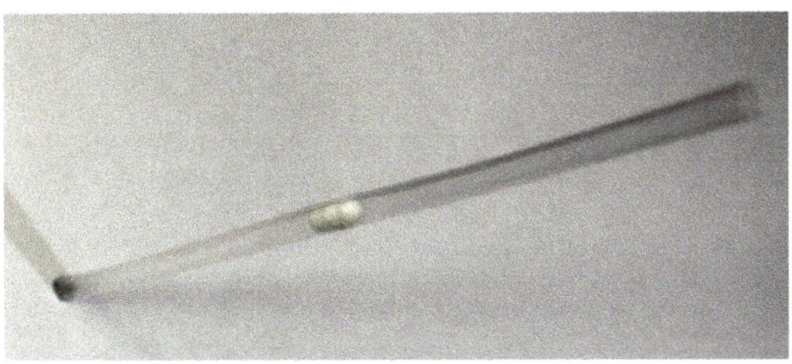

Figure 6. An illustration of a bulge formed on an S2 straw with its diameter reduced below the external diameter of a capsule #5 (for the sake of illustration a white coloured one is used). The S2 straw bulges easily due to the incision running along its length. The same incision allows to effortlessly adjust the position of the capsule, whereas the clamping force of the walls of the straw assures its alignment even during the reciprocal space option oscillatory movement in the sample chamber.

It is a matter of trial and error to find the best configuration. The authors have tested that straw S2 flawlessly holds a #5 capsule filled up to 150 mg during 20 h long 1.5 Hz and 4 cm long RSO measurement. For larger diameters, straw S1 should be used. It can house any capsule from #4 to #1. The latter has the external diameter of approximately 6.9 mm, however, in this case we suggest using a second S1 straw wrapped around the first one to increase the clamping grip.

We summarize this part by emphasizing that the single incision along the nearly entire length of the straw allows:

- A permanent reduction of its diameter (clearance) during an adequate heat treatment;
- Practically unrestricted movement along the straw of objects with effective diameters exceeding the modified diameter of the straw;
- A position-stabilizing grip sufficient to withstand acceleration and deceleration forces during the RSO measurements.

2.3. Magnetic Measurements

All measurements in the report have been taken using two commercial MPMS-XL magnetometers operating to a maximum field of $H_{\max}$ = 50 and 70 kOe, located at the Institute of Physics, Polish Academy of Sciences in Warsaw. Therefore, most of the technical remarks are strictly applicable to these units. However, the considerations presented below are so general that the required adaptive modifications could be employed to sizably improve the credibility of the research studies in other types of magnetometers, such as the SX700 of Cryogenic Ltd. (London, UK) [42] and the more popular SQUID-VSM MPMS3 of Quantum Design.

One of the main advantages of the in situ compensation method described here is that it is fully compatible with the RSO measurement mode, embedded in the MPMS-XL systems. The method has been derived to work with it, tested and validated using this improved oscillatory (mechanical) method of scanning. It allows for the collection of more scans in a given unit of time, which after the averaging procedure provide higher sensitivity and a reasonable sense of the experimental error. All the data presented here are collected by setting the scan length to 4 cm and RSO frequency to 1 Hz. Six measurements each consisting of four oscillations (24 scans all together to be averaged) are used to define the magnitude of each single experimental point. Further consideration on these setting can be found in Appendix C. Both the linear and iterative regression modes of the fitting routine (extraction method of the magnitude of the magnetic moment m) are used depending on the specific requirements. We use the sweep approach for the temperature T dependent studies.

3. The Capsule Compensation Sample Holder

The step-by-step assembly of the capsule compensated sample holder (C-CSH) is described here. Importantly, the description will provide a very good sense of the principle which makes the C-CSH so effective, robust, and easy to work with. We detail the process for the #5 polycarbonate capsules.

3.1. General Considerations

The construction process should be preceded by a careful selection of the capsules. Only capsules with very similar masses should be used. PCCs are provided by QD in two plastic bags containing separately the base parts (the containers) and the caps. This is an initial advantage as they exhibit a sizable mass distribution. The masses of the container parts range from 18 to 25 mg; the masses of caps range between 12 to 16 mg. As such, the distribution of masses of randomly assembled capsules may indeed be substantial. It therefore pays to spend the time weighing at least 30 containers and caps and arrange at least 15 capsules with the least possible spread of weights. We recommend to punch a tiny hole on the rounded part of the cap of each capsule. We use a needle from an insulin injection syringe. The hole will increase the effectiveness of the air evacuation from the interior of the capsules during purging of the sample chamber of the magnetometer. Without this hole, a tightly fitted cap can cause the capsule to slide-open when the drop of the pressure inside the capsule is much slower than in the sample chamber.

Having the weights organized, the user can start loading the capsules from any end to the already prepared straw S2, one after another, or the loading can be done as in the example detailed below. The already assembled train of 15 PCC capsules is depicted in Figure 7. The figure also documents our selection effort to prepare the set of several capsules of nearly identical weights.

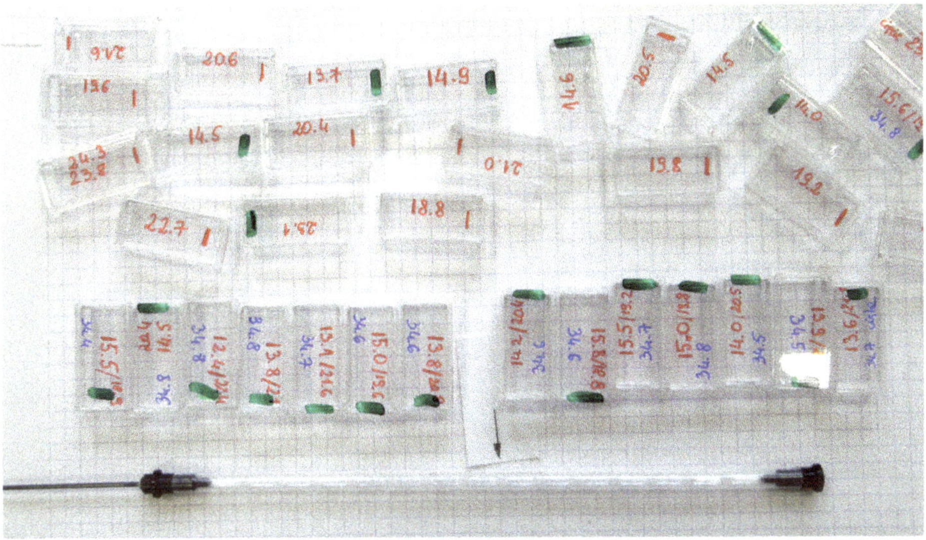

Figure 7. (**Bottom**) Fully assembled capsule compensated sample holder (C-CSH) made from 15 #5 polycarbonate capsules tightly combined in an S2 straw to make the self-compensated train. The arrow marks the position in which a small object will reside during the measurements. (**Top**) The markings on the plastic boxes located just above the C-CSH state the weight of each capsule used to assembly the train. Other boxes at the upper half of the figure are the leftovers from the capsule selection process.

3.2. The Assembly of C-CSH and the Principle of Its Effectivness

We start the assembly from placing the central capsule, the one in which the specimen will be located, at the center of the straw selected to carry the whole ensemble, as presented in Figure 8a. For the sake of presentation, we illustrate the assembly process using white-colored #5 gelatin capsules. The results of the SQUID measurements presented on the right side of Figure 8 are taken for the transparent PCCs discussed in this study, as presented in Figure 7. All of these measurements are performed at room temperature and at H = 10 kOe. The corresponding scan (the "Long Avg. Scaled Response", see Appendix B) registered by the SQUID is marked in Figure 8b as grey symbols. This scan is our reference point. It exhibits the typical shape, distorted noticeably from the ideal point object response because of the extended nature of the capsule with its length of about 11 mm. We calculate the corresponding magnetic moment using the standard point object SQUID response function (SRF). The technical aspects of this process are detailed in Appendix B. We consider two cases. First, the center of the scan window is fixed at 2.0 cm, which is at the center of the capsule. This result is marked by black line in Figure 8b. Second, the center of SRF is fixed at 2.5 cm—corresponding to the bottom of the capsule. The result of this fitting is marked by red line in Figure 8b. The former case represents the capsule completely filled with the researched material; the latter one mimics the situation when the capsule contains only a miniscule amount of the investigated material, which resides at the bottom. The obtained magnitudes of m are depicted in Figure 8d by gray-filled circles, with the edge color (black/red) corresponding to the position at which we fix the center of the SRF.

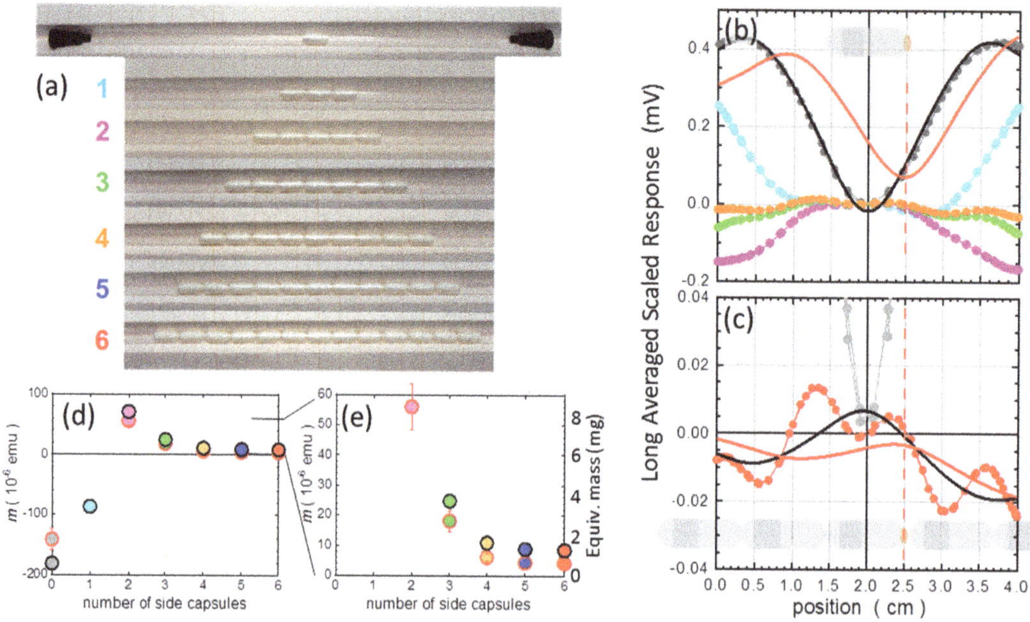

Figure 8. (**a**) The illustration of the step-by-step assembly (subpanels 1–6) of an in situ capsule compensational sample holder from capsules of size 5 and an S2 straw, that is, a straw with its diameter reduced below the external diameter of the capsules. White-colored #5 gelatin capsules are used for increased visibility only. Results presented in the rest of the panels are obtained for transparent polycrystalline capsules, as presented in Figure 7. (**b**,**c**) SQUID scans recorded for a single central capsule (grey symbols) and 1st, 2nd, 3rd, 4th, and 6th stages of the assembly (symbols of matching colors to the numerical markers in panel (**a**)). The black vertical line represents the position of the midpoint of the capsule in the scanning window of SQUID magnetometer and the red dashed one the position of the bottom of the capsule. The black and red wavy curves represent the SQUID response function (SRF) centered at these two locations. In panel (**b**) the fitting is done to the dataset corresponding to a single capsule, in panel (**c**) to the 6th stage of the assembly (red symbols). The fitting approach is described in Appendix B. (**d**) Results of the fits to the scans collected in every stage of the assembly (magnetic moment m). The fill color of each circle corresponds to the assembly stage, whereas the edge color (black or red) correspond to color marking of the alignment of the SRF: at the center of the central capsule or at its bottom, respectively. (**e**) The blown-up part of panel (**d**) for increased resolution near zero for the last stages. The added right Y-scale expresses the degree of the compensation in an equivalent mass of the central capsule of a mass of 30 mg.

The assembly continues by addition of one capsule on both sides of the central capsule, as shown in Figure 8a, case "1". The magnetometer registers this change. The form of the scan corresponding to this configuration (cyan symbol in Figure 8b) assumes a trapezoidal form and the overall amplitude is visible reduced. This is immediately reflected in the calculated magnetic moments corresponding to this configuration (cyan filled circles in Figure 8d). The main message here is that the detection unit of the magnetometer now reports a twice smaller magnetic moment despite having three times more material in the chamber. The improvement continues after placing next neighbors on each side, Figure 8a, case "2", and magenta filled circles in Figure 8d. However, an interesting thing happens now: the scan inverts (magenta symbols in Figure 8b), so the train of five diamagnetic capsules is "seen" as a blurred "paramagnetic" object.

A far more substantial change is observed when there are four compensating capsules on each side of the central one, Figure 8a, case "4", that is, when the total length of the train of capsules (about 10 cm) guarantees that neither end of the train gets very close to the pick-up coils during scanning. This makes our assembly semi-infinitely long, fulfilling the first prerequisite condition of the efficient in situ compensation. Indeed, the voltage registered during scanning is strongly reduced (orange symbols in Figure 8b), particularly in comparison to the initial case of the single capsule. The corresponding magnetic moments are marked as orange filled circles in Figure 8d. They are now very close to zero, but not strictly there. First, the train is not really very long (yet); second, it is not uniform due to its "granular" form. The capsules are not uniform themselves, they are thicker at the cap-containing half and are rounded at both ends. The resulting nonuniform mass distribution along the length of the train of capsules will always have its mark on the final form of the scan. It will be shown to be appreciably small.

We continue the assembly by adding two more capsules to each side. Actually, only five compensating gelatin #4 capsules can be added on each side of the central one. Capsules #4 are so long that 11 of them nearly fill the whole length of the straw. In the case of PCC capsules #5, one can go up to 7 capsules on each side to achieve additional positional stability with respect to the straw, and so to the SQUID pick-up coils—fulfilling yet another prerequisite condition for achieving very high sensitivity and reproducibility. The fully assembled train of 15 PCC capsules is depicted at the bottom of Figure 7, together with the documentation of the selection effort to prepare the set of several capsules of nearly identical weights. The shape of the scan obtained for the fully assembled train is marked by red bullets in Figure 8c. It certainly is very "bumpy" on its own scale, but its magnitude is truly small compared to the response of a single capsule. The SRF fitted with its centering forced to our two points of interest (black and red solid curves) yields magnetic moments indicated by black and red points in Figure 8d, and in its blown-up version, Figure 8e. By comparing these results with the starting ones for the single capsule, we obtain that magnetic signature of the in situ compensated capsule dropped about 20 times (nearly 95% of compensation) when one seeks the response at the location corresponding to the center of the capsule and about 30 times (nearly 97% of compensation) when the response at the location of the bottom of the capsule counts most. To visualize the effectiveness of the in situ compensation, the magnetic moment is again expressed in units of mass, the right scale of the graph in Figure 8e. These numbers read approximately 1.5 and 1 mg, respectively. This fact alone indicates that magnetic moments of a SRDM can be established in C-CSH with an absolute accuracy of 10 to 20% when the mass of the investigated specimen ranges between 10 and 5 mg, respectively. It is shown later in Section 4: "Discussion: validation and an example of the true potential" that one can get a much better accuracy when this residual contribution from the compensated capsules (the empty C-CSH) is considered.

Another advantage of the approach described here is that the standard mounting method of the C-CSH assembly with the long and short sample rods, required for the RSO method of measurements, can be applied. At least one black centering plug (straw adapter) has to be used to connect the straw with the rest of the sample rod. This is the reason why the incision on the straws starts a few millimeters from their one end. Here enters the connecting plug. And our experience is that for mechanical stability of the whole C-CSH assembly, one plug is acceptable. Straw S2 provides a sufficient grip to hold the train of a dozen or so #5 capsules firmly in place. However, the second plug is strongly advisable to assure a proper centering of the straw with capsules with respect to the axis of the SQUID pick-up coils. At the second (the lower) end the straw is cut and here the grip is insufficient to hold the plug. To secure the centering plug in place, we apply a small patch of a commercial surgical tape with acrylic adhesive. We elaborate a bit more on this very handy fixing agent in Appendix A.

Finally, we want to underline the usefulness of using diameter-modified straws for C-CSH-type assemblies. The key point is that the larger internal diameter of unmodified straws will allow the capsules to tilt or even rattle during the RSO measurements, an effect most likely more substantial at very low temperatures. This may add up to misalignment errors and thermal hysteresis of the results. The authors find the existing friction between the diameter-reduced straw and the capsules as advantageous during assembly and disassembly of the C-CSH. Only the bottom half of the train has to be removed from the straw to get to the central capsule to exchange the specimen. The rest of the assembly retains its location. This greatly increases the reproducibility of the assembly and so minimizes the probability of a sizable misalignment. The wall-wall friction between capsules and a straw is even more crucial in the case of gelatin capsules. As noted in Section 2.1, GCs are changing their mass when they experience the dry environment of the SQUID sample chamber (Figure 1). Similarly, their linear dimensions are changed. In particular, they shrink visibly, and so after some 10 h long measurement run an 18 cm long train of GCs gets 2–3 mm shorter. Without the radial grip provided by the wall of the diameter-reduced straw, such a contraction would result in a time-dependent 1–1.5 mm alignment shift of the specimen in the central capsule.

3.3. Examples of the Improvement

By extending the "length of the capsule" in such a harness-free manner, the capsule(s) become "stealthy". The improvement brought about by the in situ compensation is not only related to the fact that the effective mass of the capsule that houses the specimen is reduced so sizably; the gain is multiple.

At first, we note that the concept of in situ compensation makes the outcome of the experiment much less dependent on the misalignment error, that important and detrimental factor when the background measurement has to be performed and subtracted. We illustrate the improvement by performing similar z_0-dependent fitting as summarized previously in Figure 4. However, this time the SRF is fitted to the to the exemplary SQUID scan of the fully assembled empty C-CSH, red symbols in Figure 8c. The results are depicted in Figure 9, clearly indicating that misalignment errors dropped 10 to 15 times with respect to the case of a single uncompensated capsule, and that a substantial misalignment of the about 1 mm should not introduce a larger experimental uncertainty than 0.1 mg, or 1% during investigation of 10 mg of SRDM.

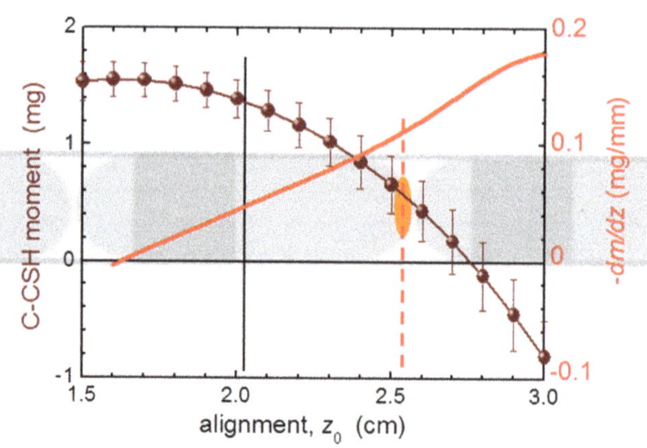

Figure 9. The position dependence of the magnetic moment (brown bullets) calculated from a scan of a fully in situ compensated train of #5 polycarbonate capsules (sketched to scale at the background) forming the capsule compensation sample holder (C-CSH) assembly. The results are related to the standard mass of a single capsule (30 mg, as in Figure 4). The magnitude of this change defines the absolute value of the misalignment error with respect to the center position (a black vertical line at z_0 = 2.0 cm). The red solid line yields the derivative of the former, that is, the magnitude of the uncertainty due to the misalignment. The orange oval mimics a small object of investigations residing at the bottom of the central capsule at the position marked by a dash red line.

Anyone who is interested in very precise magnetometry of rather large quantities of shapeless materials, such as presented previously in Figure 3, when capsule is filled completely, can stop here. Indeed, the #5 capsule with its typical volume of about 0.130 mL can accommodate approximately 100–150 mg of lightweight material (or powder), an amount that exceeds the effective mass of the in situ compensated capsule by approximately 100 times. This guarantees at least 1% absolute accuracy without any background subtraction effort. However, it has to be immediately noted here that any number evaluated in such conditions has to be corrected for the extended shape of the specimen [19,43]. The correction factor for an 11 mm long object is about 0.85 [44], although the exact value is yet to be established. Therefore, the best magnitude of m, and so of the magnetization, can be obtained in this situation after dividing the output of the magnetometer (these number are computed in the point dipole approximation) by 0.85.

Certainly, the sizably increased accuracy of the in situ compensation is not the only benefit of the method. The self-compensation of the capsules aligned in a train also indicates that other disturbing features of the capsules presented in Section 2.1 are eradicated in a similar proportion. This is first indicated in the Figure 10, where we present the full T-dependence of m measured for two C-CSHs assembled from PCC #5 and GC #4. These results have to be related to the measurements of single capsules, presented in Figure 2. The most prominent improvement seen after assembling the capsules in trains is a perfect elimination of the strong low-temperature paramagnetic component. Interestingly, in both cases a small T-dependence is seen between 100 and 300 K, corresponding to about 0.2 mg of the SRDM. This value sets the lower limit on the precision of the method when one wants to work without subtraction of the background (empty C-CSH).

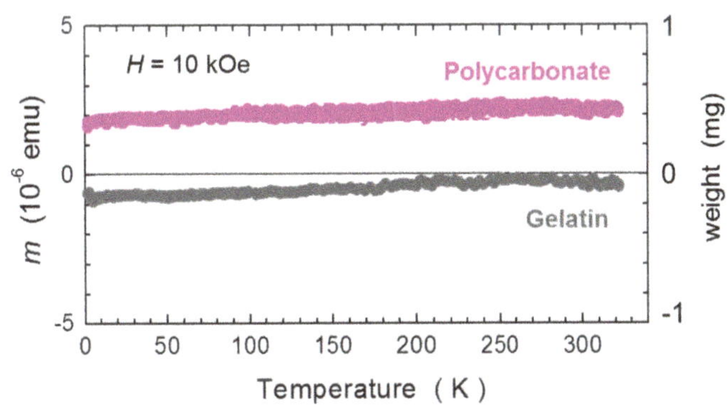

Figure 10. Temperature dependence of the magnetic moment m sensed at the location of the bottom of the central capsule for two exemplary capsule compensational sample holders assembled from gelatin (grey) and polycarbonate (magenta) capsules. Magnetic data are normalized between themselves and related to a standard mass of a 30 mg capsule. Note the complete lack of the strong low temperature upturn of m, the characteristic feature for single capsules (Figure 2).

All the test measurements are performed in $H = 10$ kOe, indicating that the improvement has occurred in all three domains: time, temperature, and H. Nevertheless, it is worth presenting a direct measurement in H. In Figure 11, the H dependence of the signal of single capsule is compared with that of a train of 14 PCC capsules, the C-CSH. In this particular case, the improvement is 45-fold, that is, the signal of the single PCC drops by so many times. Importantly, all the $m(H)$ isotherms for C-CSH show a very weak T-dependence, and particularly none at the lowest temperatures. Both $m(H)$ collected at 1.8 and 5 K go one on top of another. This confirms the results presented in Figure 10 and constitutes the base for the low-T analysis of a paramagnetic contribution in turmeric powder, presented later in the text.

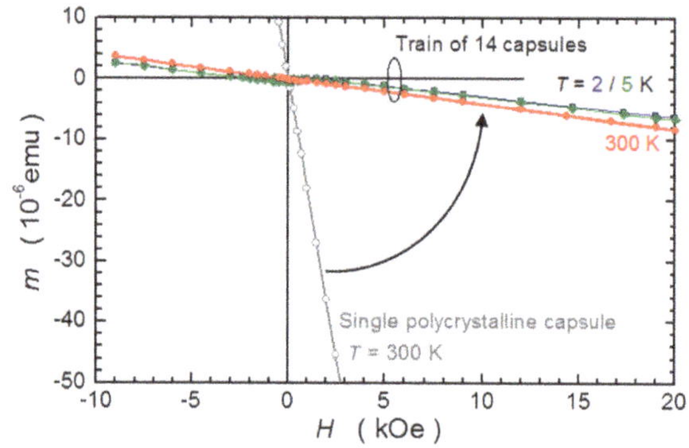

Figure 11. An example of the magnetic field H dependence of magnetic moment m for a single polycrystalline capsule (grey circles) and of the in situ compensated case of a train of nearly identical capsules loaded and stabilised in a S2 straw having its diameter reduced below the external diameter of the capsules (navy, green, and red symbols). The arrow indicates the reduction of the magnetic signal of the capsule due to the compensation effect of the neighbouring capsules.

4. Discussion: Validation and an Example of the True Potential

In the previous section we introduced the whole concept of in situ magnetic compensation of the signal of common capsules that allow the investigation of various shapeless specimens in commercial SQUID magnetometers. In the following section, we will validate to method developed here to highlight its quite remarkable experimental potential.

In order to validate the method, we investigate a very small fragment of a crystalline material that is meant to mimic other specimens of a miniscule amount. An irregular 2.1 mg crystallite of SnTe is chosen. SnTe is a narrow gap semiconductor, the sole representative of the so called topological crystalline isolators [45]. The investigated material was grown using the self-selecting vapour growth method [46]. At room temperature its magnetic susceptibility is approximately $\chi_{SnTe} = -4.6 \times 10^{-7}$ emu/g/Oe (which ties well with our arbitrary taken magnitude of χ of SRDM), but most importantly for the validation tests, χ_{SnTe} shows a very strong temperature dependence. We benchmark $\chi_{SnTe}(T)$ first. To this end we measure a 58 mg large piece of the material, indicated in Figure 12 by the grey arrow. For this measurement, this large crystallite is fixed inside the S3 narrowed straw. One of its transverse dimensions sufficiently exceeds the internal diameter of this straw so the bulged straw provides the necessary grip to stabilize its location within the straw. The results are indicated as dark gray bullets in Figure 13.

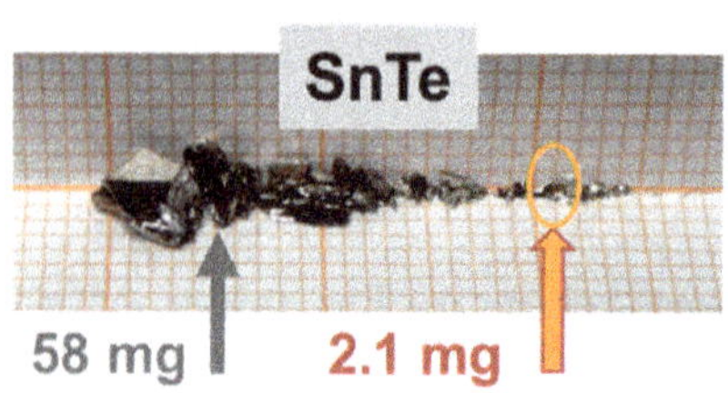

Figure 12. Selection of SnTe crystallites, leftovers from a several-cubic-centimeter large single crystal [45]. The arrows indicate two pieces used in this study: the larger for benchmarking of the temperature dependence of the magnetic susceptibility, the smaller for the validation of the in situ compensation method.

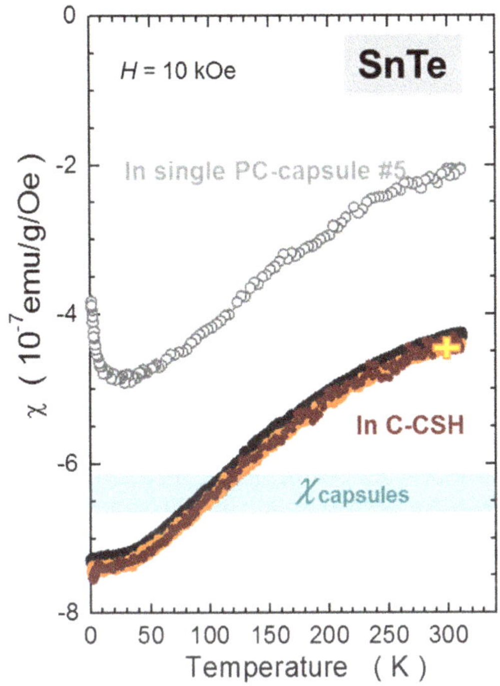

Figure 13. Temperature dependence of the magnetic susceptibility χ of SnTe established in this study by various methods. The background dark grey symbols mark the χ-benchmarking measurement of a large 58 mg crystal. Open gray symbols indicate the measurements of 2.1 mg crystallite in a single polycarbonate capsule #5. The orange symbols indicate the results obtained for the same minute crystallite mounted on a magnetically clean and uniform Si stick. The brown symbols indicate the results of the final method-validation test obtained when this small crystallite was mounted in the capsule compensated sample holder (C-CSH) elaborated in this study.

We next select the main test material, a small 2.1 mg SnTe crumb, indicated by the orange oval in Figure 12. First, we confirm that its magnetic susceptibility is identical to that of its larger neighbor. To accurately establish the absolute magnitudes of $\chi(T)$ of this tiny crystallite, we glue it to one of our standard Si sample holders. That is possible because it has two flat walls. We use an equivalently small droplet of strongly diluted GE-varnish [31]. We separately evaluate that at the same field of 10 kOe and in the whole T-range, the magnetic contribution of the solution used by us to fix the sample stays well below the error bars of the results obtained for the 2.1 mg sample, so it is disregarded in further data reduction. The results are marked by orange symbols in Figure 13, confirming the expected magnetic characteristics of the test piece.

The final two measurements are performed with this small test SnTe piece loaded (i) into a single PCC capsule, and (ii) into the central capsule of C-CSH. In both cases, we also measure the reference T-dependencies of the (i) empty capsule and (ii) empty C-CSH, which are subtracted accordingly. For both measurements the sample holder is centered on the bottom of the sample-housing capsule using the method outlined in Appendix A. All four measurements are done according to the same sequence. The results of the measurements performed in a single PCC are marked as open gray circles in Figure 13. They differ markedly from those two established previously. Obviously, the alignment error takes its toll here. We evaluated this error in Section 2.1.3 to be around 3 mg/mm for a standard diamagnetic material at the position corresponding to the bottom of the housing capsule (See Figure 4). Taking 0.5 mm as a reasonable estimation of the typical misalignment, one should expect about 1.5 mg uncertainty, which is about 70% of the mass of the test sample. Indeed, we find that the absolute values of χ_{SnTe} established in this approach differ by a factor of two at room temperature, which conforms favorably to our crude estimation. Certainly, with the identical alignment of the capsule for both these measurements. one should get a perfect reproduction of the dark grey (benchmark) or the orange results. In the authors' view, this is beyond the reach of the standard approach in the case of such minute signals. In contrast, the main part of the temperature dependence is captured quite reasonably, at least above approximately 50 K, where the T-responses of the m of the capsules are reasonably flat. Below 50 K, where the PM contribution of capsules becomes substantial (c.f. Figure 2), a small difference in alignment prevents an adequate elimination of this signal and the specific magnetic properties of the capsule mar the true properties of the investigated material. It is worth noting that one can arrive at the last conclusion only after prior knowledge of the results.

A very different outcome is obtained when this tiny crystallite is investigated in the C-CSH. Importantly, as well as in the first case, the test sample is located at the bottom of a similar housing capsule, but the several-dozen times smaller sensitivity to misalignment makes the results obtained in C-CSH practically insensitive to this issue. The expected uncertainty band in this case is only 0.1 mg/mm, which for 0.5 mm inaccuracy yields the uncertainty as low as 2.5% of the mass of the test crystallite. This final step of the validation of the in situ compensation method is marked in Figure 13 using brown symbols. It must be noted that a precise one-to-one correspondence has been achieved, which strongly underlines the practical power of the in situ compensation and its potential. It shows that it can work very effectively in a broad range of environmental parameters, providing very accurate magnitudes of magnetic moments of even minute specimens of substances exhibiting very weak magnetic signatures.

One very important technical note is appropriate here, which highlights another practical advantage of the routine usage of the in situ compensation outlined in this report. We strongly emphasize that the compensation for the unbalanced signal of the empty C-CSH has been done by a simple subtraction of the results reported by the magnetometer in the respective *.dat files. The only restrictions to be enforced on the system are: (i) to perform the T-dependence with the "Auto Tracking" option set on, and (ii) to use the "linear regression" algorithm. Apart from these, all the results reported here are obtained using the RSO method of measurements and the "Sweep" approach mode during the measurements in temperature domain. No Automatic Background Subtraction mode is necessary, neither is a cumbersome "scan-by-scan" subtraction of the reference measurements [10]. Any approach like the latter calls for excessive numerical effort and would likely cause delays in data reduction. In the method put forward here it is sufficient to operate only on the basic output *.dat files yielded by the magnetometer. In the authors' view, this very convenient property is the result of the perfectly even symmetry of the SRF and the unprecedented insensitivity to small alignment errors. The first fitting constricts channels least mean square (LMS) routines, in their fitting effort, onto specific even-parity components in the signal under processing. That is why it is impossible to correctly extract the magnitude of a weak moment located at the bottom of a single capsule, as considered in Section 2.1.3. Colloquially speaking, there is still too much of even-parity signal of the capsule itself at its bottom that this contribution takes over the searched weak source of signal. In contrast, as shown in Figure 8c, the response of the same capsule abutted by other similar capsules is highly complicated, without any dominant parity. Therefore, the same weak signal effectively breaks out and is quite correctly evaluated by LMS procedures, as exemplified in the following example.

In order to illustrate why the approach described above yields so surprisingly accurate magnitudes of very weak magnetic moments, we scrutinize the exact form of the scans recorded during these two measurements: the T-dependencies of the empty C-CSH and the same C-CSH with the 2.1 mg test piece of SnTe loaded to the central capsule. As the sole representatives, we take the very first points of these two measurements. For both measurements the C-CSH assembly was centered once, i.e., before the entire validation test, using the method described in Appendix A. There is generally no need for a second centering (correction) after inserting (removing) the sample into (from) the central capsule and reassembling the C-CSH. Just the basic care exerted during these processes allows the reconstruction of the identical enough C-CSH; a repeated centering is not needed. Moreover, the centering process on its own does not provide more accuracy than the careful reassembly of the C-CSH. Our experience gained along the development process, substantiated by the analysis of the misalignment errors (Figure 9), says that 0.5 mm reproducibility is probably the best possible result and that it is sufficient.

In Figure 14a, open light grey circles and dark grey bullets mark the shape of the response of the empty C-CSH and with 2.1 mg test crystallite loaded to the central capsule, respectively. Clearly, none of the scans resemble the typical shape of the SRF. The orange bullets represent the calculated point-by-point difference between these two scans yielding the sole contribution brought about by the insertion of the test sample. This shape exhibits a very clear correspondence to the signal expected for a diamagnetic substance, and the solid red line marks the fit of the SRF centered at the center of the scan window, i.e., where the whole assembly was centered. The match is exceptionally good and the z-shift of about 1 mm indicates the inaccuracy in alignments of the C-CSH assembly for these two measurements. The fit yields $m_{SnTe} = -9.64$ μemu. The corresponding magnitude of the magnetic susceptibility is marked by a yellow cross in Figure 13. This result corresponds to the ABS methods at its best.

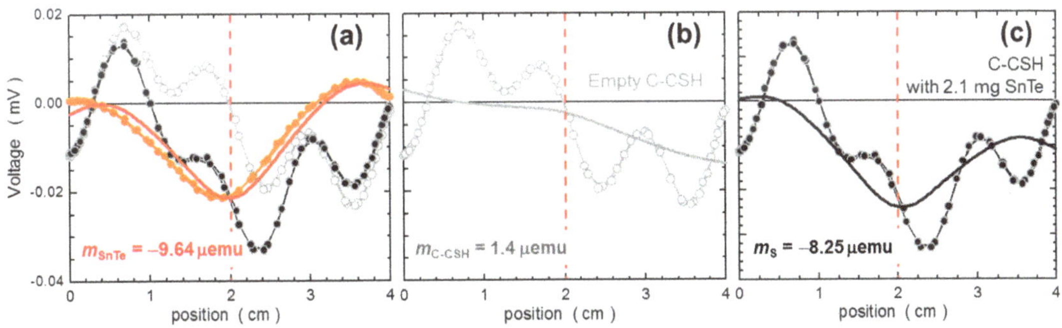

Figure 14. Exemplary scans (the long averaged scaled response voltages) representing the first experimental point of the temperature dependence of magnetic susceptibility of a 2.1 mg SnTe crystallite depicted by brown symbols in Figure 13. Light grey open symbols in (**a**,**b**) mark the scan of the empty capsule compensational sample holder (C-CSH). Dark grey symbols mark the scan obtained with the 2.1 mg SnTe piece installed in the central capsule of the C-CSH. The orange symbols in (**a**) mark the difference between scans presented in (**c**,**b**). The solid lines in matching colors in all three panels represent least mean square fit of the SQUID response function (defined in Appendix B) to the experimental points. The corresponding magnetic moments m are given in the labels.

Now we separately extract the magnetic moments from each of the two experimental scans given in Figure 14a. They are plotted in Figure 14b,c, together with the corresponding fits, marked as solid lines of matching colors. In the first case, Figure 14b, the fit yields the residual moment of the empty C-CSH $m_{C\text{-}CSH} = +1.4$ μemu. In the second case, Figure 14c, after insertion of the sample the fit yields $m_S = -8.25$ μemu. Obviously, at first glance, neither of the two fits can be classified as satisfactory. The fitted functions cut somewhat freely across the experimental points, taking rather little notice of the pattern formed by the experimental points. This is actually acceptable, as the fitting routine seeks the even-parity component, the SRF, and only these "frequencies" that correspond to $z_0 = 2$ cm. It is therefore no surprise that the difference of the magnitudes of m returned by the fittings yields the correct magnitude of $m_{SnTe} = m_S - m_{C\text{-}CSH} = -9.65$ μemu, a mere 0.1% different value than established in Figure 14a.

Finally, we exemplify the usefulness of the in situ compensation method in biological and medical studies by performing measurements of common turmeric powder. We emphasize that the investigated sample has not been prepared in any standard manner, e.g., Association of Analytical Chemists standard methods, specific for biological substances. The powder originates from a commercial herbs retailer [47]. A quantity of 2.9 mg of the powder is loaded to the #5 PCC and measured in C-CSH. This amount of material forms about a 1.5 mm thin layer occupying the bottom of the capsule; however, some of the powder spreads all over the capsule. As is shown in Figure 3, a nearly fully loaded #5 PCC accommodates approximately 70–80 mg of fresh turmeric powder. We do not focus on absolute magnitudes of m in this measurement., primarily because after loading to the SQUID sample chamber, a weak drift in the magnetic signal is observed. Similar to the gelatin, water evaporates form the fresh turmeric powder in the dry environment of the SQUID sample chamber, so for fully quantitative studies the proper preparation of such a material is mandatory. In this measurement, we aim at properties at the low temperatures end. All the data presented below are corrected for the residual weak signal of the empty C-CSH.

We start from room temperature $m(H)$ presented as orange symbols in Figure 15b. The negative slope confirms the diamagnetic character of turmeric, although a small near zero feature indicated that a small amount of ferromagnetic (FM) component is present. This may well be a contamination due to preparation of the spice at its place of origin. From the slope of this $m(H)$, we evaluate the diamagnetic susceptibility as $\chi_{turmeric} \cong -3.3 \times 10^{-7}$ emu/g/Oe. The results of the T-dependence of m presented in Figure 15a indicate that turmeric turns from a diamagnetic substance into the paramagnetic (PM) one at low temperatures. To quantify this PM component, we measure $m(H)$ at two very low temperatures: 1.8 and 5 K, shown in Figure 15b. Clearly, the two dependencies have a Brillouin-like character at high magnetic fields, with noticeable FM-like deviations around $H = 0$, similar to that at 300 K.

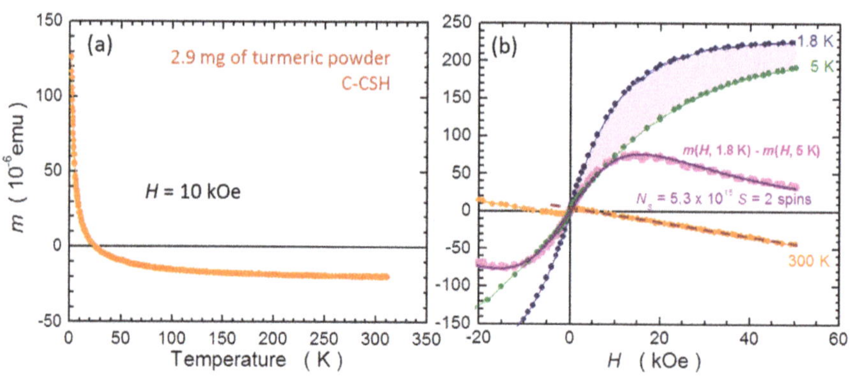

Figure 15. Results of the magnetic studies of 2.9 mg of common turmeric powder performed in capsule compensated sample holder (C-CSH). (**a**) Temperature dependence of magnetic moment m measured at external magnetic field $H = 10$ kOe. (**b**) Values of $m(H)$ at 300, 5, and 1.8 K. The brown dashed line yields the magnitude of diamagnetic susceptibility of the sample $\chi_{turmeric} \cong -3.3 \times 10^{-7}$ emu/g/Oe. The magenta area and bullets mark the difference between $m(H)$ at 2 and 5 K. The solid purple line indicates difference between Brillouin functions calculated for $S = 2$ (Equation (A2) in Appendix D). The matching factor between the theory and experiment, $N_S = 5.3(1) \times 10^{15}$, yields the number of paramagnetic $S = 2$ spins in the sample.

In order to accurately separate the PM signal in our turmeric sample, we invoke the fact that in most cases it is only the PM component that is responsible for changes of

$m(H)$ at low temperatures, a condition met in many magnetically composite systems in which the technique presented below worked with a very high precision [3,48–50]. The diamagnetism should not depend on T, so should most of the FM contaminations, as latter are typically characterized by a very high spin coupling temperatures. Indeed, the $m(T)$ dependence does not indicate any magnetic phase transition below 300 K. In such circumstances the calculated difference between the experimental points obtained at these two low temperatures should only reflect the change in the PM component—all other contributions get canceled. The magenta filled area and the corresponding symbols in Figure 15b represent this difference. The required information on the spin state is obtained from fitting of this difference by the theoretical formulae describing the difference of the two Brillouin functions taken at 1.8 and 5 K. Its particular form and relevant technical details are given in Appendix D. The best fit is obtained for $N_S = 5.3(1) \times 10^{15}$ of $S = 2$ spins in the sample, which indicates their mass concentration $x_S \cong 1.8 \times 10^{19}$ per gram of this particular powder. The other spin values of $S = 3/2$ and $5/2$ definitely yielded worse fits. The value of $S = 2$ indicates Fe^{4+} ions, as this transition metal dominates by far the other transition metals in turmeric [51]. This analysis quantifies the dominant contribution; it does not preclude the presence of other unpaired spins in the material. Commercial magnetometers are volume probes and are not element specific, so they are limited in this respect. However, although this result cannot be regarded as a standard for the turmeric powder, it is only the employment of the C-CSH that makes such as analysis quantitative. It has to be noted that, as shown in Figure 2, the capsules themselves do exhibit PM properties, and this PM of capsules would mar the results of the analysis presented above. In contrast, the self-compensation of the magnetic properties of capsules in the C-CSH assembly eradicates this detrimental property, as documented for both kinds of capsules in Figure 10 in the T-domain and in Figure 11 in the H-domain.

Where the magnetic fields are concerned, it has to be noted that the "stealthy" properties of C-CSH also lead to much smaller errors of m caused by the trapped field of the superconducting magnet [19,20,52–54]. They are also reduced 20 to 30 times, from a level of about $+4 \times 10^{-7}$ emu when the specimen is housed in a single capsule to the magnetometer noise level of 10^{-8} emu in C-CSH. Note that the shift of m due to the presence of uncompensated capsules on the return from a trip to high positive fields causes a shift of m towards positive values and it will be negative on the trip back from high negative fields, adding a FM-like hysteresis feature to the results of considerably high height of about 8×10^{-7} emu, an important value to address in the precise magnetometry of weak signals.

5. Conclusions

The construction principle, assembly method, validation measurement, and example of employment of the in situ compensation concept applied to commonly used capsules in investigations of small and irregular shape fragments, powders, and other specimens that would generally require complicated mounting in SQUID magnetometers have been presented. It has been experimentally evidenced that with incorporation of only the basic components, familiar and ubiquitously present in every magnetometry lab—common plastic straws and capsules (most preferably the polycarbonate ones)—an assembly can be completed in virtually one afternoon, which allows one to reduce the magnetic signatures of a specimen-housing capsule at least 20 times in time, temperature, and magnetic field domains. In practical terms, the reduction is equivalent to the reduction of the mass of the capsule to no more than 1.5 mg, sometimes even twice that. This permits investigation of larger amounts of materials, say between 50 and 150 mg, without any need for background signal subtraction—a procedure that is particularly prone to a large experimental inaccuracy caused by a non-ideal alignment of the sample holder. However, the effort exerted in preparation of the capsule compensated sample holder is particularly useful when only minute amounts of material are available for magnetometry. Such specimens occupy only the bottom part of the housing capsule and the background subtraction error due to misalignment can be comparable to the mass of the specimen, making the magnetometry semi-qualitative, at best. In C-CSH, the magnetic signal at the location of the bottom of the housing capsule can be readily reduced up to 30 times, with alignment errors not exceeding 0.1 mg of an equivalent mass of a common diamagnetic material. This opens a new experimental window of truly precision magnetometry of miniscule specimens of masses ranging down to a single milligram. This unprecedented property has been validated by demonstration of the ability to obtain accurate absolute values of the temperature dependence of the magnetic susceptibility of a tiny, a mere 2.1 mg, crystallite of diamagnetic SnTe.

A full applicability of the C-CSH concept to small amounts of powdered samples has been presented by investigation of a similarly minute sample of common turmeric powder. Both the magnitude of the diamagnetism of the substance and of the paramagnetic contribution due to iron ions were readily obtained from magnetic field dependencies measured at room temperature and at liquid helium temperatures, respectively. All the measurements were obtained in one experimental session, without remounting the specimen to different (specialized) sample holders. Our effort proves that the C-CSH concept works with high reproducibility rate between 1.8 and 320 K and at any magnetic field available in the magnetometer.

The concept of the in situ compensation of capsule containers has been tested in MPMS-XL SQUID magnetometers of Quantum Design. The authors are confident that after adequate modifications, a similar assembly may prove equally effective in other types of magnetometers, although for the VSM-type MPMS 3 system one would like to see even smaller polycarbonate capsules that the smallest ones of size 5 available today.

6. Patents

Relevant patent application has been registered with Polish Patent Office as P.439874.

Author Contributions: Both K.G. and M.S. contributed equally to all steps that allowed us to prepare this manuscript, that is: conceptualization, methodology, validation, formal analysis, investigation, data curation, writing—original draft preparation, writing—review and editing, and visualization; supervision, M.S.; project administration, M.S.; funding acquisition, M.S. All authors have read and agreed to the published version of the manuscript.

Funding: This study has been supported by the National Science Centre (Poland) through project OPUS (DEC–2017/27/B/ST3/02470).

Institutional Review Board Statement: Not applicable.

Informed Consent Statement: Not applicable.

Data Availability Statement: The data presented in this study are available upon reasonable request from the corresponding author.

Acknowledgments: The authors acknowledge gratefully Andrzej Szczerbakow for providing SnTe samples, Piotr Nowicki for technical support and Jacek Szczytko and Marc Kunzmann for providing us the capsules needed for this study.

Conflicts of Interest: The authors declare no conflict of interest.

Glossary

List of nonstandard terms and abbreviations used in this paper.

Abbreviation

Abbreviation	**Meaning**
ABS	automatic background subtraction
C-CSH	capsule compensational sample holder
GC	gelatin capsule
LMS	least mean square

PCC	polycarbonate capsule
QD	Quantum Design
RSO	reciprocal space option
SRDM	standard reference diamagnetic material
SRF	SQUID response function

Appendix A (Centering of C-CSH)

Setting up the position of the sample in the sample chamber—the alignment—is yet another prerequisite for the precise magnetometry. In the case of capsules housing miniscule specimens, we cannot rely on the bare voltage related to the moment of the sample as it is dwarfed by that of the capsule. Obviously, an additional magnetic marker has to be used that can be affixed temporarily to the sample holder (outside wall of the straw in this case). The authors do not recommend using pieces of magnetically strong ferrous materials as the positioning could only be done at very weak magnetic field, which calls for unnecessary frequent changing of the field in the magnetometer and related field instabilities. Our material of choice is a small, about 10 mg, crystallite of PbTe weakly doped with Cr. In a semiconductor matrix, chromium tends to precipitate to form ferromagnetic nanocrystals [55], which, as an ensemble, exhibit superparamagnetic properties even at room temperature [49], yielding a convenient, approximately 100 times stronger, response at H = 10 kOe than anything else considered in this study. An adequate amount of any similar composite material can be used for this purpose.

A narrow 2–3 cm long strip of a commercial surgical tape with acrylic adhesive serves as the bonding agent for the marker. The tape sticks firmly to the straw and comes off easily without leaving any residue. A freshly cut tape exhibits a weak magnetic response, works well down to helium temperatures, and can be considered as a very convenient "fist-aid" fixing medium in magnetometry for medium to strong magnetic materials. An example of our method is given in Figure A1.

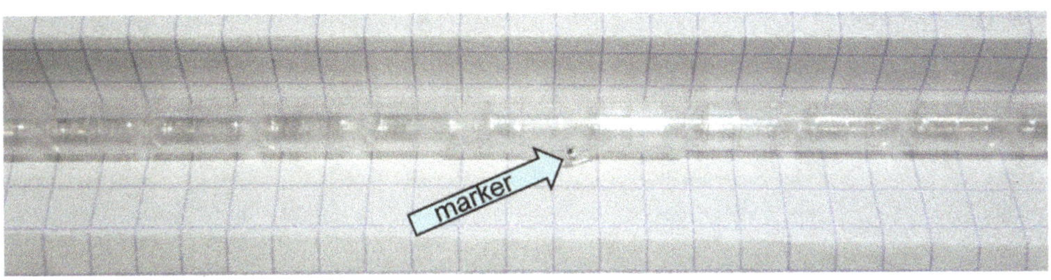

Figure A1. An example of the application of a magnetic marker for centering of the capsule compensational sample holder. As indicated, it is hard to get better reproducibility than about 0.5 mm.

Appendix B External Fitting of the SQUID Scan Voltage

The necessary position dependence of the SQUID output during scanning can be extracted from the relevant *.raw file(s), or can be taken from the *.lastscan file, both provided by the MultiVu software. For any studies focussed on precise magnetometry, we strongly advise recording the full suite of the data, that being the accompanying *.diag and *. raw files. On numerous occasions, they proved indispensable in troubleshooting or error correction. Alternatively, the required scans can be measured purposely using, for example, the [Measure] tab of MultiVu program. Among many available software packages, we use Origin's NonLinear Curve Fitting (NLCF) option, and to straightforwardly extract the magnitude of magnetic moment m we use the "Long Avg. Scaled Response", that is, the 17th column in the *.lastscan file. Our fitting function $V(z)$ comprise the second-order axial gradiometer response function (SRF with a linear term added to compensate for the offset and even-parity signal drifts) [56]:

$$V(z) = mr^3 \left\{ \frac{2}{\left(r^2 + (z - z_0)^2\right)^{\frac{3}{2}}} - \frac{1}{\left[r^2 + (z - z_0 + a)^2\right]^{\frac{3}{2}}} - \frac{1}{\left[r^2 + (z - z_0 - a)^2\right]^{\frac{3}{2}}} \right\} + A + Bz, \quad \text{(A1)}$$

where r = 0.97 cm is the radius of the gradiometer pick up coils, a = 1.519 cm is their base distance, and z_0 is the axial position of the sample (the alignment). In the coordinates used by MultiVu, z_0 should be close to 2 cm. The form given in Equation (A1) means that the numerical analysis is done in point dipole approximation. Therefore, for extended specimens the final magnitudes of m are corrected for the shape-dependent factors [19,43,44,57].

The main advantage of the external analysis of the scans is that the user possesses the ability to freely adjust with the parameters of $V(z)$. Most importantly here, one can investigate the dependence of m on the alignment.

Appendix C RSO Method Settings

As noted in the Section 2, in most cases the data presented here are collected by taking six measurements each consisting of four oscillations (24 scans all together to be averaged). There are two practical hints concerning these parameters that are worth adding to this report. First, for a given number of single scan events (4 × 6 = 24 in this case), it does not make any difference if all of them constitute one single measurement unit or, conversely, whether they are divided into 24 separate measurements containing only one scan event. The average result will be the same. However, for advanced users, particularly those who are using post-measurement method(s) to reanalyze the scans, it pays off to have more measurement units than oscillation events and to record the whole suite of the output files provided by MultiVu software (.dat, .diag, and .raw files). The advantage becomes evident when, for example, say during the ninth oscillation event, a sudden signal jump on the scan voltage occurs. This distorted voltage is "normally" averaged together with the other, correct readings (scans)—resulting in a false moment determination, accompanied by an increased magnitude of the standard deviation error. However, for the case considered here, this jump mars only the third measurement unit. A short inspection of the corresponding .diag file, manual or by a dedicated software, can eradicate such wrong readings and after recalculation of the new average, the true (or more correct) magnitude of m at the given experimental conditions is restored. This will be accompanied by a reduction of the standard deviation error down to the level of the neighboring points.

Appendix D Extraction of a Number of PM Spins

Accurate quantitative information on the paramagnetic contribution present in a composite material can be conveniently obtained from isotherms $m(H)$ measured at very low temperatures (T_1, T_2, here T_1 = 1.8 K, T_2 = 5 K) and numerical analysis of their difference $\Delta m(H, \Delta T = T_1 - T_2)$. The values of T_1 and T_2 should be much smaller than the spin–spin coupling temperature(s) of the remaining non-PM part of the sample. This allows one to treat the diamagnetic, ferromagnetic, and all other magnetic contributions as temperature independent, and so, whatever is the form of their contribution they null out in $\Delta m(H, \Delta T)$. Under such circumstances, the analysis does not have to account for them and the difference $\Delta m(H, \Delta T)$ (marked by magenta symbols in Figure 15b) can be fitted by:

$$\Delta m(H,\ \Delta T) = N_S g \mu_B S \left[B_J(H, 1.8\text{K}) - B_J(H, 5\text{K})\right] \quad \text{(A2)}$$

where $B_J(H, T)$ is the Brillouin function:

$$B_J(x) = \frac{2J+1}{2J} coth\left(\frac{2J+1}{2J} x\right) - \frac{1}{2J} coth\left(\frac{1}{2J} x\right)$$

with H and T tied in:

$$x = J\frac{g\mu_B H}{k_B T} \tag{A3}$$

Here, g = 2 is the Land'e factor, μ_B is the Bohr magneton, k_B is the Boltzmann constant, and N_S is the number of spins with $J = L + S$. For simplicity we take $L = 0$. In this approach the magnitudes of S and N_S are the only adjustable parameters. As S can only assume some discrete values, in a search of the spin state we perform three independent fittings for $J = S = 3/2$, 2, and 5/2. The best fit, obtained for $S = 2$ and $N_S = 5.3(1) \times 10^{15}$, is presented in Figure 15b.

References

1. Proselkov, O.; Sztenkiel, D.; Stefanowicz, W.; Aleszkiewicz, M.; Sadowski, J.; Dietl, T.; Sawicki, M. Thickness Dependent Magnetic Properties of (Ga,Mn)As Ultrathin Films. *Appl. Phys. Lett.* **2012**, *100*, 262405. [CrossRef]
2. Gas, K.; Kunert, G.; Dluzewski, P.; Jakiela, R.; Hommel, D.; Sawicki, M. Improved-Sensitivity Integral SQUID Magnetometry of (Ga,Mn)N Thin Films in Proximity to Mg-Doped GaN. *J. Alloy. Compd.* **2021**, *868*, 159119. [CrossRef]
3. Navarro-Quezada, A.; Gas, K.; Truglas, T.; Bauernfeind, V.; Matzer, M.; Kreil, D.; Ney, A.; Groiss, H.; Sawicki, M.; Bonanni, A. Out-of-Plane Magnetic Anisotropy in Ordered Ensembles of FeyN Nanocrystals Embedded in GaN. *Materials* **2020**, *13*, 3294. [CrossRef]
4. Navarro-Quezada, A.; Stefanowicz, W.; Li, T.; Faina, B.; Rovezzi, M.; Lechner, R.T.; Devillers, T.; d'Acapito, F.; Bauer, G.; Sawicki, M.; et al. Embedded Magnetic Phases in (Ga,Fe)N: Key Role of Growth Temperature. *Phys. Rev. B* **2010**, *81*, 205206. [CrossRef]
5. Sawicki, M.; Guziewicz, E.; Łukasiewicz, M.I.; Proselkov, O.; Kowalik, I.A.; Lisowski, W.; Dluzewski, P.; Wittlin, A.; Jaworski, M.; Wolska, A.; et al. Homogeneous and Heterogeneous Magnetism in (Zn,Co)O: From a Random Antiferromagnet to a Dipolar Superferromagnet by Changing the Growth Temperature. *Phys. Rev. B* **2013**, *88*, 085204. [CrossRef]
6. Sawicki, M.; Wang, K.Y.; Edmonds, K.W.; Campion, R.P.; Rushforth, A.W.; Foxon, C.T.; Gallagher, B.L.; Dietl, T. Coercivity Enlargement in (Ga,Mn)As Thin Films with Small Amount of MnAs Nanoclusters. *J. Magn. Magn. Mater.* **2007**, *310*, 2126–2128. [CrossRef]
7. Kaleta, A.; Kret, S.; Gas, K.; Kurowska, B.; Kryvyi, S.B.; Rutkowski, B.; Szwacki, N.G.; Sawicki, M.; Sadowski, J. Enhanced Ferromagnetism in Cylindrically Confined MnAs Nanocrystals Embedded in Wurtzite GaAs Nanowire Shells. *Nano Lett.* **2019**, *19*, 7324–7333. [CrossRef] [PubMed]
8. Yu, W.; Li, J.; Herng, T.S.; Wang, Z.; Zhao, X.; Chi, X.; Fu, W.; Abdelwahab, I.; Zhou, J.; Dan, J.; et al. Chemically Exfoliated VSe2 Monolayers with Room-Temperature Ferromagnetism. *Adv. Mater.* **2019**, *31*, 1903779. [CrossRef] [PubMed]
9. Tokmachev, A.M.; Averyanov, D.V.; Taldenkov, A.N.; Sokolov, I.S.; Karateev, I.A.; Parfenov, O.E.; Storchak, V.G. Two-Dimensional Magnets beyond the Monolayer Limit. *ACS Nano* **2021**, *15*, 12034–12041. [CrossRef]
10. Liu, Y.; Shen, Y.; Sun, L.; Li, J.; Liu, C.; Ren, W.; Li, F.; Gao, L.; Chen, J.; Liu, F.; et al. Elemental Superdoping of Graphene and Carbon Nanotubes. *Nat. Commun.* **2016**, *7*, 10921. [CrossRef]
11. Sadowski, J.; Kret, S.; Šiušys, A.; Wojciechowski, T.; Gas, K.; Islam, M.F.; Canali, C.M.; Sawicki, M. Wurtzite (Ga,Mn)As Nanowire Shells with Ferromagnetic Properties. *Nanoscale* **2017**, *9*, 2129–2137. [CrossRef]
12. Mazur, G.P.; Dybko, K.; Szczerbakow, A.; Domagala, J.Z.; Kazakov, A.; Zgirski, M.; Lusakowska, E.; Kret, S.; Korczak, J.; Story, T.; et al. Experimental Search for the Origin of Low-Energy Modes in Topological Materials. *Phys. Rev. B* **2019**, *100*, 041408. [CrossRef]
13. Sánchez-Barriga, J.; Varykhalov, A.; Springholz, G.; Steiner, H.; Kirchschlager, R.; Bauer, G.; Caha, O.; Schierle, E.; Weschke, E.; Ünal, A.A.; et al. Nonmagnetic Band Gap at the Dirac Point of the Magnetic Topological Insulator (Bi1−xMnx)2Se3. *Nat. Commun.* **2016**, *7*, 10559. [CrossRef] [PubMed]
14. Teng, J.; Liu, N.; Li, Y. Mn-Doped Topological Insulators: A Review. *J. Semicond.* **2019**, *40*, 081507. [CrossRef]
15. Grutter, A.J.; He, Q.L. Magnetic Proximity Effects in Topological Insulator Heterostructures: Implementation and Characterization. *Phys. Rev. Materials* **2021**, *5*, 090301. [CrossRef]
16. Garcia, M.A.; Fernandez Pinel, E.; de la Venta, J.; Quesada, A.; Bouzas, V.; Fernández, J.F.; Romero, J.J.; Martín González, M.S.; Costa-Krämer, J.L. Sources of Experimental Errors in the Observation of Nanoscale Magnetism. *J. Appl. Phys.* **2009**, *105*, 013925. [CrossRef]
17. Salzer, R.; Spemann, D.; Esquinazi, P.; Höhne, R.; Setzer, A.; Schindler, K.; Schmidt, H.; Butz, T. Possible Pitfalls in Search of Magnetic Order in Thin Films Deposited on Single Crystalline Sapphire Substrates. *J. Magn. Magn. Mater.* **2007**, *317*, 53–60. [CrossRef]
18. Pereira, L.M.C.; Araújo, J.P.; Bael, M.J.V.; Temst, K.; Vantomme, A. Practical Limits for Detection of Ferromagnetism Using Highly Sensitive Magnetometry Techniques. *J. Phys. D Appl. Phys.* **2011**, *44*, 215001. [CrossRef]
19. Sawicki, M.; Stefanowicz, W.; Ney, A. Sensitive SQUID Magnetometry for Studying Nanomagnetism. *Semicond. Sci. Technol.* **2011**, *26*, 064006. [CrossRef]
20. Buchner, M.; Höfler, K.; Henne, B.; Ney, V.; Ney, A. Tutorial: Basic Principles, Limits of Detection, and Pitfalls of Highly Sensitive SQUID Magnetometry for Nanomagnetism and Spintronics. *J. Appl. Phys.* **2018**, *124*, 161101. [CrossRef]

21. Pereira, L.M.C. Experimentally Evaluating the Origin of Dilute Magnetism in Nanomaterials. *J. Phys. D Appl. Phys.* **2017**, *50*, 393002. [CrossRef]
22. Gas, K.; Sawicki, M. In Situ Compensation Method for High-Precision and High-Sensitivity Integral Magnetometry. *Meas. Sci. Technol.* **2019**, *30*, 085003. [CrossRef]
23. Wang, M.; Andrews, C.; Reimers, S.; Amin, O.J.; Wadley, P.; Campion, R.P.; Poole, S.F.; Felton, J.; Edmonds, K.W.; Gallagher, B.L.; et al. Spin Flop and Crystalline Anisotropic Magnetoresistance in CuMnAs. *Phys. Rev. B* **2020**, *101*, 094429. [CrossRef]
24. Scheffler, L.; Gas, K.; Banik, S.; Kamp, M.; Knobel, J.; Lin, H.; Schumacher, C.; Gould, C.; Sawicki, M.; Kleinlein, J.; et al. Molecular Beam Epitaxy of the Half-Heusler Antiferromagnet CuMnSb. *Phys. Rev. Mater.* **2020**, *4*, 114402. [CrossRef]
25. Iskandar, N.A.J.T.N.; Guan-Yeow, Y.; Maeta, N.; Ito, M.M.; Nakamura, Y.; Gas, K.; Sawicki, M. Anisotropic and Magnetic Properties in Non-Metal and Non-Radical Organic Aggregates of Tri-Substituted Phenyl Derivatives. *New J. Chem.* **2019**, *44*, 210–217. [CrossRef]
26. Bujak, P.; Kulszewicz-Bajer, I.; Zagorska, M.; Maurel, V.; Wielgus, I.; Pron, A. Polymers for Electronics and Spintronics. *Chem. Soc. Rev.* **2013**, *42*, 8895–8999. [CrossRef]
27. Gutiérrez, L.; Vujić Spasić, M.; Muckenthaler, M.U.; Lázaro, F.J. Quantitative Magnetic Analysis Reveals Ferritin-like Iron as the Most Predominant Iron-Containing Species in the Murine Hfe-Haemochromatosis. *Biochim. Biophys. Acta Mol. Basis Dis.* **2012**, *1822*, 1147–1153. [CrossRef] [PubMed]
28. Rojas, J.M.; Gavilán, H.; del Dedo, V.; Lorente-Sorolla, E.; Sanz-Ortega, L.; da Silva, G.B.; Costo, R.; Perez-Yagüe, S.; Talelli, M.; Marciello, M.; et al. Time-Course Assessment of the Aggregation and Metabolization of Magnetic Nanoparticles. *Acta Biomater.* **2017**, *58*, 181–195. [CrossRef]
29. Menon, J.U.; Kuriakose, A.; Iyer, R.; Hernandez, E.; Gandee, L.; Zhang, S.; Takahashi, M.; Zhang, Z.; Saha, D.; Nguyen, K.T. Dual-Drug Containing Core-Shell Nanoparticles for Lung Cancer Therapy. *Sci. Rep.* **2017**, *7*, 13249. [CrossRef] [PubMed]
30. Serantes, D.; Baldomir, D. Nanoparticle Size Threshold for Magnetic Agglomeration and Associated Hyperthermia Performance. *Nanomaterials* **2021**, *11*, 2786. [CrossRef]
31. The GE/IMI 7031 Varnish Is Obtained from Cambridge Magnetic Refrigeration, Cambridge, CB5 8HR, UK. Available online: https://www.cmr-direct.com/en/cmr-gevar-25ml?filter_name=ge%20varnis (accessed on 7 January 2022).
32. Qantum Design Application Note 1500-018, Using Straw Sample Holder with DC Scan Mode. Available online: https://qdusa.com/siteDocs/appNotes/1500-018.pdf (accessed on 7 January 2022).
33. Franken, L.; Klein, M.; Spasova, M.; Elsukova, A.; Wiedwald, U.; Welz, M.; Knolle, P.; Farle, M.; Limmer, A.; Kurts, C. Splenic Red Pulp Macrophages Are Intrinsically Superparamagnetic and Contaminate Magnetic Cell Isolates. *Sci. Rep.* **2015**, *5*, 12940. [CrossRef] [PubMed]
34. Świątkowski, M.; Lanka, S.; Czylkowska, A.; Gas, K.; Sawicki, M. Structural, Spectroscopic, Thermal, and Magnetic Properties of a New Dinuclear Copper Coordination Compound with Tiglic Acid. *Materials* **2021**, *14*, 2148. [CrossRef]
35. Škrátek, M.; Dvurečenskij, A.; Kluknavský, M.; Barta, A.; Bališ, P.; Mičurová, A.; Cigáň, A.; Eckstein-Andicsová, A.; Maňka, J.; Bernátová, I. Sensitive SQUID Bio-Magnetometry for Determination and Differentiation of Biogenic Iron and Iron Oxide Nanoparticles in the Biological Samples. *Nanomaterials* **2020**, *10*, 1993. [CrossRef]
36. Dutta, R.K.; Sahu, S. Development of Oxaliplatin Encapsulated in Magnetic Nanocarriers of Pectin as a Potential Targeted Drug Delivery for Cancer Therapy. *Results Pharma Sci.* **2012**, *2*, 38–45. [CrossRef]
37. Gardias, A.; Kaszyński, P.; Obijalska, E.; Trzybiński, D.; Domagała, S.; Woźniak, K.; Szczytko, J. Magnetostructural Investigation of Orthogonal 1-Aryl-3-Phenyl-1,4-Dihydrobenzo[e][1,2,4]Triazin-4-Yl Derivatives. *Chem. A Eur. J.* **2018**, *24*, 1317–1329. [CrossRef] [PubMed]
38. Quantum Design, Inc. 10307 Pacific Center Court, San Diego, CA, USA. Available online: www.qdusa.com (accessed on 7 January 2022).
39. QD MPMS Sample Mounting Considerations 1014-201. Available online: https://qdusa.com/siteDocs/appNotes/1014-201.pdf (accessed on 7 January 2022).
40. Ney, A.; Kammermeier, T.; Ney, V.; Ollefs, K.; Ye, S. Limitations of Measuring Small Magnetic Signals of Samples Deposited on a Diamagnetic Substrate. *J. Magn. Magn. Mater.* **2008**, *320*, 3341–3346. [CrossRef]
41. Ney, A.; Harris, J.S.; Parkin, S.S.P. Temperature Dependent Magnetic Properties of the GaAs Substrate of Spin-LEDs. *J. Phys. Condens. Matter* **2006**, *18*, 4397–4406. [CrossRef]
42. Cryogenic Ltd. Unit 6, Acton Park Industrial Estate, The Vale, London W3 7QE, UK. Available online: www.cryogenic.co.uk (accessed on 7 January 2022).
43. Blott, B.H.; Daniell, G.J. The Determination of Magnetic Moments of Extended Samples in a SQUID Magnetometer. *Meas. Sci. Technol.* **1993**, *4*, 462–467. [CrossRef]
44. Hayden, M.E.; Lambinet, V.; Gomis, S.; Gries, G. Note: Evaluation of Magnetometry Data Acquired from Elongated Samples. *Rev. Sci. Instrum.* **2017**, *88*, 056106. [CrossRef] [PubMed]
45. Dziawa, P.; Kowalski, B.J.; Dybko, K.; Buczko, R.; Szczerbakow, A.; Szot, M.; Łusakowska, E.; Balasubramanian, T.; Wojek, B.M.; Berntsen, M.H.; et al. Topological Crystalline Insulator States in Pb1−xSnxSe. *Nat. Mater.* **2012**, *11*, 1023–1027. [CrossRef] [PubMed]
46. Szczerbakow, A.; Durose, K. Self-Selecting Vapour Growth of Bulk Crystals—Principles and Applicability. *Prog. Cryst. Growth Charact. Mater.* **2005**, *51*, 81–108. [CrossRef]

47. ZPH HERBES. Wiesław Sikorski, Przezmierzowo, Poland. Available online: www.herbes.com.pl (accessed on 7 January 2022).
48. Gas, K.; Sadowski, J.; Sawicki, M. Magnetic Properties of Wurtzite (Ga,Mn)As. *J. Magn. Magn. Mater.* **2021**, *533*, 168012. [CrossRef]
49. Gas, K.; Królicka, A.; Dybko, K.; Nowicki, P.; Khosravizadeh, Z.; Story, T.; Sawicki, M. Magnetic Constitution of Topologically Trivial Thermoelectric PbTe:Cr. *J. Magn. Magn. Mater.* **2021**, *537*, 168154. [CrossRef]
50. Navarro-Quezada, A.; Gas, K.; Spindlberger, A.; Karimi, F.; Sawicki, M.; Ciatto, G.; Bonanni, A. Unravelling the Local Crystallographic Structure of Ferromagnetic $Ga_yFe_{4-y}N$ Nanocrystals Embedded in GaN. *Sci. Rep.* **2021**, *11*, 2862. [CrossRef] [PubMed]
51. Restrepo-Osorio, J.; Nobile-Correa, D.P.; Zuñiga, O.; Sánchez-Andica, R.A. Determination of nutritional value of turmeric flour and the antioxidant activity of Curcuma longa rhizome extracts from agroecological and conventional crops of Valle del Cauca-Colombia. *Rev. Colomb. Química* **2020**, *49*, 26–32.
52. Mastrogiacomo, G.; Löffler, J.F.; Dilley, N.R. Soft-Magnetic Materials Characterized Using a Superconducting Solenoid as Magnetic Source. *Appl. Phys. Lett.* **2008**, *92*, 082501. [CrossRef]
53. Stefanowicz, W.; Sztenkiel, D.; Faina, B.; Grois, A.; Rovezzi, M.; Devillers, T.; d'Acapito, F.; Navarro-Quezada, A.; Li, T.; Jakieła, R.; et al. Structural and paramagnetic properties of dilute $Ga_{1-x}Mn_xN$. *Phys. Rev. B* **2010**, *81*, 235210. [CrossRef]
54. QD Application Note 1500-011: Using SQUID VSM Superconducting Magnets at Low Fields. Available online: https://qdusa.com/siteDocs/appNotes/1500-011.pdf (accessed on 7 January 2022).
55. Karczewski, G.; Sawicki, M.; Ivanov, V.; Ruester, C.; Grabecki, G.; Matsukura, F.; Molenkamp, L.W.; Dietl, T. Ferromagnetism in (Zn,Cr)Se Layers Grown by Molecular Beam Epitaxy. *J. Supercond.* **2003**, *16*, 55–58. [CrossRef]
56. Quantum Design 2002 MPMS Application Note 1014-213: Subtracting the Sample Holder Background from Dilute Samples. Available online: https://qdusa.com/siteDocs/appNotes/1014-213.pdf (accessed on 7 January 2022).
57. Stamenov, P.; Coey, J.M.D. Sample Size, Position, and Structure Effects on Magnetization Measurements Using Second-Order Gradiometer Pickup Coils. *Rev. Sci. Instrum.* **2006**, *77*, 015106. [CrossRef]

materials

MDPI

Article

Effect of Deep Cryogenic Treatment on Corrosion Behavior of AISI H13 Die Steel

Tarang Shinde [1], Catalin Pruncu [2,*], Narendra B. Dhokey [3], Anca C. Parau [4] and Alina Vladescu [4,5,*]

[1] Department of Mechanical Engineering, Finolex Academy of Management and Technology, Ratnagiri 415639, India; tarangshinde@gmail.com
[2] Department of Mechanical Engineering, Imperial College London, Exhibition Rd., London SW7 2AZ, UK
[3] Department of Metallurgy & Materials Science, College of Engineering, Shivajinagar, Pune 411005, India; nbdhokey@yahoo.co.in
[4] Department of Advanced Surface Processing and Analysis by Vacuum Technologies (ReCAST), National Institute of Research and Development for Optoelectronics INOE 2000, 409 Atomistilor St., Magurele, 077125 Bucharest, Romania; anca.parau@inoe.ro
[5] Physical Materials Science and Composite Materials Centre, Research School of Chemistry & Applied Biomedical Sciences, Tomsk Polytechnic University, Lenin's Avenue 30, 634050 Tomsk, Russia
* Correspondence: catalin.pruncu@gmail.com (C.P.); alinava@inoe.ro (A.V.); Tel.: +4021-457-57-59 (A.V.)

Abstract: AISI H13 die steel specimens were subjected to heating at 1020 °C followed by oil quenching and double tempering at 520 °C. Subsequently, these specimens were subjected to deep cryogenic treatment at −185 °C in liquid nitrogen environment for 16 h and then subjected to soft tempering at 100 °C once the specimens attained room temperature. Thereafter, the specimens were subjected to scanning electron microscopy (SEM) analysis and electron backscatter diffraction (EBSD) analysis. The electrochemical corrosion activity was investigated in 3.5% NaCl at 23 ± 0.5 °C by evaluating the evolution of open circuit potential over time and potentiodynamic curves, and electrochemical impedance spectroscopy study was also carried out. The heat-treated specimens exhibited better resistance to corrosion through more electropositive values of open circuit potential. This could be attributed to lower grain boundary area in heat-treated specimens as compared to 16 h cryogenically treated specimen as higher grain boundary areas behave as an anode in an electrochemical cell, thereby enhancing the rate of corrosion. According to electrochemical tests, the cryogenically treated surface is more resistant to corrosion, followed by heated alloy. However, both surface modification treatments improved the corrosion behavior of the untreated alloy.

Keywords: corrosion; deep cryogenic treatment; AISI H13; Nyquist; open circuit potential; tool steels; corrosion rate

Citation: Shinde, T.; Pruncu, C.; Dhokey, N.B.; Parau, A.C.; Vladescu, A. Effect of Deep Cryogenic Treatment on Corrosion Behavior of AISI H13 Die Steel. *Materials* **2021**, *14*, 7863. https://doi.org/10.3390/ma14247863

Academic Editors: Corneliu Munteanu and Ioan Doroftei

Received: 4 November 2021
Accepted: 14 December 2021
Published: 18 December 2021

1. Introduction

The tool steels are widely used in hot working industries including forging and extrusion. According to the market study and analysis, the global market size of the tool steels was USD 4.29 billion in 2020, and as per the forecast, their value is expected to cross over USD 6395.95 by 2025 [1,2]. AISI H13 hot work die steel is one of the most important materials belonging to this category. In some of the applications, AISI H13 undergoes a phase change, austenitizing some region of the component when it is under stress and heat due to friction which could promote the spalling [3]; other reasons for failure could be oxidative or abrasive wear, fatigue cracking and chipping as experienced in some of the cases [4]. The corrosion damage of H13 die steel has also been reported in the case of applications involving the manufacturing of dies for aluminum cans [5]. Since these materials are subjected to heat and mechanical stresses, their corrosion behavior needs to be studied in order to investigate their response towards changing conditions, as well as to report any possible degradation in their properties during service.

The tool steels need to possess certain characteristics in order to withstand highly challenging conditions during mechanical working operations [6]. These characteristics include hardness, resistance to wear and fatigue properties. Even though the basic combinations present in tool steels do add up these characteristics, the need of the time is to further improve them. The deep cryogenic treatment has been recommended as an add-on process to improve certain mechanical properties, especially wear resistance, as indicated by many researchers in their studies on tool steels [7–17]. This treatment increases the density of precipitation of most of the carbides in tool steels and reduces average carbide particle size [18]. Some researchers have elaborated that the deep cryogenic treatment improves the surface finish of the materials through the refinement of martensite laths [19]. This method had been beneficial for materials such as AISI H13 die steel [20–23] in order to improve the wear resistance, hardness and surface characteristics, thereby enhancing the performance of the material.

The fine carbides precipitated after cryogenic treatment assist with the improvement of these properties substantially. The cryogenic treatment even had an impact on fatigue life, as reported by some researchers [24,25]. The fine carbides could act as microcrack arresters. They delay the propagation of these microcracks, thereby enhancing the fatigue life, which is quite beneficial for applications such as mechanical working. It has also been elaborated that the deep cryogenic treatment improved the corrosion resistance for some tool steels such as 1.2080 [26]. Corrosion behavior of hot work tool steels has been studied, and it was reported by researchers that methods such as pressure vapor deposition (PVD) and nitriding could be helpful in improving the corrosion resistance of these materials [27–29]. PVD coatings assist in localizing the attack on the material surface. A good adhesiveness is essential in retaining the coating layer and thereby preventing any further attack of the chemical or any surrounding agent which could result in formation of any corrosive byproduct.

One researcher experimented with the formation of a borided layer on the surface of AISI H13 steel [30]. The borided layers formed single phase or dual phase structures and their layers were compact and crack-free, which improved the corrosion resistance of the steel, especially in the case of steam turbine applications. The deep cryogenic treatment has been beneficial in improving the corrosion characteristics of structural steels. It could certainly become a potential method for those applications where corrosion plays a very important role deciding the lifespan of the component [31]. Most of the treatments have focused upon surface modifications in order to improve the corrosion resistance of tool steels.

In the last few decades, there was an interest in increasing the wear resistance of steels by applying deep cryogenic treatment due to homogenization and stabilization of internal microstructure. The modification of other steels such as D-2, M-2 and AISI H13 by deep cryogenic treatment can be found in the literature. The aim of the paper is to investigate the corrosion resistance of the AISI H13 die steel specimens by the deep cryogenic treatment at −185 °C in liquid nitrogen environment for 16 h and to see the effect of this treatment on its properties. Electron backscatter diffraction (EBSD) analysis, grain mapping tools and Nyquist plots have been used in the current research work in order to evaluate the performance of AISI H13 die steel specimens when subjected to a corrosive atmosphere.

The study also investigates the outcomes based on the experimentation to comment upon the best treatment for H13 steel in order to withstand the corrosive atmosphere. The results have been compared to those of the untreated and conventionally heat-treated specimens.

2. Materials and Methods

2.1. Surface Modification Process

AISI H13 material was procured from Rajasthan Steels, Pune (India), to fabricate the specimens for research work. The chemical composition involves C—0.39%, Mn—0.30%, Si—0.40% Cr—5.40%, Mo—1.4%, V—1.00%, balance Fe, by weight. A set of AISI H13 specimens (diameter 8 mm and length 15 mm) was hardened at 1020 °C with a step-by-step heating (first heating to 500 °C at a rate of 7 °C/min and holding for 20 min and thereafter heating to 860 °C and soaking for 20 min) to prevent cracking due to thermal stresses. These specimens were soaked for 20 min at 1020 °C followed by oil quenching. The quenched specimens were double tempered at 520 °C for 2 h. This treatment was named conventional heat treatment for AISI H13 steel and the specimens were designated as "heated" specimens. In a cryo-chamber (Sanmar, Mumbai, India) supplied with liquid nitrogen at a cooling rate of 3 °C/min, the other set of specimens was subjected to conventional heat treatment (step-by-step heating to 1020 °C, oil quenching and double tempering at 520 °C for 2 h) and cryogenic treatment at −185 °C at a cryo soaking period of 16 h [21]. At the end of 16 h of cryo soaking period, these specimens were transferred to an insulated box (Cryobox, Sigma-Aldrich, Irvine, UK) until they attained room temperature, and thereafter the specimens were soft-tempered for 1 h at 100 °C to eliminate the cold stresses produced during the cryogenic treatment. Specimens cryogenically treated for 16 h were referred to as "cryo" specimens.

2.2. Characterization Methods of Developed Surfaces

All specimens were examined by scanning electron microscopy (SEM) (Tabletop Microscope TM3030, Hitachi, Tokyo, Japan) after mechanical polishing which involved grinding down using SiC papers (homemade), sequentially from 320 to 4000 grits. Subsequently, the ground specimens were active oxide polishing (OPS) polished for 30 min using a suspension diluted with H_2O with a ratio of 1:5 to a mirror finish.

The specimens were further characterized for grain orientation and texture direction using a Hitachi 3400 SEM-based Bruker e-flash electron backscatter diffraction (EBSD) detector (Cheshire, UK) at 20 kV acceleration voltage, 10 mA current density and 1 mm step size. The images were postprocessed using MTEX software (free Matlab toolbox, see details in https://mtex-toolbox.github.io/, accessed on 3 December 2021).

The electrochemical behavior was measured using a VERSASTAT potentiostat/galvanostat (Princeton Applied Research, Princeton, NJ, USA). The tests were performed in 3.5% NaCl solution (pH = 7.4) at 23 ± 0.5 °C. Each sample was placed in a Teflon sample holder (Princeton Applied Research, Princeton, USA) with exactly 1 cm^2 exposed to the corrosive media. A platinum electrode (XM140, Radiometer Analytical, Loveland, CO, USA) was used as the counter electrode and saturated calomel (XR110, Radiometer Analytical, Loveland, CO, USA) was used as the reference electrode. All measurements were achieved at a scanning rate of 1 mV/s. The open circuit potential (E_{OC}) was monitored for 1 h, starting right after the immersion of the sample in the 3.5% NaCl solution, and the potentiodynamic curves were recorded at −1 to +2 V vs. E_{OC}. The tests were performed according to the standard ISO 16151:2018. Electrochemical behavior of the investigated specimens was also examined by electrochemical impedance spectroscopy (EIS). Impedance measurements were performed at open circuit potential with constant perturbing AC signal amplitude of 10 mV over a frequency range extending from 0.1 to 10^4 Hz. Analysis of the spectra was performed by equivalent circuit fitting using Zview software (Princeton Applied Research, Princeton, USA). After the corrosion tests, each surface was analysed by SEM and profilometry (Dektak 150, Bruker, Billerica, MA, USA) in order to evaluate the morphology and roughness.

3. Results

3.1. Microstructural Results

The microstructural details of polycrystalline materials can be investigated with the help of electron backscatter diffraction (EBSD) analysis [32]. One of the reasons why grain boundaries must be analyzed is that the performance and integrity of the material depend upon the grain boundary networks present inside the materials [33]. The EBSD feature has made it possible to analyze more than 10^4 boundaries in order to investigate their role in properties such as corrosion resistance [34,35]. The grain size has a significant impact on the mechanical and corrosion properties of an alloy. The EBSD phase maps could be useful in identification and distribution of different phases in the alloys. Figure 1 shows the phase maps for untreated, heated and cryo-treated specimens depicting the significant precipitation of different phases in different cases. Figure 2 reports the results obtained through EBSD analysis clarifying that the heat treatment (Figure 2b) and cryogenic treatment (Figure 2c) bring about recrystallized structure in the material. A low angle grain boundary is typically the boundary between two crystal grains with a misorientation of less than 15° [36,37].

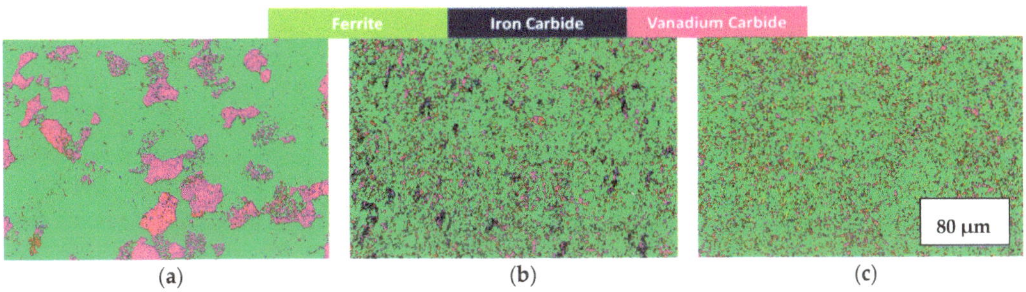

Figure 1. Maps depicting the main phase of studied material: (**a**) untreated, (**b**) heated and (**c**) cryo specimens.

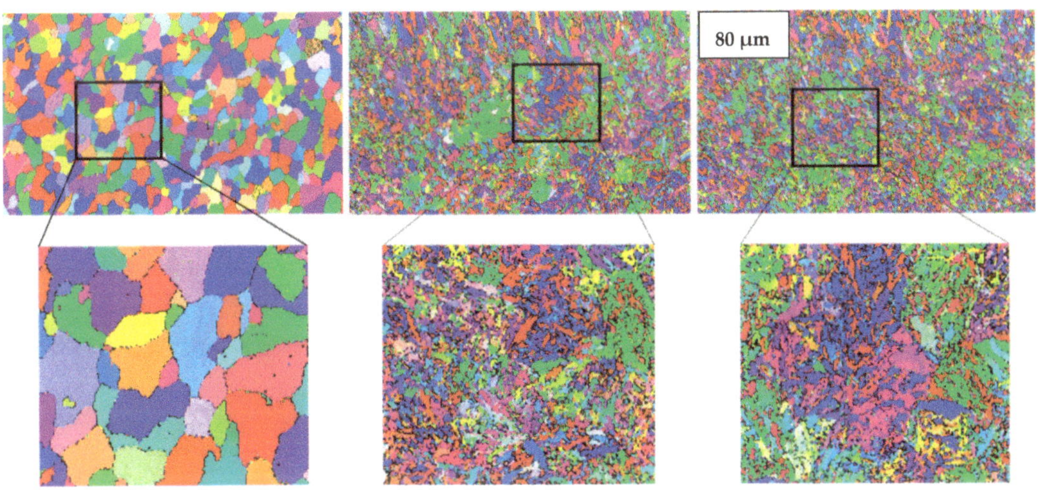

Figure 2. Inverse pole figure map obtained from EBSD analysis for (**a**) untreated, (**b**) heated and (**c**) cryo specimens.

Figure 3 elaborates the low angle grain boundary (LAGB) patterns for untreated, heated and cryo specimens which indicate that the as-received specimen has a limited number of LAGBs (noted by green) while heat-treated and cryo specimens show more LAGBs which could be attributed to the recrystallized structure in both specimens.

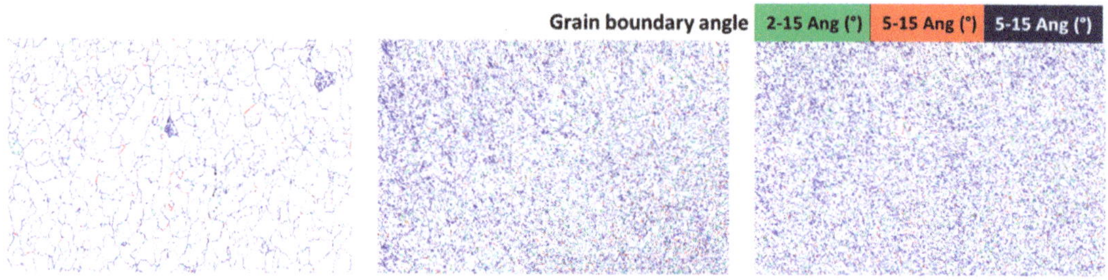

(a) (b) (c)

Figure 3. LAGBs for (**a**) untreated, (**b**) heated and (**c**) cryo specimens.

The deep cryogenic treatment of H13 steel resulted in a fine-grained structure. As geometrically necessary dislocation (GND) density measurement by EBSD has become a very popular tool in microstructural analysis [38], Figure 4 presents the geometrically necessary dislocation evolution of untreated, heated and cryo-treated specimens. It is observed from the GND maps that the dislocation density trend is untreated < heated < cryo; the highest dislocation density is found for cryo specimens (10^{14}–10^{15}), which could be attributed to increased plastic strain, resulting in accumulation of dislocations at the grain boundary areas for cryo specimens of H13 steel.

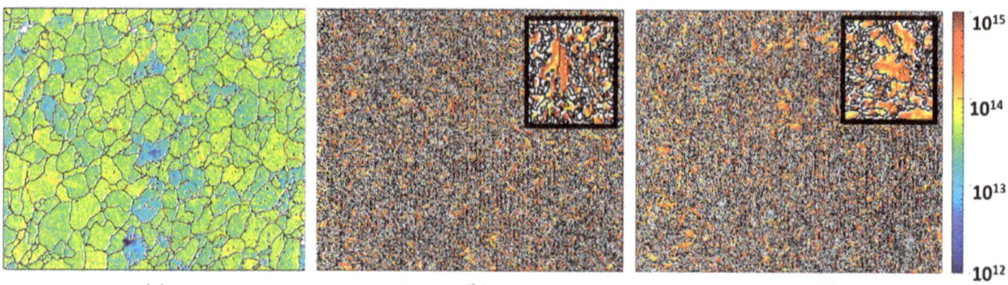

(a) (b) (c)

Figure 4. Geometrically necessary dislocation evolution of (**a**) untreated, (**b**) heated and (**c**) cryo-treated specimens.

Figure 5 represents the grain size variation for untreated, conventionally heat-treated (heated) and 16 h cryogenically treated (cryo) specimens. The graph illustrates that the grain size is minimal for 16 h cryogenically treated H13 steel for both conditions (before and after corrosion). The cryogenic treatment assists in the precipitation of fine carbides along the grain boundaries which inhibits the grain growth, resulting in a fine-grained structure [39]. Minimum grain size depicts the maximum grain boundary area inside the material. It has been shown by some authors that the grain refinement leads to increased susceptibility to corrosion [40–42], but the passivity of the film present on the surface plays a vital role in deciding the response of the material to corrosion. When untreated specimens are compared with conventionally treated (heated) ones, there is a reduction in grain size and, as stated by some authors, the corrosion resistance would be greater if the passivity of the film is maintained even for a fine-grained material [43–45].

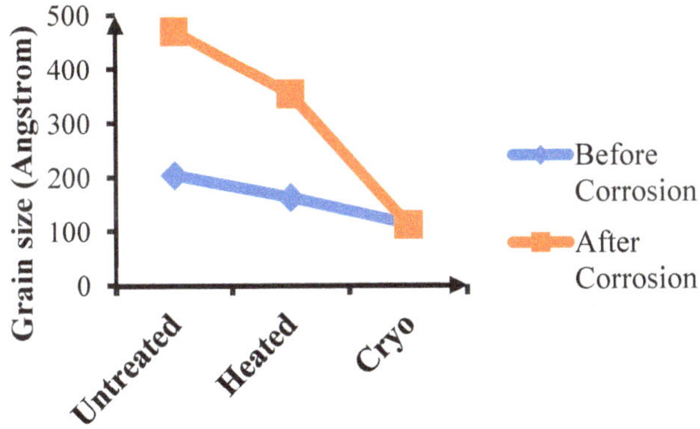

Figure 5. Grain size variation for untreated, conventionally heat-treated and 16 h cryogenically treated specimens before and after corrosion.

3.2. Electrochemical Results

The electrochemical corrosion activity was investigated in 3.5% NaCl at 23 ± 0.5 °C by evaluating the evolution of open circuit potential (OCP) over time (Figure 5) and potentiodynamic curves (Figure 6). The open circuit potential (OCP) is a parameter that is related to the protective ability of the passive film. During 1 h of immersion, the OCP value slightly changed, indicating that steady-state conditions were not reached (Figure 5). Based on the results presented in Figure 6, one may see that the "cryo" specimen has an electropositive value of E_{oc}, indicating a good resistance to NaCl attack.

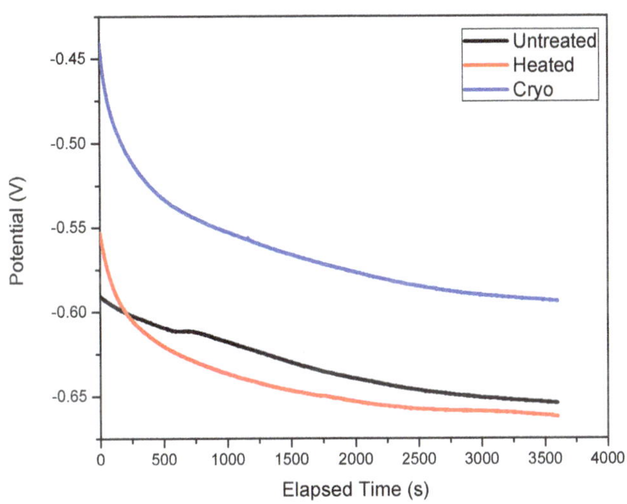

Figure 6. Evolution of the open circuit potential over time.

Figure 7 represents the polarization curves of the investigated surfaces. Based on Tafel extrapolation [46], the main corrosion parameters were extracted from Figure 6 and presented in Table 1, as frequently reported [47]. Heated specimens exhibited a more electropositive value of the corrosion potential (E_i = 0) compared to the uncoated and cryo specimens, indicating that the corrosive solution had less influence on their surfaces. The cryo specimen had a more electronegative corrosion potential, showing poor corrosion resistance to NaCl attack. Lower corrosion current density (i_{corr}) and higher polarization resistance (R_p) values were observed for heated specimens, showing good corrosion resistance. Taking into account the electrochemical parameters, it can be said that heated specimens exhibited the best corrosion resistance to 3.5% NaCl, followed by the untreated and the cryo specimens.

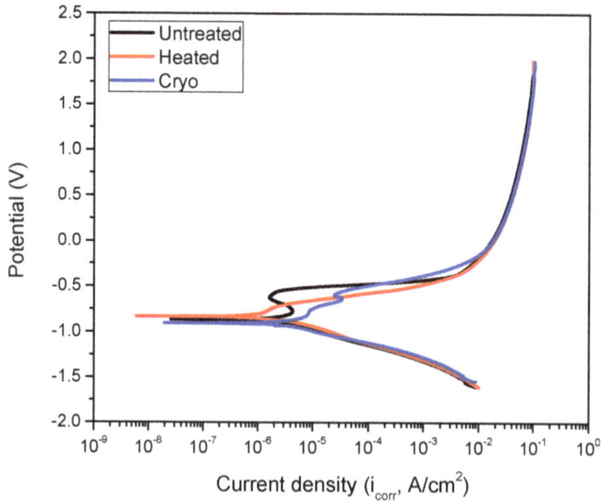

Figure 7. Potentiodynamic curves of investigated surfaces.

Table 1. Main corrosion parameters of untreated, heated and cryo specimens. $E_i = 0$: corrosion potential; i_{corr}: corrosion current density; R_p: polarization resistance; P: porosity; P_e: protective efficiency; CR: corrosion rate.

Sample	$E_i = 0$ (mV)	i_{corr} ($\mu A/cm^2$)	R_p ($k\Omega$)	P	P_e (%)	CR (μm/year)
Untreated	−865	4.387	11.644	-	-	0.051
Heated	−835	1.303	20.917	0.532	70.3	0.015
Cryo	−906	2.005	9.130	1.199	54.3	0.023

The surface porosity (P) was estimated based on Elsner's empirical equation [48,49] taking into account the values of polarization resistance (R_p) before and after applied treatments. Moreover, the protective efficiency (P_e) was also calculated based on the formula reported in [49], considering the ion corrosion densities of the untreated substrate and treated substrates. The corrosion rate (CR) has been estimated according to ASTM G102-89 standard (reapproved 2015) [50], using the following equation:

$$CR = K_i \frac{i_{corr}}{\rho} EW \quad (1)$$

where CR = corrosion rate, $K_i = 3.27 \times 10^{-3}$, ρ = materials density, i_{corr} = corrosion current density and EW = equivalent weight.

Taking into account the values of porosity, one may see that the alloy exhibited low porosity after heating, compared with the cryogenically treated alloy. Moreover, the protective efficiency in response to NaCl attack is higher in the case of the heated alloy when compared to the cryo alloy. The corrosion rate can be estimated in the following order: untreated > cryo > heated alloy. The high corrosion rate of cryogenically treated alloy can be attributed to the high porosity of the surface. However, it can be summarized that the best corrosion resistance was found for the heat-treated alloy and cryogenically treated alloy.

Electrochemical impedance spectroscopy (EIS) was used to investigate the electrochemical behavior of the investigated systems, highlighting the surface properties. For this purpose, the applied amplitude of the perturbation signal was 10 mV RMS vs. Eoc in a frequency range of 0.1–10^4 Hz. The impedance data were displayed as a Nyquist plot (Figure 8). According to this plot, one may observe that the specimens exhibit a better protection with increasing order from left to right as cryo < untreated < heated.

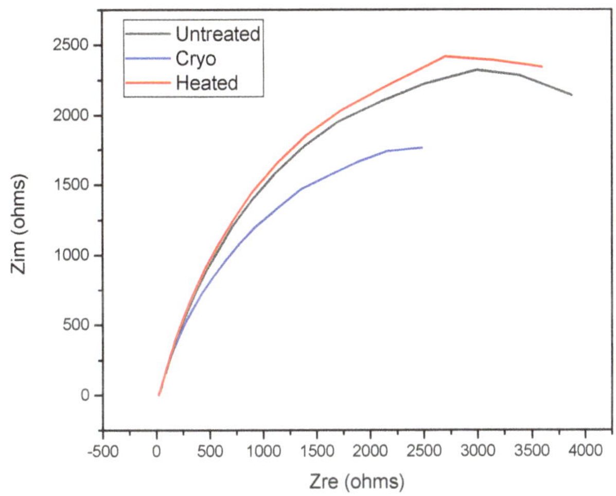

Figure 8. Nyquist plot for the investigated specimens.

3.3. Morphology before and after Electrochemical Tests

The morphology before and after electrochemical tests was evaluated by SEM analysis, and the results are presented in Figure 9. Before electrochemical tests, all surfaces appeared to be without defects or cracks. After corrosion tests, each surface was affected differently. The surface of the untreated alloy was destroyed on all corroded areas. At the magnification of 5000×, some precipitation of oxides of Al, Cr and Fe can be seen, as can the NaCl precipitates from the corrosive solution. The heated alloy is more resistant to corrosion, having a localized corrosion in some parts of the corroded region. On this surface, some oxides of Al, Cr and Fe can be seen, having a high affinity to oxygen. Some holes can also be observed, indicating that the corrosive solution penetrated more deeply. In the case of cryogenically treated alloy, many cracks appeared on the surface after the electrochemical tests. One possibility is that these cracks could be attributed to the destruction of the oxides. The heated AISI H13 steel displays better corrosion resistance compared to untreated AISI H13, and their corrosion resistance is comparable to that of the cryogenically treated alloy. In order to explain more about the corrosion process, the investigation of the roughness of the specimens was performed before and after corrosion tests.

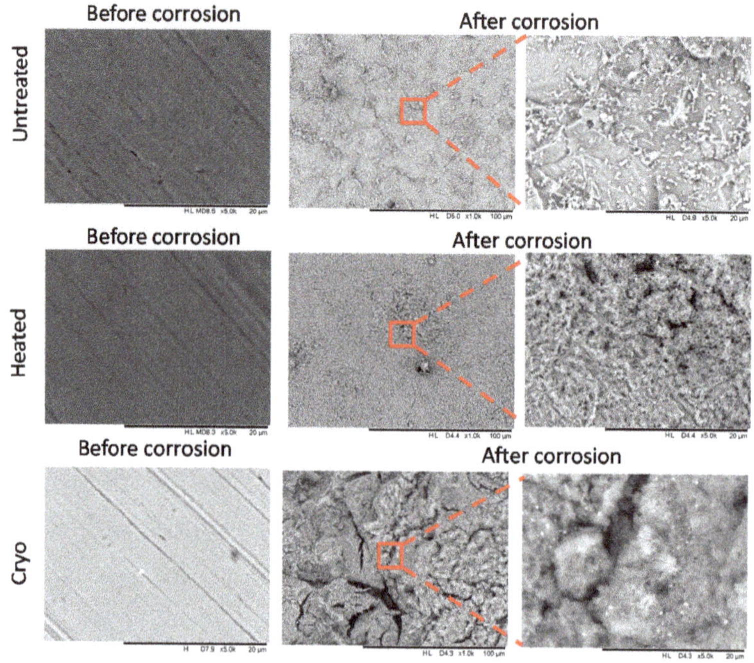

Figure 9. SEM images of the investigated specimens before and after corrosion.

3.4. Roughness before and after Electrochemical Tests

The roughness was evaluated on a length of 10 mm by a surface profilometer (Dektak 150, Bruker, Billerica, MA, USA) equipped with a 2.5 μm diameter stylus. For evaluation of roughness, two main parameters were calculated: arithmetic average deviation from the mean line (R_a) and skewness (S_{kew}). In Figure 10, the histograms of both calculated roughness parameters captured before and after electrochemical tests are presented. According to the R_a parameter, one may note that the roughness of all investigated surfaces was significantly increased after corrosion tests, indicating that all surfaces were affected by the corrosive solution. The heated and cryo alloys exhibited almost similar values of R_a, indicating that both surfaces exhibited comparable corrosion behavior. More relevant is the skewness parameter. A negative value of S_{kew} implies a surface formed of many valleys, while a positive value suggests a surface formed of mainly peaks and asperities. A S_{kew} with a value close to 0 suggests a flat surface. Taking into consideration these states, one may see that the untreated surface has an almost flat surface before corrosion and a surface with many peaks after corrosion. This finding demonstrated that the corrosive solution digs into the material and forms many pits. The heated alloy has a negative S_{kew} before corrosion tests, meaning that the surface has many valleys, which are transformed to high peaks after corrosion tests, indicating that the surface is significantly affected by the corrosive solution. The same phenomenon was also found in the case of the cryogenically treated surface. Note that the differences found for the cryo surface are small compared to those found for the heated surface, indicating that the cryo surface is more resistant to corrosion in 3.5% NaCl.

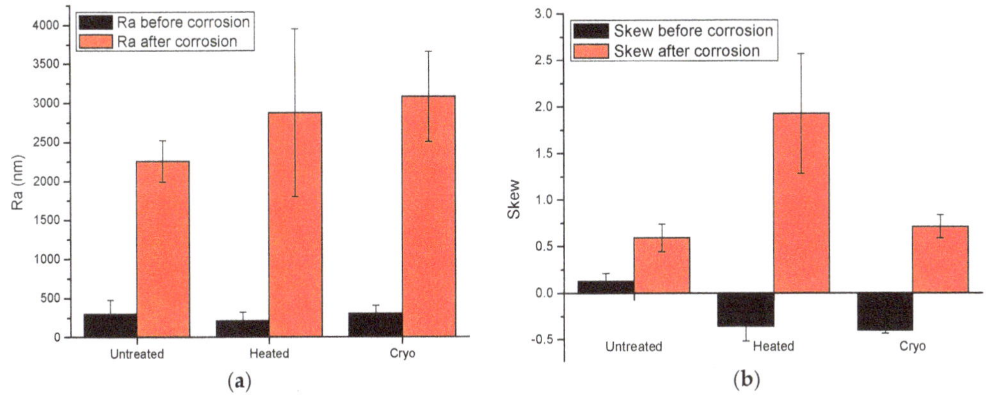

Figure 10. Roughness parameters: (**a**) R_a—arithmetic average deviation from the mean line; (**b**) S_{kew}—skewness of the investigated specimens before and after corrosion resistance.

4. Discussion

Based on the results obtained through corrosion tests, it can be said that the grain boundaries play a very important role in deciding the properties of materials under different service conditions for all types of treatments. High geometrically necessary dislocation (GND) density has been noted for 16 h cryogenically treated AISI H13 specimens (cryo specimens), which could be attributed to the increased plastic strain or strain hardening resulting in the accumulation of dislocations at the grain boundary areas. A reduction in grain size has been observed in the case of 16 h cryogenically treated specimens (cryo specimens), which could be attributed to the precipitation of fine carbides at the grain boundaries inhibiting the grain growth [12]. The highest values of open circuit potential (OCP) for AISI H13 cryo specimens indicate that these specimens are good in resisting the attack of NaCl as compared to untreated and heated specimens. The heated specimens have shown good results in the case of polarization curves with Tafel's extrapolation and estimation of surface porosity as compared to heated and cryo specimens. This indicates that the heated specimens form a protective and passive layer on the surface of the specimen. The Nyquist plots are also in agreement with the results for heated specimens. The SEM plots indicate a broken film in the case of cryo specimens, which could have resulted in increased corrosion; however, the reasons for the breakage of the film need to be determined.

Surface roughness evaluations are essential in analysing the response of materials to corrosive environments. There was a significant increase in the roughness for all types of specimens as an effect of the corrosive atmosphere. The skewness parameters evaluated for untreated, heated and cryo specimens indicate that the cryo specimens have better corrosion resistance as compared to the other two types, which could be attributed to the less significant impact of the corrosive atmosphere on cryo specimens.

One of the important aspects that need to be discussed is the influence of mechanical properties on the corrosion behavior of the materials. In this case, if the hardness is compared amongst the three categories (i.e., untreated, heated and cryo-treated), it can be seen that the hardness increases as follows: untreated (42 HRC) < heated (50 HRC) < cryo (56 HRC). The increase in hardness could be the result of the conversion of retained austenite into martensite after cryogenic treatment and the precipitation of fine carbides in the matrix of tempered martensite [12,21]. The grain refinement is also one of the reasons for the improvement in hardness. Based on these results for hardness, it can be stated that increased martensitic content indicates better corrosion resistance as in the case of cryo specimens [51].

In summary, heated and cryo specimens of AISI H13 exhibit better resistance to corrosion as compared to untreated material, and deep cryogenic treatment is useful in improving the corrosion resistance.

5. Conclusions

The current research work was aimed at investigating the corrosion behavior of cryogenically treated AISI H13 die steel. The following points summarize the major outcomes:

- The deep cryogenic treatment is beneficial in improving the corrosion resistance of AISI H13 material and could be useful in applications involving high mechanical stress and a corrosive environment.
- The grain boundaries decide the corrosion characteristics of the material; increased geometrically necessary dislocations indicate higher plastic straining. There is a reduction in grain size for AISI H13 specimens subjected to 16 h deep cryogenic treatment, which is attributed to the precipitation of fine carbides at the grain boundary areas.
- Parameters such as higher open circuit potential and skewness indicate that H13 specimens subjected to deep cryogenic treatment have a better response to a corrosive environment, whereas the surface porosity measurements and Nyquist plots confirm the superior response of heated specimens to a corrosive atmosphere.

Overall, it can be concluded that deep cryogenic treatment is useful in improving the corrosion properties of H13 die steel. Some further investigations of material surfaces exposed to a high-temperature corrosive environment could also reveal some important outcomes for H13 steel.

Author Contributions: Conceptualization, A.V., C.P. and N.B.D.; methodology, A.V., C.P., T.S. and A.C.P.; investigation, A.V., C.P., T.S. and A.C.P.; resources, A.V. and N.B.D.; writing—original draft preparation, T.S.; writing—review and editing, A.V., C.P., T.S., N.B.D. and A.C.P.; supervision, N.B.D. and A.V. All authors have read and agreed to the published version of the manuscript.

Funding: Part of this research was supported by a grant from the Romanian Ministry of Education and Research, CNCS—UEFISCDI, project number PN-III-P4-ID-PCE-2020-1264 (PCE95/2021), within PNCDI III, and by Romanian Core Program, project No. 18N/2019.

Acknowledgments: The authors are grateful to the Department of Metallurgy and Materials Science, College of Engineering, Pune (India), and the Department of Mechanical Engineering, Imperial College London (UK), for providing experimental and testing facilities. A. Vladescu is grateful to Tomsk Polytechnic University within the frame of a Tomsk Polytechnic University Competitiveness Enhancement Program grant as well as to European Regional Development Fund through Competitiveness Operational Program 2014–2020, Action 1.1.3 Creating synergies with H2020 Program; project H2020 Support Centre for European project management and European promotion, MYSMIS code 107874.

Conflicts of Interest: The authors declare that they have no known competing financial interests or personal relationships that could have appeared to influence the work reported in this paper.

References

1. Global Tool Steel Market Size by Product Group, by Application, by Geographic Scope and Forecast. 2021. Available online: https://www.verifiedmarketresearch.com/download-sample/?rid=38576 (accessed on 15 November 2021).
2. Tool Steel Market Information Report, by Format Type (Round Format, Square Format and Other Format), by Application (Automotive, Machinery and Shipbuilding), by Process (Forged and Rolled) and by Region—Forecast to 2030. 2017. Available online: https://www.marketresearchfuture.com/reports/tool-steel-market/toc (accessed on 15 November 2021).
3. Vander Voort, G.F.; Corporation, C.T. Failures of Tools and Dies. *ASM Handb. Fail. Anal. Prev.* **1986**, *11*, 563–585.
4. Kchaou, M.; Elleuch, R.; Desplanques, Y.; Boidin, X.; Degallaix, G. Failure mechanisms of H13 die on relation to the forging process—A case study of brass gas valves. *Eng. Fail. Anal.* **2010**, *17*, 403–415. [CrossRef]
5. Papageorgiou, D.; Medrea, C.; Kyriakou, N. Failure analysis of H13 working die used in plastic injection moulding. *Eng. Fail. Anal.* **2013**, *35*, 355–359. [CrossRef]

6. Mohan Lal, D.; Renganarayanan, S.; Kalanidhi, A. Cryogenic treatment to augment wear resistance of tool and die steels. *Cryogenics* **2001**, *41*, 149–155. [CrossRef]
7. Baldissera, P.; Delprete, C. Deep Cryogenic Treatment: A Bibliographic Review. *Open Mech. Eng. J.* **2008**, *2*, 1–11. [CrossRef]
8. Das, D.; Dutta, A.K.; Ray, K.K. Correlation of microstructure with wear behaviour of deep cryogenically treated AISI D2 steel. *Wear* **2009**, *267*, 1371–1380. [CrossRef]
9. Farhani, F.; Niaki, K.S.; Vahdat, S.E.; Firozi, A. Study of effects of deep cryotreatment on mechanical properties of 1.2542 tool steel. *Mater. Des.* **2012**, *42*, 279–288. [CrossRef]
10. Suchmann, P.; Jandova, D.; Republic, C.; Resistance, W. Deep Cryogenic Treatment of H11 Hot Working Tool. *Mater. Tech.* **2013**, *49*, 400–408.
11. Das, D.; Kumar, K. Structure-property correlation of sub-zero treated AISI D2 steel. *Mater. Sci. Eng. A* **2012**, *541*, 45–60. [CrossRef]
12. Das, D.; Dutta, A.K.; Toppo, V.; Ray, K.K. Effect of Deep Cryogenic Treatment on the Carbide Precipitation and Tribological Behavior of D2 Steel. *Mater. Manuf. Process.* **2007**, *22*, 474–480. [CrossRef]
13. Yan, X.G.; Li, D.Y. Effects of the sub-zero treatment condition on microstructure, mechanical behavior and wear resistance of W9Mo3Cr4V high speed steel. *Wear* **2013**, *302*, 854–862. [CrossRef]
14. Thornton, R.; Slatter, T.; Ghadbeigi, H. Effects of deep cryogenic treatment on the dry sliding wear performance of ferrous alloys. *Wear* **2013**, *305*, 177–191. [CrossRef]
15. Dhokey, N.B.; Nirbhavne, S. Dry sliding wear of cryotreated multiple tempered D-3 tool steel. *J. Mater. Process. Technol.* **2009**, *209*, 1484–1490. [CrossRef]
16. Gill, S.S.; Singh, J.; Singh, R.; Singh, H. Effect of cryogenic treatment on AISI M2 high speed steel: Metallurgical and mechanical characterization. *J. Mater. Eng. Perform.* **2012**, *21*, 1320–1326. [CrossRef]
17. Tier, M.D. Effect of heat and cryogenic treatment on wear and toughness of HSS AISI M2. *Integr. Med. Res.* **2020**, *9*, 12354–12363.
18. Jovičević-Klug, P.; Jovičević-Klug, M.; Podgornik, B. Effectiveness of deep cryogenic treatment on carbide precipitation. *J. Mater. Res. Technol.* **2020**, *9*, 13014–13026. [CrossRef]
19. Jovičević-Klug, P.; Jenko, M.; Jovičević-Klug, M.; Šetina Batič, B.; Kovač, J.; Podgornik, B. Effect of deep cryogenic treatment on surface chemistry and microstructure of selected high-speed steels. *Appl. Surf. Sci.* **2021**, *548*, 149257. [CrossRef]
20. Koneshlou, M.; Meshinchi, K.; Khomamizadeh, F. Effect of cryogenic treatment on microstructure, mechanical and wear behaviors of AISI H13 hot work tool steel. *Cryogenics* **2011**, *51*, 55–61. [CrossRef]
21. Shinde, T.; Dhokey, N.B. Influence of carbide density on surface roughness and quasi-stable wear behaviour of H13 die steel. *Surf. Eng.* **2017**, *33*, 944–952. [CrossRef]
22. Gök, M.S.; Küçük, Y.; Erdoğan, A.; Öge, M.; Kanca, E.; Günen, A. Dry sliding wear behavior of borided hot-work tool steel at elevated temperatures. *Surf. Coat. Technol.* **2017**, *328*, 54–62. [CrossRef]
23. Molinari, A.; Pellizzari, M.; Gialanella, S.; Straffelini, G.; Stiasny, K.H. Effect of deep cryogenic treatment on the mechanical properties of tool steels. *J. Mater. Process. Technol.* **2001**, *118*, 350–355. [CrossRef]
24. Shinde, T.; Dhokey, N.B. Influence of Tertiary Carbides on Improving Fatigue Limit of H13 Die Steels. *Metallogr. Microstruct. Anal.* **2017**, *6*, 398–406. [CrossRef]
25. Zhirafar, S.; Rezaeian, A.; Pugh, M. Effect of cryogenic treatment on the mechanical properties of 4340 steel. *J. Mater. Process. Technol.* **2007**, *186*, 298–303. [CrossRef]
26. Akhbarizadeh, A.; Amini, K.; Javadpour, S. Effects of applying an external magnetic field during the deep cryogenic heat treatment on the corrosion resistance and wear behavior of 1.2080 tool steel. *Mater. Des.* **2012**, *41*, 114–123. [CrossRef]
27. Molinari, A.; Pellizzari, M.; Straffelini, G.; Pirovano, M. Corrosion behaviour of a surface-treated AISI H11 hot work tool steel in molten aluminium alloy. *Surf. Coat. Technol.* **2000**, *126*, 31–38. [CrossRef]
28. Dobrzanski, L.A.; Polok, M.; Panjan, P.; Bugliosi, S.; Adamiak, M. Improvement of wear resistance of hot work steels by PVD coatings deposition. *J. Mater. Process. Technol.* **2004**, *155–156*, 1995–2001. [CrossRef]
29. Al-Ghyamah, A.; Alodan, M. Electrochemical Characterization of Heat Treated Steel. *J. King Saud Univ. Eng. Sci.* **2008**, *20*, 61–67. [CrossRef]
30. Günen, A.; Karahan, İ.H.; Karakaş, M.S.; Kurt, B.; Kanca, Y.; Çay, V.V.; Yıldız, M. Properties and Corrosion Resistance of AISI H13 Hot-Work Tool Steel with Borided B4C Powders. *Met. Mater. Int.* **2019**, *26*, 1329–1340. [CrossRef]
31. Ramesh, S.; Bhuvaneshwari, B.; Palani, G.S.; Lal, D.M.; Mondal, K.; Kumar, R. Enhancing the corrosion resistance performance of structural steel via a novel deep cryogenic treatment process. *Vacuum* **2019**, *159*, 468–475. [CrossRef]
32. Wright, S.I.; Nowell, M.M. EBSD Image Quality Mapping. *Microsc. Microanal.* **2014**, *12*, 72–84. [CrossRef] [PubMed]
33. Rohrer, G.S. Measuring and Interpreting the Structure of Grain-Boundary Networks. *J. Am. Ceram. Soc.* **2011**, *94*, 633–646. [CrossRef]
34. Adams, B.L.; Wright, S.I.; Kunze, K. Orientation Imaging: The Emergence of a New Microscopy. *Metall. Trans. A* **1993**, *24*, 819–831. [CrossRef]
35. Schwartz, A.J.; Kumar, M.; Adams, B.L.; Field, D.P. (Eds.) *Electron Backscatter Diffraction in Materials Science*; Springer: Boston, MA, USA, 2009.
36. Tochigi, E.; Nakamura, A.; Shibata, N.; Ikuhara, Y. Dislocation Structures in Low-Angle Grain Boundaries of α-Al_2O_3. *Crystals* **2018**, *8*, 133. [CrossRef]
37. Sutton, A.P.; Balluffi, R.W. *Interfaces in Crystalline Materials*; Clarendon Press: Oxford, UK, 1995.

38. Kundu, A.; Field, D.P. Geometrically Necessary Dislocation Density Evolution in Interstitial Free Steel at Small Plastic Strains. *Metall. Mater. Trans. A* **2018**, *49*, 3274–3282. [CrossRef]
39. Das, D.; Dutta, A.K.; Ray, K.K. On the refinement of carbide precipitates by cryotreatment in AISI D2 steel. *Philos. Mag.* **2009**, *89*, 55–76. [CrossRef]
40. Ralston, K.D.; Birbilis, N. Effect of Grain Size on Corrosion: A Review. *Corrosion* **2010**, *66*, 075005. [CrossRef]
41. Yun-wei, H.A.O.; Bo, D. Effect of Surface Mechanical Attrition Treatment on Corrosion Behavior of 316 Stainless Steel. *J. Iron Steel Res. Int.* **2009**, *16*, 68–72.
42. Pisarek, M.; Kędzierzawski, P.; Janik-Czachor, M.; Kurzydłowski, K.J. Effect of hydrostatic extrusion on passivity breakdown on 303 austenitic stainless steel in chloride solution. *J. Solid State Electrochem.* **2009**, *13*, 283–291. [CrossRef]
43. Afshari, V.; Dehghanian, C. Effects of grain size on the electrochemical corrosion behaviour of electrodeposited nanocrystalline Fe coatings in alkaline solution. *Corros. Sci.* **2009**, *51*, 1844–1849. [CrossRef]
44. Wang, X.Y.; Li, D.Y. Mechanical, electrochemical and tribological properties of nano-crystalline surface of 304 stainless steel. *Wear* **2003**, *255*, 836–845. [CrossRef]
45. Hadzima, B.; Estrin, Y.; Seop, H. Microstructure and corrosion properties of ultrafine-grained interstitial free steel. *Mater. Sci. Eng. A* **2007**, *462*, 243–247. [CrossRef]
46. Atrens, A. *Understanding the Corrosion of Mg and Mg Alloys*; Elsevier: Amsterdam, The Netherlands, 2018.
47. Yate, L.; Coy, L.E.; Aperador, W. Robust tribo-mechanical and hot corrosion resistance of ultra-refractory Ta-Hf-C ternary alloy films. *Sci. Rep.* **2017**, *7*, 3080. [CrossRef] [PubMed]
48. Pruncu, C.I.; Braic, M.; Dearn, K.D.; Farcau, C.; Watson, R.; Constantin, L.R.; Balaceanu, M.; Braic, V.; Vladescu, A. Corrosion and tribological performance of quasi-stoichiometric titanium containing carbo-nitride coatings. *Arab. J. Chem.* **2017**, *10*, 1015–1028. [CrossRef]
49. Vladescu, A.; Vranceanu, D.M.; Kulesza, S.; Ivanov, A.N.; Bramowicz, M.; Fedonnikov, A.S.; Braic, M.; Norkin, I.A.; Koptyug, A.; Kurtukova, M.O.; et al. Influence of the electrolyte's pH on the properties of electrochemically deposited hydroxyapatite coating on additively manufactured Ti64 alloy. *Sci. Rep.* **2017**, *7*, 16819. [CrossRef] [PubMed]
50. *ASTM G102-89(2015)e1 Standard Practice for Calculation of Corrosion Rates and Related Information from Electrochemical Measurements*; ASTM International: West Conshohocken, PA, USA, 2015.
51. Handoko, W.; Pahlevani, F.; Sahajwalla, V. Materials Enhancing Corrosion Resistance and Hardness Properties of Carbon Steel through Modification of Microstructure. *Materials* **2018**, *11*, 2404. [CrossRef] [PubMed]

materials

MDPI

Article

Diagonal Tensile Test on Masonry Panels Strengthened with Textile-Reinforced Mortar

Dragoș Ungureanu [1,*], Nicolae Țăranu [1,2], Dan Alexandru Ghiga [1], Dorina Nicolina Isopescu [1], Petru Mihai [1] and Ruxandra Cozmanciuc [1]

1 Faculty of Civil Engineering and Building Services, "Gheorghe Asachi" Technical University of Iaşi, 43 Mangeron Blvd., 700050 Iaşi, Romania; nicolae.taranu@academic.tuiasi.ro (N.Ț.); dan-alexandru.ghiga@student.tuiasi.ro (D.A.G.); dorina-nicolina.isopescu@academic.tuiasi.ro (D.N.I.); petru.mihai@academic.tuiasi.ro (P.M.); ruxandra.cozmanciuc@academic.tuiasi.ro (R.C.)

2 The Academy of Romanian Scientists, 54 Splaiul Independentei, Sector 5, 050094 Bucuresti, Romania

* Correspondence: dragos.ungureanu@staff.tuiasi.ro

Abstract: This study presents the results of an experimental and numerical program carried out on unreinforced masonry panels strengthened by textile-reinforced mortar (TRM) plastering. For this purpose, five panels were constructed, instrumented and tested in diagonal shear mode. Two panels were tested as reference. The first reference panel was left unstrengthened, while the second one was strengthened by a traditional self-supporting cement mortar matrix reinforced with steel meshes. The remaining three panels were strengthened by TRM plastering applied on one or both faces and connected with transversal composite anchors. The numerical and the experimental results evidenced a good effectiveness of the TRM systems, especially when applied on both panel facings.

Keywords: unreinforced masonry; textile-reinforced mortar; numerical and experimental study

Citation: Ungureanu, D.; Țăranu, N.; Ghiga, D.A.; Isopescu, D.N.; Mihai, P.; Cozmanciuc, R. Diagonal Tensile Test on Masonry Panels Strengthened with Textile-Reinforced Mortar. *Materials* **2021**, *14*, 7021. https://doi.org/10.3390/ma14227021

Academic Editor: Dario De Domenico

Received: 30 September 2021
Accepted: 14 November 2021
Published: 19 November 2021

1. Introduction

Unreinforced masonry (URM) has a long history of being the predominant building technique in Romania and in Europe. A large share of this existing building stock consists of historical monuments, which have been designed to resist only gravitational loads or, in many cases, have not been designed at all, but simply erected according to a few generic rules of common practice [1–7]. In Romania, the regions where the masonry monumental buildings lie are all characterized by medium-to-high levels of seismic hazard. The composite nature of the bricks and of the mortar, the stocky arrangement of the URM walls and the almost zero tensile capability of the masonry material makes this structural system potentially vulnerable to seismic actions [8–12]. Consequently, most of the existing URM buildings require strengthening interventions works, to adapt the capabilities of the structural system to the requirements of the current seismic design codes. In order to design and perform a particular strengthening technique, the structural behavior and the possible failure modes of the unstrengthened structure should be well analyzed.

During seismic excitation, URM structures can develop two possible types of dominant failure modes, the in-plane shear or the out-of-plane bending mechanisms [13,14]. The second mechanism can be avoided by improving the overall stiffness of the structure with additional transversal connections [15]. On the other hand, the in-plane shear capacity governs the global seismic performances of the URM structures, since the lateral loads from the load bearing walls are transferred through the in-plane path to the foundations [16]. The in-plane dominant failure mechanism can be divided in several characteristic modes, including failure by sliding, rocking, toe crushing and shear. Due to the high weight of the URM buildings, the in-plane failure by sliding is rarely developed. The in-plane rocking and toe crushing failures are the most ductile ones. Nonetheless, the in-plane shear is a brittle and sudden-failure mechanism, being more severe and dangerous than the other in-plane modes. Moreover, the shear failure mode can be developed during moderate or even small earthquakes and its level of severity is influenced by a large variety of parameters, including bricks and mortar quality, load bearing walls thickness and arrangement, age of the structure, type of materials, existing seismic protection, soil conditions and modal characteristics of the URM structure. For these reasons, the development and improvement of shear strengthening techniques for URM structures have been topics of interest for many researchers.

Originally, the reinforced concrete strengthening techniques were adopted for URM elements as well [17]. These traditional strengthening techniques consist in stitching or filling the cracks and plastering the masonry element with a self-supporting cement mortar matrix reinforced with steel meshes. The main disadvantage of this approach consists in the high additional weight provided by the steel reinforcements. In order to overcome this issue, various fiber-reinforced polymer (FRP) composite elements have been developed as an alternative to steel reinforcements. Despite their advantageous features, such as high strength-to-weight ratio and ease of application [18–20], using FRP composite elements in strengthening applications for URM structures is usually associated with several possible limitations. These are the potential debonding of externally applied FRP strips, the incompatibility between most of the epoxy systems and many types of substrates (due to the large stiffness discrepancies) and the unsatisfactory behavior of common resin products at high and low temperatures. Recently, besides the conventional FRP composite strengthening systems, new materials have been developed and exploited for URM structures' retrofitting. Among these materials, the textile-reinforced mortar (TRM) is considered to be highly beneficial since it has an improved compatibility with masonry substrates. This strengthening system consists in a layer of an inorganic matrix (usually a cement-based mortar) that is reinforced and applied on one or both faces of the URM walls. The reinforcements are textile fabric meshes made of fiber roving ply arranged on at least two orthogonal directions. The mechanical properties of the textile reinforcements are controlled by the amount of roving and their orientation, while the level of penetration of the mortar matrix is influenced by the mesh spacings [5,21].

According to previous studies [1,22–27], the TRM strengthening systems designed for URM structures are promising solutions that can provide significant increments in terms of shear capacity and pseudo-ductility both for in-plane and out-of-plane loading conditions. The in-plane structural behavior of the URM strengthened with TRM was studied for various wall types and configurations, including one-leaf or two-leaf brick arrangements [28–31] and tuff volcanic masonry walls [32,33]. The outcomes of some of the most relevant experimental programs related to the shear capabilities of URM walls strengthened with TRM are summarized in Table 1 [34]. The test types are abbreviated as follows: DG—diagonal test; CY—cycling test; SH—shear test. BR and VT stand for brick and volcanic tuff masonries, while the fiber meshes are denoted with S for steel, G for glass, C for carbon and B for basalt. The strength and the deformation ratios represent the capacities of the strengthened masonry specimens reported to the capacities of the URM ones.

By studying the data presented in Table 1, it can be observed that, for all the experimental campaigns, the improved behavior of the URM panels strengthened with TRM, in terms of strength and deformation capacity, has been validated. However, in some cases, the effective amount of reinforcement was not sufficient to provide significant increases in strength, but it enhanced efficiently the displacement capacity. The latter was proved to be valid even for TRM systems which consists of flax fabrics. In addition, when used as a component of TRM systems, these types of fibers do not evidence severe degradation of their mechanical properties even if they are subjected to various environmental exposures and aging protocols [35,36].

Table 1. In-plane tests of URM panels strengthened with TRM [34].

Authors	Test Type	No. of Panels	Type of Masonry	Type of Reinforcing Meshes	No. of Strengthened Sides	Failure Mechanism	Strength Ratio	Deformation Ratio
Prota et al., 2006 [33]	DG	4	TF	G	1	Tensile rupture	1.7	2.4
Prota et al., 2006 [33]	DG	4	TF	G	2	Crushing	2.2	2.3
Papanicolaou et al., 2007 [37]	SH	4	BR	C	2	Crushing	8.2	13.5
						Crushing	2.5	4.2
						Debonding and crushing	5.3	7.2
Faella et al., 2010 [38]	DG	9	TF	C	1	Debonding and crushing	5.7	—
Augenti et al., 2011 [39]	CY	1	BR	G	1	Tensile rupture	1.03	2.8
Parisi et al., 2013 [40]	DG	6	TF	G	1/2	Tensile rupture	2.4	2.7
Ismail and Ingham, 2016 [29]	CY	1	BR	S	1	Tensile rupture	1.2	1.2
Marcari et al., 2017 [32]	DG	3	TF	B	1	Vertical cracks	1.3	2.4
Marcari et al., 2017 [32]	DG	2	TF	B	2	Vertical cracks	1.6	1.6
Sagar et al., 2017 [31]	DG	15	BR	G	1	Tensile rupture	1.03	7.2
Longo et al., 2021 [41]	DG	2	BR	G	2	Vertical cracks	1.6	-
Longo et al., 2021 [41]	DG	2	BR	S	2	Tensile rupture	1.7	-
Donnini et al., 2021 [42]	DG	3	TF	G	2	Vertical cracks	2.7	4.0
Donnini et al., 2021 [42]	DG	3	BR	G	2	Vertical cracks	1.3	1

The experimental campaigns cited in Table 1 were aimed to characterize the structural behavior of the TRM systems either by testing samples made from fibers and mortar (tensile and adherence tests) or by addressing the global characterization of the systems by means of diagonal compression tests on reinforced masonry panels. By comparing these types of experimental results with the ones obtained by applying the available theoretical models, other studies reported several limitations of the theoretical models, related to the inability to consider combined or non-standard failure modes [43].

Moreover, despite the available studies carried out to investigate the shear structural behavior of the URM strengthened with TRM, which highlighted the effectiveness of this type of strengthening system, some important aspects have still not been fully investigated. The latter may refer to anchorage methods for masonry walls which enable intervention works on both faces (the side-to-side connection) or on a singular face (the middle-to-side connection); the increase in lateral capacity based on the type and configuration of the transversal connectors, pattern and effective length for various types of anchorages; the characteristic failure modes for the systems with additional reinforcement due to the radial ends of the anchorages; and the synchronization between the experimental and numerical analyses based on adapted modelling strategies.

This paper reports the results obtained through a numerical and experimental program that was carried out at the Laboratory of Heavy Structure, located at the Faculty of Civil Engineering and Building Services in Iași. The masonry materials and the strengthening systems, both traditional and modern, are rather common in Romania. In particular, five URM panels were constructed, instrumented and tested in diagonal shear mode. In addition, all the specimens were micro-modelled and subjected to a nonlinear finite element analysis using the Ansys Workbench software. Two panels were tested as reference. The first reference panel was left unstrengthened, while the second one was strengthened through a traditional self-supporting cement mortar matrix reinforced with steel meshes. The remaining three panels were strengthened by TRM plastering applied on one or both faces and connected with transversal composite anchors. After a detailed experimental set-up and testing procedures, the main experimental results and the key features of each configuration of the strengthening system are reported. In addition, the experimental results are compared to the ones obtained through a micro-detailed finite element analysis and a general good agreement is found.

2. Experimental Set-Up

2.1. URM Panel Specimens' Configurations and Materials Properties

The experimental program was performed on five 1200 × 1200 × 115 mm URM panels. The specimens were manufactured using brick clay masonry units (Figure 1) and normal mortar of M15 class [44]. All of the URM panels were constructed by qualified masons to insure a proper level of craftmanship, similar to current practice. The general configuration of the specimens is presented in Figure 2.

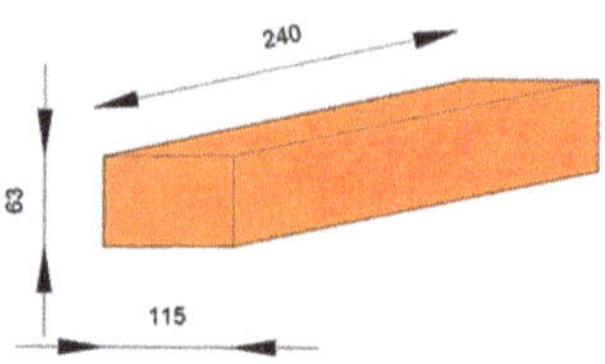

Figure 1. Brick unit (dimensions in mm) [44].

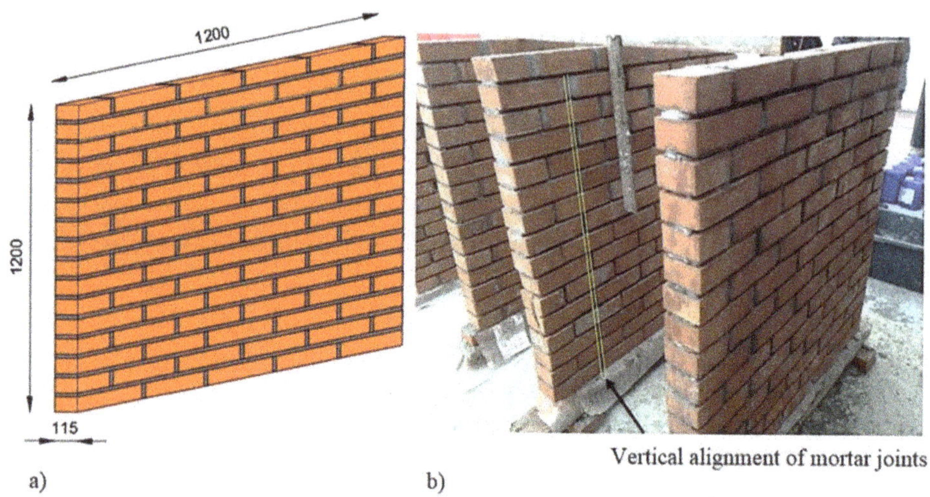

Figure 2. Specimens: (**a**) geometric configuration (values in mm) and (**b**) URM craftmanship.

The URM panels were left to cure for 28 days. Prior to strengthening and testing, the mechanical characteristics of the masonry material were determined and compared to the ones provided by the manufacturer. Bricks were the first component of the URM panels to be investigated.

Tests to determine the compressive strength were performed on five real-size specimens (Figure 3), which were loaded in compression in a Zwick Roell 1000kN universal testing machine, according to the specifications given in the Norm SR EN 772-1+A1 [45]. The value of the compressive strength provided by the manufacturer (15 MPa) was 28% lower compared to the average value that was determined experimentally (20.83 MPa). The computed standard deviation ($\sigma = 0.45$) indicates that the values tend to be close to the mean. Thus, it can be concluded that the experimentally determined values of the compressive strength of the brick units showed no statistically significant differences.

Figure 3. Brick unit tested in compression using a WAW 600 Universal Testing Machine.

The aim of the micro, non-linear analysis that is presented in Section 3 of this work is to provide a numerical model for designing purposes. Thus, the compressive strength provided by the manufacturer (the conservative value) was adopted.

Flexural and compression tests were performed to evaluate mortar strength, according to EN 1015-11 [46] (Figure 4). The flexural tests were made on a mortar prism (mortar type S) with the nominal dimensions of 40 × 40 × 160 mm, which were subjected to three-point bending loading. Each of the two parts of the mortar prisms that resulted from the flexural tests were used for the compression tests. A total number of 15 mortar prisms were subjected to the flexural test and 30 specimens were tested in compression.

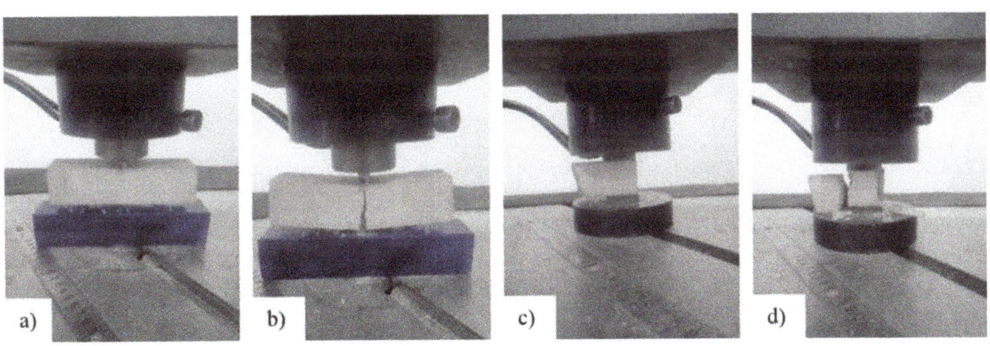

Figure 4. Flexural test: (**a**) mortar prism specimen and (**b**) failure mode. Compression test: (**c**) specimen and (**d**) failure mode.

The results were determined under the assumption of general elastic–brittle behavior of the masonry wall loaded in tension. The average value of the tensile strength—$f_{m,t}$—was approximately 3.66 MPa, while the average value of the compressive strength—$f_{m,c}$—was 18.61 MPa. The computed standard deviations ($\sigma_{flexural} = 0.80$; $\sigma_{compresion} = 1.18$) indicate that the values tend to be close to the mean. Thus, it can be concluded that the experimentally determined values showed no statistically significant differences.

The experimental values were introduced as input data for the mechanical characteristics of the micro non-linear 3D model defined in Section 3 of this work.

As mentioned before, two masonry panels were manufactured and tested as reference (noted with URM and TSM). The first one (URM) was left unstrengthened, as shown in Figure 2. The second reference panel (TSM) was strengthened using a traditional system consisting of a self-supporting cement mortar matrix reinforced with steel meshes of ϕ6. The two faces of the strengthening systems were connected by means of steel connectors of ϕ6, passing through the URM panel and fixed at the exterior in the mortar layer. The overall configuration of the traditional strengthening system is illustrated in Figure 5. The mortar used for this strengthening approach has identical characteristics with the one that was used for the URM panel assemblage.

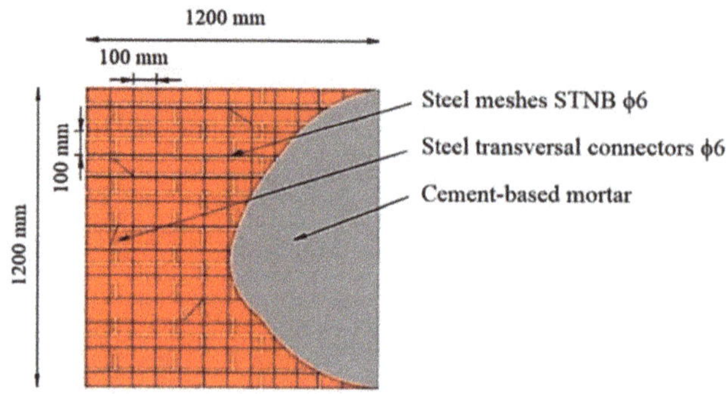

Figure 5. Configuration of the traditional strengthening system—TSM panel.

The remaining three URM panels were strengthened by plastering on one face (for one specimen—TRM1) or on both faces (for two specimens—TRM2 and TRM3) with a two-component ready-mixed, high ductility pozzolan-reaction mortar reinforced with an alkali-resistant (AR), pre-primed glass fiber mesh (Figure 6) [47,48].

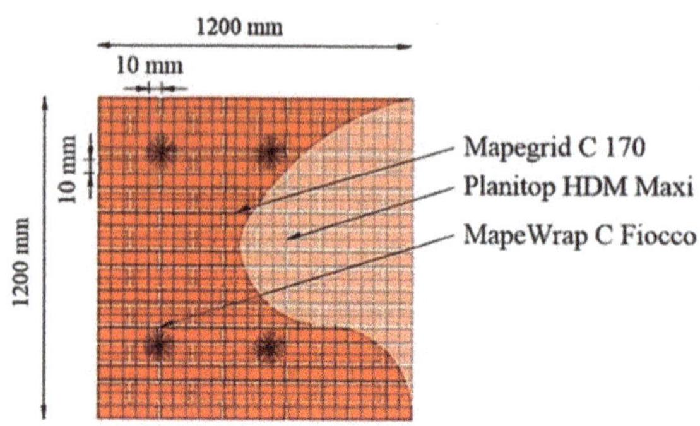

Figure 6. Configuration of the TRM strengthening system.

According to the manufacturer [47], the mortar was cement-based, made by blending special admixtures and synthetic polymers in water dispersion and reinforced with glass fibers. The physical and the mechanical characteristics of the strengthening mortar as provided by the manufacturer are listed in Table 2. This mortar has adequate adhesion strength and, when hardened, it forms a compact and uniform layer that is impermeable both to water and aggressive atmospheric gases, while remaining highly permeable to water vapors. In total, 15 mortar prisms were prepared and subjected to the flexural test and the resulted 30 specimens were tested in compression (Figure 7). The average experimental value of the tensile strength—$f_{m,t}$—was approximately 4.35 MPa, while the value given in the technical data sheet was equal with 8 MPa. Similarly, the average experimental value of the compressive strength—$f_{m,c}$—was 19.77 MPa, while the producer indicated a minimum 25 MPa strength in compression at 28 days. The computed standard deviations ($\sigma_{flexural}$ = 0.14; $\sigma_{compresion}$ = 1.21) indicate that the values tend to be close to the mean. Thus, it can be concluded that the experimentally determined values showed no statistically significant differences.

Table 2. Physical and mechanical characteristics of the strengthening mortar (provided by the manufacturer) [47].

Consistency of mix	Plastic–thixotropic
Density of wet mix (kg/m^3)	1850
Coating	Epoxy–Silicone–PFT–Titanium
Workability (h)	1
Water absorption (kg/(m^2·min$^{0.5}$)	≤0.1
Thermal conductivity (W/m·K)	0.73
Compressive strength (MPa)	25 (after 28 days)
Flexural strength (MPa)	8 (after 28 days)
Compressive modulus of elasticity (MPa)	10,000 (after 28 days)
Initial shear strength (MPa)	0.15

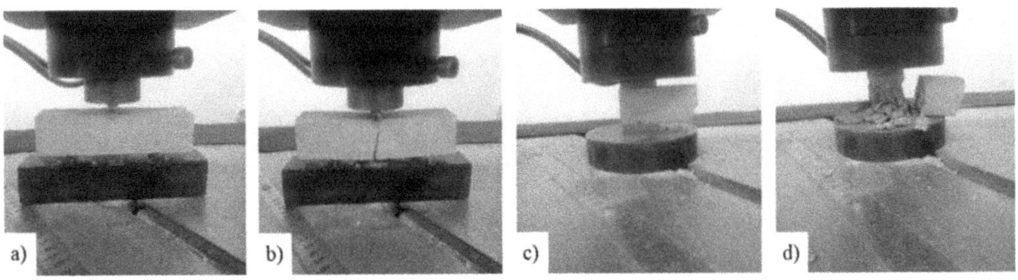

Figure 7. Flexural test: (**a**) mortar prism specimen and (**b**) failure mode. Compression test: (**c**) specimen and (**d**) failure mode.

According to the technical data sheets, the producer determined the compressive and the flexural strength according to the provisions given in the Norms EN 12190 and EN 196/1 [49,50], while the experimental values mentioned above were determined according to the specifications given in the Norm EN 1015-11 [46]. For the micro non-linear 3D model, the mechanical properties were defined according to the experimentally determined values.

The textile meshes used to reinforce the mortar are part of the same custom design system for masonry strengthening. The meshes consist in multiple dry carbon high-strength fiber roving disposed in a square pattern [48]. Due to the custom configuration of the weave, the mechanical properties of these meshes are controlled so that, when applied on URM walls together with the mortar, they make up for the masonry's lack of tensile strength and increase the overall ductility so that the shear stresses are distributed more evenly. As a result, in the event of seismic actions, the TRM strengthening system has the ability to evenly distribute the stresses and the strains over the entire surface of the URM walls, so that the dominant failure mechanism is shifted from a fragile type to a ductile one. Moreover, it should be noted that this type of TRM system has potential advantages when used on masonry historical buildings. In this frame, the system works in parallel with the existing structural elements, without influencing the distribution of the masses and the rigidity within the structures. The latter is a significant gain, particularly for seismic design, where stresses are proportional to the masses involved. The physical and mechanical properties of the mesh are listed in Table 3 [48].

Table 3. Physical and mechanical characteristics of the mesh (provided by the manufacturer) [48].

Type of Fiber	High-Strength Carbon Fiber
Weight (g/m^2)	170
Mesh size (mm)	10 × 10
Density (kg/m^3)	1830
Tensile strength (MPa)	4800
Modulus of elasticity (MPa)	230,000
Load-resistant area per unit of width (mm^2/m)	48
Equivalent thickness of dry fabric (mm)	0.048
Elongation at failure (%)	2

As it can be observed in Figure 6, the face/faces of the TRM strengthening systems were additionally fixed with four transversal composite connectors, located on the corners, so as to ensure an improved connection to the masonry substrate and to avoid the out-of-plane failure. The composite connectors were unidirectional carbon fiber threaded inside a gauge wrapping [51]. Due to this appearance, this type of composite connectors is also referred to as cords. The physical and the mechanical characteristics of the cords, as given by the manufacturer, are listed in Table 4. The configuration of the anchors was selected according to the specification given by the manufacturer [51].

Table 4. Physical and mechanical characteristics of the cords (provided by the manufacturer) [51].

Type of Fiber	High-Strength Carbon Fiber
Appearance	cord
Density (kg/m^3)	1800
Equivalent surface area of dry fabric (mm^2)	21.24
Modulus of elasticity (MPa)	230,000
Tensile strength (MPa)	4830
Elongation at failure (%)	2

In order to achieve a full interaction among the cords and the specimens, a chemical anchor was manufactured by means of an adhesive product. A bi-component styrene-free adhesive made from polyester resin was poured so as to fill the holes through which the cords were inserted. The physical and mechanical characteristics of the adhesive, as provided by the manufacturer, are listed in Table 5 [52].

Table 5. Physical and mechanical characteristics of the ϕ8 mm cords (provided by the manufacturer) [52].

Appearance	Thixotropic Paste
Density (kg/m^3)	1690
Compressive strength (MPa)	68
Flexural strength (MPa)	30
Flexural dynamic modulus of elasticity (MPa)	4025
Compressive modulus of elasticity (MPa)	6105

2.2. Construction Process

The URM panels were constructed of bricks with nominal dimensions of 1200 by 1200 by 115 mm. The latter were assembled by an experienced mason using a mortar made of Portland cement and sand, in the proportion and quality that are commonly used in the traditional Romanian masonry buildings (one part Portland cement to three parts sand). The thickness of both the horizontal and vertical mortar joints was 10 mm. Since the bricks were stored in indoor conditions, each unit was wetted before usage. After all the specimen's manufacturing processes were complete, the surfaces were cleaned of dust, pollutants or loose materials by sandblasting and low-pressure water jetting.

For the second reference panel, two steel meshes of ϕ6 were disposed on each side (Figure 8). The meshes were tied together by means of transversal steel connectors of ϕ6. The attachments between the meshes and the transversal connectors were achieved by means of thin binding steel wires. Next, the surfaces of the panel were wetted and the mortar layer was overlaid so as to cover all the reinforcements. The overall thickness of the reinforcing layer was around 12 cm.

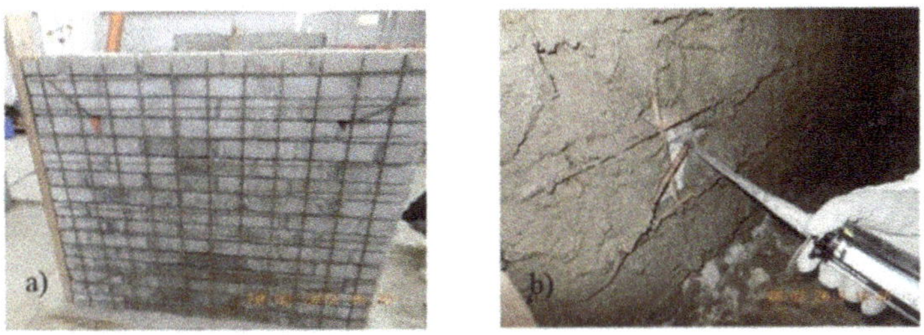

Figure 8. (**a**) TSM panel and (**b**) transversal connector.

The remaining URM panels were strengthened with TRM systems. First, all the surfaces were wetted and the first layers of the strengthening mortar were overlaid. The two constituents of the mortar, component A (powder) and component B (liquid), were mixed according to the specifications of the manufacturer [47]. The final material consisted of a plastic–thixotropic blend that was applied to the panels' surfaces with a flat metal smoothing towel.

After the first layer of the mortar was applied and while it was still wet, the meshes were placed over the surfaces of the panels. The meshes were lightly pressed down with a towel for them to adhere to the mortar. The adjacent pieces of the meshes were overlapped both widthways and lengthways by approximately 15 cm. Next, the second layer of the mortar was applied. The width of both mortar layers was approximately 5–6 mm.

In order to reduce the occurrence and development of the de-bonding phenomena and to improve the static efficiency of the system, the faces of the TRM system were tied with composite connectors [51]. For this purpose, after the mortar was cured, holes were drilled in the substrate and the loose material was removed with compressed air. Additionally, the surfaces inside the holes were also cleaned with a long-bristled brush. The protective gauze from the cords was unrolled to a length equal to the depth of the holes that were previously drilled into the panels. Before the cords were inserted into the panels, the adhesive was extruded, starting from the bottom of the holes, until the spaces were fully filled. The two components of the adhesive were automatically mixed together when the product was extruded, using a static mixer that was supplied together with the cartridges. The cords were then inserted into the adhesively filled holes and the excessive material was removed. The last remaining protective gauzes were removed and the fibers were spread over the substrate in a circular pattern. The overall diameter of the circle was approximately 25 cm. Finally, the spread fibers were embedded in a layer of mortar.

It is worth mentioning that, in the case of the URM panel strengthened only on one face, the anchoring was performed starting from the middle of the element to its face. This simulated a common situation for monumental buildings, when strengthening system needs to be applied without imposing any interventions to one side of the wall. The construction stages of the TRM1, TRM2 and TRM3 specimens are illustrated in Figure 9.

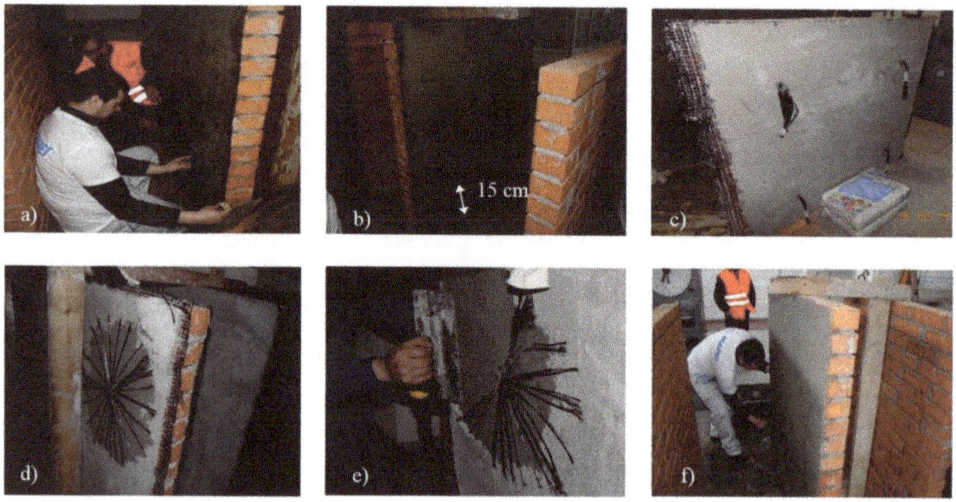

Figure 9. Construction stages: (**a**) application of the first layer of mortar and reinforcing meshes; (**b**) overlap of the reinforcing meshes; (**c**) insertion of composite cords; (**d**) cords spreading in a circular pattern; (**e**) embedment of the cords with mortar; (**f**) surface levelling and final checks.

2.3. *Instrumentation and Testing Procedures*

The tests were performed according to the specifications given in ASTM E519/E519M—15 [53]. The latter describe a test procedure designed to determine the shear strength of masonry panels with modular dimensions of 1.2 by 1.2 m, by loading them in compression along the diagonal. Due to the loading direction, a diagonal tension failure is induced within the specimen. Thus, the specimens were carefully lifted, rotated at 45^0 and positioned on the transportation carriage in such a way that no disturbance was caused either to the unstrengthened or the strengthened panels (Figure 10).

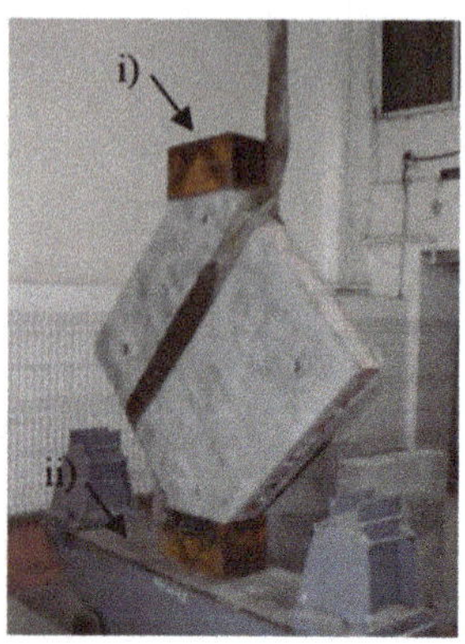

Figure 10. Specimen positioned on the transportation carriage: (i) loading shoe and (ii) transportation carriage.

The load was applied to the URM panel through a steel shoe placed at the top corner and transmitted to a similar shoe at the bottom corner (Figure 11). Each loading shoe had a loading area of 330 mm by 250 mm. A hydraulic jack with an overall capacity of 500 kN was incorporated between the clamp of the testing machine and the top loading shoe. The deformation of the specimens, elongation of diagonals and compression were monitored by four linear variable differential transducers (LVDTs), with two being placed on each face of the panel, oriented perpendicular and parallel to the loading direction (Figure 11). All the data were captured and stored using a data acquisition system (DAQ). Each test was load-controlled; thus, the load was increased monotonically at a constant speed of approximately 0.5–0.6 kN/s.

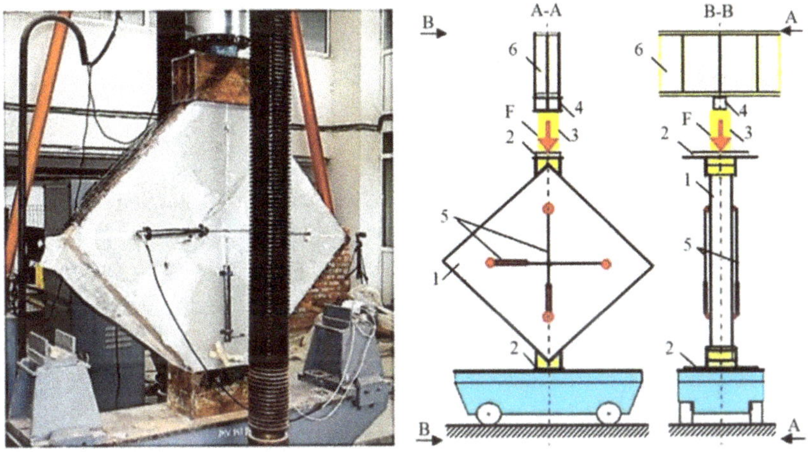

Figure 11. Specimen instrumentation and loading conditions: 1—URM wall; 2—loading shoe; 3—loading jack; 4—loading cell; 5—LVDTs; 6—steel channel.

3. Numerical Approach

3.1. Introduction

Masonry is an anisotropic composite material which consists of masonry units (bricks, stone, etc.) assembled with or without mortar. The mechanical models that were developed to evaluate the structural behavior of the URM structures fall under the category of no-tension material models. From the theoretical point of view, URM walls are considered to behave as a linear elastic material when subjected to compressive stresses. However, based on the mechanical and elastic properties of the components, the mechanical model and, by default, the numerical one may be adapted from a typical linear–elastic to an elastic–plastic one.

Generally, three distinct modelling approaches can be adopted to simulate the structural behavior of a URM wall loaded in shear [54–65]. The first one, referred to as macro-modelling, consists in a continuum homogenous element that does not distinguish between the masonry units and mortar or between the individual components of the strengthening system. The macro-model is not suitable for the analysis of the URM walls strengthened by TRM plastering since, in this case, the stresses tend to converge into a narrow region along the faces of the composite transversal connectors. More specifically, the numerical macro-models are unable to account for the micro-mechanical characteristics at the interface levels. The second approach, referred to as the simplified micro-modelling, consists in a combination between micro and macro modelling techniques. The numerical model obtained by applying this approach has both continuum elements (for masonry units and mortar) and discontinuous elements (for the interface levels). This simplification can significantly reduce the computational costs. However, when the strengthening system is made of multiple materials with various mechanical and elastic properties (e.g., reinforced mortar, textile cords, reinforcing meshes, adhesives, etc.), the simplified numerical model can undervalue the overall shear response of the strengthened URM wall. On the other hand, the detailed micro-numerical approach allows us to identify and monitor the complete stress–strain state of the strengthened URM walls. These models are designed to account for each component of the physical models and for the particular properties of each material, thus reflecting the shear structural behavior of the URM walls strengthened with TRM in a more realistic manner.

3.2. Model Definition

In this study, a detailed micro non-linear 3D model was developed to accurately simulate the laboratory test conditions (Figure 12). The non-linear numerical analyses were carried out using the commercial, multi-purpose finite element software Ansys V 15.0, Ansys, PA, USA [55–58,63,64].

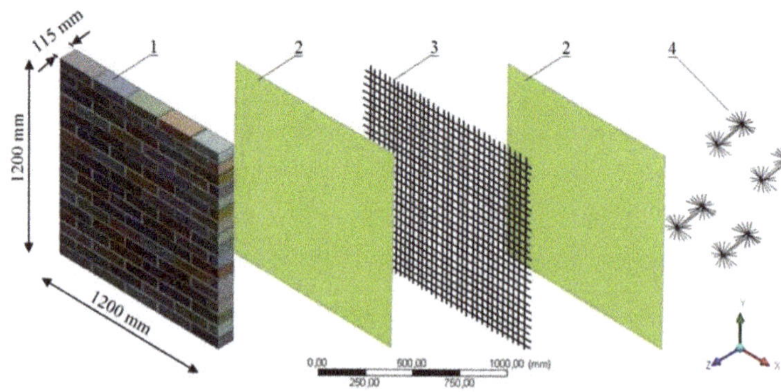

Figure 12. Three-dimensional model: 1—URM panel; 2—Planitop HDM Maxi mortar; 3—Mapegrid C 170 meshes; 4—MapeWrap C Fiocco composite connectors.

For the numerical 3D micro-models, the bricks and the mortar were simulated using SOLID187 tetrahedron finite elements (Figure 13a) [55]. The latter consist of higher-order 3D elements with 10 nodes, each of them defined by three degrees of freedom. These elements were selected based on their characteristic quadratic displacement behavior which can account for large deflections and significant strain variations. The components of the TRM system and the composite cords were modelled by SHELL 63 finite elements, as they possess both membrane and bending capabilities (Figure 13b). Moreover, the 3D micro-models were defined to account for three distinct interface levels (bricks–mortar, TRM–bricks, TRM–mortar). CONTA 174 bonded contact elements were modelled at each interface level so as to detect any possible contact separation between the TRM system and the URM panels (Figure 13c).

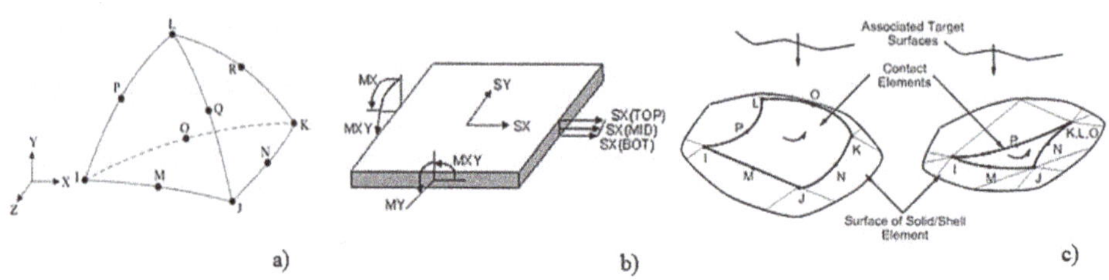

Figure 13. Finite elements: (**a**) SOLID187, (**b**) SHELL 63 and (**c**) CONTA174. *i, j, k, l, m, n, o, p, q, r—nodes; MX, Y, XY—bending moments; SX, Y—displacements. Figure adapted from [59].

The bricks were assumed to behave as linear elastic isotropic materials and the plasticity properties were entirely assessed to the mortar, which was modelled as an elastic–plastic material. In this approach, the structural characteristics of the mortar are capable to account for the global non-linear behavior of the URM module. In addition, by applying this modelling strategy, it is assured that failure occurs in a characteristic mode, which is defined by a continuous diagonal crack that passes through the head and the bed mortar joints. The plasticity of the mortar was defined using the Drucker–Prager formulation. According to the latter, the material is characterized by a specific domain of yield stresses. When the yield surface is reached, the mortar starts to deform plastically and, upon further increase in load, the plastic flow is initiated.

As mentioned before, the complete test specimens were modelled, including the interface levels. For this purpose, the augmented Lagrange formulation was used to define the contacts. The penetration points were determined for both bricks and mortar based on the integration point detection.

Since the stresses vary with a high gradient between distinct components of the 3D model, the mesh size was set differently for bricks, mortar, TRM and composite cords. For the bricks, the maximum mesh size was limited to 30 mm; for the mortar and the composite cords, the mesh size was limited to maximum 2 mm; whereas, for the TRM system, a maximum length of 80 mm was imposed in the model (Figure 14).

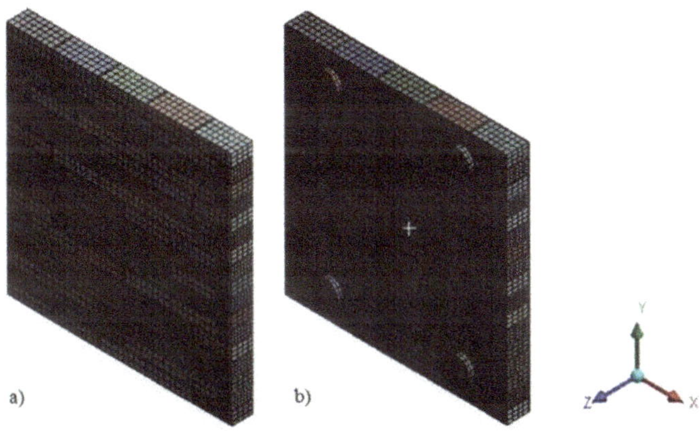

Figure 14. Finite element mesh of (**a**) URM panel and (**b**) URM panel strengthened with TRM.

The nonlinear 3D micro-detailed models were analyzed by using the Newton–Raphson iteration method. More specifically, the method was applied by gradually increasing the load, up to the limit where the convergence criterion was satisfied. Furthermore, whenever required, various sub-steps were defined so as to divide the load ratio. Thus, the final step/sub-step corresponded to the ultimate load. This value was reached at the end of the elastic branch, immediately before an abrupt change in displacement was recorded. This change also indicated that the convergence criterion was no longer satisfied.

4. Results and Discussion

In the case of the URM reference panel, the failure occurred in the mortar layer by splitting the specimen along the horizontal and vertical joints (Figure 15). Besides the typical propagation path of the crack (through the mortar layer), shortly before the force reached its ultimate value, the crack propagated even through the bricks.

By analyzing the load–displacement curve (on both faces of the panel) of the URM reference panel, depicted based on both experimentally and numerical results, it can be observed that, in the first stage, the specimen's response was quasi-linear–elastic with almost constant stiffness value. In the second plateau of the graph, the load–displacement curve illustrates a plastic phase which represents the degraded stiffness state of the URM panel (Figure 16). The plastic behavior was recorded immediately after the first cracks had started to develop. At the end of the plastic branch, all the cracks indicated by arrows in Figure 15 were joined in a continuous crack that spanned, in the vertical direction, between the loaded corners of the masonry panel. At the end of the test, the masonry panel was separated into two parts along the direction of the crack.

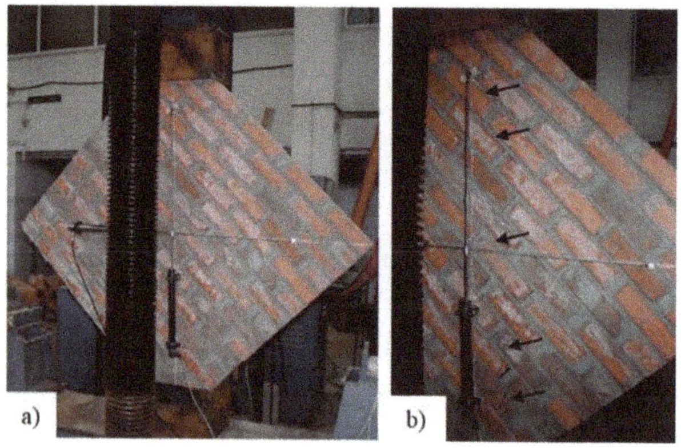

Figure 15. (**a**) URM panel loaded into the testing machine; (**b**) Failure mode—crack propagation pattern through the mortar joints.

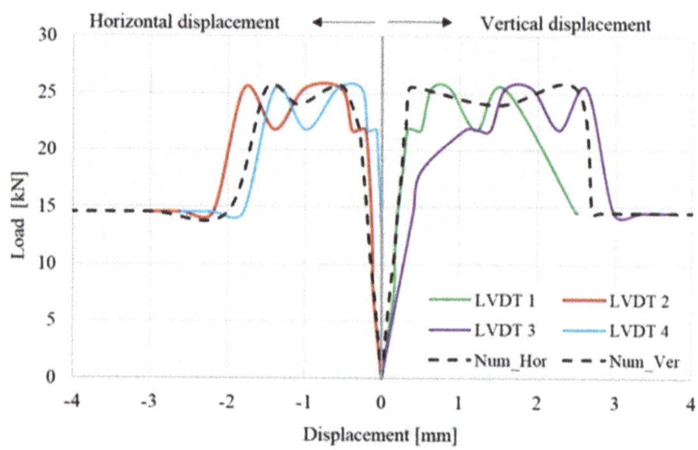

Figure 16. Load–displacement distribution for the URM panel.

As it can be observed, the numerical and the experimental results are in good agreement both for ultimate values and distributions. It should also be mentioned that a sign convention was adopted in order to represent, in the same graph, both the horizontal and vertical displacements. For the evaluation of the shear stress–shear strain state, all the data were taken as absolute values.

In the case of the TSM panel, even though the overall thickness of the strengthening systems is around 12 cm, the typical pattern for crack development (across the masonry mortar joints) can be easily identified (Figure 17). When the load reached its ultimate value, large surfaces of mortar started to expel, exposing the steel reinforcement meshes (Figure 17b). Moreover, the failure by panel splitting is evidenced by the distribution of the horizontal displacement, where it can be observed that, at a relatively constant load, or even slightly decreasing, the displacements increase largely. Two local failures were also recorded in the ranges 38–42 KN and 50–52 KN, respectively (Figure 18).

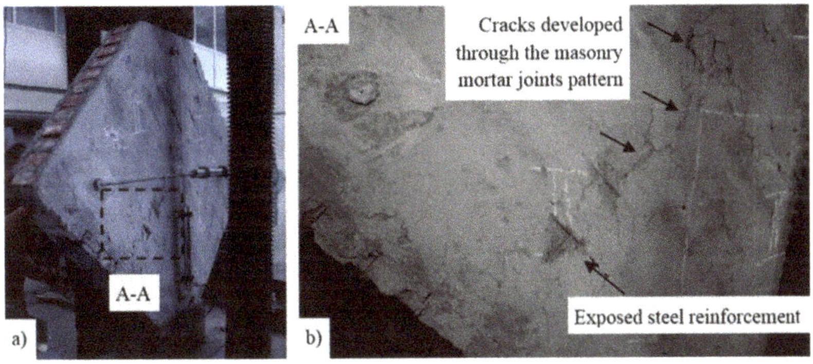

Figure 17. (**a**) Failure mode of the TSM panel and (**b**) detail of A-A.

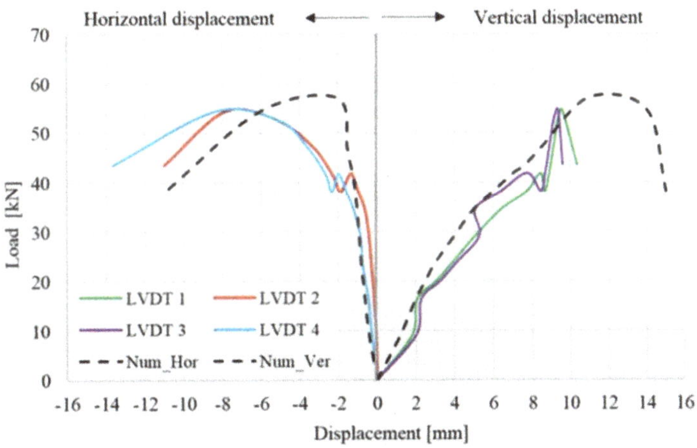

Figure 18. Load–displacement distribution for the TSM panel.

Unlike the reference URM panels, the TRM-strengthened ones (TRM1, TRM2 and TRM3) evidenced a gradual failure mode, which is highly advantageous for the masonry structures located in earthquake-vulnerable areas. For the TRM1 panel, the crack pattern is similar to the ones developed for the reference panels (Figure 19a). However, on the plastered face, the crack pattern is slightly shifted toward the borders of the circular regions where the composite connectors were embedded with the mortar (Figure 19b).

The TRM 2 and TRM3 panels developed a particular crack pattern which followed the border line of the composite anchors (Figure 20). Besides preventing the de-bonding phenomena and improving the static efficiency of the TRM system (the main functions of the composite anchors), the exterior part of the cords provided an additional reinforcement for the system. More specifically, the radial orientation of these fibers modified the crack pattern by shifting its original vertical direction towards the border region of the exterior side of the composite anchors (Figure 20b,c). Moreover, the sliding effect at the joints was negligible due to the presence of the end parts of the anchors (also referred as *diatones*). On the contrary, in the case of similar experimental works, that were performed on masonry panels tied with punctual transversal connectors, the sliding phenomena was more intense, as reported in [66].

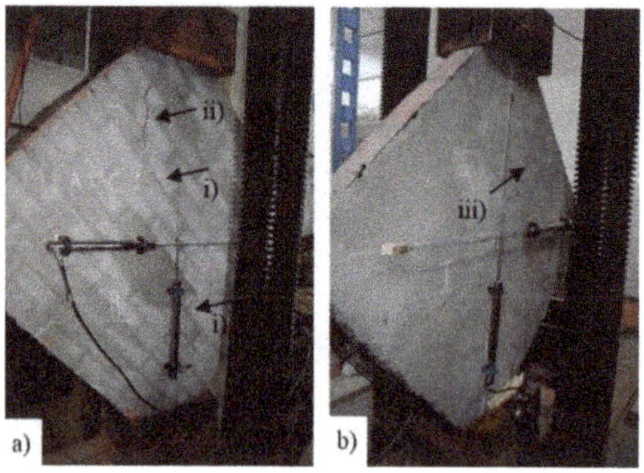

Figure 19. URM panel strengthened with TRM on one face (TRM1 panel). (**a**) Unstrengthened face: (i) crack propagation through mortar joints and (ii) crack propagation through brick units; (**b**) Strengthened face: (iii) crack propagation close to the composite anchors.

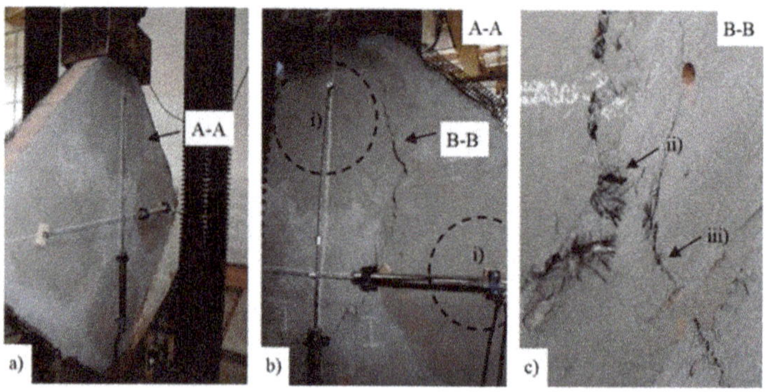

Figure 20. TRM2 and TRM 3 panels. (**a**) Crack pattern. (**b**) Detail of A-A: (i) exterior side of the composite anchors. (**c**) Detail of B-B: (ii) exposed composite cords and (iii) failure of the TRM by fiber mesh rupture.

As it can be observed in Figures 21–25, beside the linear–elastic plateau, for all the specimens (including the URM panel strengthened only on one face), a characteristic plateau of constant load and increasing displacement was recorded. This plateau is similar to the plastic range (yield plateau) representation of structural carbon steel elements. As shown in the load–displacement curves, in the plastic stage, the load remains constant while the displacements continue to increase, meaning that the elongation and contraction continue as long as the load is not removed. In addition, this behavior indicates that the TRM systems provide significant additional ductility that enables the URM panels to largely deform before failure. Nevertheless, this behavior demonstrates the efficiency of the TRM systems that are applied only on one face of the panel, a case which often comes across in the strengthening applications of monumental masonry buildings. The anchoring techniques (both the side-to-side and the middle-to-side method) provide sufficient lateral capacity to avoid the possible out-of-plane failure.

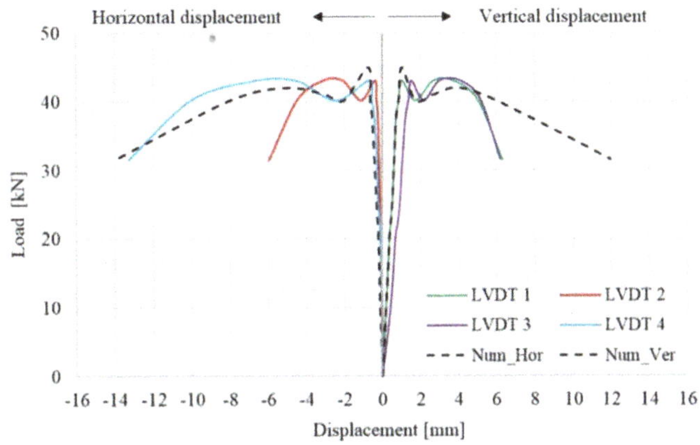

Figure 21. Load–displacement distribution for the URM panel strengthened with TRM on one face (TRM1 panel).

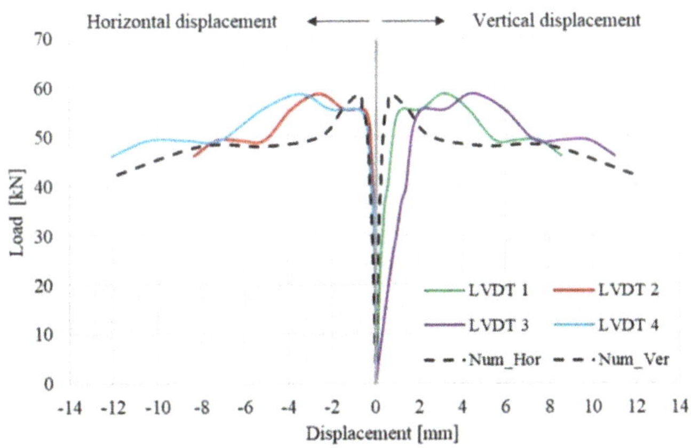

Figure 22. Load–displacement distribution for the URM specimen strengthened with TRM on both faces (TRM2 panel).

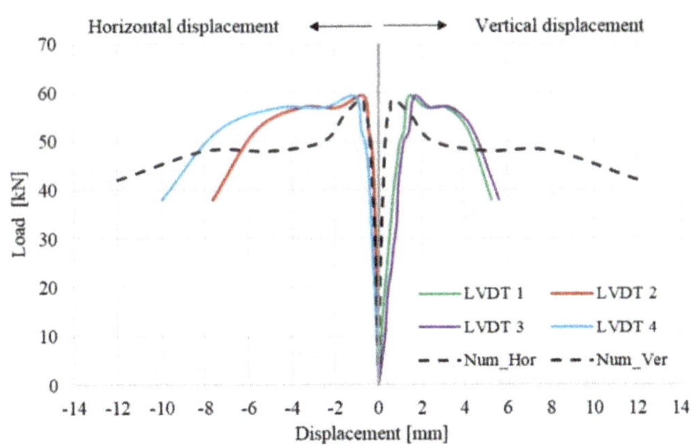

Figure 23. Load–displacement distribution for the URM specimen strengthened with TRM on both faces (TRM3 panel).

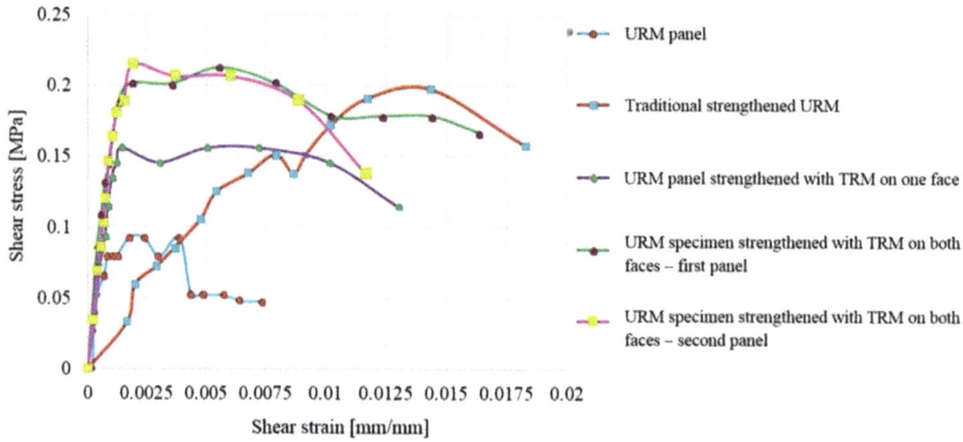

Figure 24. Shear stress–shear strain distributions: experimental results.

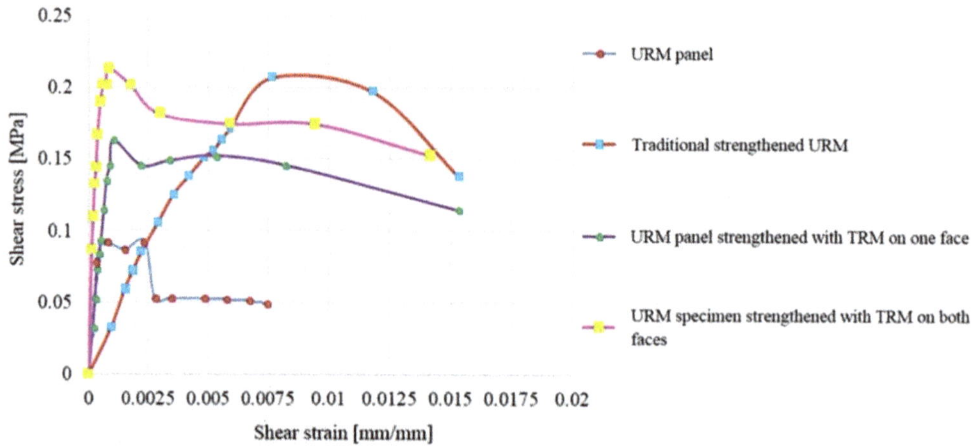

Figure 25. Shear stress–shear strain distributions: numerical results.

The shear stress–shear strain distributions of all panels are presented in Figures 24 and 25. According to ASTM E519/E519M—15 the shear stress is computed using Equation (1) [53].

$$S_s = \frac{0.707P}{A_n} \tag{1}$$

where S_s—shear stress (MPa); P—load measured along the diagonal pattern; A_n—net area of the panel; $A_n = \frac{w+h}{2} t \times n$; w—width of the panel (mm); h—height of the panel (mm); t—thickness of the panel; n—the percentage of the gross area that is solid (expressed as a decimal).

According to ASTM E519/E519M—15, the shear strain is computed using Equation (2) [53].

$$\gamma = \frac{\Delta V + \Delta H}{g} \tag{2}$$

where γ —shear strain (mm/mm); ΔV—shortening on vertical direction; ΔH—extension on horizontal direction; g—monitoring length.

As it can be observed in Figures 24 and 25, for all the specimens, the shear stress–shear strain distribution curves start with a relatively steep slope and increase linearly until the beginning of the plastic range.

In the cases of the URM panel and of the traditional, strengthened panel (URM and TSM panels), in the plastic phase, a substantial degraded stiffness can be observed. On the other hand, in the cases of the TRM-strengthened panels (TRM1, TRM2 and TRM3), the deformation capacity is considerably improved and the characteristic (yield plateau) that was previously mentioned for the load–displacement distribution can be easily identified. Thus, TRM systems provide significant additional ductility that enables the shear stresses to remain constant while the shear strains continue to increase, meaning that the elongation and contraction continues as long as the load is not removed.

The maximum shear stress recorded for the URM panels strengthened on both faces (TRM2 and TRM3) were 20% higher than the one measured for the URM panel strengthened only on one face. When it comes to the numerical values, similarly, an increase of 33% was recorded for the maximum shear stress for the panels strengthened on both faces when compared to the one of the panel strengthened on one face. The experimentally obtained value for the shear strain was 23% higher for the TRM2 and TRM3 specimens when compared to the one measured for the TRM1 specimen. However, the numerical values of the shear strains were almost equal in both cases.

The ultimate values of both the shear stress and shear strain that were experimentally recorded for the TSM panel were almost equal to the ones recorded for the TRM2 and TRM3 panels. However, the structural improvement of the TSM panel was obtained by increasing the specimen overall thickness by approximately 12 cm, while the overall thickness of the TRM2 and TRM3 panels was increased by only 2 cm. Thus, the total weight of the traditional strengthened system was considerably higher than the one of the modern TRM system.

According to ASTM E519/E519M—15 [53], the modulus of rigidity—G—is computed as the ratio between the shear stress—S_s—and the shear strain—γ—[53]. The stiffness of the URM panels can be evaluated by quantifying the modulus of elasticity—E. The latter can be calculated using Equation (3), where υ was adopted as 0.25, as reported in previous studies relating to similar URM panels [14,60].

$$E = 2G(1 + \upsilon) \tag{3}$$

The ductility of the URM panels can be quantified by the drift ratio—δ—Equation (4).

$$\delta_u = \frac{\Delta u}{H} \tag{4}$$

where Δu—the displacement measured along the diagonal, corresponding to the ultimate load; H—height of the URM specimens.

The experimental and the numerical results are listed in Table 6.

As presented in Figures 21–25, the response of all the specimens was non-linear and the post-peak behavior in the cases of the strengthened panels was characterized by a relatively gradual strain softening. The shear stiffness of the TSM panel, represented by the modulus of rigidity (G), increased by 34%, when compared to the experimentally recorded value for the URM panel. Similarly, the numerical results show an increase by 43% from the previously mentioned values. In the case of the TRM1 panel, the shear stiffness was increased by 36% (experimentally determined value) and 31% (numerically determined value), when compared to the values determined for the URM panel. The largest increase in the shear stiffness was recorded for the TRM2 and TRM3 panels (approximately 42% with respect to the experimental value and 62% with respect to the numerical value).

Table 6. Summary of experimental and numerical results.

Characteristics	URM	TSM	TRM1	TRM2	TRM3
P_{ult_exp} (kN)	25.432	54.378	43.024	58.695	59.364
P_{ult_num} (kN)	24.900	57.210	45.155	58.345	58.345
S_{s_exp} (MPa)	0.043	0.157	0.114	0.167	0.137
S_{s_num} (MPa)	0.042	0.138	0.114	0.152	0.152
γ_{exp} (mm/mm)	0.007	0.017	0.012	0.015	0.012
γ_{num} (mm/mm)	0.008	0.015	0.015	0.014	0.014
G_{exp} (MPa)	6.142	9.235	9.500	11.133	11.417
G_{num} (MPa)	5.250	9.200	7.600	10.857	10.857
E_{exp} (MPa)	15.355	23.088	23,750	27.833	28.542
E_{num} (MPa)	13.125	23.000	19.000	27.143	28.843
δ_{u_exp} (%)	0.220	0.609	0.473	0.650	0.331
δ_{u_num} (%)	0.236	0.884	0.707	0.707	0.707

The ductility of the URM panel, expressed as the drift ratio, is considerably lower when compared to the values determined for the strengthened panels. Thus, increases in ductility of 64% for the TSM panel, 54% for the TRM1 panel, 66% for the TRM2 panel and 34% for the TRM3 panel were recorded with respect to the experimentally determined value of the URM panel. Similarly, increases in ductility of 73% for the TSM panel and 67% for the TRM1, TRM2 and TRM 3 panels were recorded with respect to the numerically determined value of the URM panel.

5. Conclusions

This paper presents the main outcomes of a numerical and experimental study which relates to the structural efficiency of a modern strengthening technique designed for URM panels. The latter consists of a TRM jacketing that can be applied either on one side of the wall or on both sides. The structural TRM layers can be anchored both by the side-to-side method and the middle-to-side one.

From the findings summarized above, the following can be concluded:

- The observed damaged patterns both for the URM and TSM specimens matched the characteristic crack path, which consisted in a continuous-propagation step-like line that spread along the compressed diagonal, through the horizontal and vertical mortar joints.
- The TRM2 and TRM3 panels developed a particular crack pattern which followed the border line of the composite anchors.
- The TRM systems evidenced a general and significant increase in the original shear capability and ductility of the URM panel. They also enabled the URM panels to largely deform before failure.
- The selected type of anchorage system provided sufficient lateral capacity to avoid the possible out-of-plane failure.
- The ultimate loads that were experimentally determined increased by 53% for the TSM panel, 41% for the TRM1 panel, 56% and 57% for the TRM2 and TRM3 panels, when compared to the ultimate load recorded for the URM panel.
- Similarly, the numerical results evidenced an increase of 56% for the ultimate load of the traditional strengthened panel, 45% for the panel plastered with TRM on one face and 57% for the 3D micro-model with TRM jacketing on both faces, when compared to the ultimate load recorded for the URM reference panel.
- The ultimate values of the mechanical properties that were experimentally and numerically investigated were almost equal for the TSM panel and the TRM2 and TRM3 panels.
- The numerical and the experimental results validate the efficiency of the TRM systems that are applied only on one face of the wall, a case which often comes across in the strengthening applications of monumental masonry buildings.

Although further experimental studies are required for an adequate assessment of the investigated strengthening techniques (e.g., position, dimension and number of the composite anchors) and for the validation of the micro-numerical model, the studies discussed in this paper represent a background for the general structural efficiency of the TRM strengthening systems.

Author Contributions: Conceptualization, N.Ț.; methodology, P.M.; validation, N.Ț.; formal analysis, D.U.; investigation, D.N.I. and D.U.; resources, D.N.I.; data curation, D.U. and N.Ț.; writing—original draft preparation, D.U.; writing—review and editing, N.Ț. and D.N.I.; supervision, N.Ț.; project administration, D.A.G. and R.C. All authors have read and agreed to the published version of the manuscript.

Funding: This research project was funded by "ANAGENNISI—Innovative Reuse of All Tyre Components in Concrete", project FP7-ENV-2013-603722, PNII.

Institutional Review Board Statement: Not applicable.

Informed Consent Statement: Not applicable.

Data Availability Statement: The data underlying this article will be shared on reasonable request from the corresponding author.

Conflicts of Interest: The authors declare no conflict of interest.

References

1. Papanicolaou, C.G.; Triantafillou, T.C.; Papathanasiou, M.; Karlos, K. Textile reinforced mortar (TRM) versus FRP as strengthening material of URM walls: Out-of-plane cyclic loading. *Mater. Struct.* **2008**, *41*, 143–157. [CrossRef]
2. Casolo, S.; Sanjust, C.A. Seismic analysis and strengthening design of a masonry monument by a rigid body spring model: The "Maniace Castle" of Syracuse. *Eng. Struct.* **2009**, *31*, 1447–1459. [CrossRef]
3. Bothara, J.K.; Dhakal, R.P.; Mander, J.B. Seismic performance of an unreinforced masonry building: An experimental investigation. *Earthq. Eng. Struct. Dyn.* **2010**, *39*, 45–68. [CrossRef]
4. Bui, T.-L.; Si Larbi, A.; Reboul, N.; Ferrier, E. Shear behaviour of masonry walls strengthened by external bonded FRP and TRC. *Compos. Struct.* **2015**, *132*, 923–932. [CrossRef]
5. Shabdin, M.; Zargaran, M.; Attari, N.K.A. Experimental diagonal tension (shear) test of Un-Reinforced Masonry (URM) walls strengthened with textile reinforced mortar (TRM). *Constr. Build. Mater.* **2018**, *164*, 704–715. [CrossRef]
6. Brandonisio, G.; Mele, E.; De Luca, A. Closed form solution for predicting the horizontal capacity of masonry portal frames through limit analysis and comparison with experimental test results. *Eng. Fail. Anal.* **2015**, *55*, 246–270. [CrossRef]
7. Cundari, G.A.; Milani, G.; Failla, G. Seismic vulnerability evaluation of historical masonry churches: Proposal for a general and comprehensive numerical approach to cross-check results. *Eng. Fail. Anal.* **2017**, *82*, 208–228. [CrossRef]
8. Yardim, Y.; Lalaj, O. Shear strengthening of unreinforced masonry wall with different fiber mortar jacketing. *Constr. Build. Mater.* **2016**, *102*, 149–154. [CrossRef]
9. Mosoarca, M.; Onescu, I.; Onescu, E.; Anastasiadis, A. Seismic vulnerability assessment methodology for historic masonry buildings in the near-field areas. *Eng. Fail. Anal.* **2020**, *115*, 104662. [CrossRef]
10. Sayin, B.; Yildizlar, B.; Akcay, C.; Gunes, B. The retrofitting of historical masonry buildings with insufficient seismic resistance using conventional and non-conventional techniques. *Eng. Fail. Anal.* **2019**, *115*, 454–463. [CrossRef]
11. Micelli, F.; Cascardi, A. Structural assessment and seismic analysis of a 14th century masonry tower. *Eng. Fail. Anal.* **2020**, *107*, 104198. [CrossRef]
12. Vlachakis, G.; Cerveza, M.; Barbat, G.B.; Saloustros, S. Out-of-plane seismic response and failure mechanism of masonry structures using finite elements with enhanced strain accuracy. *Eng. Fail. Anal.* **2019**, *97*, 534–555. [CrossRef]
13. Kalali, A.; Kabir, M.Z. Experimental response of double-wythe masonry panels strengthened with glass fiber reinforced polymers subjected to diagonal compression tests. *Eng. Struct.* **2012**, *39*, 24–37. [CrossRef]
14. Mustafaraj, E.; Yardim, Y. Retrofitting damaged unreinforced masonry using external shear strengthening techniques. *J. Build. Eng.* **2019**, *26*, 100913. [CrossRef]
15. Wood, W.A. *Load Distribution of Masonry Walls Subject to Concentrated Loads over Opening*; University of Pittsburgh: Pittsburgh, PA, USA, 1998.
16. Gattesco, N.; Boem, I. Experimental and analytical study to evaluate the effectiveness of an in-plane reinforcement for masonry walls using GFRP meshes. *Constr. Build. Mater.* **2015**, *88*, 94–104. [CrossRef]
17. Țăranu, N.; Cozmanciuc, R.; Ențuc, I.; Budescu, M.; Munteanu, V.; Isopescu, D. The behaviour of the interface between carbon fibre reinforced polymer composite plates and concrete. *Rom. J. Mater.* **2015**, *45*, 232–239.
18. Bejan, L.; Țăranu, N.; Sârbu, A. Advanced polymeric composites with hybrid reinforcement. *J. Optoelectron. Adv. Mater.* **2010**, *12*, 1930–1934.

19. Ţăranu, N.; Banu, C.; Oprişan, G.; Budescu, M.; Munteanu, V.; Ioniţă, O. Tensile characteristics of glass fibre reinforced polymeric bars. *Rom. J. Mater.* **2010**, *40*, 323–331.
20. Ungureanu, D.; Ţăranu, N.; Lupăşteanu, V.; Isopescu, D.N.; Oprişan, G.; Mihai, P. Experimental and numerical investigation on adhesively bonded single lap and thick adherents joints between pultruded GFRP composite profiles. *Compos. Part B Eng.* **2018**, *146*, 49–59. [CrossRef]
21. Bernat, E.; Gil, L.; Roca, P.; Escrig, C. Experimental and analytical study of TRM strengthened brickwork walls under eccentric compressive loading. *Constr. Build. Mater.* **2013**, *44*, 35–47. [CrossRef]
22. Torres, B.; Ivorra, S.; Baeza, F.J.; Estevan, L.; Varona, B. Textile reinforced mortars (TRM) for repairing and retrofitting masonry walls subjected to in-plane cyclic loads. An experimental approach. *Eng. Struct.* **2021**, *231*, 111742. [CrossRef]
23. Carozzi, F.G.; Milani, G.; Poggi, C. Mechanical properties and numerical modeling of fabric reinforced cementitious matrix (FRCM) systems for strengthening of masonry structures. *Compos. Struct.* **2014**, *107*, 711–725. [CrossRef]
24. Papanicolaou, C.; Triantafillou, T.; Lekka, M. Externally bonded grids as strengthening and seismic retrofitting materials of masonry panels. *Constr. Build. Mater.* **2011**, *25*, 504–514. [CrossRef]
25. da Porto, F.; Mosele, F.; Modena, C. In-plane cyclic behaviour of a new reinforced masonry system: Experimental results. *Eng. Struct.* **2011**, *33*, 2584–2596. [CrossRef]
26. Wang, X.; Lam, C.C.; Lu, V.P. Bond behaviour of steel-TRM composites for strengthening masonry elements: Experimental testing and numerical modelling. *Constr. Build. Mater.* **2020**, *253*, 119157. [CrossRef]
27. Blanksvärd, T.; Täljsten, B. Strengthening of concrete structures with cement based bonded composites. *Nord. Concr. Res.* **2008**, *2*, 133–154.
28. Koutas, L.; Triantafillou, T.; Bousias, S. Analytical modeling of masonry-infilled RC frames retrofitted with textile-reinforced mortar. *J. Compos.* **2014**, *19*, 1–14. [CrossRef]
29. Ismail, N.; Ingham, J.M. In-plane and out-of-plane testing of unreinforced masonry walls strengthened using polymer textile reinforced mortar. *Eng. Struct.* **2016**, *118*, 167–177. [CrossRef]
30. Ismail, N.; Ingham, J.M. Polymer textiles as a retrofit material for masonry walls. *Proc. Inst. Civ. Eng.-Struct. Build.* **2014**, *167*, 15–25. [CrossRef]
31. Sagar, S.L.; Singhal, V.; Rai, D.C.; Gudur, P. Diagonal shear and out-of-plane flexural strength of fabric-reinforced cementitious matrix-strengthened masonry wallets. *J. Compos. Constr.* **2017**, *21*, 04017016. [CrossRef]
32. Marcari, G.; Basili, M.; Vestroni, F. Experimental investigation of tuff masonry panels reinforced with surface bonded basalt textile-reinforced mortar. *Compos. Part B Eng.* **2017**, *108*, 131–142. [CrossRef]
33. Prota, A.; Marcari, G.; Fabbrocino, G.; Manfredi, G.; Aldea, C. Experimental in plane behaviour of tuff masonry strengthened with cementitious matrix-grid composites. *J. Compos. Constr.* **2006**, *10*, 1081–1097. [CrossRef]
34. Kouris, L.A.S.; Triantafillou, T.C. State-of-the-art on strengthening of masonry structures with textile reinforced mortar (TRM). *Constr. Build. Mater.* **2018**, *188*, 1221–1233. [CrossRef]
35. Ferrara, G.; Coppola, B.; Di Maio, L.; Incarnato, L.; Martinelli, E. Tensile strength of flax fabrics to be used as reinforcement in cement-based composites: Experimental tests under different environmental exposures. *Compos. Part B Eng.* **2019**, *168*, 511–523. [CrossRef]
36. Ferrara, G.; Martinelli, E. Tensile behaviour of Textile Reinforced Mortar composite systems with flax fibres. In Proceedings of the 12th Fib International PhD Symposium in Civil Engineering, Prague, Czech Republic, 29–31 August 2018; Volume 29, pp. 1–7.
37. Papanicolaou, C.G.; Triantafillou, T.C.; Karlos, K.; Papathanasiou, M. Textile reinforced mortar (TRM) versus FRP as strengthening material of URM walls: In-plane cyclic loading. *Mater. Struct.* **2007**, *40*, 1081–1097. [CrossRef]
38. Faella, C.; Martinelli, E.; Nigro, E.; Paciello, S. Shear capacity of masonry walls externally strengthened by a cement-based composite material: An experimental campaign. *Constr. Build. Mater.* **2010**, *24*, 84–93. [CrossRef]
39. Augenti, N.; Parisi, F.; Prota, A.; Manfredi, G. In-plane lateral response of a full scale masonry subassemblage with and without an inorganic matrix-grid strengthening system. *J. Compos. Constr.* **2011**, *15*, 578–590. [CrossRef]
40. Parisi, F.; Iovinella, I.; Balsamo, A.; Augenti, N.; Prota, A. In-plane behaviour of tuff masonry strengthened with inorganic matrix-grid composites. *Compos. Part B Eng.* **2013**, *45*, 1657–1666. [CrossRef]
41. Longo, F.; Cascardi, A.; Lassandro, P.; Aielo, M.A. Energy and seismic drawbacks of masonry: A unified retrofitting solution. *J. Build. Pathol. Rehabil.* **2021**, *6*, 1–24. [CrossRef]
42. Donnini, J.; Maracchini, G.; Lenci, S.; Corinaldesi, V.; Quagliarini, E. TRM reinforced tuff and fired clay brick masonry: Experimental andanalytical investigation on their in-plane and out-of-plane behavior. *Constr. Build. Mater.* **2021**, *272*, 121643. [CrossRef]
43. Ferrara, G.; Caggegi, C.; Martinelli, E.; Gabor, A. Shear capacity of masonry walls externally strengthened using Flax-TRM composite systems: Experimental tests and comparative assessment. *Construct. Build. Mater.* **2020**, *261*, 120490. [CrossRef]
44. Sistem Complet de Zidărie Ceramic. Available online: https://www.wienerberger.ro/content/dam/wienerberger/romania/marketing/documents-magazines/brochures/Wienerberger_Catalog_Porotherm_2021.pdf (accessed on 19 November 2021).
45. Partea 1: Determinarea rezistenţei la compresiune. In *Metode de Încercare a Elementelor Pentru Zidărie*; SR EN 772-1+A1:2016; SAI Global Standards: Chicago, IL, USA, 2016. (In Romanian)
46. *Methods of Test for Mortar for Masonry. Determination of Flexural and Compressive Strength of Hardened Mortar*; EN 1015-11:2019; BSI Standards: London, UK, 2019.

47. Planitop HDM Maxi—Technical Data Sheet, 1065–3. 2019. Available online: https://www.mapei.com/ae/en/products-and-solutions/products/detail/planitop-hdm-maxi (accessed on 14 November 2021).
48. Mapegrid C 170—Technical Data Sheet, 1025–1026. 2019. Available online: https://www.mapei.com/it/en/products-and-solutions/products/detail/mapegrid-c-170 (accessed on 14 November 2021).
49. Products and Systems for the Protection and Repair of Concrete Structures. Test Methods. Determination of Compressive Strength of Repair Mortar. EN 12190:1999. Available online: https://standards.iteh.ai/catalog/standards/cen/4191a705-4001-4412-ae57-f21e7994bb79/en-12190-1998 (accessed on 14 November 2021).
50. Methods of Testing Cement. Determination of Strength. EN 196-1:2016. Available online: https://standards.iteh.ai/catalog/standards/cen/37b8816e-4085-4dcc-a642-a383d9bddd6c/en-196-1-2016 (accessed on 14 November 2021).
51. MapeWrap C Fiocco—Technical Data Sheet, 1012/1017/1026-8. 2019. Available online: https://www.mapei.com/nz/en/products-and-solutions/products/detail/mapewrap-c-fiocco (accessed on 14 November 2021).
52. Mapefix PE Wall—Technical Data Sheet, 5803/10. 2012. Available online: https://www.mapei.com/it/en/products-and-solutions/products/detail/mapefix-pe-sf (accessed on 14 November 2021).
53. *Standard Test Method for Diagonal Tension (Shear) in Masonry Assemblages*; ASTM E519/E519M-15; ASTM International: West Conshohocken, PA, USA, 2021.
54. Ghiga, D.A.; Țăranu, N.; Ungureanu, D.; Isopescu, D.N.; Scutaru, M.C.; Hudișteanu, I. Numerical modelling of structural behaviour of URM panels strengthened with cement matrix composites. *IOP Conf. Ser. Mater. Sci. Eng.* **2019**, *591*, 012037. [CrossRef]
55. Drougkas, A.; Roca, P.; Molins, C. Experimental analysis and detailed micro-modeling of masonry walls subjected to in-plane shear. *Eng. Fail. Anal.* **2019**, *95*, 82–95. [CrossRef]
56. Hudisteanu, I.; Țăranu, N.; Isopescu, D.N.; Bejan, L.; Axinte, A.; Ungureanu, D. Improving the mechanical properties of composite laminates through the suitable selection of the corresponding materials and configurations. *Rom. J. Mater.* **2017**, *47*, 252–266.
57. Ungureanu, D.; Țăranu, N.; Isopescu, D.N.; Lupășteanu, V.; Mihai, P.; Hudișteanu, I. Analytical and numerical study of adhesively bonded composite pultruded elements. *Rom. J. Mater.* **2017**, *47*, 522–531.
58. Ghiga, D.A.; Țăranu, N.; Ungureanu, D.; Isopescu, D.N.; Oprișan, G.; Hudișteanu, I. *IOP Conference Series—Materials Science and Engineering, Modtech International Conference—Modern Technologies in Industrial Engineering VIII*; IOP Publishing Ltd.: Bristol, UK, 2020; Volume 916, p. 012041.
59. Halabian, A.M.; Mirshahzadeh, L.; Hashemol-Hosseini, H. Non-linear behaviour of unreinforced masonry walls with different Iranian traditional brick-work settings. *Eng. Fail. Anal.* **2014**, *44*, 46–65. [CrossRef]
60. Karaton, M.; Aksoy, H.S.; Sayin, E.; Calayir, Y. Nonlinear seismic performance of a 12th century historical masonry bridge under different earthquake levels. *Eng. Fail. Anal.* **2017**, *79*, 408–421. [CrossRef]
61. Scacco, J.; Grillanda, N.; Milani, G.; Lourenco, P.B. Novel non-linear static numerical model for curved masonry structures based on a combined adaptive limit analysis and discrete FE computations. *Int. J. Solids Struct.* **2022**, *236–237*, 111265. [CrossRef]
62. Pineda, P. Collapse and upgrading mechanisms associated to the structural materials of a deteriorated masonry tower. Nonlinear assessment under different damage and loading levels. *Eng. Fail. Anal.* **2016**, *63*, 72–93. [CrossRef]
63. Nelson, T.; Wang, E. Reliable FE-Modeling with ANSYS (Ansys Support Data Base). Available online: https://support.ansys.com/staticassets/ANSYS/staticassets/resourcelibrary/confpaper/2004-Int-ANSYS-Conf-24.PDF (accessed on 1 June 2021).
64. Ansys Courses. Available online: https://www.sharcnet.ca/Software/Ansys/15.0.7/enus/help/ans_thry/thy_el174.html (accessed on 1 June 2021).
65. Mustafaraj, E.; Yardim, Y. Usage of ferrocement jacketing for strengthening of damaged unreinforced masonry (URM) walls. In Proceedings of the 3rd International Balkans Conference on Challenges of Civil Engineering, Tirana, Albania, 19–21 May 2016; pp. 153–163.
66. Casardi, A.; Leone, M.; Aielo, M.A. Transversal joining of multi-leaf masonry through different types of connector: Experimental and theoretical investigation. *Constr. Build. Mater.* **2020**, *272*, 120733. [CrossRef]

materials

Article

Structural Response of Bonded Joints between FRP Composite Strips and Steel Plates

Vlad Lupăşteanu [1], Dragoş Ungureanu [1,*], Nicolae Ţăranu [1,2], Dorina Nicolina Isopescu [1], Radu Lupăşteanu [1] and Petru Mihai [1]

[1] Faculty of Civil Engineering and Building Services, "Gheorghe Asachi" Technical University of Iaşi, 43 Mangeron Blvd., 700050 Iaşi, Romania; vlad.lupasteanu@academic.tuiasi.ro (V.L.); nicolae.taranu@academic.tuiasi.ro (N.Ţ.); dorina-nicolina.isopescu@academic.tuiasi.ro (D.N.I.); radu.lupasteanu@academic.tuiasi.ro (R.L.); petru.mihai@academic.tuiasi.ro (P.M.)
[2] The Academy of Romanian Scientists, 54 Splaiul Independentei, Sector 5, 050094 Bucuresti, Romania
* Correspondence: dragos.ungureanu@staff.tuiasi.ro

Abstract: This paper presents the outcomes of an experimental and numerical study performed on epoxy-bonded single lap joints (SLJs) between carbon fiber-reinforced polymer (CFRP) composite strips and steel elements. For the experimental program, 34 specimens were prepared by varying the type of the composite strip and the type of adhesives and their thicknesses; all specimens were loaded in axial tension up to failure. The specific failure mechanisms were identified and commented on the basis of the performed tests, and the load–displacement curves were plotted. Additionally, the strain distributions along the bond lengths at different load stages, the shear stress–displacements (slip) variations and the stress–strain distributions for the CFRP strips were plotted and investigated. The numerical simulations, based on 3D finite element method (FEM) analysis, provided consistent results, in good agreement with the experimental ones for all parameters that were investigated and discussed in this paper.

Keywords: polymer–matrix composites; bonded joints; epoxy adhesives; finite element analysis (FEA)

Citation: Lupăşteanu, V.; Ungureanu, D.; Ţăranu, N.; Isopescu, D.N.; Lupăşteanu, R.; Mihai, P. Structural Response of Bonded Joints between FRP Composite Strips and Steel Plates. *Materials* **2021**, *14*, 6722. https://doi.org/10.3390/ma14216722

Academic Editor: Aleksander Muc

Received: 11 October 2021
Accepted: 5 November 2021
Published: 8 November 2021

1. Introduction

Over the past decades, composite materials have been widely used to strengthen existing structural elements made of traditional materials [1–3]. The use of fiber-reinforced polymer (FRP) composite products (laminates, plates, fabrics and sheets) to repair and strengthen reinforced concrete and masonry structures is well established [4–6]. On the other hand, the application of FRP materials to steel structures remains understudied since traditional methods of repairing and strengthening are still highly used. These methods consist in either adding supplementary metallic elements to increase the cross-sectional dimensions or embedding the existing steel elements into concrete sections [7]. However, the traditional strengthening methods of steel structures have two main disadvantages. The first one refers to the fact that the added elements have considerable weight, and the installing technologies may further affect the damaged element (the case of the residual stresses that develop due to the welding process). The second disadvantage refers to the time needed to carry out the intervention works, which is often quite long, thus affecting the serviceability of the construction.

Compared to the traditional methods used for strengthening, the techniques based on FRP composite materials have remarkable advantages that arise from their superior mechanical characteristics and corrosion resistance properties [7–10]. The structural efficiency of these methods was analyzed by various research teams by conducting experimental programs based on numerical analysis methods and by adjusting and applying the existent theoretical models specific to the hybrid consolidation solutions [11–14]. However, the progress in this research area has been achieved mainly through the relatively recent development of FRP composite products that are structurally compatible with steel elements.

Most of the FRP composite products used to repair and strengthen steel elements are attached to their surfaces by an adhesive layer. The main function of the adhesives is to ensure the contact between the steel element and the FRP one, thus guaranteeing an efficient transfer of the stresses. Therefore, the interface behavior of the FRP composite products bonded to steel surfaces is a key aspect in assuring an efficient strengthening solution for structural or non-structural steel elements [15].

This paper describes an experimental and numerical study performed at the Faculty of Civil Engineering and Building Services of Iasi. The study included 34 single lap joints (SLJs) specimens, each of them being composed of a carbon fiber-reinforced polymer (CFRP) strip-bonded to a steel element. The single-lap joints are among the most common types of bonded connections, being utilized in various engineering applications [16,17]. Each specimen was instrumented so that the load, the slip between the adherents and the axial strains of the un-bounded part of the CFRP strip could be monitored during the tension tests. All specimens were tested up to failure using the load control procedure. The same typologies of bonded connections were numerically studied using the finite element method (FEM) analysis. A good agreement was obtained between the results provided by the experimental program and the ones obtained through numerical analysis.

The main objectives of the research program consist in studying, based on simplified models, the suitable configuration of bonded joints between common CFRP strips and steel elements, with respect to the type and the thickness of the adhesive. The bonding behavior is one of the key parameters that influence the structural response of steel elements which are strengthened with CFRP products.

2. Experimental Program

2.1. Specimens Description

The main aim of the carried out experimental program consists in the investigation of the specific characteristics of the interface behavior between steel elements and CFRP strips bonded with adhesives. During the design and preparation stages of the specimens, the following parameters were taken as variables: the axial stiffness of the CFRP composite strips, the moduli of elasticity and the thicknesses of the adhesives and the surface preparation method for the steel substrates. Thus, two types of CFRP composite strips (Sika Carbodur S512 and Sika Carbodur M514) [18,19] were bonded to either prepared or unprepared steel surfaces using two types of adhesives (Sikadur 30 and Sikadur 330) [20,21], having three different thicknesses (1 mm, 2 mm, 3 mm). For each combination, at least two identical specimens were prepared. For all specimens, the bond length was 350 mm, being higher than the effective one, calculated using the analytical model given by [22,23].

2.2. Materials Properties

The steel plates used to assemble the specimens were cut from a 10 mm thick plate using a high-speed circular metal sawing machine. After the cutting process, the edges of the steel plates were rectified and polished in order to obtain the design dimensions of the specimens. The geometrical features and the mechanical properties of the steel plates are given in Table 1.

Table 1. Properties of steel plates according to EN 1993-1-1:2006 [24].

Table S235.	S235 JR
Thickness, t_s (mm)	10
Width, b_s (mm)	120
Length, l_s (mm)	500
Yielding strength, $f_{y,s}$ (N/mm^2)	235

Table 1. *Cont.*

Table S235.	S235 JR
Ultimate strength in tension, $f_{u,s}$ (N/mm^2)	360
Modulus of elasticity, E_s (GPa)	210
Shear modulus of elasticity, G_s (GPa)	81
Poisson's ratio, ν_s	0.3
Thermal coefficient of expansion, α (1/°C)	12×10^{-6}

The CFRP strips used to assemble the specimens belong to the Sika's Carbodur composite products range. The selection of the two types, S512 and M514, was made by taking into account one of the main objectives of the experimental program, which is to determine the influence of the axial stiffness of the CFRP strip on the shear structural response of the bonded specimens. Thus, the two types of CFRP strips have similar tensile strength but different moduli of elasticity, namely 165 GPa for S512 and 210 GPa for M514 [18,19]. The most relevant properties of the CFRP composite strips, as provided by the manufacturer, are presented in Table 2.

Table 2. Properties of the CFRP composite strips [18,19].

Properties	Sika Carbodur S512	Sika Carbodur M514
Density (kg/m^3)	1600	
Fibre volume fraction (%)	≥68	
Thickness, $t_{CFRP,512/514}$ (mm)	1.2	1.4
Width, b_{CFRP} (mm)	50	50
Length, l_{CFRP} (mm)	550	550
Cross sectional area (mm^2)	60	70
Longitudinal modulus of elasticity, $E_{CFRP,512/514}$ (GPa) (average values)	165	210
Tensile strength, $f_{t,CFRP,512/514}$ (N/mm^2)	3100	3200
Elongation at break, $\varepsilon_{u,CFRP,512/514}$ (%) (minimum values)	>1.70	>1.35

The adhesives used for bonding the CFRP composite strips to the surface of the steel plates, Sikadur 30 and Sikadur 330, are provided by the same manufacturer and are adequate for this kind of hybrid bonding system. Both products are thixotropic, consisting of a combination of epoxy resins, additives and hardening agents. In addition, from a mechanical point of view, both adhesives have linear-elastic behavior. As it can be observed in Table 3, the tensile strengths and the ultimate strains of the adhesives have similar values, while the modulus of elasticity of Sikadur 30 is almost three times higher than that of Sikadur 330 [20,21]. The values presented in Table 3 are given by the producer, being valid for specific conditions of preparing and curing the adhesives, which were carefully ensured during the experimental program.

Table 3. Properties of the adhesives [20,21].

Type	Density (kg/m^3) (Mixed)	Compressive Strength, $f_{c,a}$ (N/mm^2)	Tensile Strength, $f_{t,a}$ (N/mm^2)	Modulus of Elasticity, E_a (GPa)	Elongation at Break, $\varepsilon_{u,adh}$ (%)
Sika 30	1650	70–80 (7 days, +10 °C)	25–28 (7 days, +15 °C)	12.8 (7 days, +15 °C)	1.0%
Sika 330	1300	30 (7 days, +23 °C)	33.8 (7 days, +23 °C)	4.5 (7 days, +23 °C)	0.9%

2.3. Experimental Set-Up

After all the necessary materials for assembling the specimens were shaped to their nominal dimensions, the first step consisting in preparing the steel surface was performed.

According to the Mechanical Theory, the degree of adhesion that may be obtained for an adhesively bonded joint is directly linked to the roughness of the substrates [25,26]. Thus, 30 of the 34 steel plates were sandblasted to increase the specific roughness of the surface that comes into contact with the adhesive [27]. The remaining four steel plates were not sandblasted so as to determine the effect of this surface treatment method. The sandblasting process was performed in a special room with a controlled environment using GL-80 steel grits made by W Abrasive that was shot onto the surface of the steel plates from a distance of 5 cm and at an inclination angle of 45° (Figure 1, Table 4) [28]. After sandblasting, the steel plates were vacuumed and wrapped in a plastic foil to prevent future contamination.

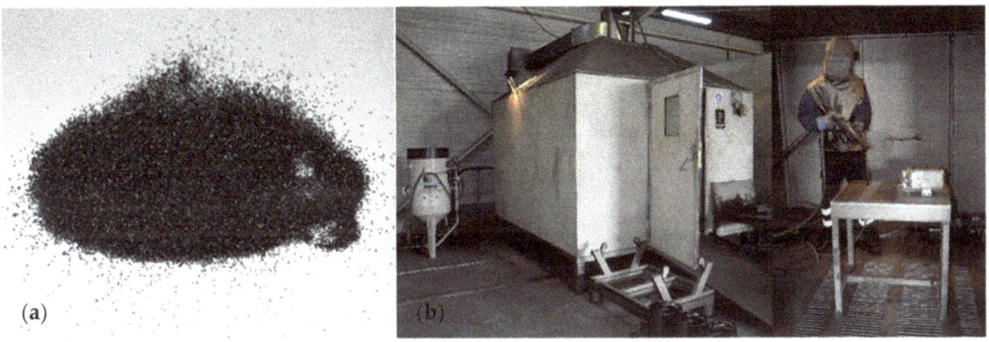

Figure 1. (**a**) GL-80 steel grits; (**b**) sandblasting chamber [28].

Table 4. GL-80 steel grits produced by W Abrasive [28].

Grading	0.180 mm—99.8% 0.425 mm—0.2%
Hardness (Rockwell scale)	55.7 HRC
Density (kg/m^3)	7600
Chemical composition	$C \geq 0.85\%/P \leq 0.05\%/S \leq 0.05\%/0.40\% < Si < 1.20\%$

The bonding of the CFRP strips was performed within a maximum of 24 h of sandblasting the steel plates. For each specimen, the CFRP strips were cleaned with a special solvent (Sika ColmaCleaner) [29] to remove all the protective coatings that could affect the quality of the bonds. The adhesive was then prepared by mixing the components, as indicated in their technical data sheets [20,21].

One of the most applied techniques of bonding composite products on different substrates recommends that the adhesive should be applied initially to the composite strip and then pressed on the base element [30,31]. However, this technique is particularly suitable for site applications where the technological phases have to be carried out within a limited time. In addition, when the bond length is relatively small (close to the effective one) and the thickness of the adhesive layer is fixed, it is difficult to determine the amount of adhesive that should be applied to the composite strip. In this case, two factors with a negative impact on the mechanical performances of the bond may occur. The first refers to a bond area that is insufficiently covered with adhesive, which generally results in a lower thickness than the one desired. The second consists in applying the excess adhesive which slumps during the pressing of the composite element on the base member. On the other hand, when adhesive joints are assembled under laboratory conditions and their geometrical tolerances are minimal, it is necessary to use specific control methods to ensure that the designed geometrical characteristics are obtained. For this purpose, in order to obtain a constant thickness of the adhesive layer, chrome-steel spherical spacers of 1, 2 or 3 mm were fixed on the surface of the steel plate, depending on the type of the specimen. These spacers do not affect the structural behavior of the joint since the voids that may be formed near them collapse to a negligible volume when the specimens are fixed with clamps. Additionally, the spherical steel spacers (and/or the calibrated wires) are recommended by practice guides as some of the most suitable methods to control the adhesive thickness [32]. A special rig was designed (Figure 2b) to control the quantity and the homogeneity of the adhesive that is applied on the surface of the CFRP strip. Figure 2a presents the laboratory setup of the rig. After the assembling stage had ended, the specimens were fixed with clamps and stored in constant temperature and humidity conditions for 14 days until the curing process was completed.

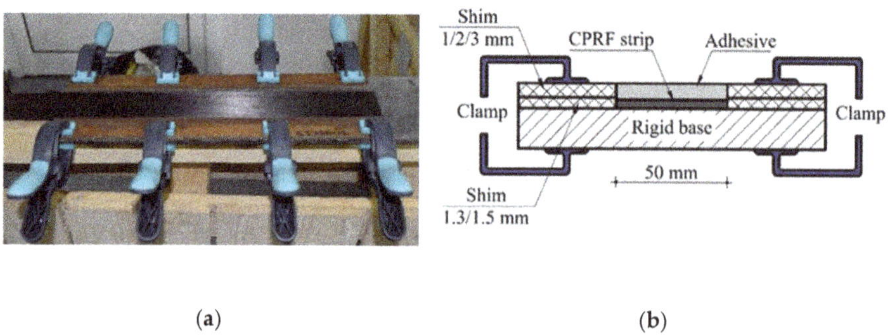

Figure 2. Rig for control of the adhesive thickness. (**a**) Laboratory setup, (**b**) Designed rig.

Each specimen was marked with a code (i.e., S1 – S512-330-1-II), indicating the type of the CFRP composite strip, the type of adhesive and its thickness. The last term stands for identifying specimens with identical configurations. The general layout of the specimens is presented in Figure 3. In Table 5, the first five columns show the configurations of the specimens, while the last two columns list their corresponding failure loads (P_{ult}) and failure mechanisms recorded during the experimental tests.

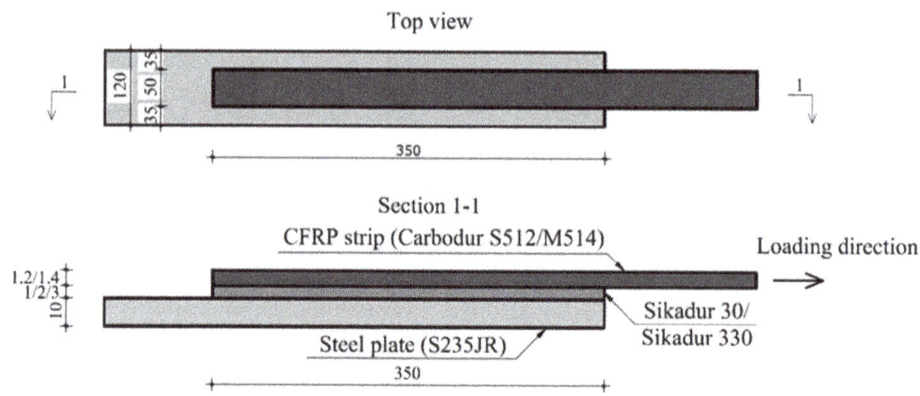

Figure 3. General layout of the specimens.

Table 5. Specimen details, ultimate loads and failure modes.

No.	Type	Adhesive Type	Adhesive Tensile Strength, $f_{t,a30/330}$ (N/mm^2)	Adhesive Modulus of Elasticity, $E_{a,30/330}$ (GPa)	Adhesive Thickness, $t_{a,30/330}$ (mm)	Ultimate Load, P_{ult} (kN)	Failure Mechanism
1	S-S512-30-1-I	Sikadur 30	25	12.80	1	42.10	C+D
2	S-S512-30-1-II	Sikadur 30	25	12.80	1	38.30	C+D
3	S-S512-30-1-III	Sikadur 30	25	12.80	1	39.40	C+D+SAI
4	S-S512-30-2-I	Sikadur 30	25	12.80	2	37.85	C+D+SAI
5	S-S512-30-2-II	Sikadur 30	25	12.80	2	38.65	C+D
6	S-S512-30-2-III	Sikadur 30	25	12.80	2	34.87	C+D+SAI
7	S-S512-330-1-I	Sikadur 330	30	4.50	1	74.80	CAI
8	S-S512-330-1-II	Sikadur 330	30	4.50	1	78.80	CAI
9	S-S512-330-1-III	Sikadur 330	30	4.50	1	48.00	CAI+D
10	S-S512-330-2-I	Sikadur 330	30	4.50	2	70.20	CAI
11	S-S512-330-2-II	Sikadur 330	30	4.50	2	61.25	CAI+SAI
12	S-M514-30-1-I	Sikadur 30	28	12.80	1	22.41	C+SAI
13	S-M514-30-1-II	Sikadur 30	28	12.80	1	51.06	C+SAI+D
14	S-M514-30-1-III	Sikadur 30	28	12.80	1	46.55	C+SAI+D
15	S-M514-30-2-I	Sikadur 30	28	12.80	2	10.00	C+D
16	S-M514-30-2-II	Sikadur 30	28	12.80	2	42.85	C+D
17	S-M514-330-1-I	Sikadur 330	30	4.50	1	80.80	SAI+CAI +D
18	S-M514-330-1-II	Sikadur 330	30	4.50	1	99.25	CAI
19	S-M514-330-1-III	Sikadur 330	30	4.50	1	85.00	CAI+SAI
20	S-M514-330-2-I	Sikadur 330	30	4.50	2	56.90	CAI+D
21	S-M514-330-2-II	Sikadur 330	30	4.50	2	53.35	CAI+SAI
22	S-S512-30-3-I	Sikadur 30	28	12.80	3	39.30	C+D+SAI
23	S-S512-30-3-II	Sikadur 30	28	12.80	3	38.75	C
24	S-S512-30-3-III	Sikadur 30	28	12.80	3	36.43	C+D
25	S-S512-330-3-I	Sikadur 330	30	4.50	3	23.10	CAI+D
26	S-S512-330-3-II	Sikadur 330	30	4.50	3	51.25	CAI
27	S-M514-30-3-I	Sikadur 30	28	12.80	3	39.80	C+D+SAI
28	S-M514-30-3-II	Sikadur 30	28	12.80	3	42.95	C+D
29	S-M514-330-3-I	Sikadur 330	30	4.50	3	49.65	CAI+D +SAI
30	S-M514-330-3-II	Sikadur 330	30	4.50	3	44.60	CAI+SAI
31	NS-S512-30-1-I	Sikadur 30	28	12.80	1	32.1	C+SAI
32	NS-S512-30-1-II	Sikadur 30	28	12.80	1	31.5	C+SAI
33	NS-M514-330-1-I	Sikadur 330	30	4.50	1	54.05	SAI+CAI
34	NS-M514-330-1-II	Sikadur 330	30	4.50	1	50.6	SAI+CAI

Failure mechanisms notations: C—Cohesive adhesive layer failure, D—CFRP strip delamination, SAI—Steel–adhesive interface debonding, CAI—CFRP–adhesive interface debonding.

2.4. Specimens Instrumentation and Loading Conditions

The parameters that were monitored during the shear pull-of tests are the applied tensile force, the relative displacements between the CFRP strip and the steel plate and the variation of the strains along the composite element. The shearing effect in the adhesive layer was obtained by loading all specimens in tension at their free ends in a Zwick/Roell 100 kN hydraulic test machine (ZwickRoell, Ulm, Germany) (Figure 4), located in the Composite Materials Laboratory of the Faculty of Civil Engineering and Building Services, Iași.

(a) (b)

Figure 4. Test machine and specimen's instrumentation. (**a**) Front view, (**b**) Lateral view.

All tests were performed under force-control by imposing a loading rate of 5 kN/min. The ultimate forces were recorded by the testing machine connected to the acquisition system. The displacement between the CFRP strip and the steel plate was recorded using a linear variation displacement transducer (LVDT) mounted on the two elements outside the bond area. The LVDTs have been attached after the specimens were fixed into the testing machine. The strains along the CFRP strips were measured by attaching 6 strain gauges on their top faces, 5 within the overlap area and 1 outside. Three specimens (S-M514-30-1-I/II/III) were instrumented with 9 strain gauges in order to investigate the variation of the strains at smaller intervals. The instrumented configuration of the specimens is presented in Figure 5.

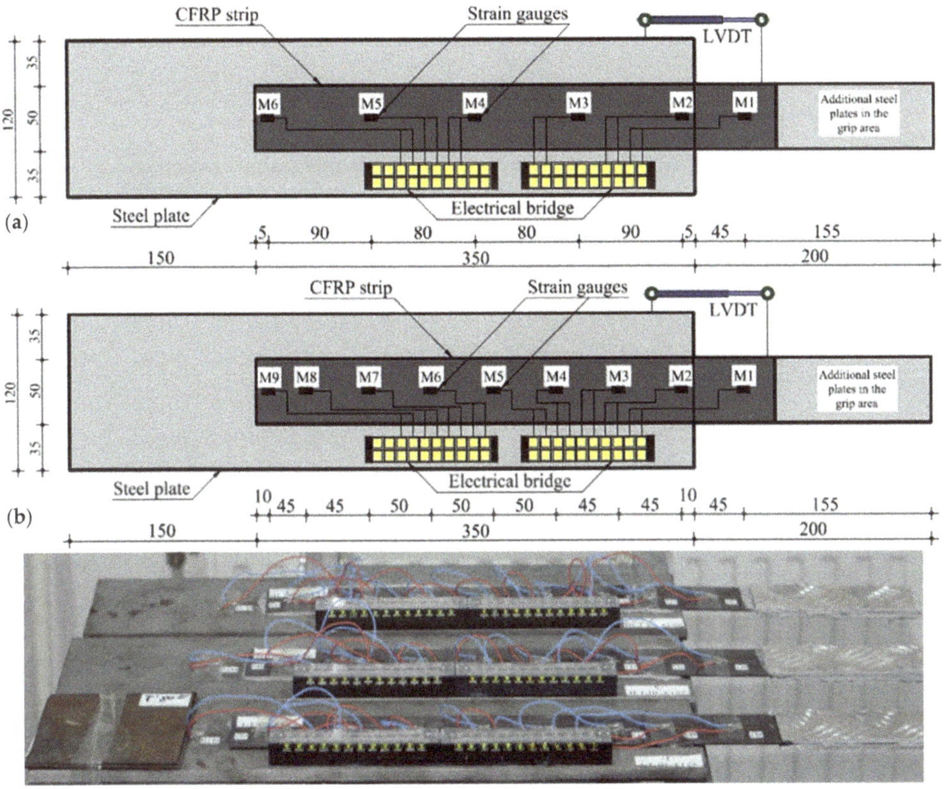

(c)

Figure 5. Specimens instrumentation with strain gauges and LVDT (dimensions in mm). (**a**) Graphical representation of specimens with 5 strain gauges, (**b**) Graphical representation of specimens with 9 strain gauges, (**c**) Real specimens with 9 strain gauges.

3. Experimental Results

All 34 specimens were loaded in tension up to failure. Based on the monitored parameters mentioned above, the following characteristics of the bonded joint were analyzed: failure modes, load–displacement behavior, strain distribution along the bond length at different loading stages, bond–slip relations and shear stress–strain variations of the CFRP strips.

3.1. Characteristic Failure Modes and the Corresponding Ultimate Loads

All failure mechanisms that were identified correspond to those already presented by other authors in previous research works [33–39]. However, only 6 of the 34 specimens failed by a single dominant mechanism, while the other 28 developed combined failure modes, characterized by two or three failure patterns. The failure mechanisms of some representative specimens are illustrated in Figure 6.

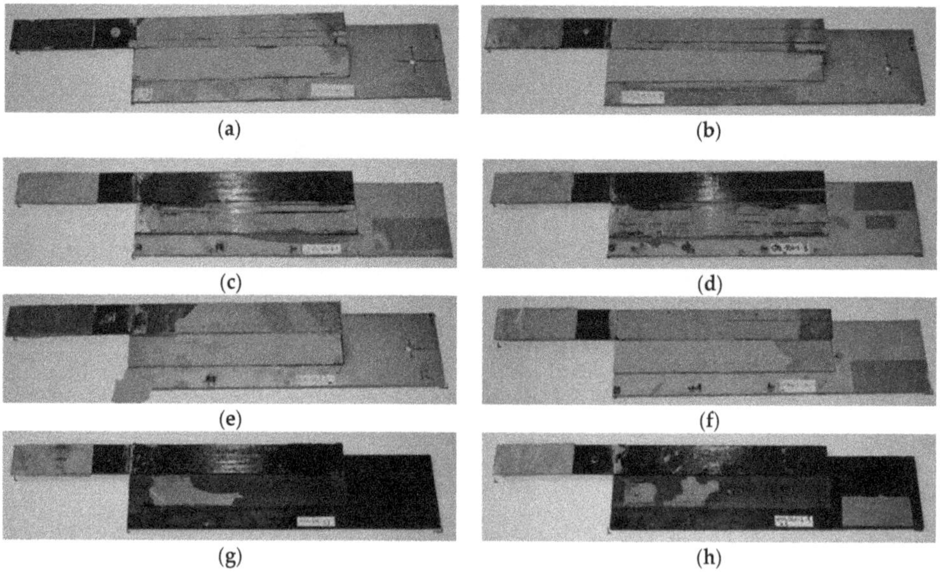

Figure 6. (**a**,**b**) Specimens 1 and 2, Failure mechanism: Cohesive adhesive layer failure (dominant) and CFRP strip delamination (secondary); (**c**,**d**) Specimens 8 and 10, Failure mechanism: CFRP–adhesive interface debonding; (**e**,**f**) Specimens 6 and 27, Failure mechanism: Cohesive adhesive layer failure (dominant), CFRP strip delamination (secondary) and steel–adhesive interface debonding (tertiary); (**g**,**h**) Specimens 33 and 34, Failure mechanism: steel–adhesive interface debonding (dominant) and CFRP–adhesive interface debonding (secondary).

For the specimens bonded with the Sikadur 30 adhesive, the dominant failure mechanism was the cohesive one, while for those bonded with Sikadur 330 adhesive, the interface debonding mostly between the CFRP strip and the adhesive was the most common failure mode.

By analyzing the failure mechanisms of the specimens with identical configurations, it can be concluded that in most cases, when the predominant failure (cohesive or debonding at the composite–adhesive interface) is accompanied by secondary modes, such as the delamination of the CFRP strip or debonding at the steel–adhesive interface, the ultimate forces decrease considerably (especially for S-S512-330 specimens, where the ultimate forces are up to 55% lower).

For the specimens bonded with Sikadur 30 adhesive, the dominant failure mechanism was the cohesive one, which had a progressive behavior, starting from the loaded end and propagating gradually towards the opposite one. For the specimens made with Sikadur 330 adhesive, no cohesive failures were recorded. The most frequent failure modes were initiated at the CFRP strip–adhesive interface and occurred in a brittle and sudden manner.

The specimens with steel surfaces that were not sandblasted failed only by interface debonding between the steel plate and the adhesive at considerably lower ultimate loads when compared to the specimens having identical configurations but mechanically treated steel surfaces.

For the specimens having S512 CFRP strips, bonded with Sikadur 30 adhesive, it was observed that the ultimate forces were not directly influenced by the variations in the thickness of the adhesive layer. Thus, the highest values of the ultimate forces were obtained for the 1 mm adhesive thickness, and those recorded for thicknesses of 2 and 3 mm were smaller by 7% and 5%, respectively. For the same specimen configurations but a different type of adhesive (Sikadur 330), the lowest values of the ultimate forces were recorded for the bonded joints with 3 mm adhesive thickness.

Moreover, as it was expected, the failure mechanism strongly influenced the ultimate forces, as significant variations of the failure forces were recorded for specimens with identical configurations but a different failure pattern. The premature delamination of the CFRP strip caused an important decrease of the ultimate force: 37.5% for the 1 mm adhesive layer and 55% for the 3 mm adhesive layer. For the specimens with 2 mm adhesive thickness, the additional failure mode, developed at the steel–adhesive interface, led to an additional 13% decrease of the ultimate force, compared to the singular mode exhibited by the specimen that failed only at the CFRP–adhesive interface.

The same remarks are valid for the specimens made with M514 CFRP composite strip and bonded with Sikadur 30 adhesive. The changes in the thickness of the adhesive layer do not produce significant variations of the ultimate forces. However, for the M514 CFRP specimens bonded with Sikadur 330 adhesive, the maximum ultimate forces were recorded for the 1 mm adhesive layer, while the specimens having 2 and 3 mm adhesive thicknesses exhibited ultimate forces that were 38% and 47% smaller, respectively.

The effectiveness of the surface-treatment method was also demonstrated through the significant differences between the ultimate forces. All specimens that were not surface-treaded failed prematurely (predominantly at the steel–adhesive interface) at ultimate forces that were 33.65%–98.1% lower than the corresponding surface-treated configurations.

3.2. *Load–Displacement Behaviour*

The load–displacement distribution curves were plotted based on the values recorded during the testing process. Figures 7–9 illustrate the load–displacement curves for some representative specimens.

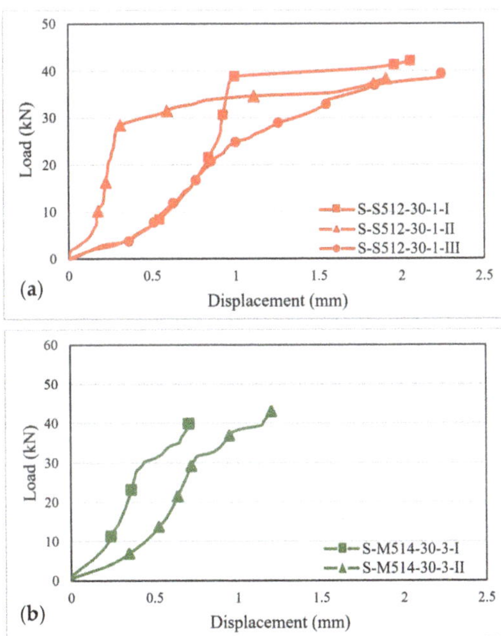

Figure 7. Load–displacement curves for: (**a**) S-S512-30-1 and (**b**) S-M514-30-3 specimens.

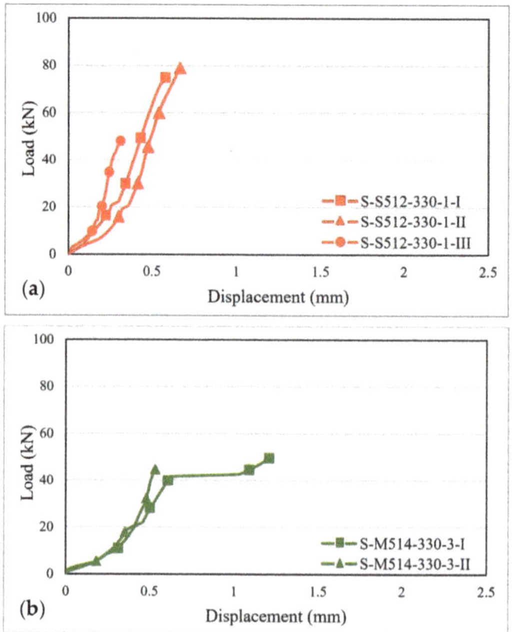

Figure 8. Load–displacement curves for: (**a**) S-S512-330-1 and (**b**) S-M514-330-3 specimens.

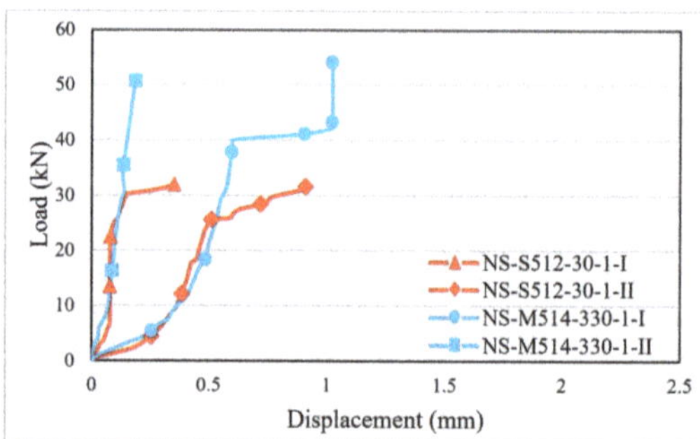

Figure 9. Load–displacement curves for specimens made with non-sandblasted steel plates.

As it can be observed in Figure 7a, for the S-S512-30 samples, two types of behavior characterized by different shapes and slopes of the load–displacement curves were obtained. In the first case, the curve displays an upward plateau followed by a second region of constant forces but rapidly increasing displacements. In the second case, the initial stiffness of the joint is smaller, and the slope of the curve does not undergo substantial changes during the loading stage. For specimens of the S-M514-30 type, the superior stiffness of the CFRP strip led to the diminishing of the ultimate displacements, which had maximum values of about 1.3 mm. The changes in the slope of the load–displacement curves can be correlated with the local failures that occurred along the bond length or with the transition from one failure mechanism to another.

For specimens of the S-S512-330 series (Figure 8a), the load–displacement curves are characterized by a single upward plateau with a slope that is significantly higher when compared to the one of the specimens of the S-S512-30 series. The ultimate displacements are in the range of 0.55–0.65 mm and do not vary significantly with respect to the thickness of the adhesive.

For the S-M514-330 specimens (Figure 8b), the load–displacement curves have similar shapes and slopes to those belonging to the S-S512-330 series. However, it can be observed that the maximum displacements decrease with respect to the stiffness of the CFRP strip, especially for the specimens bonded with the 1 mm adhesive layer, where the maximum displacements ranged between 0.3–0.35 mm.

For the specimens made with non-sandblasted steel plates (Figure 9), the load–displacement curves are similar to those of the corresponding surface-treated specimens. However, due to the premature interface debonding, the curves are characterized by only one ascending branch that stops at maximum displacements of 1 mm ended by the premature debonding. This proves that by sand-blasting the contact surface of the steel plates, the mechanical interlocking between the adhesive and the steel is significantly improved, especially due to the higher specific contact surface area.

3.3. Strain Distribution

The variation of the strains distribution along the bond length is one of the most important parameters in describing the specific characteristics of the bonding mechanism between steel elements and FRP composite strips [40]. During the experimental tests, the strains' variations were recorded using the strain gauges fixed on the surface of the CFRP profiles. These values were later compared to the ones provided by the numerical analysis of the models. Figures 10–12 show the strain distribution curves along the bond length at different load application rates for some representative specimens. The intervals between the fractions of the ultimate force at which the strain distribution curves were drawn are greater in the range 0.1–0.7 P_{ult} and smaller in the 0.85 P_{ult} - P_{ult} range in order to accurately identify the changes that occur before the failure of the specimens.

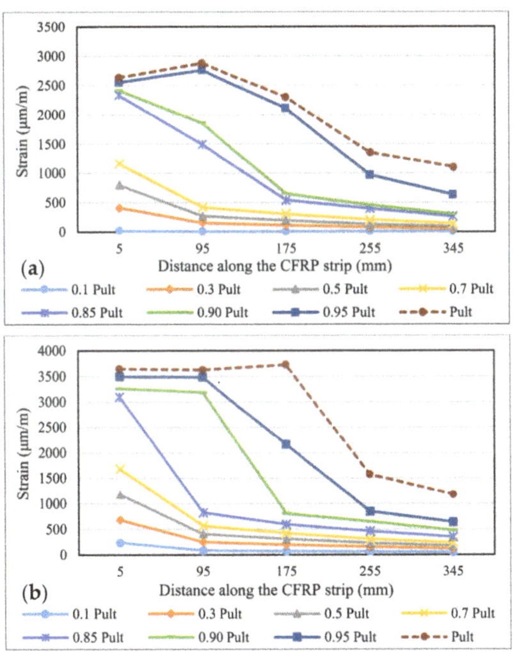

Figure 10. Strains distribution along the bond length for specimens: (**a**) S-S512-30-2-I and (**b**) S-M514-30-1-II.

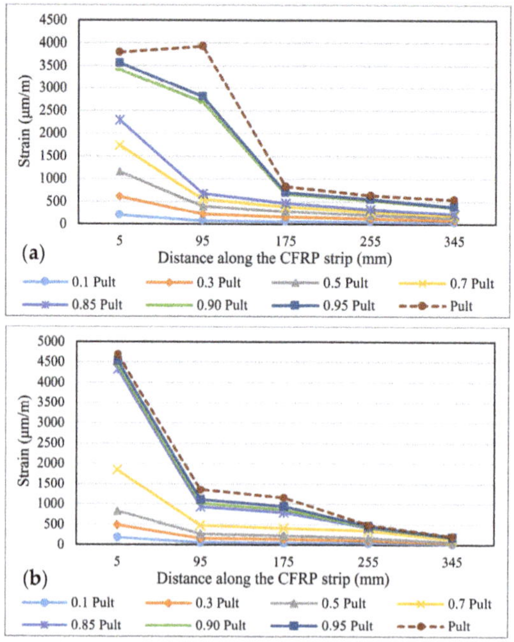

Figure 11. Strains distribution along the bond length for specimens: (**a**) S-S512-330-1-I and (**b**) S-M514-330-3-I.

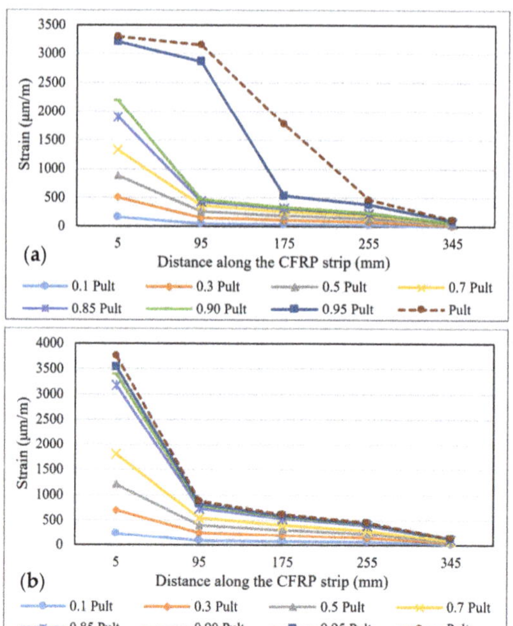

Figure 12. Strains distribution along the bond length for specimens: (**a**) NS-S512-30-1-I and (**b**) S-M514-330-3-I.

By comparing the strain distribution curves obtained for the specimens made with the Sikadur 30 adhesive to those of the specimens made with Sikadur 330, it is observed that in the first case, the values are much more scattered, thus demonstrating the progressive behavior of the cohesive failure mechanism. Furthermore, the ultimate values of the strains recorded for the specimens made with Sikadur 30 are around 4000 μm/m, while the ones recorded for the joints made with Sikadur 330 are greater, being close to 8000 μm/m for the S-S512-330-1 series.

It can be observed that with the increase in the axial stiffness of the composite strips, in the case of specimens made with M514 strips, the plateau of constant strain distributions extends more towards the free end of the composite elements, thus mobilizing a larger portion of the bond length. However, with respect to the thickness of the adhesive layer, no significant variations in the distribution and ultimate values of the strains were observed.

By comparing the strain distributions of the specimens made with non-sandblasted steel plates (NS-S512-330-1 and NS-M514-330-1 series) with the ones obtained for specimens with identical configuration, but with steel-treated surfaces (S-S512-30-1 and NS-M514-330-1 series), it can be concluded that for the first, the bond length is mobilized in a lower fraction, which is demonstrated by the narrow area on which the range of constant strain distribution extends.

3.4. Shear Stress–Displacement (Slip) Distributions

The shear stress–displacement curves, also referred to as bond–slip curves, establish a direct relationship between the values of the shear stresses and the corresponding displacements between the steel plate and the CFRP strip calculated at various key points located along the bond line [41]. In the case of SLJ specimens, the values of these two parameters can be obtained based on the strain variations recorded by the strain gauges by applying Equations (1) and (2) for the intervals between the positions of the strain gauges (Δ1–Δ5).

$$\tau_{i/2} = \frac{(\varepsilon_i - \varepsilon_{i+1})}{(L_{i+1} - L_i)} E_{CFRP} t_{CFRP} \quad (1)$$

$$\delta_{i/2} = \frac{(\varepsilon_i - \varepsilon_{i+1})}{4}(L_{i+1} - L_i) + \sum_{i-1}^{n} \frac{(\varepsilon_{i+1} + \varepsilon_{i+2})}{2}(L_{i+2} - L_{i+1}) \quad (2)$$

where:

ε_i = the strain value recorded by the strain gauge "*i*";

n = number of strain gauges;

L_i = the distance measured from the loaded end to the strain gauge "*i*";

E_{CFRP},$_{FRP}$ = elasticity modulus and width of the CFRP strip;

$\tau_{i/2}$ = the shear stress value computed at the middle distance between the strain gauge "*i*" and "*i*+1";

$\delta_{i/2}$ = the displacement value computed at the middle distance between the strain gauge "*i*" and "*i*+1".

Figures 13–15 show the shear stress–displacement (bond–slip) curves for some representative specimens. The main conclusion resulting from the analysis of the shear stress–displacement curves is that the shape and the slope of the curves are directly influenced by the adhesive characteristics and by the particularities of the dominant failure mode. In the case of specimens made with Sikadur 30 adhesive, the graph has a bi-linear pattern, characterized by an upward plateau and a descending one that is initiated immediately after reaching the ultimate shear stress. This behavior is common for the adhesive joints that fail in a cohesive manner since this specific failure mode is characterized by progressive development of stresses (mainly shear) up to the ultimate limit that corresponds to the shear failure of the adhesive.

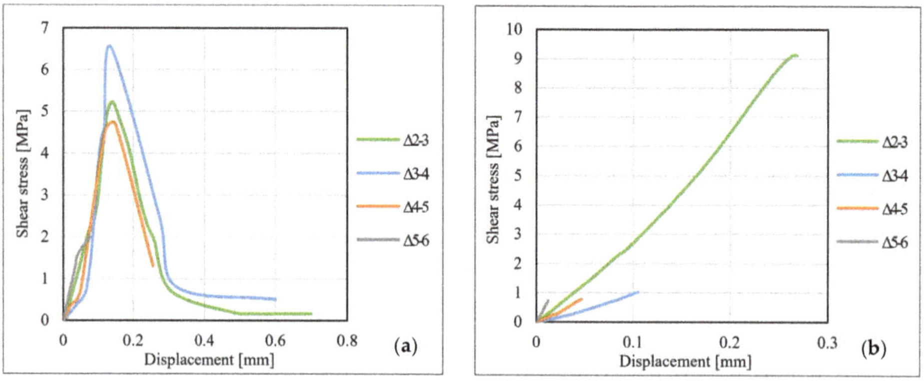

Figure 13. Bond–slip curves for specimens: (**a**) S-S512-30-2-I and (**b**) S-S512-330-1-II.

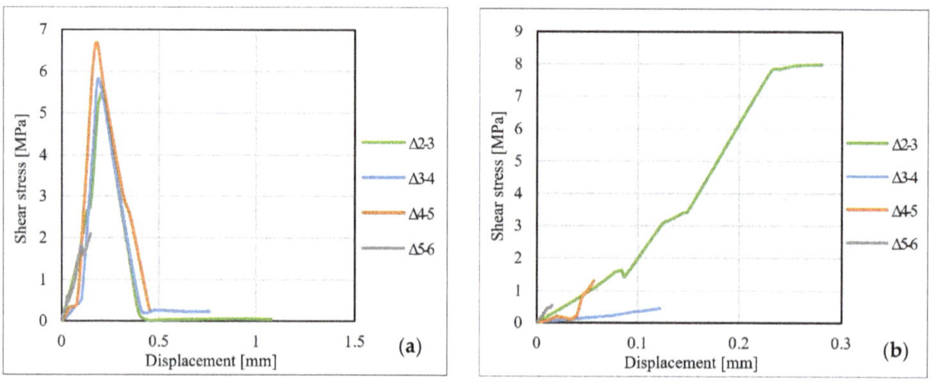

Figure 14. Bond–slip curves for specimens: (**a**) S-M514-30-3-I and (**b**) S-M514-330-3-I.

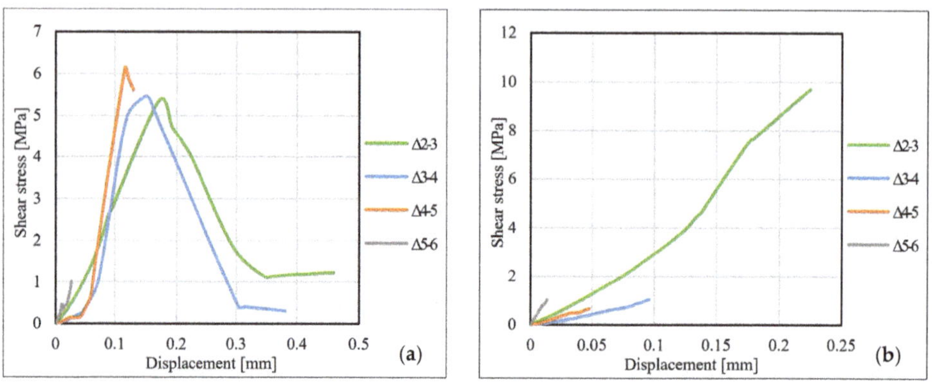

Figure 15. Bond–slip curves for specimens: (**a**) NS-S512-30-1-II and (**b**) NS-M514-330-1-I.

For the specimens made with Sikadur 330 adhesive, in most of the cases, the shear stress–displacement curves are characterized by a single upward plateau. The failure mechanisms that were identified in the case of these joints occurred mostly at the adhesive–CFRP strip interface in a brittle and sudden manner (non-cohesive).

For the specimens made with non-sandblasted steel plates, the shear stress–displacement curves kept the same shape as those of the other specimens. For joints made with adhesive Sikadur 30, premature failure of the steel–adhesive interface led to a weaker mobilization of the bond length. The curves corresponding to the values recorded for the strain gauges no. 4–6 are suddenly interrupted, as can be observed in Figure 15b. The ultimate values of the shear stress for the specimens made with non-sandblasted steel plates are lower than of the corresponding specimens made with sandblasted steel plates, by approximately 28% for the S512 CFRP strip and 30% in the case M514 CFRP strip, respectively.

3.5. Stress–Strain Distributions for CFRP Strips

The strain variations in the free area of the CFRP strip were monitored during the loading stage using strain gauge no. 1 fixed to the upper side of the strip. The stresses were computed by dividing the load to the cross-sectional area of the CFRP strip, which has values of 60 mm^2 for those of type S512 and 70 mm^2 for those of type M514. Figures 16 and 17 illustrate the stress–strain curves for some representative specimens.

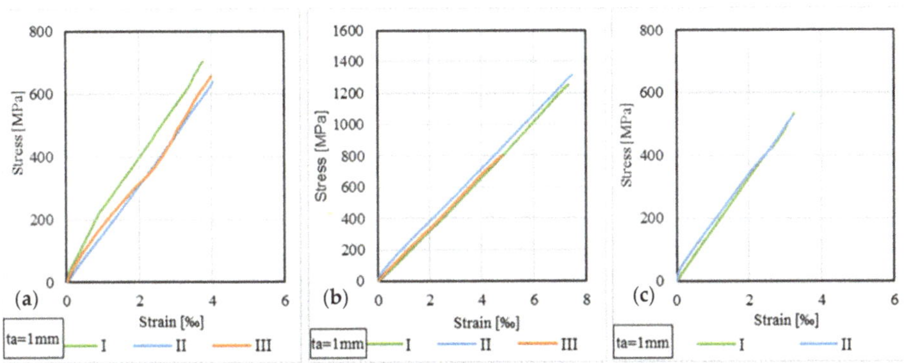

Figure 16. Stress–strain curves for S512 strip: (**a**) S-S512-30-1-I, II, III; (**b**) S-S512-330-1-I, II, III; (**c**) NS-S512-30-1-I, II.

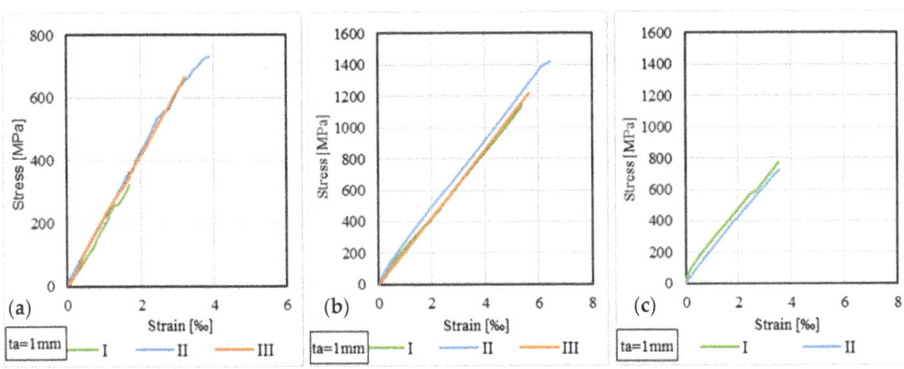

Figure 17. Stress–strain curves for M514 strip: (**a**) S-M514-30-1-I, II, III; (**b**) S-M514-330-1-I, II, III; (**c**) NS-M514-30-1-I, II.

By analyzing the stress–strain distributions at the level of the CFRP strips, it can be observed that, in general, the specimens that have the same configuration (i.e., CFRP strip, adhesive type and thickness), the stress–strain curves have similar shapes and ultimate values. For example, in the case of S512 CFRP strips, the maximum tensile stresses recorded in the free (unbounded) area of the CFRP are considerably smaller than the tensile strength of the adhesives (21% for specimens bonded with Sikadur 30 and 42% for Sikadur 330). The same remark is also valid for M514 CFRP strips, where the maximum tensile stresses were 23% and 44% of the tensile strength of Sikadur 30 and Sikadur 330 adhesives, respectively.

4. Numerical Modelling

4.1. Finite Element Model Description

The specimens were modeled using ANSYS Workbench software, the Static Structural module [42]. The 3D models were built as solid bodies, and the contact between the elements was considered as fully bonded using three interconnected primitives of parallelepiped shapes with eight nodes. Each node is characterized by three degrees of freedom. The parameters of connectivity and position were defined for each node to match the SLJ geometric configurations.

The adequacy of the finite element mesh was verified by the authors by progressively increasing the mesh density until the stresses, strain and displacements values converged by 1%. Thus, the final models were meshed using tetrahedral elements of 0.15 mm for the adhesive layer and hexahedral elements of 0.25 mm and 2.0 mm for the CFRP strip and steel plates, respectively. The size of the finite elements assigned for the steel plates was selected based on the results provided by the preliminary analysis, which had shown that no significant effects were produced at this level. A mesh refinement with the level set to 0.1 (0.1 mm maximum length of the mesh elements) was performed for the edges of the bond area since these regions are susceptible to stress concentrations [43,44]. The transition between the meshing areas with different elements and distinct refinement levels has been progressively achieved, with a reduced gradient, through specific zones, also called smooth transition regions. By modeling these regions, the risk of developing discontinuities in the mesh network is avoided. Additionally, by applying a smooth mesh and a fine refinement level, it is ensured that any variation in stresses and strains in the adhesive layer will be accurately determined [45]. Figure 18 presents the meshed models at the extremities of the bond length.

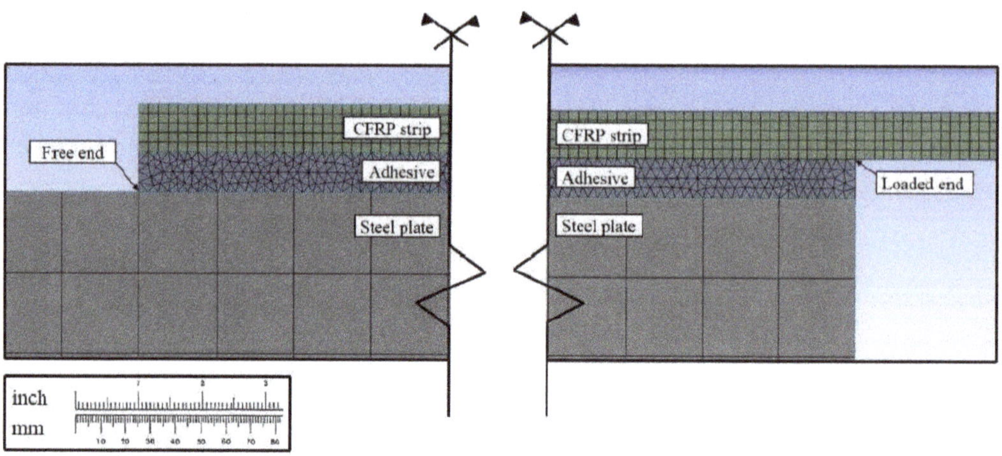

Figure 18. FE discretization at the extremities of the bond length.

The metallic plates were modeled as S235 steel grade, with mechanical properties identical to those given in Eurocode 3 - EN 1993-1-1 (2006) [24]. The CFRP composite profiles were modeled as being linear-elastic, and orthotropic materials and their mechanical properties were applied based on the values presented in Table 2. The epoxy Sikadur 30 and Sikadur 330 adhesives were assumed to be linear-elastic, and isotropic materials and their physical and mechanical properties were taken from Table 3.

The boundary conditions were assessed in order to simulate the ones applied in the experimental program. Each model was considered to be fixed at the left bottom extremity, and the tensile load was applied at the opposite free end of the CFRP strip. The fracture criterion approach, valid for singular failure mechanisms, was not considered suitable to be applied in the numerical simulation since the failures that were recorded during the experimental program consisted of combined mechanisms and patterns at both the interface and adhesive levels. Therefore, the numerical models were loaded with the corresponding ultimate forces obtained during the experimental tests so as to determine the consistency between the experimental and numerical results with respect to the investigated parameters.

4.2. Results

The models were loaded in tension, and the variations in total displacement, shear stresses and strain along the bond area were monitored and compared to the experimental values. Figure 19 shows the total displacement maps for two representative specimens at the level of the CFRP strip.

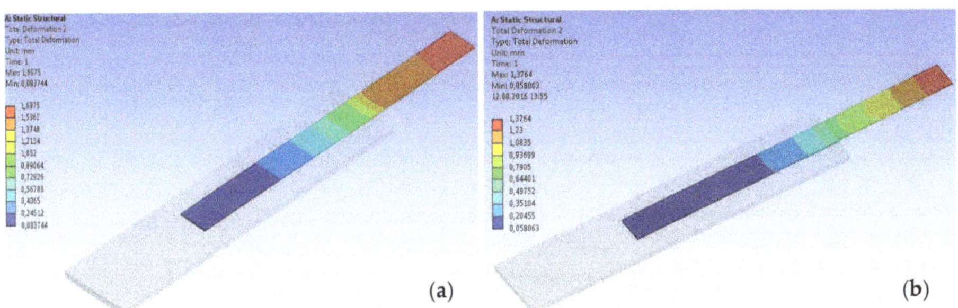

Figure 19. Displacement map at the level of the CFRP strip: (**a**) S-S512-30-1 (**b**) S-M514-30-2.

The displacement variations obtained through the numerical analysis were compared to those recorded during the experimental testing. As can be observed from Figure 20, the ultimate values obtained through both analysis approaches are in good agreement. However, the numerical approach provides a linear load–displacement curve without displaying the changes in stiffness produced by local, intermediate debonding failures. Moreover, based on the experimental load–displacement curves, it can be observed that the specimens exhibited a local non-linear behavior contrarily to the linear elastic behavior provided by the manufacturer of the adhesives [20,21].

Based on the results of numerical modeling, the distributions of the shear stresses along the bond area were investigated. These distributions are presented in the form of chromatic maps, for two representative specimens (Figure 21), at the level of the adhesive layer.

By analyzing the shear stress variations based on the chromatic maps presented above, it can be seen that the maximum values developed near the loaded end, and their intensity gradually decreases towards the free end. Moreover, the ultimate values of the shear stresses, determined through experimental and numerical analysis approaches, are in good agreement.

Another parameter that was investigated based on the numerical analysis approach is the strain distributions along the contact area between the CFRP strips and the steel plates. These distributions are presented in the form of chromatic maps for two representative specimens (Figure 22) at the level of the CFRP strip. Additionally, Figure 23 presents the graphical representation of the strain distribution for both experimental and numerical approaches.

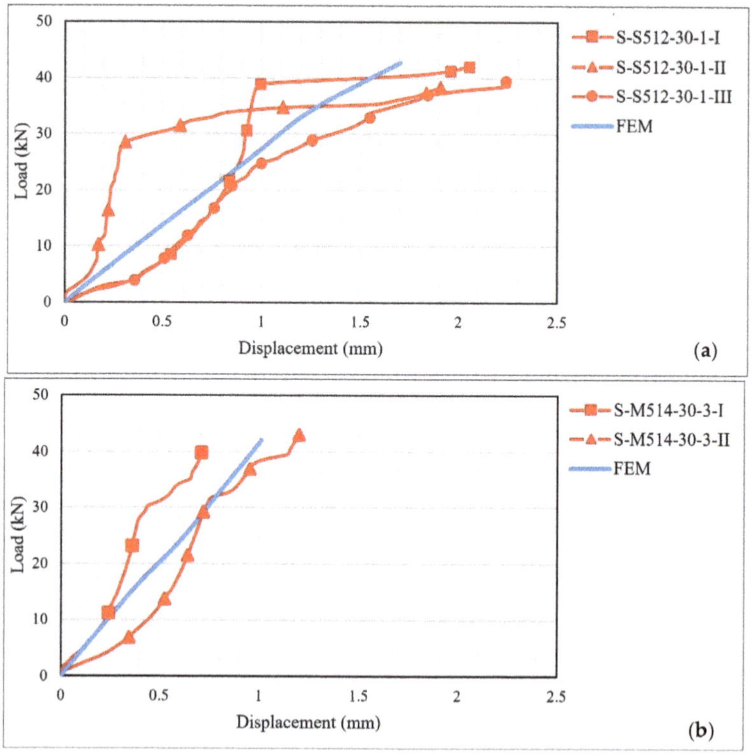

Figure 20. Load–displacement curves for specimens: (**a**) S-S512-30-1 and (**b**) S-M514-330-3 types.

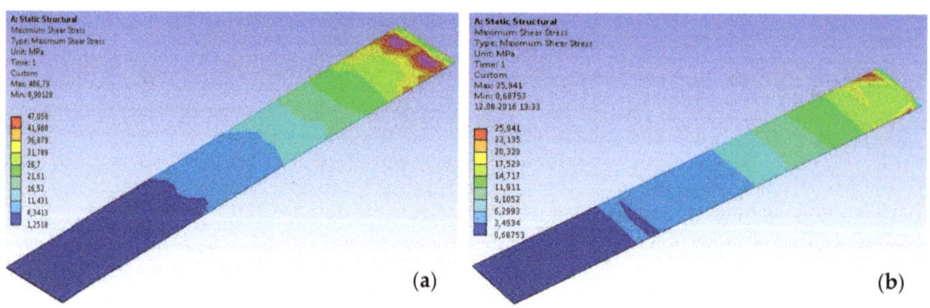

Figure 21. Typical shear stress distribution maps at the level of the adhesive layer: (**a**) S-S512-30-1 and (**b**) S-M514-30-1 types (only the adhesive layer is presented).

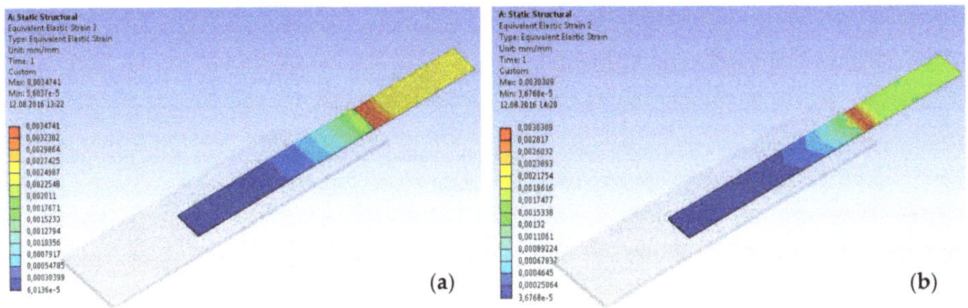

Figure 22. Typical strain distribution maps at the level of the CFRP strip: (**a**) S-S512-30-3 and (**b**) S-M514-30-3 types.

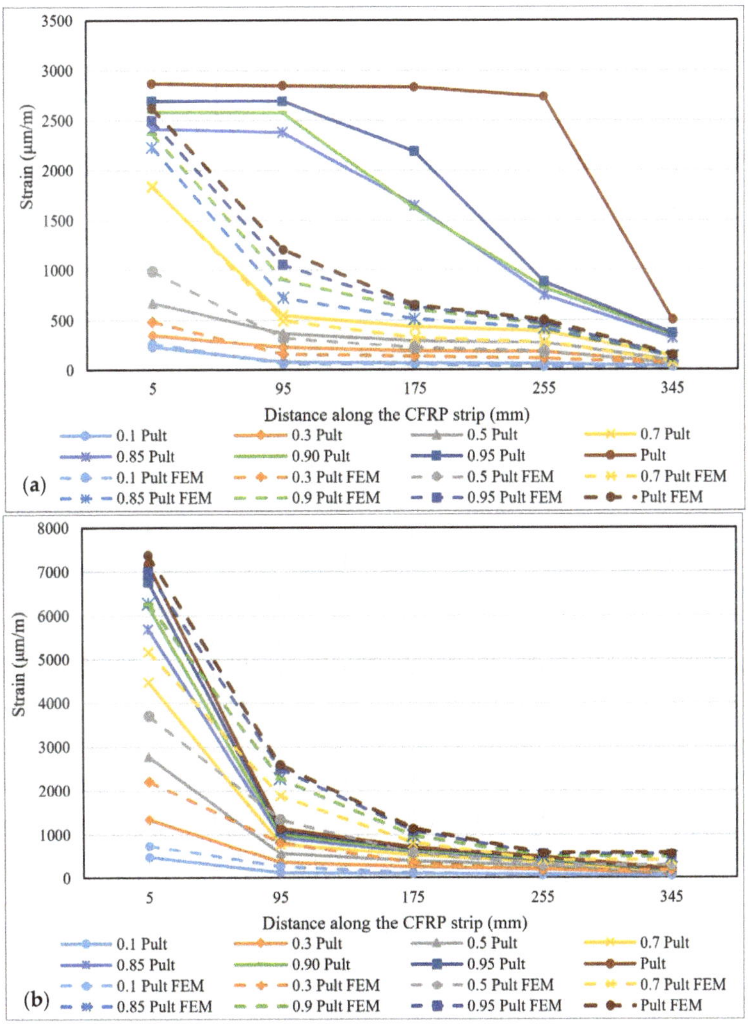

Figure 23. Strain distributions along the contact area between the CFRP strips and the steel plates for specimens: (**a**) S-M514-30-2 and (**b**) S-S512-330-1.

The comparative analysis of the strain distributions along the CPAF strip at different loading steps shows that for both the specimens made with adhesive Sikadur 30 and the specimens made with adhesive Sikadur 330, the ultimate strains calculated through numerical modelling have similar values to those obtained by experimental testing. Moreover, the shape and the slopes of the curves are similar in both cases. However, the local, intermediate failures that were recorded during the experimental tests were not precisely captured by the numerical simulation.

5. Discussions

Based on the results obtained by applying both experimental and numerical analysis approaches, the following aspects can be commented:

- The failure modes recorded during the experimental program correspond to the typologies already presented in previous research papers. The following failure typologies were identified: cohesive (shearing of the adhesive layer), CFRP strip delamination, steel –adhesive interface debonding and CFRP–adhesive interface debonding. However, only six specimens failed under a singular failure pattern, while the other 28 specimens failed under a combination of two or three failure mechanisms at both the interface and adhesive layers.
- By analyzing the variation of the ultimate forces, it was observed that for both S512/M514–30 and S512/M514–330 series, by increasing the thickness of the adhesive layer from 1 mm to 2 mm or 3 mm, the ultimate forces decrease up to 47% (the case of the S-M514-330-3 specimen).
- The load–displacement response is characterized by two distinct patterns. The first pattern, specific to the cohesive failures, consists of two distinct branches: one of constant variation between load and displacement, demonstrating the initial linear behavior, and the other, where the load is relatively constant, but the displacements are rapidly increasing. The second pattern, specific to failures at the interface levels, is characterized by a single branch of linear behavior interrupted by the sudden and brittle failure. This response was commonly specific to the specimens bonded with Sikadur 330 adhesive. The maximum displacements are directly influenced by the axial rigidity of the CPRF strip and by the thickness of the adhesive layer since the maximum displacements were recorded for the S512 strip and Sikadur 30 adhesive with 3 mm thickness.
- By analyzing the shear stresses–displacement (bond–slip curves) distributions, it can be observed that both for the joints made with S512 CFRP strips and for those made with M514 CFRP strips, the displacements corresponding to the ultimate shear stresses are varying with respect to the type and the thickness of the adhesive. Thus, in most of the cases, the ultimate values of the shear stress and displacements decrease with the increase in the adhesive layer thickness. Additionally, the ultimate values corresponding to the specimens bonded with Sikadur 330 adhesive are significantly higher than those of the specimens made with Sikadur 30 adhesive. However, the bond–slip curves can be used as a design tool only for the cohesive failures since the latter is controlled by the shear strength of the adhesive.
- By analyzing the strain variations along the bond line at different loading steps, it was observed that their distributions were initially concentrated near the loaded end. However, as the load increases, local debonding of the CFRP strip occurred and the strains distribution extended along the bond length. The limited extent of the strains demonstrates that the adhesive bonds are characterized by an effective bond length.
- Based on the numerical analysis of the finite element models, the chromatic maps were obtained in relation to the total displacement, shear stresses and shear strains. In addition, the strain distribution along the bond line at different loading steps was investigated. The numerical results are generally in good agreement with the experimental ones. Thus, the values of the ultimate displacements obtained by the two approaches are matched, both of them decreasing in relation to the axial rigidity of

the CFRP strips. However, the non-linear behavior, characterized by the experimentally obtained distributions, cannot be simulated by numerical modelling since this approach involves a linear analysis. The shear stresses and strain distributions that were numerically computed confirmed that the bond length is loaded only within a limited region, validating the previous theories referring to the effective bond length. Thus, the maximum values develop near the loaded end of the CFRP strip, and their intensity gradually decreases towards the free end.

- Based on the results that were obtained through this combined experimental and numerical program, it can be concluded that the variation in the axial rigidity of the CFRP strip does not produce significant changes in the general behavior of the bonded system. On the other hand, the elastic and geometrical properties of the adhesive have a substantial impact on the structural response of the bond. The cohesive failures were obtained only for Sikadur 30 adhesive, characterized by a higher modulus of elasticity, while all specimens bonded with Sikadur 330, with a modulus of elasticity almost three times smaller than Sikadur 30, failed prematurely, by interface debonding or by delamination. This demonstrates that the shear strength of the adhesive is only partially used. Moreover, for SLJ bonded with adhesives having lower stiffness, the adhesion at the CFRP–adhesive interface should be further improved in order to avoid premature debonding. Generally, by increasing the thickness of the adhesive, the ultimate tensile forces decrease, probably due to the changes in the stress state, when the peeling stresses lead to premature debonding.

6. Conclusions

This paper presents the outcomes of a complex experimental and numerical study referring to the stress–strain behavior and failure particularities of single lap joints (SLJs) specimens composed of CFRP strips bonded to steel elements.

By analyzing different geometrical and constitutive configurations of the bonded specimens, it has been observed that the failure mechanisms are of critical importance when assessing the efficiency of the joint. The most suitable one consists of cohesive failure since it mobilizes the largest percentage of the adhesive's strength. However, the cohesive failures were obtained only for the adhesive with a high modulus of elasticity, while in the case of the other type of adhesive, having a modulus of elasticity almost three times smaller, all failures were premature by either debonding or delamination.

As was expected, the axial rigidity of the CFRP element does not produce important changes in the structural response of the bonded joint since it is mostly controlled by the adhesive's characteristics and by the bonding effect. However, the adhesion mechanism is strongly influenced by the degree of preparing the contact surface. Thus, the sand-blasting method of surface treating the steel plates was efficient and produced a higher specific contact surface area.

The results of the numerical analysis were in generally good agreement with the experimental ones. The non-linear real behavior of the adhesives and the localized debonding effects that lead to premature failures cannot be accurately simulated by linear analysis. However, the distribution of the shear stresses and strains that were numerically computed confirmed that the bond length is loaded only within a limited region, thus validating the previous theories referring to the effective bond length.

Author Contributions: Conceptualization, N.Ţ.; methodology, V.L.; validation, P.M. and R.L.; formal analysis, D.U. and V.L.; investigation, D.N.I. and D.U.; resources, V.L. and R.L.; data curation, V.L.; writing—original draft preparation, D.U.; writing—review and editing, V.L., N.Ţ. and D.N.I.; supervision, N.Ţ.; project administration, D.U. and V.L. All authors have read and agreed to the published version of the manuscript.

Funding: This research was funded by "Ecoinnovative Products and Technologies for Energy Efficiency in Constructions—EFECON", project ID P_40_295/105524, a program co-financed by the European Regional Development Fund through Operational Program Competitiveness 2014–2020.

Institutional Review Board Statement: Not applicable.

Informed Consent Statement: Not applicable.

Data Availability Statement: The data underlying this article will be shared on reasonable request from the corresponding author.

Conflicts of Interest: The authors declare no conflict of interest.

References

1. Justo, J.; Paris, F. Experimental mechanical characterization of composite-concrete joints. *Compos. Part B Eng.* **2018**, *154*, 148–156. [CrossRef]
2. Gil, L.; Bernat-Maso, E. Experimental and analytical flexural performances of reinforced concrete beams strengthened with post-tensioned near surface mounted basalt composite laminates. *Compos. Part B Eng.* **2019**, *157*, 47–57. [CrossRef]
3. Shalaby, H.A.; Maha, M.H.; Safar, S.S. Parametric study of shear strength of CFRP strengthened end-web panels. *Steel Comp. Struct.* **2019**, *31*, 159–172.
4. Țăranu, N.; Banu, C.; Oprișan, G.; Budescu, M.; Munteanu, V.; Ioniță, O. Tensile characteristics of glass fibre reinforced polymeric bars. *Rom. J. Mater.* **2010**, *40*, 323–331.
5. Oprișan, G.; Țăranu, N.; Munteanu, V.; Budescu, M.; Cozmanciuc, C.; Oltean, R. Improvement of concrete strength through confining with composite membranes. *Rom. J. Mater.* **2011**, *41*, 302–315.
6. Kazem, H.; Rizkalla, S.; Kobayashi, A. Shear strengthening of steel plates using small-diameter CFRP strands. *Compos. Struct.* **2018**, *184*, 78–91. [CrossRef]
7. Siwowski, T.W.; Siwowska, P. Experimental study on CFRP-strengthened steel beams. *Compos. Part B Eng.* **2018**, *149*, 12–21. [CrossRef]
8. Bejan, L.; Țăranu, N.; Sârbu, A. Advanced polymeric composites with hybrid reinforcement. *J. Optoelectron. Adv. Mater.* **2010**, *12*, 1930–1934.
9. Wang, H.T.; Wu, G. Bond-slip models for CFRP plates externally bonded to steel substrates. *Compos. Struct.* **2018**, *184*, 1204–1214. [CrossRef]
10. Bejan, L.; Sârbu, A.; Țăranu, N. Component properties influence upon the elasticity modules of the cloth-reinforced polymeric composites. *Revista de Materiale Plastice* **2007**, *44*, 22–25.
11. Ghiga, D.A.; Țăranu, N.; Ungureanu, D.; Isopescu, D.N.; Oprișan, G.; Hudișteanu, I. A detailed micro-modelling approach for the diagonal compression test of strengthened stone masonry walls. *IOP Conf. Ser. Mater. Sci. Eng.* **2020**, *916*, 012041. [CrossRef]
12. Hartanto, D.; Hamed, E.; Agarwal, A.; Foster, S.J. Behaviour of steel-CFRP lap joints under hygrothermal cycles and sustained loadings. *Compos. Struct.* **2018**, *203*, 740–749. [CrossRef]
13. Ghiga, D.A.; Țăranu, N.; Ungureanu, D.; Isopescu, D.N.; Scutaru, M.C.; Hudișteanu, I. Numerical modelling of structural behaviour of URM panels strengthened with cement matrix composites. *IOP Conf. Ser. Mater. Sci. Eng.* **2019**, *591*, 012037. [CrossRef]
14. Hudișteanu, I.; Țăranu, N.; Isopescu, D.N.; Bejan, L.; Axinte, A.; Ungureanu, D. Improving the mechanical properties of composite laminates through the suitable selection of the corresponding materials and configurations. *Rom. J. Mater.* **2017**, *47*, 252–266.
15. Al-Emrani, M.; Linghoff, D. Performance of steel beams strengthened with CFRP laminate—Part 2: FE analyses. *Compos. Part B Eng.* **2010**, *41*, 516–522.
16. Ungureanu, D.; Țăranu, N.; Isopescu, D.N.; Lupășteanu, V.; Mihai, P.; Hudișteanu, I. Analytical and numerical study of adhesively bonded composite pultruded elements. *Rom. J. Mater.* **2017**, *47*, 522–531.
17. Lupășteanu, V.; Țăranu, N.; Mihai, P.; Oprișan, G.; Lupășteanu, R.; Ungureanu, D. Behaviour of CFRP-to-steel interfaces in adhesively bonded joints. *Rom. J. Mater.* **2016**, *46*, 512–522.
18. Sika CarboDur S, Product data sheet – CARBODUR S Section, Identification no. 02 04 01 01, 01.2008, Brasov, Romania. Available online: https://rou.sika.com/content/dam/dms/ro01/f/Lamele%20Sika%20CarboDur.pdf (accessed on 5 May 2021).
19. Sika CarboDur M, Product data sheet – CARBODUR M Section, Identification no. 02 04 01 01, 01.2008, Brasov, Romania. Available online: https://rou.sika.com/content/dam/dms/ro01/f/Lamele%20Sika%20CarboDur.pdf (accessed on 5 May 2021).
20. Sikadur 30, Product data sheet, Identification no: 02 04 01 04 001 0 000001, 2008, Brasov, Romania. Available online: http://www.concordoptim.ro/wp-content/uploads/2013/08/Sikadur%C2%AE-30-Adeziv-pentru-lamele-din-fibre-de-carbon.pdf (accessed on 5 May 2021).
21. Sikadur 330, Product data sheet, Identification no: 02 04 01 04 001 0 000006, 2007, Brasov, Romania. Available online: https://rou.sika.com/dms/getdocument.get/9e195986-57ec-3922-8ab3-562c82a20a26/Sikadur%20300.pdf (accessed on 5 May 2021).
22. Xia, S.H.; Teng, J.G. Behaviour of FRP-to-steel bonded joints. *Proc. Int. Symp. Bond. Behav. FRP Struct. BBFS* **2005**, 411–418.
23. Fernando, N.D. Bond Behaviour and Debonding Failures in CFRP-Strengthened Steel Members. Ph.D. Thesis, The Hong Kong Polytechnic University, Department of Civil and Structural Engineering, Hong Kong, 2010.
24. *EN 1993-1-1:2005 Design of steel structures—Part 1-1: General rules and rules for buildings*; European Committee for Standardization: Brussels, Belgium, 2005.

25. Ungureanu, D.; Ţăranu, N.; Lupăşteanu, V.; Isopescu, D.N.; Oprişan, G.; Mihai, P. Experimental and numerical investigation of adhesively bonded single lap and thick adherents joints between pultruded GFRP composite profile. *Compos. Part B Eng.* **2018**, *146*, 49–59. [CrossRef]
26. Ungureanu, D.; Ţăranu, N.; Isopescu, D.N.; Lupăşteanu, V.; Scutaru, M.C.; Hudişteanu, I. Failure particularities of adhesively bonded joints between pultruded GFRP composite profiles. *IOP Conf. Ser. -Mater. Sci. Eng.* **2018**, *400*, 032011. [CrossRef]
27. Zeng, J.J.; Gao, W.Y.; Liu, F. Interfacial behavior and debonding failures of full-scale CFRP-strengthened H-section steel beams. *Compos. Struct.* **2018**, *201*, 540–552. [CrossRef]
28. Available online: https://wabrasives.com/en/standard-products/angular-steel-grit-blast-cleaning-surface-preparation.html (accessed on 12 May 2021).
29. Sika Colma Cleaner, Product data sheet, Identification no. 020516020000000001, 04.2015, Romania. Available online: https://i.dedeman.ro/media/file/file/s/i/sika_colma_cleaner.pdf (accessed on 14 May 2021).
30. Ţăranu, N.; Ciobanu, P.; Popoaiei, S.; Mihai, P.; Budescu, M.; Lupăşteanu, V. Improving the structural response of reinforced concrete beams strengthened with near surface mounted composite strips. *Rom. J. Mater.* **2016**, *44*, 325–340.
31. Oprişan, G.; Ţăranu, N.; Budescu, M.; Enţuc, I. Structural behaviour of reinforced concrete beams strengthened by CFRP plate bonding. *Rom. J. Mater.* **2012**, *42*, 387–398.
32. Da Silva, L.F.M. Preparing lap joints with flat adherends. In *Testing Adhesive Joints. Best Practices*; da Silva, L.F.M., Dillard, D.A., Blackman, B., Adams, R.D., Eds.; Wiley—VCH Verlag GmbH & Co. KGaA: Weinheim, Germany, 2012; Volume 2012, pp. 15–21.
33. Al-Azzawi, Z.; Stratford, T.; Rotter, M.; Bisby, L. FRP strengthening of web panels of steel plate girders against shear buckling Part-I: Static series of tests. *Compos. Struct.* **2018**, *206*, 728–738. [CrossRef]
34. Peiris, A.; Harik, I. FRP-steel bond study of IM and UHM CFRP strips. *Constr. Build. Mater.* **2018**, *185*, 628–637. [CrossRef]
35. Hudişteanu, I.; Ţăranu, N.; Isopescu, D.N.; Enţuc, I.; Oprişan, G.; Ungureanu, D. Numerical analysis of intralaminar damage evolution on various composite laminates. *IOP Conf. Ser. -Mater. Sci. Eng.* **2018**, *400*, 042031. [CrossRef]
36. Chandrathilaka, E.R.K.; Gamage, J.C.P.H.; Fawzia, S. Mechanical characterisation of CFRP/steel bond cured and tested at elevated temperature. *Compos. Struct.* **2019**, *207*, 471–477. [CrossRef]
37. Jiang, X.; Kolstein, M.H.; Bijlaard, F.S.K. Experimental and numerical study on mechanical behavior of an adhesively-bonded joint of FRP–steel composite bridge under shear loading. *Compos. Struct.* **2014**, *108*, 87–399. [CrossRef]
38. Jiang, X.; Kolstein, M.H.; Bijlaard, F.S.K. Study on mechanical behaviors of FRP-to-steel adhesively-bonded joint under tensile loading. *Compos. Struct.* **2013**, *98*, 192–201. [CrossRef]
39. Jiang, X.; Kolstein, M.H.; Bijlaard, F.S.K. Experimental investigation on mechanical behavior of FRP-to-steel adhesively-bonded joint under combined loading-part 1: Before hygrothermal aging. *Compos. Struct.* **2015**, *125*, 672–686. [CrossRef]
40. Al-Mosawe, A.; Al-Mahaidi, R.; Zhao, X.L. Bond behaviour between CFRP laminates and steel members under different loading rates. *Compos. Struct.* **2016**, *148*, 236–251. [CrossRef]
41. He, J.; Xian, G. Debonding of CFRP-to-steel joints with CFRP delamination. *Compos. Struct.* **2016**, *153*, 12–20. [CrossRef]
42. *ANSYS Workbench User's Guide*; ANSYS: Canonsburg, PA, USA, 2009.
43. Stein, N.; Rosendahle, P.L.; Becker, W. Homogenization of mechanical and thermal stresses in functionally graded adhesive joints. *Compos. B Eng.* **2017**, *111*, 279–293. [CrossRef]
44. Razavi, S.M.J.; Ayatollahi, M.R.; Majidi, H.R.; Berto, F. A strain-based criterion for failure load prediction of steel/CFRP double strap joints. *Compos. Struct.* **2018**, *206*, 116–123. [CrossRef]
45. Ribeiro, F.M.F.; Campilho, R.D.S.G.; Carbas, R.J.C.; da Silva, L.F.M. Strength and damage growth in composite bonded joints with defects. *Compos. B Eng.* **2016**, *100*, 91–100. [CrossRef]

materials

Review

Magnetic Nanoparticles Used in Oncology

Veronica Manescu (Paltanea) [1,2], Gheorghe Paltanea [2,*], Iulian Antoniac [1,3] and Marius Vasilescu [1]

1 Faculty of Material Science and Engineering, University Politehnica of Bucharest, 313 Splaiul Independentei, District 6, RO-060042 Bucharest, Romania; veronica.paltanea@upb.ro (V.M.); antoniac.iulian@gmail.com (I.A.); marius.vasilescu@upb.ro (M.V.)
2 Faculty of Electrical Engineering, University Politehnica of Bucharest, 313 Splaiul Independentei, District 6, RO-060042 Bucharest, Romania
3 Academy of Romania Scientist, 54 Splaiul Independentei, RO-050094 Bucharest, Romania
* Correspondence: gheorghe.paltanea@upb.ro

Abstract: Recently, magnetic nanoparticles (MNPs) have more and more often been used in experimental studies on cancer treatments, which have become one of the biggest challenges in medical research. The main goal of this research is to treat and to cure advanced or metastatic cancer with minimal side effects through nanotechnology. Drug delivery approaches take into account the fact that MNPs can be bonded to chemotherapeutical drugs, nucleic acids, synthetized antibodies or radionuclide substances. MNPs can be guided, and different treatment therapies can be applied, under the influence of an external magnetic field. This paper reviews the main MNPs' synthesis methods, functionalization with different materials and highlight the applications in cancer therapy. In this review, we describe cancer cell monitorization based on different types of magnetic nanoparticles, chemotherapy, immunotherapy, magnetic hyperthermia, gene therapy and ferroptosis. Examples of applied treatments on murine models or humans are analyzed, and glioblastoma cancer therapy is detailed in the review. MNPs have an important contribution to diagnostics, investigation, and therapy in the so called theranostics domain. The main conclusion of this paper is that MNPs are very useful in different cancer therapies, with limited side effects, and they can increase the life expectancy of patients with cancer drug resistance.

Keywords: magnetic nanoparticles; cancer therapy; drug delivery; radiosensitizing properties; glioblastoma; magnetic hyperthermia; ferroptosis

Citation: Manescu, V.; Paltanea, G.; Antoniac, I.; Vasilescu, M. Magnetic Nanoparticles Used in Oncology. *Materials* **2021**, *14*, 5948. https://doi.org/10.3390/ma14205948

Academic Editor: Paulo José Gomes Coutinho

Received: 1 September 2021
Accepted: 8 October 2021
Published: 10 October 2021

1. Introduction

Magnetic nanoparticles (MNPs) have been used in nanomedicine and the manufacturing of nanosensors for a long time. It is well known that MNPs have a controlled interaction with an external magnetic field, generated by using permanent magnets; therefore, as a consequence, the position of an MNP could be used to exactly establish where a medical problem exists. An alternating magnetic field induces a thermal heating process in the MNPs, and they are very useful as a contrast substance in nuclear magnetic resonance imaging (MRI) and as drug delivery vehicle for different diseases. The alloys of FeCo, magnetite (Fe_3O_4) and maghemite (γ—Fe_2O_3) are the most used materials, because a high value of the spin magnetic moment is present [1]. In the case of Fe_3O_4 the iron and the oxygen atoms are arranged in an inversed spinel structure, and the γ—Fe_2O_3 is characterized by the fact that the iron is bivalent. The bivalent iron could catalyze the oxidation of the hydrogen peroxide, resulting in toxicity due to the reactive species of oxygen. Usually, maghemite particles must have an outer shell, made with biocompatible materials such as gold or other polymers or resins, to prevent the toxicity of this alloy [2,3].

This paper focuses on magnetic nanoparticles used in oncology, and it provides a comprehensive review on this topic, taking into account specific aspects of the neoplasm treatment. The preparation methods of the MNPs used in medicine, classified as chemical, physical, and biological technologies, are presented.

Magnetic nanoparticles' functionalization is a very important topic, and it will also be taken into account in the discussion. The synthesis process of the MNPs with gold (Au@MNPs) implies two important steps: the synthesis of the iron oxide and the application of a gold shell. The gold shell can be directly applied onto the magnetic core [4–7] or indirectly, by adding a glue-like material layer between the iron oxide and gold shell [8]. Other inorganic and organic compounds such as silica, silver and polymers are also a viable candidate for MNP functionalization [9]. Today, superparamagnetic nanoparticles of iron oxide with a biocompatible shell (SPIONs) are being used more and more often in personalized medicine for the diagnosis and treatment of different types of cancers [2,10–13]. With the help of SPIONs, a real-time monitorization of the treatment is possible, by combining SPIONs with different medicines and then by following the changes that appear in the T1 or T2 relaxivity parameters [14]. The multifunctional SPIONs are a very important factor in immunotherapy, because through SPIONs the modulation of the macrophage polarization is possible, and in the case of a human monocyte cellular line, they can be transformed from the M_2 state (pro-tumorigenic) to the M_1 state (anti-tumorigenic). As an intermediate effect, antigen transport through lymph nodes is possible and activates the dendritic cells that can cause, as a final result, an important response from the T cells [15].

In the case of "in vivo" application, the hydrodynamic dimension of magnetic nanoparticles becomes one of the most important parameters, being restricted by the anatomical characteristics of the human body and by the fact that this type of materials is delivered in an aqueous solution. Nanoparticles smaller than 150 nm could pass the endothelial barrier, but the blood–brain barrier is permeable for magnetic nanoparticles with reduced dimensions and electric neutral lipidic molecules with a concentration lower than 500 g/mol [16–18]. The area to volume ratio is a limitative factor, which is taken into consideration in the nanoparticle synthesis process.

Magnetic nanoparticles tend to form aggregates that are very important if the material is used as a contrast substance in magnetic resonance investigations [19–21]. MNPs also possess numerous advantages, such as a large surface area to volume ratio, pore sizes, functionalized surfaces, multiple sites for interaction or binding, and low mass transfer resistance, which makes them very suitable to be used in different disease treatments.

In this review, we will present different approaches to treat cancer by combining chemotherapy, radiotherapy, immunotherapy, and gene therapy with MNPs or SPIONs, as depicted in Figure 1.

Magnetic-induced hyperthermia uses SPIONs to destroy the malignant tissues, which are difficult to treat through conventional therapies, and it will be detailed in Section 5. The tumoral tissues are targeted with SPIONs, whose magnetic properties are enhanced by using an external alternating magnetic field that induces a heating process in the neoplasm area. It is well known that by increasing the tissue temperature above 42 °C, the structural and functional proteins alter and the necrosis of the tumoral tissue begins.

The ferroptosis process will also be discussed and the side effects of the MNPs or SPIONs treatment will be highlighted in the final section of this review. The challenges of these innovative treatments will be presented.

Today, these treatments are successfully applied in murine animal models, but these approaches are considered very useful in the case of cerebral cancers such as glioblastoma, and they could be lifesaving in patient cases, where classical treatments provide poor results.

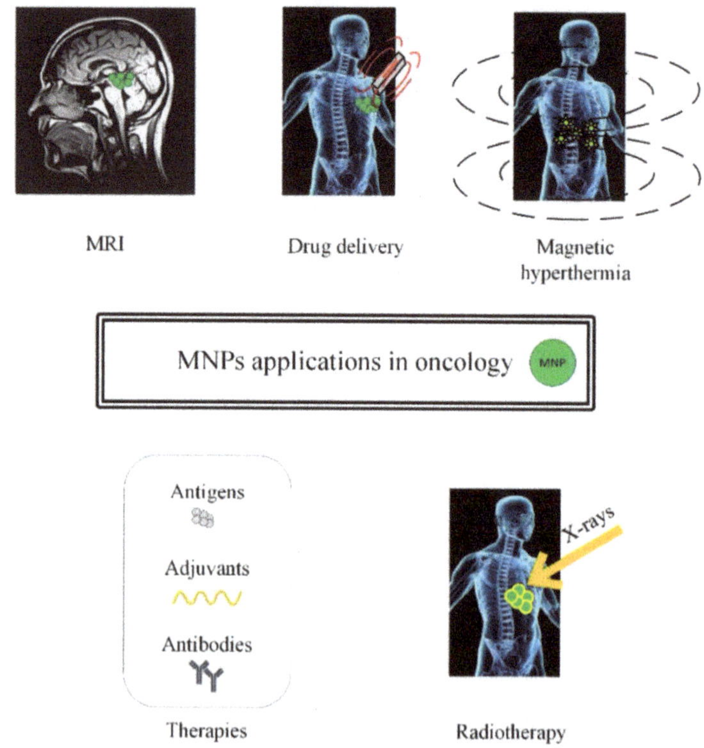

Figure 1. Schematic representation of magnetic nanoparticles' applications in oncology (MRI investigations, drug delivery in chemotherapy, magnetic hyperthermia, immunotherapy and radiotherapy).

2. Magnetic Nanoparticles' Synthesis

In the case of biomedical applications, there are some important restrictions, which must be taken into account. Each medical investigation and treatment requires special and dedicated nanoparticles. The surface effects, the crystallinity grade, and the link between the magnetic properties and the dimensions or shapes of the nanoparticles should be considered. Currently, the ideal shape of a magnetic nanoparticle is spherical or quasi-spherical.

The synthesis process can be divided into three principal categories as a function of the aggregation state of the precursors (gaseous, aqueous, or solid). Another classification can be performed as a function of the process type (chemical, physical, or biological). The most important synthesis methods presented in the literature are co-precipitation, sol-gel procedure, and hydrothermal synthesis [22]. Additionally, new methods were developed, such as microemulsion, thermal decomposition, sonolysis, and biosynthesis [23]. In Table 1, the main synthesis procedures that are used in practice are presented.

One of the most commonly used chemical methods is co-precipitation. The synthesis process of the nanoparticles used in this method can be completed in an open or closed environment. The open medium includes the following steps: co-precipitation, hot organic solvent use, a gaseous route or solid chemical reactions, and the closed medium involves oil- or water-based microemulsions and porous inorganic or organic templates. In Figure 2 are presented scanning electron microscopy (SEM) images of MNPs obtained through the co-precipitation method at different magnification scales.

Table 1. Synthesis methods for magnetic nanoparticles.

	Synthesis Method	Conditions	Size (nm)	Particle Morphology	Reference
Chemical methods	Aqueous co-precipitation	Temperature 20–90 °C Water solvent	15–200	Spherical, rhombic	[24]
	Sol-gel synthesis	Temperature 200–400 °C Organic solvent	20–200	Spherical	[23]
	Hydrothermal synthesis	Temperature 220 °C Water-ethanol solvent	520	Spherical	[25]
	Microemulsions	Temperature 20–50 °C Organic solvent	4–12	Spherical, cubic	[26]
	Electrochemical deposition	Temperature 70–100 °C Organic solvent	3–8	Spherical	[22]
	Polyol method	Temperature 120–280 °C Organic solvent	5–40	Spherical	[22]
Physical methods	Sonochemical method	Temperature 25 °C Water	10–30	Spherical, rod	[22]
	Thermal decomposition	Temperature 100–320 °C Organic	3–20	Spherical	[25]
	Spray pyrolysis	Temperature 400–700 °C Organic	5–60	Spherical with aggregates	[27]
	Laser-induced pyrolysis	Temperature 1100 °C Organic	5–30	Spherical less large	[27]
Biological methods	Biomimetic synthesis	-	50–100	Spherical cluster, Cubo-octahedral	[28]
	Microbial synthesis	Virus Tobacco mosaic virus	5	Disordered, prismatic and film	[28,29]
	Plant-mediated synthesis	Eucalyptus Leaf	20–80	Spheroidal	[28]

The co-precipitation in an aqueous solution consists of using Fe^{2+} and Fe^{3+} precursors, dissolved in an aqueous solution and then mixed with reactive substances, in order to precipitate in the spinelic system. In the case of the Massart method [30], a degassed solution of chlorides based on Fe^{2+} and Fe^{3+} is added to an ammonia solution, washed several times, and then stabilized in basic or acid suspensions. The aqueous medium is kept at a low temperature, and thus the co-precipitation method is considered to be one of the most commonly used technologies [31,32].

Other methods are based on hot organic solvents instead of water. When the dimensions of the particles are between 15 and 20 nm, through the co-precipitation method, using a Fe_2SO_4 precursor combined with a weak oxidant, high-quality nanoparticles are obtained [33–37].

In the case of hot organic solvents, MNPs with a high grade of crystallinity and a good monodispersing property are synthetized. The main mechanisms are nucleation and growth, and the coordination ligands are used to impose a good dispersion grade of the magnetic nanoparticles, which have a passivate surface. The particle dimension depends on the boiling temperature of the solvent. This technology consists of mixing the substances at room temperature; then an increase in the temperature is supposed to be applied at a value at which magnetic nanoparticles are rapidly formed. These particles have hydrophobic properties, and they are transferred to an aqueous medium and used in biomedical applications. Some FePt or FePt-Au magnetic nanoparticles are obtained based on crystalline iron oxide, formed from an iron precursor such as $Fe(acac)_3$, lauric acid or lauryl ammine [37].

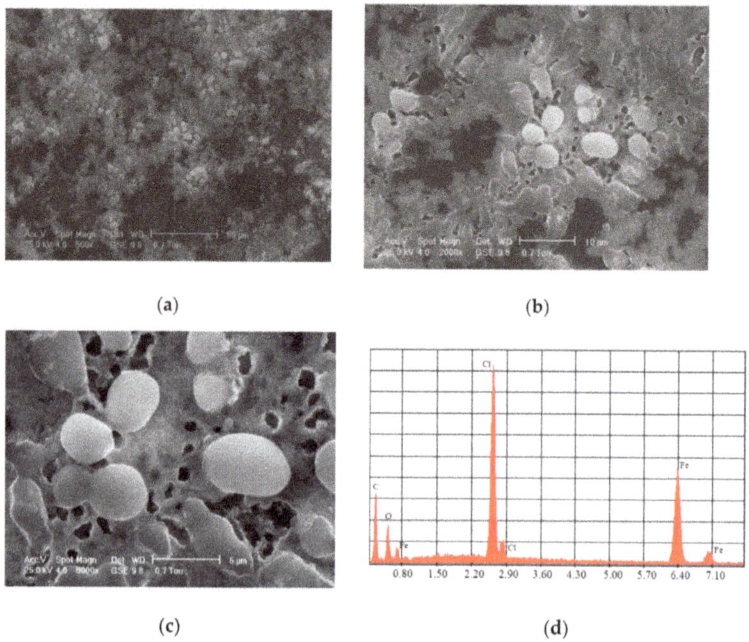

Figure 2. SEM analysis made in the case of Fe_3O_4 MNPs obtained through the co-precipitation method at three different magnification scales: (**a**) 500×; (**b**) 200×; (**c**) 5000×; and the resulting elemental analysis (**d**).

Another important method is the growth of magnetic nanoparticles in limitative conditions (growth under confinement) and the dimensions and shape of the nanoparticles are controlled [38,39]. This closed domain could be considered a matrix that represents a barrier used to control the material nucleus growth. This matrix could be beneficial for obtaining special properties of the magnetic material. Organized surfactants such as microemulsions, defined as thermodynamic stable dispersions, and composed of two non-miscible liquids, are usually employed. The surfactant determines microcavity formation, which represents the closure area. In this way, nucleation, growth and particle agglomeration are avoided. A classical case is the Al_2O_3 matrix with cylindrical pores [40].

The sol-gel process is based on the hydrolysis and condensation chemical reactions of metal precursors at low temperatures, resulting in the formation of an inorganic network [41]. The conversion of monomers into a colloidal solution determines the sol, and the gel is produced after the evaporation of the solvent, and by introducing particles in the inorganic network. Through this method, single-phase, fine, dense, and homogenous MNPs can be obtained. Although the sol-gel process is environmentally friendly and it is not an expensive method, it presents with a limited efficiency, and the synthesis process is longer than in the case of the co-precipitation method.

The hydrothermal method consists of MNPs' generation through a mixture between aqueous bivalent metals in acetate solutions with iron nitrate and a carbon nano template at a pH higher than 7. The final product is washed, centrifuged and annealed at 500 °C, in order to eliminate the carbon compounds. Ethylenediamine or citric acid can be used in the synthesis process [42].

Another important method is based on polyol. The metal ions are reduced in a polyol solution and precipitate as nanoparticles with dimensions lower than 100 nm. [43]. MNPs with small sizes can be obtained on dendrimer platforms or by using ultrasound waves [44]. Electrochemical deposition of metal is used to synthetize MNPs, by reducing the metallic ions dissolved by anode and by depositing them at the cathode. Pyrolysis, Fe^{3+} spray solution or the CO_2 laser heating process of gases that contain iron precursors were also reported as synthesis methods for MNPs [43].

The biological synthesis technologies are based on plants (bark, leaf, root, fruit, seed, flower), fungus, bacteria, algae, virus, and marine organisms, and they are considered simple, low-cost, non-toxic and eco-friendly. Two examples, applied in the case of iron-based MNPs, are presented in Table 1.

Another preparation method consists of metal–organic material structures, inside which the metallic ions are combined with organic molecules. The inorganic matrices have superior chemical and mechanical properties. The most used inorganic matrix is that based on silica, but carbon nanotubes or materials with a structure similar to graphene or titania nanotubes are also seen in practice. The magnetic nanoparticles are encapsulated in silica, by using ammonia hydroxide through the Stoeber method [45]. Carbon nanotubes were treated with nitric acid in an $FeCl_3$ suspension [46].

The gaseous state synthesis is very suitable for the manufacture of magnetic nanoparticles with dimensions lower than 5 nm. This method is based on the formation of gaseous substances oversaturated with vapors due to chemical reactions. The oversaturated vapor state contains atoms or molecules of a material, set at a higher pression than vapor pression. New particles are nucleated through laser pyrolysis or using carbonyl precursors. The pyrolysis is performed using the heating process of a gaseous mixed compound with the help of a continuous laser wave, which initiates and maintains the decomposition reaction. Once an energy level is reached, the metal atoms are produced through a fast coalescence procedure, resulting in nanoparticles which can supplementarily react with another gas components [47,48].

Through solid synthesis, MNPs with sizes lower than 20 nm can be obtained, but this method is rarely used.

The chemical and physical methods are considered conventional technologies and they are characterized by high costs and complex procedures. Co-precipitation and thermal/hydrothermal decomposition are the most commonly used methods because they are easy and secure.

3. Functionalization of Magnetic Nanoparticles

The encapsulation and coating of magnetic nanoparticles represent two important steps, which could be adopted in the MNPs synthesis for medical applications. Usually, hydrophilic substances are used, which are suitable for blood circulation and protein adhesion. Sometimes, thin polymers can be utilized [49,50]. If an organic matrix is involved, the human body does not present an immune response. Synthetic or natural polymer matrices made from polysaccharides or polyesters using the co-precipitation method are also a suitable solution.

The type of the coating which is applied on an MNP is important, because the shell layer could enhance or reduce the magnetic properties of the core, the shell thickness, and the size of the nanoparticles. The main methods used for MNPs' functionalization are performed in situ or post synthesis, involving mechanisms such as ligand addition, ligand exchange or encapsulation.

3.1. Innorganic Compounds

In the case of inorganic compounds, a specific group of ligands should be chosen, in order to manifest a high affinity for MNPs. The inorganic materials used in functionalization process are gold, silica, or silver.

Gold is an inert metal, and it is resistive to oxidation and corrosion. Nowadays, there are a high number of techniques associated with gold deposition, such as sputtering, thermal evaporation, and chemical deposition.

Magnetic nanoparticles for medical applications, functionalized with gold, can be considered as a magnetic core covered with a gold shell (Fe_3O_4@Au).

The synthesis procedure of Fe_3O_4@Au nanoparticles includes two steps: firstly, the iron oxide synthesis and secondly the gold shell application. There are two principal methods used to obtain a gold shell, as shown in Figure 3. The first one is called the direct method, where the gold shell is directly formed on the magnetic material; in the second case—the indirect method—a glue-like intermediary material is used [51].

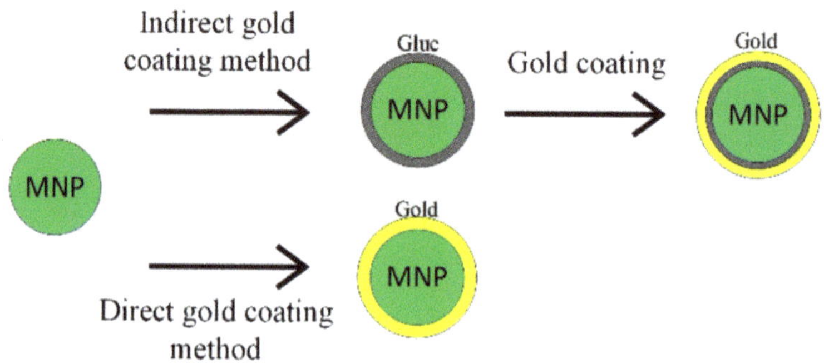

Figure 3. Gold coating methods' schematic representation.

The gold shell can be applied when the magnetite particles are placed in an aqueous solution or in an inorganic material. For the aqueous solution case, the gold shell is created by reducing the Au^{3+} ions through ammonium citrate or borohydride [52]. A disadvantage of this method consists of the fact that if the experimental conditions are not entirely respected and controlled, individual gold nanoparticles appear, and outer gold shell formation becomes impossible. When the magnetic core is included in an organic phase such as oleic acid, the gold outer shell can be formed [53]. The synthesis of Au@MNPs is possible when a non-polar solvent is present, and the obtained gold shell is thick.

Another method in gold functionalization is inverse micelle technology, which is a complicated procedure in which centyltrimethylammonium bromide, octan and 1-butan are used as surfactants and oil is used as a co-surfactant [52].

The gold shell can also be obtained through the indirect method, interposing a glue layer between the magnetic core and the gold outer shell, as presented in Figure 4 [52,53]. The gold shell must completely cover the magnetic core, in order to ensure a high chemical and physical stability. It has to contain a large number of chemical groups, such as poly-L-histidina, cross-linked polymer (cyclotriphosphazene-co-4, 4′-sulfonyldiphenol) (PZS), polyethylene-imine or silica that can chelate with metallic ions.

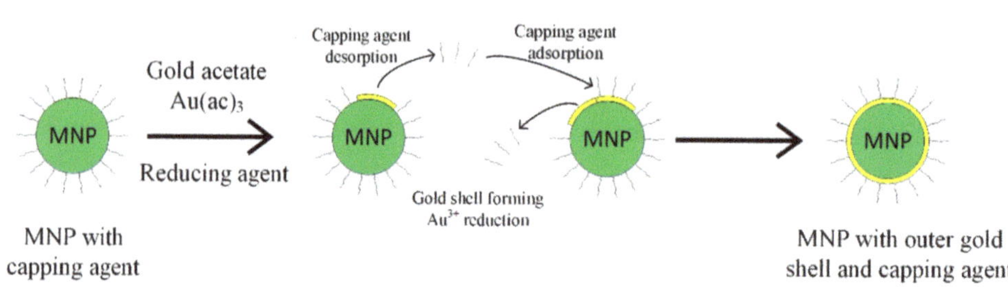

Figure 4. Capping agent desorption and gold coating principle.

When the gold shell was formed on a substrate as silica, the co-precipitation method was applied to synthesize the magnetic core and then sol-gel technology, based on the hydrolysis process of tetraethyl orto-silicate, was used to obtain the silica cover. Three layers of positive charged poli-electrolytes are formed on the magnetic nanoparticles covered with silica. The Au ions are negatively charged, and a gold shell can be obtained through the self-assembly method. Gold shells thicker than 30 nm were reported in [54,55].

The functionalization of MNPs with gold presents several advantages in biomedical applications, because it has a positive impact on the specific sensitivity and compatibility. Au@MNPs can form aggregates due to the magnetic attraction forces developed between particles. The surface changes play an important role in the stability increase, which is affected by the electrostatic exchanges on the material's surface. The gold shell prevents the metal ion emission from the magnetic core, increasing the MNPs' biocompatibility [56,57].

Silver coatings facilitate the germicidal action of the MNPs, and they are very similar to gold coatings. Silver can generate core–shell structures and heterostructures. Hybrid nanoparticles of Fe_3O_4@C@Ag were used for drug release, MRI and two-photon fluorescence imaging, presenting advantages due to having on their surface both carbon and silver.

Silica (SiO_2) is functionalized with MNPs through a similar process to ligand absorption. A silica coating inhibits the aggregation of the MNPs and improves the chemical stability and biocompatibility of the shell–core assembly. Two main methods have been developed to stabilize the silica coating. The first one consists of shielding the magnetic dipole interaction with the silica coating, and the second one is based on charging the MNPs with negative charge, so an electric repulsion phenomenon appears between the shell and the core. SiO_2@MNPs could be prepared according to the Stoeber method, in which SiO_2 is generated through hydrolysis and condensation of a sol-gel precursor (tetraethyl orthosilicate, TEOS). As an example, in [58,59], 200 mg of Fe_3O_4 was introduced in an ethanol suspension under ultrasonic irradiation for 30 min in nitrogen atmosphere. Then, 11 mL of NH3 (28%) and 400 μL of tetra-ortho-silicate (TEOS) in the Fe_3O_4 suspension were sonicated for 2 h. The final compound, Fe_3O_4@SiO_2, was washed in ethanol and then collected through a magnetic separation procedure.

The second method is the silica deposition from silicic acid solution, and through this method, the particle size can be effectively controlled. Additionally, a higher percentage of the MNPs' surface can be covered.

The emulsion method consists of controlling the silica shell using micelles or inverse micelles. This procedure is based on the separation of the core–shell MNPs from surfactants linked to the emulsion system.

A main advantage of silica coatings is the presence of surface silanol groups that react with coupling agents, and ligands can be attached to MNPs. In this way, amine groups were introduced on the surface of Fe_3O_4@SiO_2 through hydrolysis and condensation of an organosilane (silanization method). Biomolecules could be easily added by using alkoxysilanes with active groups.

Another method for obtaining potential vectors for drug-delivery treatments involving magnetic nanostructures is based on silica mesoporous structured materials (MCMs). Usually, nonmagnetic or magnetic silica nanoparticles are prepared using the sol-gel procedure based on silica xerogels, hydrothermal synthesis and laser pyrolysis. MCMs present a large surface zone and small pores with dimensions between 15 and 100 Å. The pore openings are hexagonal (MCM-14), cubic (MCM-48) and lamellar (MCM-50) ordered. In [60], the drug miltefosine (hexadecyl 2-(trimethylazaniumyl) ethyl phosphate), used for different fungi, bacteria, cancers and parasite species, was loaded onto a new composite mesoporous silica drug carrier 20–30 nm in size, which contains magnetic nanoparticles modified with NH_2 and PEG chains. It was concluded that about 11–25% of the drug content was deposited inside the pores of the silica carrier. Miltefosine loaded onto PEGylated magnetic silica magnetite nanoparticles exhibits a very good antiproliferative activity and antineoplastic potential against human tumor HUT-78 cell line (cutaneous T lymphocytes). This material exhibits important properties as a nanocarrier of miltefosine.

In the nanoscale range, nanoporous silica particles can also be used that can cover the MNPs and embed different oncological drugs on the silica support. A SO_3H-functionalized magnetic nanoporous silica/polymer nanocomposite was developed in [61]. Spherical

100 nm particles were synthesized that had the anticancer drug mitoxantrone (MTX) loaded in the magnetic nanoporous silica support, covered with a chitosan layer infused with prednisolone (PRD). The final coating layer consisted of alginate. The cytotoxic activity was assessed against HD-MY-Z (Hodgkin lymphoma), EJ (urinary bladder carcinoma) and K-562 (myelogenous leukemia) cellular lines, and this innovative material shows great potential in pharmacological applications.

3.2. Organic Compounds

MNPs can be functionalized with organic compounds after or during the synthesis process. Organic functional materials present a good biocompatibility and biodegradability and do not have an important influence on the magnetic properties of MNPs. The functional groups are carboxyl, aldehyde, hydroxyl and amino, which, after the functionalization of MNPs with organic compounds, can be linked to deoxyribonucleic acid (DNA), proteins, enzymes or antibodies. After the functionalization process, the final structure is classified into three types: core–shell, matrix (mosaic or shell-core) and $shell_x$–core–$shell_y$ (Figure 5).

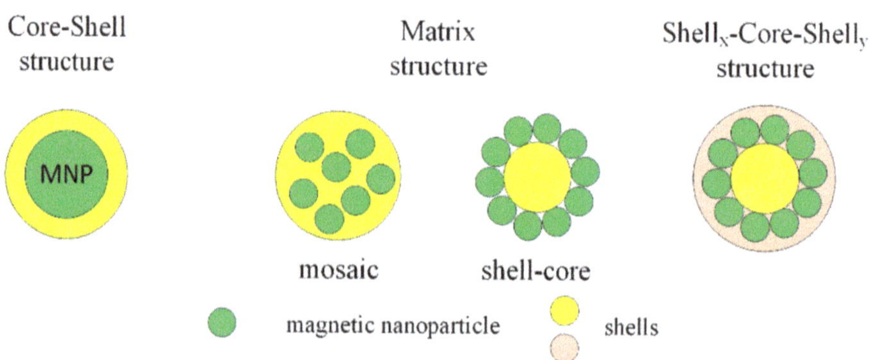

Figure 5. Classification of functionalized structure types with organic compound.

The most important organic compounds that can be used in the functionalization of MNPs will be detailed and explain in the review.

The first type of organic compound is the *monomer* class that includes *citrates* and *phosphates*. The surface of MNPs is stabilized in an aqueous solution through the adsorption of citric acid, which have carboxyl groups. The acid is adsorbed on the nanoparticle surface with the help of carboxylate functionalities. This process is dependent on the size and the shape of the particle. MNPs functionalized with citrates are characterized by a negative charge on the particle surface, generating hydrophilic properties. In Figure 6, transmission electron microscopy (TEM) images of Fe_3O_4 nanoparticles synthetized through the co-precipitation method are presented. It can be observed that the unfunctionalized MNPs tend to form aggregates, since the citrate-functionalized MNPs present a spherical shape and a relatively uniform dispersion.

Regarding the phosphates group, alkanesulfonic and alkanephosphonic acids can be used as surfactants. After the functionalization with phosphates, the dispersions of MNPs are thermodynamically stable. This method was used to prepare SPIONs, based on magnetite, through co-precipitation of bivalent and trivalent iron ions, using poly(vinylalcohol phosphate) (PVAP). The size of the particles is strongly dependent on the PVAP concentration [50]. From Figure 6c, it can be seen that MNPs functionalized with phosphates tend to form aggregates, a fact that makes them very useful in MRI investigations.

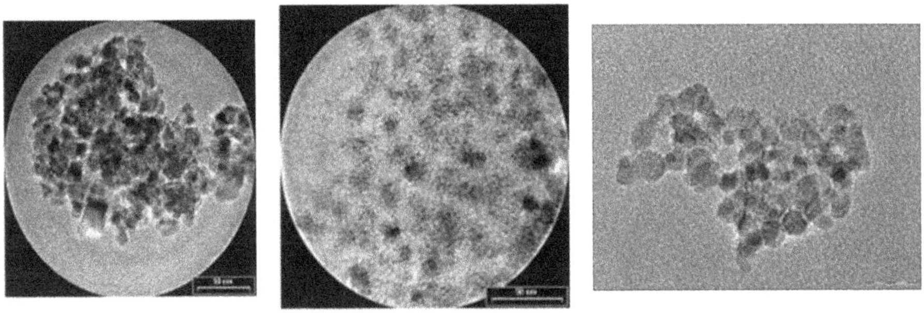

(a) (b) (c)

Figure 6. TEM images of Fe_3O_4 MNPs: (**a**) unfunctionalized MNPs (average medium size of particles 8.5 ± 2 nm); (**b**) citrate-functionalized MNPs (average medium size of particles 9.5 ± 1 nm); (**c**) phosphate-functionalized MNPs.

If *small molecules* are employed in MNP functionalization, they are characterized by a low value of the hydrodynamic radius, a fact that is beneficial for nanotechnology. *Short chain aminosilanes* or *amines* are usually used to coat the MNPs' surface. Magnetite nanoparticles, functionalized with these substances, were considered as being highly water-stable. Another type of small molecule is thiol as an organosulfur compound, 2,3-meso dimercaptosuccinic acid (DMSA), which is characterized through two thiol and two carboxyl groups. It was observed that thiol-functionalized nanoparticles are stable under physiological pH. The stability of the coating could be increased if DMSA is protected through a polyethylene glycol (PEG) chain.

In the *macromolecules* class, natural or synthetic polymers are intensively added for MNPs' functionalization. A polymer coating balances the van der Waals and magnetic attraction forces, which acts on MNPs. Coating approaches, performed in situ and post-synthesis, are based on reaction mixtures that are simultaneously combined with polymer functionalization. Ligand exchange and ligand adsorption are used together to provide a stable structure. In the case of MNPs' functionalization with polymers, the agglomeration of the particles is prevented, and different surface properties are obtained.

The most commonly used *natural polymers* are *dextran* and *chitosan*. Dextran increases the MNPs' biocompatibility and is a polysaccharide composed of R-D glucopyranosyl. This polymer's presence limits the particle size, and it has the advantage that it can be aminated using ammonia, which permits conjugation with complementary chemical groups. Chitosan is alkaline-based, biodegradable, biocompatible, and hydrophilic. MNPs functionalized with chitosan have shown a high MRI contrast and they exhibit very good properties as carriers in different types of treatments [62]. Figure 7 shows SEM analysis performed for Fe_3O_4 functionalized with chitosan.

Polyethylene glycol (PEG) is a synthetic, hydrophilic, and biocompatible polymer. PEG-coated magnetite nanoparticles present solubility and stability in physiological and aqueous media. Post-synthesis coating technology is preferred in the case of PEG functionalization. *Polyvinil alcohol* (PVA) is a synthetic polymer that can be transformed into gel, and it is highly biocompatible and hydrophilic. A PVA coating prevents self-agglomeration, and the colloidal stability of the PVA-coated particles is good. *Alginate* is another type of synthetic polymer, and it can be characterized as an electrolytic polysaccharide with carboxyl groups, which interact with iron, resulting in the stable assembly of alginate MNPs due to electrostatic repulsion.

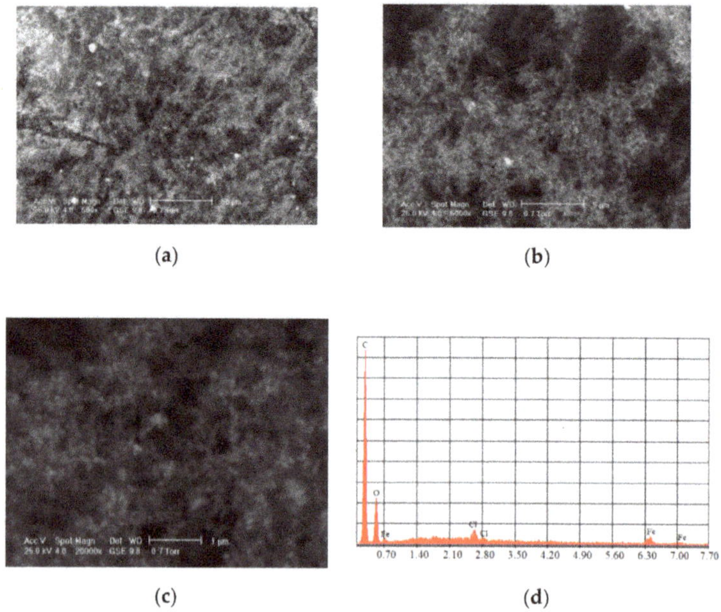

Figure 7. SEM analysis made in the case of Fe_3O_4 MNPs obtained through the co-precipitation method functionalized with chitosan at different magnification scales: (**a**) 500×; (**b**) 5000×; (**c**) 20,000×; and the resulting elemental analysis (**d**).

A special class of synthetic polymers is the *liposomes*. In magnetic drug targeting therapy, the liposomes play the role of carriers, because they present a large encapsulation domain, and biomolecules such as DNA or proteins can be encapsulated by liposomes as one unit [63,64]. The liposomes can be combined with different types of magnetic particles, using the dialysis process of a single unilamellar vesicle or through the extrusion of a combination between phospholipids and magnetic particles, as presented in Figure 8.

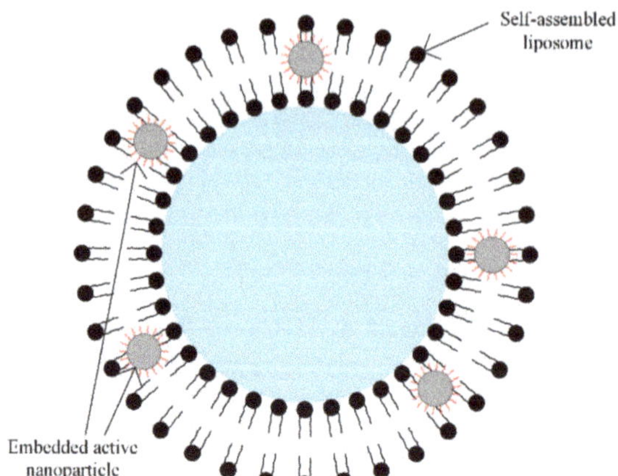

Figure 8. Self-assembled liposome combined with nanoparticles. Solid or water-soluble drugs can be included inside the core region (blue zone) and between interfaces (black spheres); embedded active nanoparticles or oil-soluble drugs in the lipid bilayer can be included.

Micelles or *micellar polymers* can be used as a coating layer for MNPs, and they are made from self-assembled amphiphilic polymers, which contain both hydrophobic and hydrophilic groups. The hydrophobic core helps to facilitate the hydrophobic components' movements, and the hydrophilic shell determines the MNP's high stability in the biologic medium [65,66]. The micelles are benefic for MNP aggregation, in the case of contrast substances used in nuclear magnetic resonance [67].

Hydrogels are considered one of the most promising substances, used in magnetic nanoparticle synthesis for medical applications (Figure 9). These are networks of hydrophilic polymers that contain a high amount of water or physiological fluids. The hydrogels exhibit a high biocompatibility and important therapeutical properties [40,68–72]. When nanogels are used, interactions between materials and living cells are possible, and different types of drugs can be delivered.

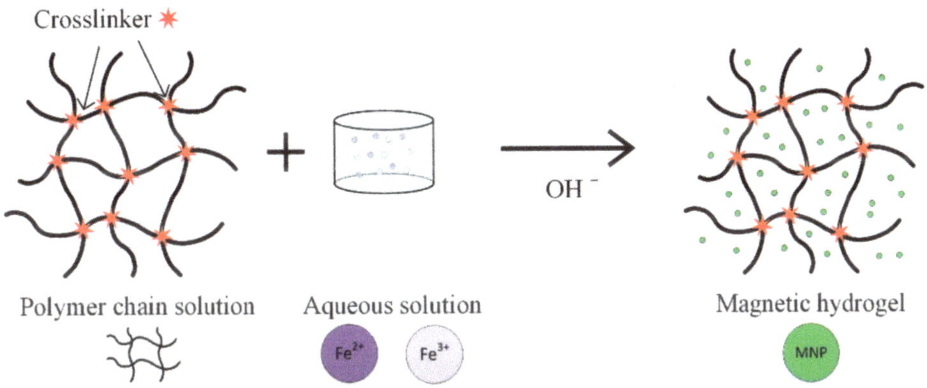

Figure 9. Schematic representation of the in situ co-precipitation method used to prepare a magnetic hydrogel.

Polymeric nanogels in aqueous medium are obtained through the co-polymerization of N-vinylcaprolactam or vinylimidazole in aqueous solution in the presence of a link agent and a reaction initiator, soluble in water [70]. The high grade of porosity of these nanogels permits their use as templates for MNPs, through the hydrolysis reaction of $FeCl_2$ or $FeCl_3$.

The *epoxy resin* group is useful for different material functionalization processes. SU8 UV photosensitive materials are biocompatible, and they are suitable for the preparation of biochemical coupling structures. The epoxy resin group reacts with nucleophile substances (NH_2) through a special mechanism. It is also possible to perform the transformation of an epoxy group in a hydroxyl group through a hydrolysis chemical reaction that is based on using an acid, basic or capping agent (i.e., ethanolamine). A primary amine reacts with a carboxylic acid or with an epoxide. The basic characteristics of the solution (with a pH of 9) are suitable for this reaction, and the process is followed by deprotonation [73,74].

The side effects of the hydrolysis reaction can be avoided by using an organic solvent such as dimethylformamide (DMF). Regarding the chemical link agents, there are electrophile or nucleophile substances. The electrophile materials accept electrons, and they are epoxy resins, NHS esters, aldehyde R-CHO and carboxylic acid R-COOH, and the nucleophile materials release electrons (primary amine R-NH_2, thiol R-SH, alcohol R-OH and water H_2O) [58,59,75,76].

The material morphology could be investigated through infrared spectroscopy, FT-IR transform, X-ray diffraction (XRD), X-ray energy dispersive spectroscopy (EDX), thermogravimetric analysis (TGA), dynamic light scattering (DLS) and zeta potential measurements. The dependence between magnetization and the applied magnetic field is usually measured with a SQUID or vibrating sample (VSM) magnetometers [62]. In Figure 10, different types of analysis made in the case of Fe_3O_4 nanoparticles are given as examples.

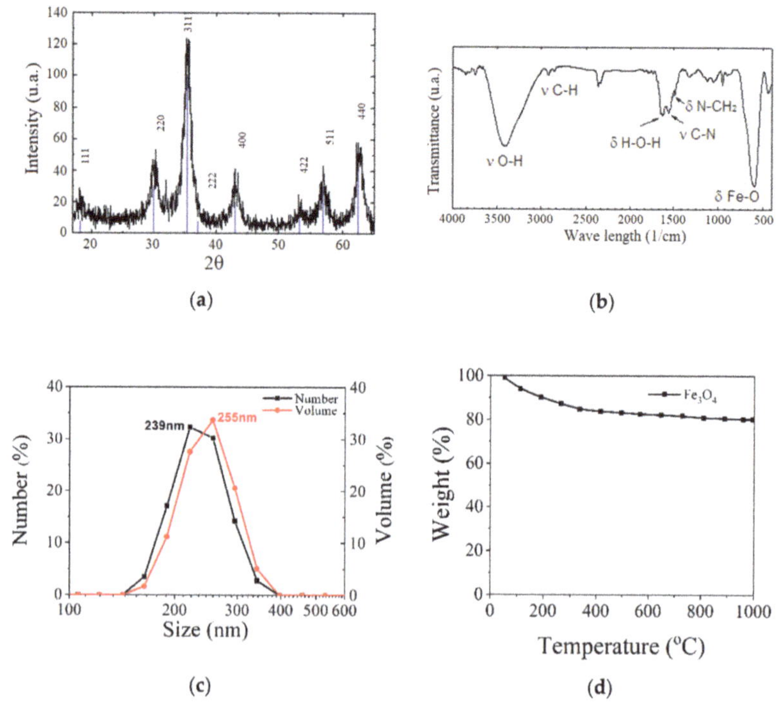

Figure 10. Different types of analysis made on Fe_3O_4 nanoparticles: (**a**) X-ray diffraction (XRD); (**b**) FTIR analysis; (**c**) DLS investigation; and (**d**) TGA analysis.

4. Magnetic Properties of MNPs

Usually, MNPs are considered to exhibit a ferrimagnetic, ferromagnetic or superparamagnetic behavior. Below a given size of the nanoparticle, which depends on the material type, MNPs are characterized through a zero value of the magnetic moment, in the absence of an external magnetic field, and they are called superparamagnetic particles (SPIONs). The magnetic moment rapidly increases when a magnetic field is acting on MNPs, and it is oriented along the field direction.

SPIONs are very important in medical applications such as magnetic drug targeting, MRI and different cancer treatments due to the fact that when the external magnetic field is removed, SPIONs do not exhibit any magnetic properties and the aggregate formation is prevented. Superparamagnetic nanoparticles present a strong response to external fields, and so they provide a high control of the medical application. In Figure 11, an example of a magnetization curve for iron-based nanoparticles in a superparamagnetic state is shown.

MNPs are characterized through saturation magnetization M_s (maximum induced magnetization), remanent magnetization M_r (the value of the magnetization, which can be measured after the magnetic field is removed) and coercivity H_c (magnetic field strength that is necessary to reduce the magnetization value to zero). In the case of superparamagnetic materials, the dependence $M(H)$ follows a sigmoidal function, and it presents no hysteresis phenomena [77].

The saturation magnetization is a function of the temperature, and it is well known that above a so-called blocking temperature, T_B ferrimagnetic and ferromagnetic materials have a superparamagnetic character. This behavior is due to the difference between measuring time and relaxation time as follows: if the relaxation time is lower than the measuring time, the particles are in a superparamagnetic state; otherwise they are in ferrimagnetic or ferromagnetic working conditions [78,79]. When the MNPs' diameter is kept below a critical value during the synthesis procedure, they are characterized by a single domain structure, and at the lowest diameter value, they are in a superparamagnetic state. If the diameter increases after this critical value, the particle tends to develop a multi-domain magnetic structure (Figure 12). In the case of Fe_3O_4 particles, the value of the diameter at which superparamagnetic behavior appears (D_s) is around 25 nm, and the critical diameter for the mono-domain magnetic structure (D_c) is 82 nm. In the case of γ—Fe_2O_3, these values are D_s = 30 nm and D_c = 90 nm [80].

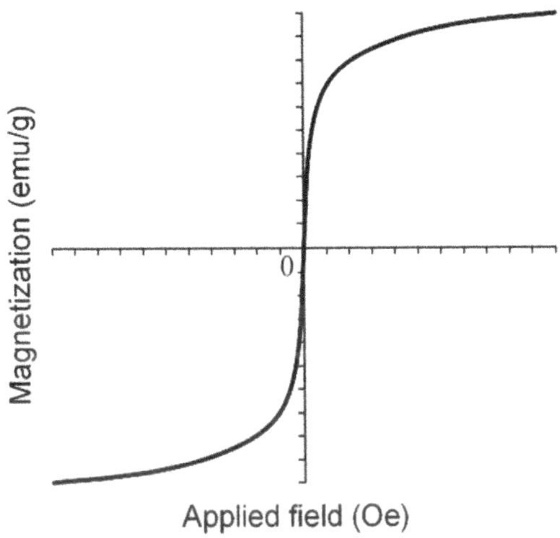

Figure 11. Magnetization curve for SPIONs can be modeled using sigmoidal functions. These types of particles do not present hysteresis phenomena, and so after external magnetic field removal, they do not tend to form aggregates.

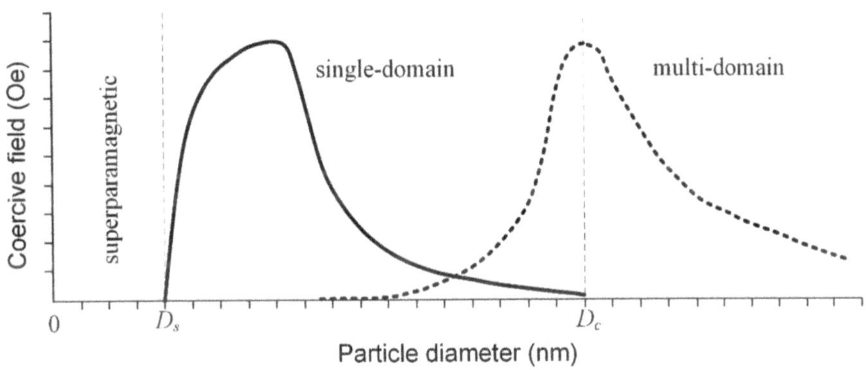

Figure 12. Coercive field variation as a function of the particle diameter, considering the transition between superparamagnetic state to single-domain and to multi-domain state. For the single-domain case, the coercivity variation follows the solid graphical representation in the case of particles that do not interact between them and the dashed line, for the coupled particles.

The size of the MNPs and the synthesis method are directly linked to the saturation magnetization. The value of saturation magnetization M_s proportionally increases with particle size, and finally it becomes equal to the bulk magnetization value, independent of the synthesis method. Regarding the coercivity field value H_c, it has a similar behavior, but after a specific size value is reached, a transition from a mono-domain to a multi-domain structure occurs, and so, as a consequence, a decrease in coercivity is observed [81–83]. In Table 2, the abovementioned magnetic properties in the case of Fe_3O_4 nanoparticles are summarized.

Table 2. Main magnetic properties of Fe_3O_4 nanoparticles.

Synthesis Method	Size (nm)	M_s (emu/g)	H_c (Oe)	Reference
Aqueous co-precipitation	4.9 6.3 8.6	60 65 58		[81]
Thermal decomposition	4.2 7.4 8.1 17 45	75 70 65 82 92	318 270 70 364 640	[82]
Non-aqueous homogeneous solution of polyols	6.6 11.6 17.8	71 77 83	16 15 3	[83]

The shape of the magnetic nanoparticles is also an important parameter. Refs. [84,85] analyzed the influence of this parameter, and it was concluded that Fe_3O_4 spherical nanoparticles have a higher value of T_B than cubic and spherical $Zn_{0.4}Fe_{2.6}O_4$ MNPs, and cubic particles exhibit a higher saturation point when compared with other shapes of nanoparticles.

The magnetic properties of MNPs appear due to the presence or absence of unpaired valence electrons, which are placed on the metal atoms or metal ions. Chemical composition is a very important parameter, and it depends on precursor concentration, synthesis method and dopant nature. The solvent could have an influence on the magnetic state, and so in [78], it was observed that when benzyl alcohol was used, Mn-doped ZnO became paramagnetic and Co-doped MNPs exhibited ferromagnetic properties.

The Curie temperature T_c of the MNPs, defined as the temperature above which magnetic nanoparticles have a zero value of magnetization, can be variated through chemical composition. This parameter should be carefully controlled, in order to avoid the overheating phenomenon in magnetic hyperthermia. As an example, in the case of aluminum content variation in $Y_3Fe_{5-x}Al_xO_{12}$, with x between 0 and 2, the T_c was measured between −40 and 280 °C [78].

5. Applications of Magnetic Nanoparticles in Cancer Therapy

The iron oxides with a crystalline core (Fe_3O_4) present great potential for use in oncological medicine due to their high compatibility [86], biodegradability [19] and easy synthesis [69]. Additionally, spherical magnetite nanoparticles with a diameter lower than 20 nm exhibit a superparamagnetic behavior that is a very important quality in medical imaging. An iron superparamagnetic particle (SPION) is made from a magnetite magnetic core, which provides the property of contrast in magnetic resonance imaging (MRI). This material has a biocompatible shell with a high number of functional groups, making it very useful for cancerous tumor follow up. Usually, a SPION is a magnetic single-domain structure, and it has high magnetic susceptibility. When such a special nanoparticle is placed in an alternating magnetic field, it can generate heat through Neel and Brown relaxation processes [87]. SPIONs do not have remanence and coercivity, as was explained in Section 4, and their agglomeration in human body is prevented.

Magnetic nanoparticles have received Food and Drug Administration (FDA) approval for the iron treatment deficiency, as a contrast agent in MRI and as a carrier for magnetic drug targeting therapy in different parts of the human body, biotherapeutical transport, magnetic hyperthermy, photodermal ablation, and photodynamic therapy [88].

Magnetic nanoparticles are also used in personalized medicine, termed theranostics. The diagnosis and monitoring of the cancer are performed based on medical imaging and through chemotherapeutical or biotherapeutical agents such as medicines, nucleic acids and proteins. A clear representation of the tumoral mass is very useful for doctors, because they can anticipate the patient's response when a tumor is surgically removed and the future treatment is established. As medical imagining methods, X-ray radiography, echography, computerized tomography (CT scan), nuclear magnetic resonance (MRI) and positron emission tomography (PET) are analyzed.

The accumulation of magnetic nanoparticles at the tumor sites consists of two types: active or passive targeting. Passive targeting is based on the enhanced permeability and retention (EPR) phenomenon. A characteristics of the tumoral tissue represent the perfored blood vessels, and abnormal lymph drainage can be observed inside the tumor. Nanoparticles with dimensions lower than 100 nm can pass through these broken vessels, and they can migrate inside the tumor and remain there for a longer period of time, as compared with nanoparticle clearance in the blood (EPR phenomenon). Passive targeting is limited, because not all the cancerous tumors in the human body exhibit the EPR phenomenon and the permeability grade of the tumoral vascular system is not homogenous [89,90].

Active targeting is performed through nanoparticle modification, by attaching different groups to the MNP surface. These ligands have the possibility to recognize the biological structure's unicity and neoplasm cells, which accumulate at the tumor site. The magnetic nanoparticles can be guided using an external magnetic field.

It has been proven that the iron oxide nanoparticles can generate microscopic holes inside endothelial cells. This phenomenon is called NanoEL. In order to not block the magnetic nanoparticles in the endothelial tissue, they are delivered through transcellular endocytosis or diffusion transport through the cellular membrane, because junctions between cells appear. The nanoparticles enter into vascularized organs such as the kidneys, liver or spleen at a high speed [91]. The NanoEL effect is much faster in comparison with endocytosis or diffusion through the cellular junction, and it is described in the existence of a intracellular link between the nanoparticle and VE-cadherin [92].

5.1. Cancer Cell Monitorization Using SPIONs

Today, the efficacity of the cancer treatment is estimated by taking into account the biopsy tissue analysis or using the medical images of the anatomical structures before and after the treatment. The time between the treatment's initiation and the results' apparition could be weeks or months. With the help of SPION nanoparticles, a real-time monitorization of the treatment is initiated, drugs are delivered, the response of the healthy tissue is evaluated, and the treatment adjustment becomes possible. A successful treatment was obtained in the case of murine dendritic cells with a combination of SPIONs, by increasing the anti-tumorigenic response through T cells stimulation. If dendritic cells are combined with SPIONs, it becomes possible to visualize their trajectories [93].

Other successful studies followed the pharmacokinetics of different active substances using SPIONs in different organs [94–97]. Another application of the SPIONs in drug delivery is made by detecting the changes in the two relaxivity parameters, T1 and T2. The transformation between a non-load state and a load state can be observed. In [98], studies were carried out in which the nanoparticles were used as a carrier for chemotherapeutical medicines, such as doxorubicin combined with Flutax 1 and DiR. The magnetic nanoparticles have a polymeric pH shell that degrades when an acid medium is present. The iron oxide carriers are called "nanophores", and the loading with different active substances increases the SPIONs' transverse T2 and longitudinal T1 nuclear magnetic resonance proton relaxation times. This type of chemotherapeutic treatment has proven to be more effective than classical routes, because the medicines do not have any chemical modification, and they are directly delivered at the tumor site. Drugs such as doxorubicin were modified and combined with polymers and targeting moieties, using the EPR effect. Nutrient receptors such as folate or transferrin receptors were also added to improve the drug delivery. Fluorochrome was used to trace the distribution of molecular groups. Doxorubicin was delivered in liposomal formulations, and clinical studies proved that higher doses of medicine could be released at the tumor sites, without many significant side effects. Ferumoxytol was approved by the FDA in 2009 to treat iron deficiency. In [98], these MNPs were analyzed as carriers for hydrophobic drugs, which were loaded inside the nanoparticle through the co-incubation process. This innovative delivery system could be approved faster by the FDA, because ferumoxytol is already accepted, and thus developing a way to load the active substances in a clinically approved vehicle will soon be easier in clinical studies. A great advantage of using magnetic nanoparticles consists of the fact that these materials could be followed through MRI and changes in relaxivity parameters.

Ferumoxytol is considered an ideal nanophore in drug delivery, and it was coated with carboxymethyl dextran [99]. After the dialysis process, performed in order to remove accidental cargo substances, the group was loaded with Flutax 1, which is a fluorescent taxol that can link to the taxol microtubule, a site with high affinity. It has been proven that ferumoxytol can also carry a near-infrared fluorophore DiR and the chemoterapeutical active substance doxorubicin. Through the dynamic light scattering method, it was observed that medicine loading has no effect on the average diameter of ferumoxytole, proving that no interactions occur between the nanoparticles and their cargo substances. The prepared colloidal solution has to be stable in a physiological medium (i.e., basic pH), and when the acidic conditions are met, the active substance is released. Ferumoxytol keeps doxorubicin at a pH of 7.2, and then, since pH decreases, it slowly releases the drug [98].

Loading or unloading the iron oxide particles can be observed, without using fluorophores, by analyzing the changes in T2 and T1 parameters and using an MRI device. Slight changes in these parameters were observed at a pH of 6.8. As the pH decreased, rapid decreases were also measured for T2 and T1, suggesting that the active substance is released at the tumor site.

5.2. Chemotherapy

The main aim of chemotherapy treatment is to avoid tumor growth with the help of chemical agents, which inhibit cellular function. The DNA chain can be interrupted by using these agents, and repair of the DNA is prohibited. Sometimes, different types of proteins can be administrated to stop the multiplication of damaged cells [100]. The chemotherapeutical drugs do not demonstrate selectivity, and they exhibit a high grade of toxicity in the human body, even acting on healthy cells. To avoid this activity, chemotherapeutic medicines are loaded in magnetic nanoparticles and directly delivered, as was already presented in Section 5.1 at the tumor site (see Figure 13). In this way, the overall toxicity of the treatment is decreased, and the active substance does not demonstrate chemical or physical degradation. In recent studies, TMZ, doxorubicin, paclitaxel and 5-fluorouracil were combined with SPIONs [101–103].

The alkylating agent temozolomide (TMZ) is an imidazotetrazine derivative of dacarbazine and the drug Temodar. TMZ has a lipophilic structure, and it can be orally administrated, because the human blood–brain barrier (BBB) is permeable to this substance. The compound presents stability at a pH lower than 5 and at pH higher than 7 if it is hydrolyzed [104].

Temozolomide was approved in 2005 to treat glioblastoma (GMB) patients. Glioblastoma multiforme is grade IV astrocytoma cancer [105]. Firstly, studies were conducted in which a classical treatment approach was taken into consideration. In the first phase of these clinical studies, the patients were administered drugs at doses between 750 and 1000 mg/m^2, 5 days a week for 4 months. TMZ was combined with radiotherapy and the GBM patients took 75 $mg/m^2/day$ for 6 weeks, and at the same time focal radiotherapy was applied. Then, they took 150 mg/m^2 TMZ 5 days/week followed by at least 20 days free of cancer treatment [106].

Due to the high toxicity of this drug, poor TMZ penetration across BBB, short half-life, the non-specific nature of the active substance and its non-targeted character, it is be better to functionalize it with different ligands (i.e., folic acid) and then load it in magnetite nanoparticles coated with a polymeric shell, composed of poly(ethylene glygol)-poly(butylene adipate)-poly(ethylene glycol) (TMZ-SPION-PEG-PBA-PEG) [107].

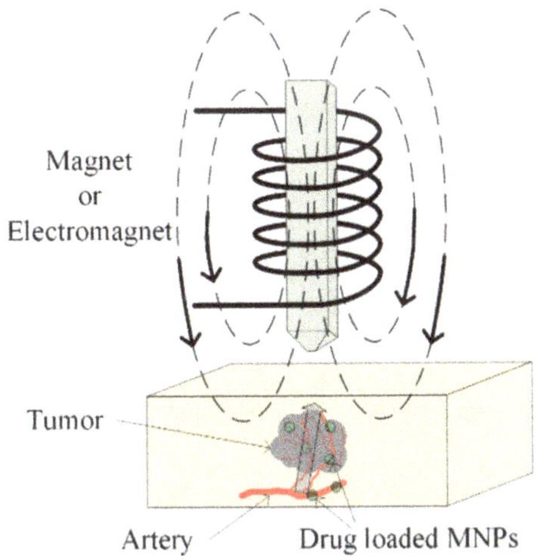

Figure 13. Magnetic drug targeting application. Drug-loaded MNPs are directly injected in the arterial blood of the tumor. A magnetic field is applied, and it causes MNPs' enhancement in the tumoral zone.

This shell, formed from tree layers of biodegradable and biocompatible polymers, helps to improve the therapeutic efficacity. This hydrophilic shell inhibits the reticuloendothelial system's capacity to recognize and eliminate the MNPs. A higher stability in the physiological fluids observed, and increased blood circulation is obtained. The magnetite hydrophobic core is suitable to be loaded with lipophilic medicines, and their release at the tumor site is controlled.

The magnetic nanoparticles have dimensions between 10 and 100 nm, and thus the accumulation of SPIONs at the tumor zone is performed via the passive targeting process, as well as the EPR effect. Folic acid (FA) can be linked to the folate receptor (FR), which is present in cancerous cells such as GBM, breast and ovarian unhealthy tissues [107].

Another way to prepare an efficient treatment for GBM was proposed in [108], by preparing lipid magnetic nanovectors (LMNVs) loaded with SPIONs and functionalized with transferrin receptor antibodies (anti-TfR Ab). An outer shell made from biotin and m-PEG was analyzed, exploiting a biotin–streptavidin interaction.

Another study described aminosilane-coated SPIONs and took into account the use of magnetic hyperthermia to treat the GBM [109]. If a SPION is loaded with TMZ, its efficacity is increased by using a chlorotoxin (CTX), which that has a beneficial effect on the final product's stability [110].

Some studies [111,112], in the case of bladder cancer, have shown that the treatment with mitomicyn-C decreased the recurrence rate after 10 years by 40%.

5.3. Immunotherapy

The immunotherapy method is an old one, but classical treatments are effective only in a small group of patients. Factors that influence the treatment success are previous treatments, the heterogeneity of the tumor and its stage, and the immunosuppressive phenotype of the cancer [113]. A higher grade of immunogenicity of the tumor ensures a better response to the immunotherapy. This type of therapy increases the immune system response to the cancerous tumor, by stimulating it or by hindering the cancerous cell signal. Artificial antibodies were developed that fight against the tumor, using different routes, such as antigens (Herceptin-targeting HER-2/neu for breast cancer), antibodies, which sever the blood supply of the tumor, cytokines such as interferon α and β and immune inhibitors (PD-1 inhibitors) [114,115].

Superparamagnetic nanoparticles are used together with immunotherapy, and the particle dimensions are very important. SPIONs with dimensions lower than 1000 nm are used for liver treatments, and particles smaller than 10 nm can be filtered by the kidneys. Due to lympho-hepatic tropism, SPIONS with intermediate dimensions can form agglomerations in the spine, liver or spleen, and in lymph nodes. These nanoparticles demonstrate an interaction with macrophage monocytes, dendritic cells and tumoral cells, which can be embedded through phagocytosis or micropinocytosis, depending on the surface nature [116,117].

The macrophage cells are phagocytic cells of the innate immune system. They represent the first response of the human body to infections, acting in pathogen recognition, the degradation of cellular intruders, and phagocytosis. Macrophages work well in the presence of T cells, and they induce tissue homeostasis, repair, and reconstruction of the damaged tissue. They also have a great disadvantage, because they are usually associated with autoimmune disease, cancers, angiogenesis, metastasis, and tumor development [118].

Dendritic cells are also a cellular line that detect the existence of pathogens and play an important role in the activation of the immune system. They have antigenic peptides on the major histocompatibility complex (MHC), and induce T-cell activation and differentiation. Dendritic cells are usually called sentinel cells, and they are located inside the lymphoid organs or at interfaces such as the intestines or skin [119].

The human immune system has different mechanisms for transport and integration, in which the magnetic nanoparticles are used in tumor treatment and differential medical imaging. A particular case is lymphatic drainage, realized using magnetic nanoparticles. MNPs are delivered in the lymphatic system, or they are transported by the antigen cells and guided through the lymphatic system to lymph nodes, where MNPs localize in different types of cells, and can regulate the immune system's response. In order to increase the embedding process of the nanoparticles, different types of receptors and antibodies can be linked. Antibodies were used which control the epithelial cells' growth for the encapsulation of the carcinoma cells [120]. If the corresponding cells are not vital for the human body, the antigen differentiation process can be successfully applied.

The T anticancer cells are considered indispensable tools for the eradication of different types of neoplasm. If, at the tumor site, an increased number of T cells is detected, the survival rate of the patient increases. T cells have receptors that can recognize pathological cells at the tumor site. By blocking the proteins CTLA-4 and PD-1 through specific antibodies, which are present in drugs, they can restore the entire immune system [121–124]. Pembrolizumab recently received FDA approval for different types of cancers, such as esophageal or gastroesophageal carcinoma, melanoma, applicable for patients with inoperable tumors or metastatic cases, non-small cell lung cancer, small cell lung cancer, head and neck squamous cell cancer, classical Hodgkin Lymphoma, etc. This drug is a monoclonal antibody directed against the PD-1 receptor on the cell surface, and it prevents the binding and activation of PD-L1 and PD-L2 and activates the T cells. Another monoclonal antibody is nivolumab, and it received FDA approval in 2015. Its action route is similar to pembrolizumab, and it can be combined with different chemotherapeutic approaches. It is very suitable for use with MNPs, but the PD-1 monoclonal antibodies target ligands and they can break the link between PD-L1 on cancerous cells and the immune system [125,126]. The first experiments consisted of loading the T cells with SPIONS as vehicles for immune modulatory substances and injecting and guiding them into the tumor sites by applying an external magnetic field [125]. Magnetic nanoparticles have an influence on macrophage polarization modulation, because it is already well known that the macrophage monocytes present two polarization states: non-tumorigenic M1 and pro-tumorigenic M2. In the case of a human monocyte line, it is possible to transform the phenotype M2 into phenotype M1, when SPIONs are used [126–140].

5.4. Radiotherapy

Radiotherapy is one of the most important treatments used to fight cancer. Tumor radiosensitization is an important phenomenon, and it can be enhanced by combining classical radiotherapy with different drugs or even with magnetic nanoparticles. Magnetic nanoparticles present a high value of the atomic number Z and exhibit important X-ray photon absorption, leading to an increase in secondary electrons that produce human DNA deterioration when the radiation dose is increased [141]. Nanoparticles, which are characterized by a high proton number, are usually linked to increased oxidative stress, expressed by reactive oxygen species (ROS) such as hydroxyl radicals that damages the cells' DNA.

SPIONs were used for a long time as contrast agents in MRI investigations, but they increase the radiosensitivity of the tumor, making the radiotherapy treatments more effective [142–144]. In [145], it was found that the radiosensitizing properties of SPIONs have an influence on the MCF7 breast cancer cells, a fact that was indicated by the production of reactive oxygen species. This effect was analyzed in [146,147] in the case of colon cancer cells, in which SPIONs were combined with a 150 MeV proton beam. When 13 nm iron oxide nanoparticles were concomitantly used with 7.1 keV monochromatic X-rays, the colon tumor cells in murine animal model were drastically reduced [148]. In the radiotherapy treatment domain, energy sources in the MeV range are used, and the Compton effect is present [149].

In [150], the radiosensitizing efficacity of Fe_3O_4–OA, Ag and Fe_3O_4@Au nanoparticles was compared on a U251 cellular line (GBM line). A schematic representation of the process is presented in Figure 14.

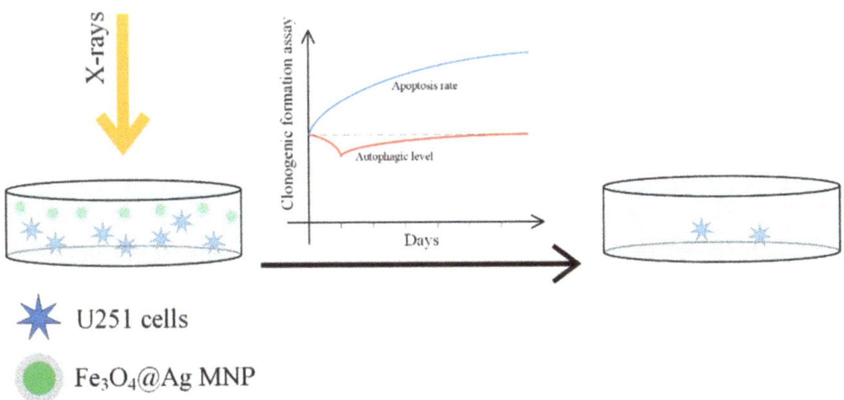

Figure 14. Fe_3O_4@Ag nanoparticles acting as a nano-radiosensitizer for GBM cellular culture treatment. The MNP radiosensitivity magnification effect is due to a decrease in the autophagy at the beginning of the treatment and followed by an increase in the apoptosis.

When SPIONs functionalized with gold or silver are used, the surface properties are changed. The gold nanoparticles increase the cellular death rate due to radiation, by interacting with thiols in the cytoplasm. Silver nanoparticles increase radiosensitivity in glioma by inducing cell apoptosis. This enhancement is stronger than in the case of gold nanoparticles [151–155]. SPIONs are guided to the tumor sites with the help of an external magnetic field, they can produce an increase in the local temperature due to magnetic hyperthermia, and through a Fenton reaction they can lead to ROS production. The best results were obtained in the case of Fe_3O_4@Ag, and it can be concluded that this material has a highly efficient radiosensitivity that adds to the radiation effect and destroys the cancer cell. The classical cell death processes due to radiation are necrosis, apoptosis, and autophagy [156].

Some studies have reported that a combination between growth factor receptor (C225) and Fe_3O_4@Ag can increase the sensitivity of nasopharyngeal carcinoma cells to radiotherapy, and the radiosensitivity of Fe_3O_4@Ag is higher than gemcitabine (Gemzar) or 5-fluoroacil [157].

In [158], 38 clinical studies were compared, obtained between 1987 and 2014. This clinical research took into account different types of neoplasm, such as head, neck, esophagus, skin, lungs, etc. Radiotherapy was applied as a singular treatment to 1717 persons, and a combined treatment of radiotherapy and hyperthermy (thermoradiotherapy) was used for 1761 patients. In these cases, an increased rate of survival of 54.9% was recorded, as compared with the rate of 39.8% that was determined for single therapy. The hyperthermic effect of SPIONs increases the radiotherapy's effect [158,159].

5.5. Magnetic Hyperthermia

Magnetic hyperthermia has been known about since 1957, and it was first presented in [160], where cancerous tumors were heated with the help of magnetic particles placed in an external magnetic field (Figure 15).

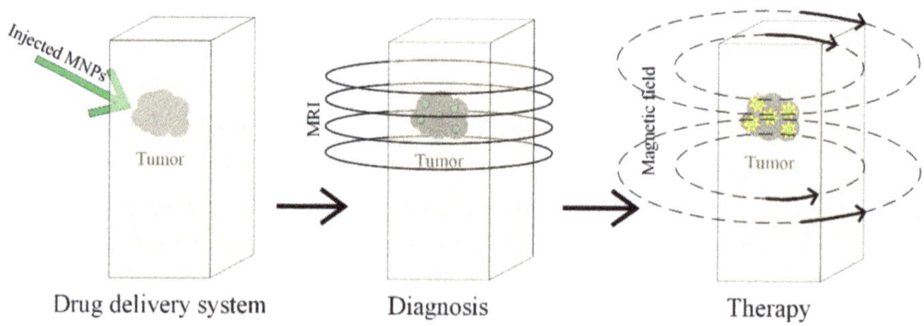

Figure 15. Diagnosis and treatment of the tumor based on MNPs, using magnetic hyperthermia.

When magnetic nanoparticles are mediated with magnetic hyperthermia, the treatment is very efficient, because a lot of cancerous cells are killed, and the surrounding healthy tissues are left undamaged [160–165]. This combined treatment is based on intracellular hyperthermia, and it is possible to heat only the harmful cells and the link between cells and targeting ligands with MNPs becomes possible. It can be observed that the local and homogenous heat presents a high grade of selectivity. This type of treatment is in clinical trials for GBM cancer [166].

Magnetic hyperthermia that uses MNPs or SPIONs (SPIONs-MH) is based on locally increasing the temperature above 42 °C, with an impact on cancer cell physiology, by altering the structural proteins and generating the necrosis or apoptosis of the cells [162,165,167].

The treatment's success depends on the following factors: (1) the cancerous cells must be selectively targeted; (2) the SPIONs' penetration inside the cancerous cells should be facile; (3) the encapsulation of the SPIONs in the unhealthy cells has to be completed; (4) SPIONs must be biocompatible and non-toxic; (5) the treatment should be feasible from the clinical point of view; (6) the treatment has to be a general one, and different types of tumors can be treated, even metastatic ones [168,169].

The SPIONs-MH have strict limitations regarding the effective value H and the frequency f of the alternating applied magnetic field, so the product $H \times f$ must be strictly lower than 5×10^9 due to biomedical conditions [170]. When SPIONs are used, it must be taken into account that sometimes, their low thermal conversion efficiency is due to a decreased value of magnetic susceptibility, and this is an important problem in medical applications. In order to address this limitation and to obtain improved induction heating properties, modulation of the particle size [171], a strict control of the chemical composition [172–179], biofunctionalization of the SPIONs' surface, and modification of the particle shape [86,180] should be taken into account. It is already well-known that an accumulation of SPIONs in lysosomes, which are known as the "suicidal bags" of the cell, because they can destroy their own cells through specific hydrolytic enzymes, leads to lysosomal cell death by increasing the local temperature and the production of ROS inside the lysosome [181,182].

The heating effect of SPIONs is due to the magnetic hysteresis phenomenon, Neel relaxation, induced eddy currents and Brownian relaxation. The time interval between two random flips of the SPION magnetization due to temperature is known as the Neel relaxation time. SPIONs are ferrimagnetic or ferromagnetic nanoparticles, because they contain iron, and so a small hysteretic dependence between magnetic flux density and magnetic field strength is expected to appear. The energy losses, necessary to magnetize such a material, are usually called hysteresis losses. The heat generated by SPIONs is equal to the hysteresis cycle area, and in the case of a very small nanoparticle, the magnetization of the material is due to spin magnetic moment rotations. When an alternating magnetic field is applied, the spins' rotation must overcome the friction phenomenon, which appears between the easy magnetization axis of the particles and atomic lattices due to the Neel relaxation phenomenon. The friction effect can also be generated between the SPIONs and the area near them in the case of Brownian relaxation. In nano-magnetism, the Neel and Brown relaxation mechanisms are classical processes which completely characterize the single-domain nanoparticle magnetization [77].

If the time necessary to measure the magnetization of the nanoparticle is higher than the Neel relaxation time, in the absence of an external magnetic field, the average magnetization of SPIONs, dispersed in different aqueous solution, is equal to zero [183].

The main experimental technique used for measuring MNPs' magnetic heating is calorimetry, which measures the specific heat absorption rate (SAR), specific loss power (SLP) and intrinsic loss power (ILP) [181,184]. The hysteresis loop of SPIONs or MNPs is determined through alternating current magnetometry. The heat dissipated during a cycle is equal to the hysteresis area, and SAR can be computed as in [171,181], but this type of experimental measurement is performed at different values for the magnetic field strength and frequency that are not in the biological range, and so the magnetic characterization is not entirely accurate for SAR detection [185]. In Table 3, experimental data, obtained in the case of Fe_3O_4 magnetic nanoparticles with different shape and sizes, are presented [184,185].

Table 3. Shape and size effect on MNPs' magnetic heating parameters for Fe_3O_4 nanoparticles [184,185].

Shape	Size (nm)	f (kHz)	M_s (emu/g)	H (Oe)	SAR/SLP (W/g)	ILP (nHm^2/kg)
sphere	22	500	65	195	716	5.96
disc	125	488	435	600	5000	4.48
ring	70	400	-	744	3050	2.18

When the shell of an MNP has been functionalized with a fluorophore through a bond, which can break at a given temperature [186], it is possible to measure the fluorophore content after the heating process, and the bond breaking temperature is determined through fluorescence thermometry.

In the case of the application of an alternating magnetic field, the MNPs develop due to the Joule effect and eddy current losses, which contribute to an increase in the total heating effect. In [187], the thermogenesis of MNPs when an external magnetic field is applied was presented. Films of Au@MNPs present a magnetothermal effect at a high frequency in the order of 10^2 kHz.

MH is a promising therapeutic strategy in GBM treatment, as depicted in Figure 16.

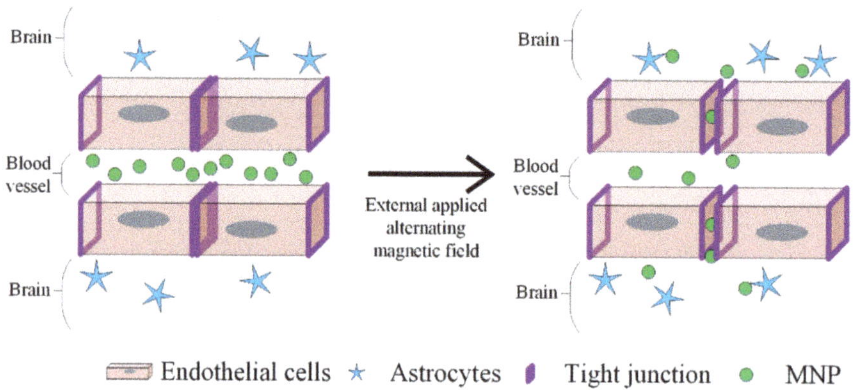

Figure 16. BBB disruption using MNPs and applying an alternative magnetic field. The heat generated by MNPs opens up the tight junction to deliver the drugs. The endothelial cells, astrocytes and MNPs are represented by different symbols, as indicated in the figure legend.

SPIONs can be coated with aminosilane, a polymer that improves protein and cell adhesion. This material was used in preclinical investigations [188–190] and in some phase I and II clinical trials [191–196]. The first clinical study was described in [195], and it was applied in the case of 14 GBM patients. The entire group received a 15 nm aminosilane-coated MNP suspension that was neuro-navigationally guided through the tumor site, followed by 4 to 10 thermotherapy applications. At a maximum field value of 14 kA/m in 90% tumor volumes, a temperature between 39 and 46 °C was obtained.

In 2009, a second postmortem study on three GBM patients was developed, and histological analysis showed that MNPs were confined near the tumor sites, demonstrating that it is necessary to administer MNPs in multiple places in order to obtain a uniform heat production [193].

Another clinical study performed on 66 patients using MH in combination with radiotherapy was published in [196]. Although an increased survival rate was obtained, serious side effects such as increased intracranial pressure or necrosis [197] and moderate side effects such as headache, fatigue, thrombocytopenia, leukopenia, vomiting were reported.

The general characteristics of the SPIONs used in MH is that the SAR depends on the particle size, shape, magnetization, chemical composition, and concentration of SPIONs, magnetic field intensity and frequency. It is better that magnetic fields with high oscillation frequencies are avoided due to eddy current generation [198].

Magnetic hyperthermia mediated through SPIONs has proven its efficacity, by causing about 50% cell death of the U87-MG cellular line (GBM—malignant glioma) compared with only 20% cell death in the HUVEC line (healthy endothelial cells) [199]. Zinc-doped ferrite nanoparticles have a higher saturation point than in the case of classical particles and a constant anisotropy energy, and they have proven to be effective for U87-MG by inducing a hyperthermia effect for a frequency of 700 kHz at 1.7 kA/m [200]. The achieved temperature was 41.5°C, and it was high enough to produce cell death. Human-like collagen protein-coated SPIONs have a rapid heating capacity, by comparing them with uncoated particles, because the MNPs' agglomeration is reduced, and they have an enhanced surface stability [194]. Another approach consists of using iron oxide-coated cationic liposomes, because they present a high cellular affinity and electrostatic interactions with the anionic phospholipids from the cellular membrane. Their effect was investigated on T-9 glioma cells, and a temperature of 42 °C was achieved in 20 min, and as a final result, 100% cellular cancerous death was obtained in 40 min [174,194,199,201,202]. In [194,203], poly-ethylene-glycol-based hydrogel with MNPs, which induces MH, was analyzed on M059K glioblastoma cells, and the results were satisfactory.

Magnetic hyperthermia has a great potential to become a first line treatment for GBM patients and for other cancers. Other studies have incorporated MNPs and methotrexat (MTX) to treat breast cancer [159,204–206]. Other chemotherapeutical medicines, such as doxorubicin, can be combined with MH treatments, making this approach more effective. Another advantage consists of the damage to the extracellular matrix of cancer cells, due to heat effect that facilitates drug elution at the tumor site. These types of studies were performed on a murine animal model.

5.6. Gene Therapy

Gene therapy is similar to chemotherapy, and the main difference between them is that instead of delivering small drugs, guided biological agents such as DNA, RNA or siRNA, proteins, and peptides are delivered to the tumor site, and in this way the cellular death can be induced. When cancer appears in one body part, the deteriorated human DNA is characterized through the appearance of atypical proteins, and the disease can be treated through defect gene replacement inside the cancerous cells. Alternatively, siRNA works by suppressing the protein actions, which are due to the cancerous cells. As a classical definition, siRNA is a double-stranded interfering RNA formed from a 21–23-nucleotide sequence. siRNA is combined with nanoparticles that can carry and deliver it to cells, because medical RNA has the property to be very degradable when it is directly injected into the bloodstream [207]. From siRNA, RNA-induced silencing complex (RISC) is produced, which is formed in the cell cytoplasm. The RISC binds the mRNA of the cell. siRNA is usually used to target oncogenes. When siRNA is carried by an MNP, the particle becomes almost unrecognizable for the human body, the cell uptake will be improved, and the circulation time becomes longer. The MNPs will accumulate at the tumor sites and due to tumor vascularization, the particle can easily enter inside the tumor, and the poor lymphatic drainage permits particle accumulation.

The magnetic nanoparticles, which encapsulate or bind DNA or RNA, are covered with a cationic polymer, leading to a positively charged layer, favoring an electrostatic interaction between negatively charged nucleic acid and the MNPs. Other studies showed that for a successful DNA delivery, it is necessary that the MNPs interact with the cell nucleus [208,209]. Recent approaches involved the development of "intelligent" MNPs that use specific coatings, which prevent the delivery carrier from being degraded and permit drug release after entering the tumoral cell. Some initial coatings were performed with disulfide-containing polyethylimine (PEI), and in recent years, polymer shell coatings based on chitosan, PEI, and PEG were developed [210].

In the case of GBM cancer, the tumor presents many genetic aberrations, and cationic-coated MNPs as non-viral vehicles for gene drugs were able to deliver tumor necrosis factor-related apoptosis-inducing ligand (hTRAIL), encoding plasmid open reading frame (pORF-hTRAIL) with high efficiency and survival rate in a rodent glioma model [211]. In the near future, gene delivery with the help of MNPs could become a safe option to treat GBM, because MNPs represent a new way to address infiltrating cells of GBM.

5.7. Ferroptosis

Ferroptosis is classified as an oxidative regulated cell death (RCD), and the main involved mechanisms are iron-dependent peroxidation of lipids and plasma membrane rupture [212]. Inside the living cells, the mitochondria represent the most important reactive oxygen species (ROS) source. Ferroptosis can be carried out in two ways: extrinsic or intrinsic. The extrinsic modality consists of cell membrane transporters' (cystine/glutamate) inhibition or by stimulating iron transporters such as serotransferrin and lactotransferrin [213]. The intrinsic pathway is characterized by a blocking of intracellular antioxidant enzymes such as glutathione peroxidase (GPX4). In the case of an increased intracellular amount of lipid reactive oxygen species (L-ROS) that exceed the GPX4, it results in damage to the cellular redox homeostasis [214].

Ferroptosis has three important characteristics, as follows: the oxidation reaction of polyunsaturated fatty acid (PUFA) from the phospholipids of the cellular membrane, the presence of iron as a result of redox chemical reaction, and finally the loss of lipid hydroperoxide's (LOOH) repair properties [215]. It can be concluded that ferroptosis is usually due to a deadly metabolic imbalance, and it can represent an adequate solution for cancers which are resistant to other therapies [216]. Changes that appear in the cell mitochondria, such as volume decrease, bilayer membrane density increase, external mitochondrial membrane disruption, and the absence of mitochondrial cristae [217,218], are important parameters and favor the ferroptosis process. In [219], it was shown that if a clear and spherical cell which contains an empty cytosol is microscopically detected, is considered to be a ferroptotic cell.

From a biochemical point of view, the ferroptotic cell is characterized by the peroxidation of PUFAs in membrane phospholipide due to an increased quantity of redox active divalent iron Fe^{2+}. This process is controlled through GPX4, which converts LOOH into a lipid alcohol [220].

Different genes connected with iron metabolism in the human body (transferrin receptor TFR1, iron response element binding protein 2, IREB2 and ferritin heavy chain FtH), lipid synthesis and oxidative stress pathways are also involved in controlling the ferroptosis process [221,222].

Ferroptosis can be considered an useful tool, because the FDA have approved the so-called ferroptosis inducers (erastin, sorafenib, sulfasalazine (SAS), cisplatin, 1-buthionine sulfoximine (BSO), artesunate (ART), lanperisone, RSL3, altretamine, iron ionophores, doxorubicin into mesoporous carbon nanoparticles/withaferin A/poly(ethylene glycol)-coated (PEGylated) silica nanoparticles, etc.).

SPIONs are very important in this RCD process, because they induce cellular death through an increase in iron level and ROS generation [223]. ROS include superoxide, oxygen peroxide, hydroxyl radical, singular oxygen species, peroxyl radical, alkoxyl radical, hypochlorous acid, and ozone. In [223], cisplatin with SPIONs as a carrier based on FePt had proven efficacity through the hydrolysis of iron oxide contained in SPIONs. Cancer cells have an acidic pH, which favors iron intracellular release and Fenton-like reaction generation, which produce ROS. The chemical Fenton-like reaction consists of divalent iron Fe^{2+}'s reaction with hydrogen peroxide (H_2O_2), generating ROS [224–226].

When SPIONs, considered as ROS generators through a Fenton-like reaction, are combined with anticancer drugs, the treatment will be more efficient. The toxicity of SPIONs is shown in Figure 17 and consists of cell membrane deterioration, cytoskeleton structural modifications, oxidative damage of DNA chains, lysosome working condition modifications, ROS generation, mitochondria damage, and the generation of inflammatory factors [227,228].

Mitochondria-mediated ferroptosis has a potential use in cancer treatments, but supplementary in vivo models to fulfill this theory are necessary at this moment of time [213,221,222,224–227].

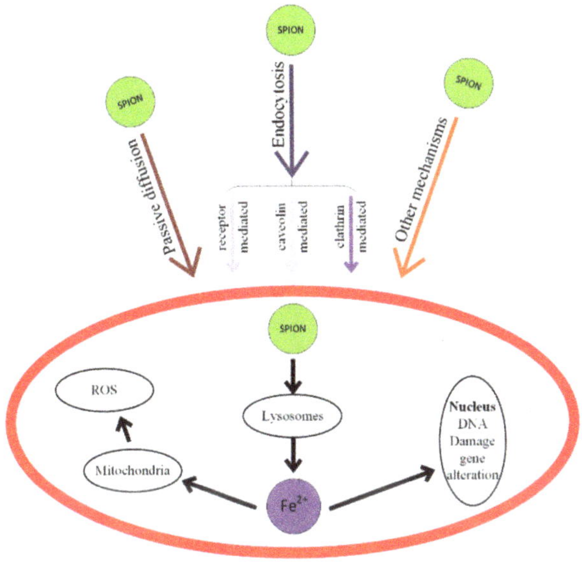

Figure 17. Mechanisms of SPIONs' interaction with cells and types of induced toxicity. SPIONs mainly interact with cells through passive diffusion or endocytosis. The endocytosis pathways are: clathrin-mediated, caveolin-mediated and other receptor-mediated.

6. Challenges and Discussions Regarding the MNPs Use in Cancer Therapy

Nanoparticles are used more and more oftne in many medical fields as orthopedy, stomatology and neurosurgery. In [229–232], composites consisting of a combination between magnetic nanoparticles, calcium phosphates and biopolymers are prepared through a biomimetic co-precipitation method for bone tissue engineering. In dental restauration, metallic nanoparticles are usually involved in designing adhesives used in coherence tomography evaluation. Graphene-based nanomaterials are another innovative solution for tissue engineering in the dental field [233–235]. Selenium nanoparticles were involved in surface modification of the titanium mesh for cranioplasty [236–238]. In the nanoparticle field, magnetic nanoparticles represent a special class, because they demonstrate an intrinsic association with oncological domain.

Although some MNPs and SPIONs have received FDA approval, their use as treatments in cancer therapy should be carefully conducted and controlled, because a high amount of these materials could be harmful. MNPs and SPIONs can exhibit toxicity, and thus they sometimes are functionalized with different inorganic or organic compounds in order to obtain a passivate surface. Modern vectors for drug-delivery applications are also silica-based mesoporous structured materials, due to their specific textural characteristics, consisting of uniform pores with laboratory-controllable sizes, a high specific surface area and high chemical and thermal stability [60,61]. Different studies, presented in the paper, provide evidence that MNPs have a very low level of toxicity, when their concentration in solutions is lower than 100 μg/mL. The analysis made in vivo or in vitro proves that SPIONs can alter living cells due to cytoskeletal deterioration, ROS generation, mitochondria damage, and DNA structure change. It is supposed that the SPIONs' toxicity is not due only to their ferromagnetic or ferrimagnetic cores, but also to nanoparticle agglomerations, which can become instable in biological media or different solutions, similar to body fluids. In the case of SPIONs, the shape (spherical, rod or beam), the dimensions, the outer shell, and the iron/material ratio used for the outer shell (polymer or gold) are important parameters that must be taken into account.

If laboratory-developed magnetic nanoparticles are used in animal models, their toxicity has to be carefully analyzed. A high number of MNPs or SPIONs, FDA-approved or laboratory-developed, were successfully applied on murine animal models. In 2007, clinical studies were initiated regarding the SPIONs used in the treatment of GBM patients, through magnetic hyperthermia. If MNPs are addressed for clinical investigations, the physic-chemical properties, toxicity, and the possibility to become carriers for different drugs have to be considered. The shell of the magnetic core must be biocompatible, and the liver and kidneys have to synthetize them fast. The degraded components must also be smaller than 5 nm.

When SPIONs are used in magnetic hyperthermia, their distribution is uniform and the therapeutical temperature has to be achieved inside the tumor volume, and the surrounding tissue must present minimal damage [228].

MNPs and SPIONs have different side effects, such as liver, brain, skin, stomach function deterioration, with visible symptoms such as hair loss, nausea and vomiting and even nerve damage that could unfortunately be persistent for 10 years [213].

MNPs are monitored with the help of MRI techniques, and this is a very important step in cancer monitorization and drug delivery treatments. Today, MNPs are of great importance, because they are considered very useful in early-stage cancer detection, and in the near future, they may be a viable treatment for patients who suffer from aggressive types of cancer such as GBM. It is expected that this technology will applied at a medium scale for patients at a global level.

Some studies [213,221,222,224] claim that the side effects of the MNPs used in chemotherapy, radiotherapy, gene therapy, immunotherapy and magnetic hyperthermia can be successfully addressed by the self-defense system of the body.

The most important characteristic of the MNPs is that they can be linked to drugs, antibodies, proteins, enzymes and can directly target a specific organ or tumor site using an external magnetic field. MNPs or SPIONs can be heated in alternating magnetic fields and they can destroy the cancer cells.

An important collaborative work between medical specialists and nanomaterial science researchers is important, because together, they can generate steps forwards to ameliorate or cure cancer. As future perspectives, important progress has been made in the development of new MNPs for pre-clinical cancer theranostics applications. Notable steps were made in the synthesis of efficient magnetite/mesoporous media as drug delivery treatments, and in the near future, complex platform could be developed for cancer treatments [60,61,239]. Few nanomaterials have shown success in clinical cases, because supplementary research must be carried out in order to understand nanomaterial–human cell interactions and biological barriers in different types of cancers [240,241]. There are also industrial barriers because good laboratory practice has to be developed. An increase in the number of preclinical trials with MNPs has been observed, and it is expected that this will evolve into clinical studies, by taking into account the great potential of MNPs to diagnose and treat cancer. The in vivo behavior of MNPs must be carefully investigated. In order to obtain a high efficiency of MNP treatments, a size reduction and the use of different bio-compatible shells have to be taken into account. In this way, the blood circulation and the time taken to reach the cancerous tissues is greatly reduced. As we have underlined in the paper, despite MNPs' success in murine models as theranostic agents, their loading capacity and their affinity to target cancer cells must be analyzed and improved if possible. The topic of MNPs is a new and important research domain, which is in continuous development, and synthesis of magnetic drug delivery systems should be investigated. Two very important future perspectives consist of MNPs' elimination from the living body and overcoming the reported long-term toxicity cases.

Author Contributions: V.M. and G.P. conceived and designed the study, V.M., M.V. and I.A. acquired the data, V.M. and G.P. wrote the paper, I.A. provide original experimental results, I.A. and M.V. critically revised the article. All authors have read and agreed to the published version of the manuscript.

Funding: This research received no external funding.

Institutional Review Board Statement: Not applicable.

Informed Consent Statement: Not applicable.

Data Availability Statement: Data sharing not applicable.

Conflicts of Interest: The authors declare no conflict of interest.

References

1. Seo, W.; Lee, J.; Sun, X.; Suzuki, Y.; Mann, D.; Liu, Z.; Terashima, M.; Yang, P.C.; McConnel, M.; Nishimura, D.; et al. FeCo/graphitic-shell nanocrystals as advanced magnetic-resonance-imaging and near-infrared agents. *Nat. Mater.* **2006**, *5*, 971–976. [CrossRef]
2. Laurent, S.; Saei, A.A.; Behzadi, S.; Panahifar, A.; Mahmoudi, M. Superparamagnetic iron oxide nanoparticles for delivery of therapeutic agents: Opportunities and challenges. *Expert Opin. Drug Deliv.* **2014**, *11*, 1449–1470. [CrossRef]
3. Cornell, R.M.; Schwertmann, U. *The Iron Oxides: Structure, Properties, Reactions, Occurrences and Uses*, 2nd ed.; Wiley-VCH: Weinheim, Germany, 2003.
4. Ahmad, T.; Bae, H.; Rhee, I.; Chang, Y.; Jin, S.U.; Hong, S. Gold-coated iron oxide nanoparticles as a T2 contrast agent in magnetic resonance imaging. *J. Nanosci. Nanotechnol.* **2012**, *12*, 5132–5137. [CrossRef]
5. Zhu, N.; Ji, H.; Yu, P.; Niu, J.; Farooq, M.U.; Akram, M.W.; Udego, I.O.; Li, H.; Niu, X. Surface Modification of Magnetic Iron Oxide Nanoparticles. *Nanomaterials* **2018**, *8*, 810. [CrossRef]
6. Ghorbani, M.; Hamishehkar, H.; Arsalani, N.; Entezami, A.A. A novel dual-responsive core-crosslinked magnetic-gold nanogel for triggered drug release. *Mater. Sci. Eng. C Mater. Biol. Appl.* **2016**, *1*, 436–444. [CrossRef]
7. Lo, C.K.; Xiao, D.; Choi, M.M.F. Homocysteine-protected gold-coated magnetic nanoparticles: Synthesis and characterization. *J. Mater. Chem.* **2007**, *17*, 2418–2427. [CrossRef]
8. Hu, Y.; Meng, L.; Niu, L.; Lu, Q. Facile Synthesis of Superparamagnetic Fe_3O_4@polyphosphazene@Au Shells for Magnetic Resonance Imaging and Photothermal Therapy. *ACS Appl. Mater. Interfaces* **2013**, *5*, 4586–4591. [CrossRef]
9. Hoang Thi, T.T.; Nguyen Tran, D.H.; Bach, L.G.; Vu-Quang, H.; Nguyen, D.C.; Park, K.D.; Nguyen, D.H. Functional Magnetic Core-Shell System-Based Iron Oxide Nanoparticle Coated with Biocompatible Copolymer for Anticancer Drug Delivery. *Pharmaceutics* **2019**, *11*, 120. [CrossRef] [PubMed]
10. Kumari, P.; Ghosh, B.; Biswas, S. Nanocarriers for cancer-targeted drug delivery. *J. Drug Target* **2016**, *24*, 179–191. [CrossRef] [PubMed]
11. ud Din, F.; Aman, W.; Ullah, I.; Qureshi, O.S.; Mustapha, O.; Shafique, S.; Zeb, A. Effective use of nanocarriers as drug delivery systems for the treatment of selected tumors. *Int. J. Nanomed.* **2017**, *12*, 7291. [CrossRef] [PubMed]
12. Yalcin, S.; Unsoy, G.; Mutlu, P.; Khodadust, R.; Gunduz, U. Polyhydroxybutyrate-Coated Magnetic Nanoparticles for Doxorubicin Delivery: Cytotoxic Effect Against Doxorubicin-Resistant Breast Cancer Cell Line. *Am. J. Ther.* **2014**, *21*, 453–461. [CrossRef]
13. Fazilati, M. Anti-neoplastic Applications of Heparin Coated Magnetic Nanoparticles Against Human Ovarian Cancer. *J. Inorg. Organomet. Polym. Mater.* **2014**, *24*, 551–559. [CrossRef]
14. Kostevšek, N. A Review on the Optimal Design of Magnetic Nanoparticle-Based T2 MRI Contrast Agents. *Magnetochemistry* **2020**, *6*, 11. [CrossRef]
15. Persano, S.; Das, P.; Pellegrino, T. Magnetic Nanostructures as Emerging Therapeutic Tools to Boost Anti-Tumour Immunity. *Cancers* **2021**, *13*, 2735. [CrossRef] [PubMed]
16. Hajal, C.; Campisi, M.; Mattu, C.; Chiono, V.; Kamm, R.D. In vitro models of molecular and nano-particle transport across the blood-brain barrier. *Biomicrofluidics* **2018**, *12*, 042213. [CrossRef]
17. Barua, S.; Mitragotri, S. Challenges associated with Penetration of Nanoparticles across Cell and Tissue Barriers: A Review of Current Status and Future Prospects. *Nano Today* **2014**, *9*, 223–243. [CrossRef] [PubMed]
18. D'Agata, F.; Ruffinatti, F.A.; Boschi, S.; Stura, I.; Rainero, I.; Abollino, O.; Cavalli, R.; Guiot, C. Magnetic Nanoparticles in the Central Nervous System: Targeting Principles, Applications and Safety Issues. *Molecules* **2018**, *23*, 9. [CrossRef] [PubMed]
19. Sun, C.; Lee, J.S.; Zhang, M. Magnetic nanoparticles in MR imaging and drug delivery. *Adv. Drug Deliv. Rev.* **2008**, *60*, 1252–1265. [CrossRef] [PubMed]
20. Ebadi, M.; Bullo, S.; Buskaran, K.; Hussein, M.Z.; Fakurazi, S.; Pastorin, G. Dual-Functional Iron Oxide Nanoparticles Coated with Polyvinyl Alcohol/5-Fluorouracil/Zinc-Aluminium-Layered Double Hydroxide for a Simultaneous Drug and Target Delivery System. *Polymers* **2021**, *13*, 855. [CrossRef]
21. Hayashi, K.; Tokuda, A.; Nakamura, J.; Sugawara-Narutaki, A.; Ohtsuki, C. Tearable and Fillable Composite Sponges Capable of Heat Generation and Drug Release in Response to Alternating Magnetic Field. *Materials* **2020**, *13*, 3637. [CrossRef]
22. Hosu, O.; Tertis, M.; Cristea, C. Implication of Magnetic Nanoparticles in Cancer Detection, Screening and Treatment. *Magnetochemistry* **2019**, *5*, 55. [CrossRef]
23. Chen, Z.; Wu, C.; Zhang, Z.; Wu, W.; Wang, X.; Yu, Z. Synthesis, functionalization, and nanomedical applications of functional magnetic nanoparticles. *Chin. Chem. Lett.* **2018**, *29*, 1601–1608. [CrossRef]
24. Cristea, C.; Tertis, M.; Galatus, R. Magnetic nanoparticles for antibiotics detection. *Nanomaterials* **2017**, *7*, 119. [CrossRef]

25. Xu, J.K.; Zhang, F.F.; Sun, J.J.; Sheng, J.; Wang, F.; Sun, M. Bio and nanomaterials based on Fe_3O_4. *Molecules* **2014**, *19*, 21506–21528. [CrossRef]
26. Liu, C.; Zou, B.; Rondinone, A.J.; Zhang, Z.J. Reverse micelle synthesis and characterization of superparamagnetic $MnFe_2O_4$ spinel ferrite nanocrystallites. *J. Phys. Chem. B* **2000**, *104*, 1143–1145. [CrossRef]
27. Abd Elrahman, A.A.; Mansour, F.R. Targeted magnetic iron oxide nanoparticles: Preparation, functionalization and biomedical application. *J. Drug Deliv. Sci. Technol.* **2019**, *52*, 702–712. [CrossRef]
28. Ahmad, F.; Ashraf, N.; Ashraf, T.; Zhou, R.B.; Yin, D.C. Biological synthesis of metallic nanoparticles (MNPs) by plants and microbes: Their cellular uptake, biocompatibility, and biomedical applications. *Appl. Microbiol. Biotechnol.* **2019**, *103*, 2913–2935. [CrossRef]
29. Karade, V.C.; Dongale, T.D.; Sahoo, S.C.; Kollu, P.; Chougale, A.D.; Patil, P.S.; Patil, P.B. Effect of reaction time on structural and magnetic properties of green-synthesized magnetic nanoparticles. *J. Phys. Chem. Solids* **2018**, *120*, 161–166. [CrossRef]
30. Massart, R. Preparation of aqueous magnetic liquids in alkaline and acidic media. *IEEE Trans. Magn.* **1981**, *17*, 1247–1248. [CrossRef]
31. Montaseri, H.; Alipour, S.; Vakilinezhad, M.A. Development, evaluation and optimization of superparamagnetite nanoparticles prepared by co-precipitation method. *Res. Pharm. Sci.* **2017**, *12*, 274–282. [CrossRef]
32. Slimani, S.; Meneghini, C.; Abdolrahimi, M.; Talone, A.; Murillo, J.P.M.; Barucca, G.; Yaacoub, N.; Imperatori, P.; Illés, E.; Smari, M.; et al. Spinel Iron Oxide by the Co-Precipitation Method: Effect of the Reaction Atmosphere. *Appl. Sci.* **2021**, *11*, 5433. [CrossRef]
33. Hollingsworth, J.A.; Klimov, V.I. Soft chemical synthesis and manipulation of semiconductor nanocrystals. In *Nanocrystal Quantum Dots*, 2nd ed.; Taylor and Francis: Boca Raton, FL, USA, 2010; pp. 1–62.
34. Shouheng, S.; Murray, C.B.; Weller, D.; Folks, L.; Moser, A. Monodisperse FePt Nanoparticles and Ferromagnetic FePt Nanocrystal Superlattices. *Science* **2000**, *287*, 1989–1992. [CrossRef]
35. Park, J.; An, K.; Hwang, Y.; Park, J.G.; Noh, H.J.; Kim, J.Y.; Park, J.H.; Hwang, N.M.; Hyeon, T. Ultra-large-scale syntheses of monodisperse nanocrystals. *Nat. Mater.* **2004**, *3*, 891–895. [CrossRef]
36. Hyeon, T. Chemical synthesis of magnetic nanoparticles. *Chem. Commun.* **2003**, *8*, 927–934. [CrossRef]
37. Choi, J.S.; Jun, Y.W.; Yeon, S.I.; Kim, H.C.; Shin, J.S.; Cheon, J. Biocompatible heterostructured nanoparticles for multimodal biological detection. *J. Am. Chem. Soc.* **2006**, *128*, 15982–15983. [CrossRef]
38. Sanchez-Dominguez, M.; Pemartin, K.; Boutonnet, M. Preparation of inorganic nanoparticles in oil-in-water microemulsions: A Soft and Versatile Approach. *Curr. Opin. Colloid Interface Sci.* **2012**, *17*, 297–305. [CrossRef]
39. Lopez-Quintela, M.A. Synthesis of nanomaterials in microemulsions: Formation Mechanisms and Growth Control. *Curr. Opin. Colloid Interface Sci.* **2003**, *8*, 137–144. [CrossRef]
40. Song, M.M.; Song, W.J.; Bi, H.; Wang, J.; Wu, W.L.; Sun, J.; Yu, M. Cytotoxicity and cellular uptake of iron nanowires. *Biomaterials* **2010**, *31*, 1509–1517. [CrossRef]
41. Dippong, T.; Levei, E.A.; Cadar, O. Recent Advances in Synthesis and Applications of MFe2O4 (M = Co, Cu, Mn, Ni, Zn) Nanoparticles. *Nanomaterials* **2021**, *11*, 1560. [CrossRef]
42. Jalalian, M.; Mirkazemi, S.M.; Alamolhoda, S. The effect of poly vinyl alcohol (PVA) surfactant on phase formation and magnetic properties of hydrothermally synthesized $CoFe_2O_4$ nanoparticles. *J. Magn. Magn. Mater.* **2016**, *419*, 363–367. [CrossRef]
43. Mulens, V.; Morales, M.P.; Barber, D. Development of magnetic nanoparticles for cancer gene therapy: A comprehensive review. *Int. Sch. Res. Not.* **2013**, *2013*, 646284. [CrossRef]
44. Morel, A.L.; Nikitenko, S.I.; Gionnet, K.; Wattiaux, A.; Lai-Kee-Him, J.; Labrugere, C.; Chevalier, B.; Deleris, G.; Petibois, C.; Brisson, A.; et al. Sonochemical approach to the synthesis of Fe(3)O(4)@SiO(2) core-shell nanoparticles with tunable properties. *ACS Nano* **2008**, *2*, 847–856. [CrossRef] [PubMed]
45. Lee, D.C.; Mikulec, F.V.; Pelaez, J.M.; Koo, B.; Korgel, B.A. Synthesis and magnetic properties of silica-coated FePt nanocrystals. *J. Phys. Chem. B.* **2006**, *110*, 11160–11166. [CrossRef] [PubMed]
46. Chen, X.; Klingeler, R.; Kath, M.; El Gendy, A.A.; Cendrowski, K.; Kalenczuk, R.J.; Borowiak-Palen, E. Magnetic silica nanotubes: Synthesis, drug release, and feasibility for magnetic hyperthermia. *ACS Appl. Mater. Interfaces* **2012**, *4*, 2303–2309. [CrossRef] [PubMed]
47. Gutsch, A.; Muhlenweg, H.; Kramer, M. Tailor-Made Nanoparticles via Gas-Phase Synthesis. *Small* **2005**, *1*, 30–46. [CrossRef] [PubMed]
48. Kodas, T.T.; Hampden-Smith, M. *Aerosol Processing of Materials*, 1st ed.; John Wiley & Sons Inc.: New York, NY, USA, 1999.
49. Patel, T.; Zhou, J.; Piepmeier, J.M.; Saltzman, W.M. Polymeric nanoparticles for drug delivery to the central nervous system. *Adv. Drug Deliv. Rev.* **2012**, *64*, 701–705. [CrossRef]
50. Liu, D.; He, C.; Poon, C.; Lin, W. Theranostic nanoscale coordination polymers for magnetic resonance imaging and bisphosphonate delivery . *J. Mater. Chem. B* **2014**, *2*, 8249–8255. [CrossRef] [PubMed]
51. Silva, S.M.; Tavallaie, R.; Sandiford, L.; Tilley, R.D.; Gooding, J.J. Gold coated magnetic nanoparticles: From preparation to surface modification for analytical and biomedical applications. *Chem. Commun.* **2016**, *52*, 7528–7540. [CrossRef]
52. Kumar, S.; Gandhi, K.S.; Kumar, R. Modeling of formation of gold nanoparticles by citrate method. *Ind. Eng. Chem. Res.* **2007**, *46*, 3128–3136. [CrossRef]

53. Mills, A.; Britton, M. NMR study of the influence of n-alkanol co-surfactants on reverse micelles in quaternary microemulsions of cetyltrimethylammonium bromide (CTAB). *Magn. Reson. Chem.* **2017**, *55*, 425–432. [CrossRef]
54. Dong, W.J.; Li, Y.S.; Niu, D.C.; Ma, Z.; Gu, J.L.; Chen, Y.; Zhao, W.R.; Liu, X.H.; Liu, C.S.; Shi, J.L. Facile Synthesis of Monodisperse Superparamagnetic Fe_3O_4 Core@hybrid@Au Shell Nanocomposite for Bimodal Imaging and Photothermal Therapy. *Adv. Mater.* **2011**, *23*, 5392–5397. [CrossRef]
55. 55. Kim, J.; Park, S.; Lee, J.E.; Jin, S.M.; Lee, J.H.; Lee, I.S.; Yang, I.; Kim, J.S.; Kim, S.K.; Cho, M.H.; et al. Designed fabrication of multifunctional magnetic gold nanoshells and their application to magnetic resonance imaging and photothermal therapy. *Angew. Chem.* **2006**, *45*, 7754–7758. [CrossRef] [PubMed]
56. Salgueirino-Maceira, V.; Correa-Duarte, M.A.; Farle, M.; Lopez-Quintela, A.; Sieradzki, K.; Diaz, R. Bifunctional gold-coated magnetic silica spheres. *Chem. Mater.* **2006**, *18*, 2701–2706. [CrossRef]
57. Lin, C.-W.; Chen, J.-M.; Lin, Y.-J.; Chao, L.-W.; Wei, S.-Y.; Wu, C.-H.; Jeng, C.-C.; Wang, L.-M.; Chen, K.-L. Magneto-Optical Characteristics of Streptavidin-Coated Fe_3O_4@Au Core-Shell Nanoparticles for Potential Applications on Biomedical Assays. *Sci. Rep.* **2019**, *9*, 16466. [CrossRef] [PubMed]
58. Ajinkya, N.; Yu, X.; Kaithal, P.; Luo, H.; Somani, P.; Ramakrishna, S. Magnetic Iron Oxide Nanoparticle (IONP) Synthesis to Applications: Present and Future. *Materials* **2020**, *13*, 4644. [CrossRef]
59. Pham, X.H.; Hahm, E.; Kim, H.M.; Son, B.S.; Jo, A.; An, J.; Tran Thi, T.A.; Nguyen, D.Q.; Jun, B. Silica-Coated Magnetic Iron Oxide Nanoparticles Grafted onto Graphene Oxide for Protein Isolation. *Nanomaterials* **2020**, *10*, 117. [CrossRef] [PubMed]
60. Popova, M.; Koseva, N.; Trendafilova, I.; Lazarova, H.; Mitova, V.; Mihaly, J.; Momekova, D.; Konstantinov, S.; Koleva, I.Z.; Petkov, P.S.; et al. Design of PEG-modified magnetic nanoporous silica based miltefosine delivery system: Experimental and theoretical approaches. *Micropor. Mesopor. Mat.* **2021**, *310*, 110664. [CrossRef]
61. Popova, M.; Trendafilova, I.; Szegedi, A.; Momekova, D.; Mihaly, J.; Momekov, G.; Kiss, L.F.; Lazar, K.; Koseva, N. Novel SO_3H functionalized magnetic nanoporous silica/polymer nanocomposite as a carrier in a dual-drug delivery system for anticancer therapy. *Micropor. Mesopor. Mat.* **2018**, *263*, 96–105. [CrossRef]
62. Lee, H.S.; Kim, E.H.; Shao, H.; Kwak, B.K. Synthesis of SPIO-chitosan microspheres for MRI-detectable embolotherapy. *J. Magn. Magn. Mater.* **2005**, *293*, 102–105. [CrossRef]
63. Cortés, H.; Hernández-Parra, H.; Bernal-Chávez, S.A.; Prado-Audelo, M.L.D.; Caballero-Florán, I.H.; Borbolla-Jiménez, F.V.; González-Torres, M.; Magaña, J.J.; Leyva-Gómez, G. Non-Ionic Surfactants for Stabilization of Polymeric Nanoparticles for Biomedical Uses. *Materials* **2021**, *14*, 3197. [CrossRef]
64. Al-Jamal, W.T.; Kostarelos, K. Liposomes: From a clinically established drug delivery system to a nanoparticle platform for theranostic nanomedicine. *Acc. Chem. Res.* **2011**, *44*, 1094–1104. [CrossRef]
65. Essa, M.L.; El-Kemary, M.A.; Ebrahem Saied, E.M.; Leporatti, S.; Nemany Hanafy, N.A. Nano targeted Therapies Made of Lipids and Polymers have Promising Strategy for the Treatment of Lung Cancer. *Materials* **2020**, *13*, 5397. [CrossRef]
66. Su, W.; Wang, H.; Wang, S.; Liao, Z.; Kang, S.; Peng, Y.; Han, L.; Chang, J. PEG/RGD-modified magnetic polymeric liposomes for controlled drug release and tumor cell targeting. *Int. J. Pharm.* **2012**, *426*, 170–181. [CrossRef]
67. Ai, H.; Flask, C.; Weinberg, B.; Shuai, X.-T.; Pagel, M.; Farrell, D.; Duerk, J.; Gao, J. Magnetite-Loaded Polymeric Micelles as Ultrasensitive Magnetic-Resonance Probes. *Adv. Mater.* **2005**, *17*, 1949–1952. [CrossRef]
68. Wang, J.; Shah, Z.H.; Zhang, S.; Lu, R. Silica-based nanocomposites via reverse microemulsions: Classifications, preparations, and applications. *Nanoscale* **2014**, *6*, 4418–4437. [CrossRef] [PubMed]
69. Stephen, Z.R.; Kievit, F.M.; Zhang, M. Magnetite Nanoparticles for Medical MR Imaging. *Mater. Today* **2011**, *14*, 330–338. [CrossRef]
70. Frachini, E.; Petri, D. Magneto-Responsive Hydrogels: Preparation, Characterization, Biotechnological and Environmental Applications. *J. Braz. Chem. Soc.* **2019**, *30*, 2010–2028. [CrossRef]
71. Wong, H.L.; Wu, X.Y.; Bendayan, R. Nanotechnological advances for the delivery of CNS therapeutics. *Adv. Drug Deliv. Rev.* **2012**, *64*, 686–700. [CrossRef] [PubMed]
72. Omwoyo, W.N.; Ogutu, B.; Oloo, F.; Swai, H.; Kalombo, L.; Melariri, P.; Mahanga, G.M.; Gathirwa, J.W. Preparation, characterization, and optimization of primaquine-loaded solid lipid nanoparticles. *Int. J. Nanomed.* **2014**, *11*, 3865–3874. [CrossRef]
73. Cavalli, G.; Banu, S.; Ranasinghe, R.T.; Broder, G.R.; Martins, H.F.; Neylon, C.; Morgan, H.; Bradley, M.; Roach, P.L. Multistep synthesis on SU-8: Combining microfabrication and solid-phase chemistry on a single material. *J. Comb. Chem.* **2007**, *9*, 462–472. [CrossRef]
74. Marie, R.; Schmid, S.; Johansson, A.; Ejsing, L.; Nordström, M.; Häfliger, D.; Christensen, C.B.; Boisen, A.; Dufva, M. Immobilisation of DNA to polymerised SU-8 photoresist. *Biosens. Bioelectron.* **2006**, *21*, 1327–1332. [CrossRef]
75. Tsakos, M.; Schaffert, E.; Clement, L.; Villadsen, N.; Poulsen, T. Ester coupling reactions–an enduring challenge in the chemical synthesis of bioactive natural products. *Nat. Prod. Rep.* **2015**, *32*, 605–632. [CrossRef]
76. Azizi, N.; Kamrani, P.; Saadat, M. A magnetic nanoparticle-catalyzed regioselective ring opening of epoxides by aromatic amines . *Appl. Organomet. Chem.* **2016**, *30*, 431–434. [CrossRef]
77. Gavrila, H.; Chiriac, H.; Ciureanu, P.; Ionita, V.; Yelon, A. *Magnetism Tehnic si Aplicat*, 1st ed.; Editura Academiei: Bucharest, Romania, 2000.
78. Kolhatkar, A.G.; Jamison, A.C.; Litvinov, D.; Willson, R.C.; Lee, T.R. Tuning the magnetic properties of nanoparticles. *Int. J. Mol. Sci.* **2013**, *14*, 15977–16009. [CrossRef] [PubMed]

79. Knobel, M.; Nunes, W.C.; Socolovsky, L.M.; Biasi, E.; Vargas, J.M.; Denardin, J.C.J. Superparamagnetism and other magnetic features in granular materials: A review on ideal and real systems. *Nanosci. Nanotech.* **2008**, *8*, 2836–2857. [CrossRef]
80. Ferreira, M.; Sousa, J.; Pais, A.; Vitorino, C. The Role of Magnetic Nanoparticles in Cancer Nanotheranostics. *Materials* **2020**, *13*, 266. [CrossRef] [PubMed]
81. Pereira, C.; Pereira, A.M.; Fernandes, C.; Rocha, M.; Mendes, R.; Fernandez-Garcia, M.; Guedes, A.; Tavares, P.B.; Greneche, J.-M.; Araujo, J.P.; et al. Superparamagnetic MFe_2O_4 (M = Fe, Co, Mn) nanoparticles: Tuning the particles size and magnetic properties through a novel one-step coprecipitation route. *Chem. Mater.* **2012**, *24*, 1496–1504. [CrossRef]
82. Guardia, P.; Labarta, A.; Batlle, X. Tuning the size, the shape, and the magnetic properties of iron oxide nanoparticles. *J. Phys. Chem. C* **2011**, *115*, 390–396. [CrossRef]
83. Caruntu, D.; Caruntu, G.; O'Connor, C.J.J. Magnetic properties of variable-sized Fe_3O_4 nanoparticles synthesized from non-aqueous homogeneous solutions of polyols. *Phys. D Appl. Phys.* **2007**, *40*, 5801–5809. [CrossRef]
84. Noh, S.H.; Na, W.; Jang, J.T.; Lee, J.H.; Lee, E.J.; Moon, S.H.; Lim, Y.; Shin, J.S.; Cheon, J. Nanoscale magnetism control via surface and exchange anisotropy for optimized ferrimagnetic hysteresis. *Nano Lett.* **2012**, *12*, 3716–3721. [CrossRef] [PubMed]
85. Zhen, G.; Muir, B.W.; Moffat, B.A.; Harbour, P.; Murray, K.S.; Moubaraki, B.; Suzuki, K.; Madsen, I.; Agron-Olshina, N.; Waddington, L.; et al. Comparative study of magnetic behavior of spherical and cubic superparamagnetic iron oxide nanoparticles. *J. Phys. Chem. C* **2011**, *115*, 327–334. [CrossRef]
86. Weissleder, R.; Stark, D.D.; Engelstad, B.L.; Bacon, B.R.; Compton, C.C.; White, D.L.; Jacobs, P.; Lewis, J. Superparamagnetic iron oxide: Pharmacokinetics and toxicity. *Am. J. Roentgenol.* **1989**, *152*, 167–173. [CrossRef] [PubMed]
87. Wahajuddin, A.S. Superparamagnetic iron oxide nanoparticles: Magnetic nanoplatforms as drug carriers. *Int. J. Nanomed.* **2012**, *7*, 3445–3471. [CrossRef]
88. Mukherjee, S.; Liang, L.; Veiseh, O. Recent Advancements of magnetic nanomaterials in cancer therapy. *Pharmaceutics* **2020**, *12*, 147. [CrossRef]
89. Silva, F.; Cabral Campello, M.P.; Paulo, A. Radiolabeled Gold Nanoparticles for Imaging and Therapy of Cancer. *Materials* **2021**, *14*, 4. [CrossRef]
90. Peer, D.; Karp, J.M.; Hong, S.; Farokhzad, O.C.; Margalit, R.; Langer, R. Nanocarriers as an emerging platform for cancer therapy. *Nat. Nanotechnol.* **2007**, *2*, 751–760. [CrossRef]
91. Setyawati, M.I.; Tay, C.Y.; Leong, D.T. The gap between endothelial cells: Key to the quick escape of nanomaterials? *Nanomedicine* **2014**, *9*, 1591–1594. [CrossRef]
92. Huang, Y.; Wang, S.; Zhang, J.-Z.; Wang, H.-X.; Wu, L. Stealthy nanoparticles protecting endothelium barrier from leakiness by resisting the absorption of VE-cadherin. *Nanoscale* **2021**, *13*, 12577–12586. [CrossRef]
93. Sehl, O.C.; Gevaert, J.J.; Melo, K.P.; Knier, N.N.; Foster, P.J. A Perspective on Cell Tracking with Magnetic Particle Imaging. *Tomography* **2020**, *6*, 315–324. [CrossRef]
94. Ardeshirpour, Y.; Chernomordik, V.; Hassan, M.; Zielinski, R.; Capala, J.; Gandjbakhche, A. In vivo fluorescence lifetime imaging for monitoring the efficacy of the cancer treatment. *Clin. Cancer Res.* **2014**, *20*, 3531–3539. [CrossRef] [PubMed]
95. Wáng, Y.X.; Idée, J.M. A comprehensive literatures update of clinical researches of superparamagnetic resonance iron oxide nanoparticles for magnetic resonance imaging. *Quant. Imaging Med. Surg.* **2017**, *7*, 88–122. [CrossRef]
96. Hong, S.C.; Lee, J.H.; Lee, J.; Kim, H.Y.; Park, J.Y.; Cho, J.; Lee, J.; Han, D.W. Subtle cytotoxicity and genotoxicity differences in superparamagnetic iron oxide nanoparticles coated with various functional groups. *Int. J. Nanomed.* **2011**, *6*, 3219–3231. [CrossRef]
97. Basuki, J.S.; Duong, H.T.; Macmillan, A.; Erlich, R.B.; Esser, L.; Akerfeldt, M.C.; Whan, R.M.; Kavallaris, M.; Boyer, C.; Davis, T.P. Using fluorescence lifetime imaging microscopy to monitor theranostic nanoparticle uptake and intracellular doxorubicin release. *ACS Nano* **2013**, *7*, 10175–10189. [CrossRef] [PubMed]
98. Kaittanis, C.; Shaffer, T.M.; Ogirala, A.; Santra, S.; Perez, J.M.; Chiosis, G.; Li, Y.; Josephson, L.; Grimm, J. Environment-responsive nanophores for therapy and treatment monitoring via molecular MRI quenching. *Nat. Commun.* **2014**, *5*, 3384. [CrossRef] [PubMed]
99. Wang, G.; Serkova, N.J.; Groman, E.V.; Scheinman, R.I.; Simberg, D. Feraheme (Ferumoxytol) is recognized by proinflammatory and anti-inflammatory macrophages via scavenger receptor type AI/II. *Mol. Pharm.* **2019**, *16*, 4274–4281. [CrossRef] [PubMed]
100. Chow, A.Y. Cell cycle control by oncogenes and tumor supressors: Driving the transformation of normal cells into cancerous cells. *Nat. Educ.* **2010**, *3*, 7.
101. Janko, C.; Ratschker, T.; Nguyen, K.; Zschiesche, L.; Tietze, R.; Lyer, S.; Alexiou, C. Functionalized superparamagnetic iron oxide nanoparticles (SPIONs) as platform for the targeted multimodal tumor therapy. *Front. Oncol.* **2019**, *9*, 59. [CrossRef]
102. Khaledian, M.; Nourbakhsh, M.S.; Saber, R.; Hashemzadeh, H.; Darvishi, M.H. Preparation and evaluation of Doxorubicin-oloaded PLA–PEG–FA copolymer containing superparamagnetic iron oxide nanoparticles (SPIONs) for cancer treatment: Combination therapy with hyperthermia and chemotherapy. *Int. J. Nanomed.* **2020**, *15*, 6167–6182. [CrossRef]
103. Raviraj, V.; Pham, B.T.T.; Kim, B.J.; Pham, N.T.H.; Kok, L.F.; Painter, N.; Delic, N.C.; Jones, S.K.; Hawkett, B.S.; Lyons, J.G. Non-invasive transdermal delivery of chemotherapeutic molecules in vivo using superparamagnetic iron oxide nanoparticles. *Cancer Nano* **2021**, *12*, 6. [CrossRef]
104. Zhang, J.; Stevens, M.F.; Bradshaw, T.D. Temozolomide: Mechanism of action, repair and resistance. *Curr. Mol. Pharmacol.* **2012**, *5*, 102–114. [CrossRef]

105. Brooks, L.J.; Clements, M.P.; Burden, J.J.; Kocher, D.; Richards, L.; Devesa, S.C.; Zakka, L.; Woodberry, M.; Ellis, M.; Jaunmuktane, Z.; et al. The white matter is a pro-differentiative niche for glioblastoma. *Nat. Commun.* **2021**, *12*, 2184. [CrossRef]
106. Brandes, A.A.; Basso, U.; Reni, M.; Vastola, F.; Tosoni, A.; Cavallo, L.; Ferreri, A.J.; Panucci, M.G.; Monfardini, S.; Ermani, M. First-line chemotherapy with cisplatin plus fractionated temozolomide in recurrent glioblastoma multiforme: A phase II study of the gruppo italiano cooperativo di neuro-oncologia. *Clin. Oncol.* **2004**, *22*, 1598–1604. [CrossRef] [PubMed]
107. Emamgholizadeh Minaei, S.; Khoei, S.; Khoee, S.; Karimi, M.R. Tri-block copolymer nanoparticles modified with folic acid for temozolomide delivery in glioblastoma. *Int. J. Biochem. Cell Biol.* **2019**, *108*, 72–83. [CrossRef] [PubMed]
108. Caffery, B.; Lee, J.S.; Alexander-Bryant, A.A. Vectors for Glioblastoma Gene Therapy: Viral & Non-Viral Delivery Strategies. *Nanomaterials* **2019**, *9*, 105. [CrossRef]
109. Rego, G.N.A.; Nucci, M.P.; Mamani, J.B.; Oliveira, F.A.; Marti, L.C.; Filgueiras, I.S.; Ferreira, J.M.; Real, C.C.; Faria, D.P.; Espinha, P.L.; et al. Therapeutic Efficiency of Multiple Applications of Magnetic Hyperthermia Technique in Glioblastoma Using Aminosilane Coated Iron Oxide Nanoparticles: In Vitro and In Vivo Study. *Int. J. Mol. Sci.* **2020**, *21*, 958. [CrossRef]
110. Zhao, M.; van Straten, D.; Broekman, M.; Préat, V.; Schiffelers, R.M. Nanocarrier-based drug combination therapy for glioblastoma. *Theranostics* **2020**, *10*, 1355–1372. [CrossRef]
111. Zargar, H.; Aning, J.; Ischia, J.; So, A.; Black, P. Optimizing intravesical mitomycin C therapy in non-muscle-invasive bladder cancer. *Nat. Rev. Urol.* **2014**, *11*, 220–230. [CrossRef]
112. Eldin, O.S.; Fouda, A.M.; Youssef, A.R.; Hamilton, P.; Maxwell, P.; Williamson, K.E. Reduction of mitomycin C resistance in human bladder cancer T24 cells by knocking-down ras oncogene. *Cancer Drug Resist.* **2018**, *1*, 59–71. [CrossRef]
113. Scotté, F.; Ratta, R.; Beuzeboc, P. Side effects of immunotherapy: A constant challenge for oncologists. *Curr. Opin. Oncol.* **2019**, *31*, 280–285. [CrossRef]
114. Baselga, J.; Albanell, J. Mechanism of action of anti-HER2 monoclonal antibodies. *Ann. Oncol.* **2001**, *12* (Suppl. 1), S35–S41. [CrossRef]
115. Filin, I.Y.; Solovyeva, V.V.; Kitaeva, K.V.; Rutland, C.S.; Rizvanov, A.A. Current Trends in Cancer Immunotherapy. *Biomedicines* **2020**, *8*, 621. [CrossRef] [PubMed]
116. Mahmoudi, M.; Sant, S.; Wang, B.; Laurent, S.; Sen, T. Superparamagnetic iron oxide nanoparticles (SPIONs): Development, surface modification and applications in chemotherapy. *Adv. Drug Deliv. Rev.* **2011**, *63*, 24–46. [CrossRef] [PubMed]
117. Crețu, B.E.; Dodi, G.; Shavandi, A.; Gardikiotis, I.; Șerban, I.L.; Balan, V. Imaging Constructs: The Rise of Iron Oxide Nanoparticles. *Molecules* **2021**, *26*, 3437. [CrossRef] [PubMed]
118. ThermoFisher SCIENTIFIC. Available online: https://www.thermofisher.com/ro/en/home/life-science/cell-analysis/cell-analysis-learning-center/immunology-at-work/macrophage-cell-overview.html (accessed on 8 July 2021).
119. ThermoFisher SCIENTIFIC. Available online: https://www.thermofisher.com/ro/en/home/life-science/cell-analysis/cell-analysis-learning-center/immunology-at-work/dendritic-cell-overview.html (accessed on 8 July 2021).
120. Gires, O.; Pan, M.; Schinke, H.; Canis, M.; Baeuerle, P.A. Expression and function of epithelial cell adhesion molecule EpCAM: Where are we after 40 years? *Cancer Metastasis Rev.* **2020**, *39*, 969–987. [CrossRef]
121. Shapiro, C.L.; Recht, A. Side effects of adjuvant treatment of breast cancer. *N. Engl. J. Med.* **2001**, *344*, 1997–2008. [CrossRef]
122. Ceeraz, S.; Nowak, E.C.; Burns, C.M.; Noelle, R.J. Immune checkpoint receptors in regulating immune reactivity in rheumatic disease. *Arthritis Res. Ther.* **2014**, *16*, 1–12. [CrossRef]
123. Freeman, G.J.; Long, A.J.; Iwai, Y.; Bourque, K.; Chernova, T.; Nishimura, H.; Fitz, L.J.; Malenkovich, N.; Okazaki, T.; Byrne, M.C.; et al. Engagement of the PD-1 immunoinhibitory receptor by a novel B7 family member leads to negative regulation of lymphocyte activation. *J. Exp. Med.* **2000**, *192*, 1027–1034. [CrossRef] [PubMed]
124. Leach, D.R.; Krummel, M.F.; Allison, J.P. Enhancement of antitumor immunity by CTLA-4 blockade. *Science* **1996**, *271*, 1734–1736. [CrossRef] [PubMed]
125. Schmid, D.; Park, C.G.; Hartl, C.A.; Subedi, N.; Cartwright, A.N.; Puerto, R.B.; Zheng, Y.; Maiarana, J.; Freeman, G.J.; Wucherpfennig, K.W.; et al. T cell-targeting nanoparticles focus delivery of immunotherapy to improve antitumor immunity. *Nat. Commun.* **2017**, *8*, 1747. [CrossRef]
126. Mühlberger, M.; Janko, C.; Unterweger, H.; Schreiber, E.; Band, J.; Lehmann, C.; Dudziak, D.; Lee, G.; Alexiou, C.; Tietze, R. Functionalization of T lymphocytes for magnetically controlled immune therapy: Selection of suitable superparamagnetic iron oxide nanoparticles. *J. Magn. Magn. Mater.* **2019**, *473*, 61–67. [CrossRef]
127. Shaik, A.P.; Shaik, A.S.; Al Majwal, A.; Al Faraj, A. Blocking Interleukin-4 Receptor α Using Polyethylene Glycol Functionalized Superparamagnetic Iron Oxide Nanocarriers to Inhibit Breast Cancer Cell Proliferation. *Cancer Res. Treat.* **2017**, *49*, 322–329. [CrossRef]
128. Lunov, O.; Uzhytchak, M.; Smolková, B.; Lunova, M.; Jirsa, M.; Dempsey, N.M.; Dias, A.L.; Bonfim, M.; Hof, M.; Jurkiewicz, P.; et al. Remote Actuation of Apoptosis in Liver Cancer Cells via Magneto-Mechanical Modulation of Iron Oxide Nanoparticles. *Cancers* **2019**, *11*, 1873. [CrossRef] [PubMed]
129. Hao, H.; Ma, Q.; He, F.; Yao, P. Doxorubicin and Fe_3O_4 Loaded Albumin Nanoparticles with Folic Acid Modified Dextran Surface for Tumor Diagnosis and Therapy. *J. Mater. Chem. B* **2014**, *2*, 7978–7987. [CrossRef] [PubMed]
130. Comes Franchini, M.; Baldi, G.; Bonacchi, D.; Gentili, D.; Giudetti, G.; Lascialfari, A.; Corti, M.; Marmorato, P.; Ponti, J.; Micotti, E.; et al. Bovine Serum Albumin-Based Magnetic Nanocarrier for MRI Diagnosis and Hyperthermic Therapy: A Potential Theranostic Approach against Cancer. *Small* **2010**, *6*, 366–370. [CrossRef] [PubMed]

131. Yu, S.-M.; Gonzalez-Moragas, L.; Milla, M.; Kolovou, A.; Santarella-Mellwig, R.; Schwab, Y.; Laromaine, A.; Roig, A. Bio-Identity and Fate of Albumin-Coated SPIONs Evaluated in Cells and by the C. elegans Model. *Acta Biomater.* **2016**, *43*, 348–357. [CrossRef] [PubMed]
132. Vidawati, S.; Barbosa, S.; Taboada, P.; Villar, E.; Topete, A.; Mosquera, V. Study of Human Serum Albumin-SPIONs Loaded PLGA Nanoparticles for Protein Delivery. *Adv. Biol. Chem.* **2018**, *8*, 91–100. [CrossRef]
133. Subbiahdoss, G.; Sharifi, S.; Grijpma, D.W.; Laurent, S.; Van Der Mei, H.C.; Mahmoudi, M.; Busscher, H.J. Magnetic Targeting of Surface-Modified Superparamagnetic Iron Oxide Nanoparticles Yields Antibacterial Efficacy against Biofilms of Gentamicin-Resistant Staphylococci. *Acta Biomater.* **2012**, *8*, 2047–2055. [CrossRef]
134. Dadfar, S.M.; Roemhild, K.; Drude, N.I.; von Stillfried, S.; Knüchel, R.; Kiessling, F.; Lammers, T. Iron Oxide Nanoparticles: Diagnostic, Therapeutic and Theranostic Applications. *Adv. Drug Deliv. Rev.* **2019**, *138*, 302–325. [CrossRef] [PubMed]
135. Fang, C.; Wang, K.; Stephen, Z.R.; Mu, Q.; Kievit, F.M.; Chiu, D.T.; Press, O.W.; Zhang, M. Temozolomide Nanoparticles for Targeted Glioblastoma Therapy. *ACS Appl. Mater. Interfaces* **2015**, *7*, 6674–6682. [CrossRef] [PubMed]
136. Hu, S.H.; Liao, B.J.; Chiang, C.S.; Chen, P.J.; Chen, I.W.; Chen, S.Y. Core-Shell Nanocapsules Stabilized by Single-Component Polymer and Nanoparticles for Magneto-Chemotherapy/Hyperthermia with Multiple Drugs. *Adv. Mater.* **2012**, *24*, 3627–3632. [CrossRef]
137. Sun, C.; Du, K.; Fang, C.; Bhattarai, N.; Veiseh, O.; Kievit, F.; Stephen, Z.; Lee, D.; Ellenbogen, R.G.; Ratner, B.; et al. PEG Mediated Synthesis of Highly Dispersive Multifunctional Superparamagnetic Nanoparticles: Their Physicochemical Properties and Function in Vivo. *ACS Nano* **2010**, *4*, 2402–2410. [CrossRef]
138. Alupei, L.; Peptu, C.A.; Lungan, A.M.; Desbrieres, J.; Chiscan, O.; Radji, S.; Popa, M. New Hybrid Magnetic Nanoparticles Based on Chitosan-Maltose Derivative for Antitumor Drug Delivery. *Int. J. Biol. Macromol.* **2016**, *92*, 561–572. [CrossRef]
139. Shakeri-Zadeh, A.; Khoee, S.; Shiran, M.B.; Sharifi, A.M.; Khoei, S. Synergistic Effects of Magnetic Drug Targeting Using a Newly Developed Nanocapsule and Tumor Irradiation by Ultrasound on CT26 Tumors in BALB/c Mice. *J. Mater. Chem. B* **2015**, *3*, 1879–1887. [CrossRef]
140. Shakeri-Zadeh, A.; Shiran, M.B.; Khoee, S.; Sharifi, A.M.; Ghaznavi, H.; Khoei, S. A New Magnetic Nanocapsule Containing 5-Fluorouracil: In Vivo Drug Release, Anti-Tumor, and pro-Apoptotic Effects on CT26 Cells Allograft Model. *J. Biomater. Appl.* **2014**, *29*, 548–556. [CrossRef] [PubMed]
141. Russell, E.; Dunne, V.; Russell, B.; Mohamud, H.; Ghita, M.; McMahon, S.J.; Butterworth, K.T.; Schettino, G.; McGarry, C.K.; Prise, K.M. Impact of superparamagnetic iron oxide nanoparticles on in vitro and in vivo radiosensitisation of cancer cells. *Radiat. Oncol.* **2021**, *16*, 104. [CrossRef] [PubMed]
142. Dulińska-Litewka, J.; Łazarczyk, A.; Hałubiec, P.; Szafrański, O.; Karnas, K.; Karewicz, A. Superparamagnetic Iron Oxide Nanoparticles-Current and Prospective Medical Applications. *Materials* **2019**, *12*, 617. [CrossRef]
143. Lee, S.H.; Kim, B.H.; Na, H.B.; Hyeon, T. Paramagnetic inorganic nanoparticles as T1 MRI contrast agents. *Wiley Interdiscip. Rev.: Nanomed. Nanobiotechnol.* **2014**, *6*, 196–209. [CrossRef] [PubMed]
144. Hu, P.; Fu, Z.; Liu, G.; Tan, H.; Xiao, J.; Shi, H.; Cheng, D. Gadolinium-Based Nanoparticles for Theranostic MRI-Guided Radiosensitization in Hepatocellular Carcinoma. *Front. Bioeng. Biotechnol.* **2019**, *7*, 368. [CrossRef]
145. Klein, S.; Sommer, A.; Distel, L.V.; Neuhuber, W.; Kryschi, C. Superparamagnetic iron oxide nanoparticles as radiosensitizer via enhanced reactive oxygen species formation. *Biochem. Biophys. Res. Commun.* **2012**, *425*, 393–397. [CrossRef]
146. Kirakli, E.; Takan, G.; Hoca, S.; Muftuler, F.Z. Superparamagnetic iron oxide nanoparticle (SPION) mediated in vitro radiosensitization at megavoltage radiation energies. *J. Radioanal. Nucl. Chem.* **2018**, *9*, 595–602. [CrossRef]
147. Abdul Rashid, R.; Zainal Abidin, S.; Khairil Anuar, M.A.; Tominaga, T.; Akasaka, H.; Sasaki, R.; Kie, K.; Razak, K.A.; Pham, B.T.T.; Hawkett, B.S.; et al. Radiosensitization effects and ROS generation by high Z metallic nanoparticles on human colon carcinoma cell (HCT116) irradiated under 150 MeV proton beam. *OpenNano* **2019**, *4*, 100027. [CrossRef]
148. Choi, G.H.; Seo, S.J.; Kim, K.H.; Kim, H.T.; Park, S.H.; Lim, J.H.; Kim, J.K. Photon activated therapy (PAT) using monochromatic synchrotron X-rays and iron oxide nanoparticles in a mouse tumor model: Feasibility study of PAT for the treatment of superficial malignancy. *Radiat. Oncol.* **2012**, *7*, 184. [CrossRef] [PubMed]
149. Mesbahi, A. A review on gold nanoparticles radiosensitization effect in radiation therapy of cancer. *Rep. Pract. Oncol. Radiother.* **2010**, *15*, 176–180. [CrossRef]
150. Zhang, X.; Liu, Z.; Lou, Z.; Chen, F.; Chang, S.; Miao, Y.; Zhou, Z.; Hu, X.; Feng, J.; Ding, Q.; et al. Radiosensitivity enhancement of Fe_3O_4@Ag nanoparticles on human glioblastoma cells. *Artif. Cells Nanomed. Biotechnol.* **2018**, *46*, 975–984. [CrossRef] [PubMed]
151. Xu, R.; Ma, J.; Sun, X.; Chen, Z.; Jiang, X.; Guo, Z.; Huang, L.; Li, Y.; Wang, M.; Wang, C.; et al. Ag nanoparticles sensitize IR-induced killing of cancer cells. *Cell Res.* **2009**, *19*, 1031–1034. [CrossRef]
152. Liu, P.; Huang, Z.; Chen, Z.; Xu, R.; Wu, H.; Zang, F.; Wang, C.; Gu, N. Silver nanoparticles: A novel radiation sensitizer for glioma? *Nanoscale* **2013**, *5*, 11829–11836. [CrossRef]
153. Wu, H.; Lin, J.; Liu, P.; Huang, Z.; Zhao, P.; Jin, H.; Ma, J.; Wen, L.; Gu, N. Reactive oxygen species acts as executor in radiation enhancement and autophagy inducing by AgNPs. *Biomaterials* **2016**, *101*, 1–9. [CrossRef]
154. Wu, H.; Lin, J.; Liu, P.; Huang, Z.; Zhao, P.; Jin, H.; Wang, C.; Wen, L.; Gu, N. Is the autophagy a friend or foe in the silver nanoparticles associated radiotherapy for glioma? *Biomaterials* **2015**, *62*, 47–57. [CrossRef]
155. Liu, P.; Jin, H.; Guo, Z.; Ma, J.; Zhao, J.; Li, D.; Wu, H.; Gu, N. Silver nanoparticles outperform gold nanoparticles in radiosensitizing U251 cells in vitro and in an intracranial mouse model of glioma. *Int. J. Nanomed.* **2016**, *11*, 5003–5014. [CrossRef]

156. Eriksson, D.; Stigbrand, T. Radiation-induced cell death mechanisms. *Tumour. Biol.* **2010**, *31*, 363–372. [CrossRef]
157. Zhao, D.; Sun, X.; Tong, J.; Ma, J.; Bu, X.; Xu, R.; Fan, R. A novel multifunctional nanocomposite C225-conjugated Fe_3O_4/Ag enhances the sensitivity of nasopharyngeal carcinoma cells to radiotherapy. *Acta Biochim. Biophys. Sin.* **2012**, *44*, 678–684. [CrossRef]
158. Datta, N.R.; Ordóñez, S.G.; Gaipl, U.S.; Paulides, M.M.; Crezee, H.; Gellermann, J.; Marder, D.; Puric, E.; Bodis, S. Local hyperthermia combined with radiotherapy and-/or chemotherapy: Recent advances and promises for the future. *Cancer Treat. Rev.* **2015**, *41*, 742–753. [CrossRef]
159. Bañobre-López, M.; Teijeiro, A.; Rivas, J. Magnetic nanoparticle-based hyperthermia for cancer treatment. *Rep. Pract. Oncol. Radiother.* **2013**, *18*, 397–400. [CrossRef] [PubMed]
160. Gilchrist, R.K.; Medal, R.; Shorey, W.D.; Hanselman, R.C.; Parrott, J.C.; Taylor, C.B. Selective inductive heating of lymph nodes. *Ann. Surg.* **1957**, *146*, 596–606. [CrossRef] [PubMed]
161. Blanco-Andujar, C.; Teran, F.J.; Ortega, D. Current outlook and perspectives on nanoparticle-mediated magnetic hyperthermia. In *Iron Oxide Nanoparticles for Biomedical Applications*; Mahmoudi, M., Laurent, S., Eds.; Elsevier: Amsterdam, The Netherlands, 2018; pp. 197–245.
162. Kumar, C.S.; Mohammad, F. Magnetic nanomaterials for hyperthermia-based therapy and controlled drug delivery. *Adv. Drug Deliv. Rev.* **2011**, *63*, 789–808. [CrossRef] [PubMed]
163. Shubayev, V.I.; Pisanic, T.R., 2nd; Jin, S. Magnetic nanoparticles for theragnostics. *Adv. Drug Deliv. Rev.* **2009**, *61*, 467–477. [CrossRef] [PubMed]
164. Yoo, D.; Lee, J.H.; Shin, T.H.; Cheon, J. Theranostic magnetic nanoparticles. *Acc. Chem. Res.* **2011**, *44*, 863–874. [CrossRef] [PubMed]
165. Ho, D.; Sun, X.; Sun, S. Monodisperse magnetic nanoparticles for theranostic applications. *Acc. Chem. Res.* **2011**, *44*, 875–882. [CrossRef]
166. Espinosa, A.; Curcio, A.; Cabana, S.; Radtke, G.; Bugnet, M.; Kolosnjaj-Tabi, J.; Péchoux, C.; Alvarez-Lorenzo, C.; Botton, G.A.; Silva, A.K.A.; et al. Intracellular Biodegradation of Ag Nanoparticles, Storage in Ferritin, and Protection by a Au Shell for Enhanced Photothermal Therapy. *ACS Nano* **2018**, *12*, 6523–6535. [CrossRef]
167. Hildebrandt, B.; Wust, P.; Ahlers, O.; Dieing, A.; Sreenivasa, G.; Kerner, T.; Felix, R.; Riess, H. The cellular and molecular basis of hyperthermia. *Crit. Rev. Oncol. Hematol.* **2002**, *43*, 33–56. [CrossRef]
168. Ong, Y.S.; Bañobre-López, M.; Costa Lima, S.A.; Reis, S. A multifunctional nanomedicine platform for co-delivery of methotrexate and mild hyperthermia towards breast cancer therapy. *Mater. Sci. Eng. C Mater. Biol. Appl.* **2020**, *116*, 111255. [CrossRef] [PubMed]
169. Datta, N.R.; Krishnan, S.; Speiser, D.E.; Neufeld, E.; Kuster, N.; Bodis, S.; Hofmann, H. Magnetic nanoparticle-induced hyperthermia with appropriate payloads: Paul Ehrlich's "magic (nano)bullet" for cancer theranostics? *Cancer Treat. Rev.* **2016**, *50*, 217–227. [CrossRef]
170. Jordan, A.; Scholz, R.; Wust, P.; Fähling, H.; Felix, R. Magnetic fluid hyperthermia (MFH): Cancer treatment with AC magnetic field induced excitation of biocompatible superparamagnetic nanoparticles. *J. Magn. Magn. Mater.* **1999**, *201*, 413–419. [CrossRef]
171. Rosenweig, R.E. Heating magnetic fluid with alternating magnetic field. *J. Magn. Magn. Mater.* **2002**, *252*, 370–374. [CrossRef]
172. Lee, J.H.; Jang, J.T.; Choi, J.S.; Moon, S.H.; Noh, S.H.; Kim, J.W.; Kim, J.G.; Kim, I.S.; Park, K.I.; Cheon, J. Exchange-coupled magnetic nanoparticles for efficient heat induction. *Nat. Nanotechnol.* **2011**, *6*, 418–422. [CrossRef]
173. Liu, X.L.; Ng, C.T.; Chandrasekharan, P.; Yang, H.T.; Zhao, L.Y.; Peng, E.; Lv, Y.B.; Xiao, W.; Fang, J.; Yi, J.B.; et al. Synthesis of Ferromagnetic $Fe_{0.6}$ $Mn_{0.4}$ O Nanoflowers as a New Class of Magnetic Theranostic Platform for In Vivo T1 -T2 Dual-Mode Magnetic Resonance Imaging and Magnetic Hyperthermia Therapy. *Adv. Healthc. Mater.* **2016**, *5*, 2092–2104. [CrossRef]
174. Jang, J.T.; Nah, H.; Lee, J.H.; Moon, S.H.; Kim, M.G.; Cheon, J. Critical enhancements of MRI contrast and hyperthermic effects by dopant-controlled magnetic nanoparticles. *Angew. Chem. Int. Ed.* **2009**, *48*, 1234–1238. [CrossRef]
175. Chen, R.; Christiansen, M.G.; Anikeeva, P. Maximizing hysteretic losses in magnetic ferrite nanoparticles via model-driven synthesis and materials optimization. *ACS Nano* **2013**, *7*, 8990–9000. [CrossRef] [PubMed]
176. Jang, J.T.; Lee, J.; Seon, J.; Ju, E.; Kim, M.; Kim, Y.I.; Kim, M.G.; Takemura, Y.; Arbab, A.S.; Kang, K.W.; et al. Giant Magnetic Heat Induction of Magnesium-Doped γ-Fe_2 O_3 Superparamagnetic Nanoparticles for Completely Killing Tumors. *Adv. Mater.* **2018**, *30*, 1704362. [CrossRef]
177. Liu, X.; Peng, M.; Li, G.; Miao, Y.; Luo, H.; Jing, G.; He, Y.; Zhang, C.; Zhang, F.; Fan, H. Ultrasonication-Triggered Ubiquitous Assembly of Magnetic Janus Amphiphilic Nanoparticles in Cancer Theranostic Applications. *Nano Lett.* **2019**, *19*, 4118–4125. [CrossRef]
178. Du, Y.; Liu, X.; Liang, Q.; Liang, X.J.; Tian, J. Optimization and Design of Magnetic Ferrite Nanoparticles with Uniform Tumor Distribution for Highly Sensitive MRI/MPI Performance and Improved Magnetic Hyperthermia Therapy. *Nano Lett.* **2019**, *2019 19*, 3618–3626. [CrossRef]
179. Sugumaran, P.J.; Liu, X.L.; Herng, T.S.; Peng, E.; Ding, J. GO-Functionalized Large Magnetic Iron Oxide Nanoparticles with Enhanced Colloidal Stability and Hyperthermia Performance. *ACS Appl. Mater. Interfaces* **2019**, *11*, 22703–22713. [CrossRef] [PubMed]
180. Liu, X.L.; Yang, Y.; Ng, C.T.; Zhao, L.Y.; Zhang, Y.; Bay, B.H.; Fan, H.M.; Ding, J. Magnetic vortex nanorings: A new class of hyperthermia agent for highly efficient in vivo regression of tumors. *Adv. Mater.* **2015**, *27*, 1939–1944. [CrossRef] [PubMed]

181. Clerc, P.; Jeanjean, P.; Hallali, N.; Gougeon, M.; Pipy, B.; Carrey, J.; Fourmy, D.; Gigoux, V. Targeted Magnetic Intra-Lysosomal Hyperthermia produces lysosomal reactive oxygen species and causes Caspase-1 dependent cell death. *J. Control. Release* **2018**, *270*, 120–134. [CrossRef]
182. Domenech, M.; Marrero-Berrios, I.; Torres-Lugo, M.; Rinaldi, C. Lysosomal membrane permeabilization by targeted magnetic nanoparticles in alternating magnetic fields. *ACS Nano* **2013**, *7*, 5091–5101. [CrossRef] [PubMed]
183. de la Fuente, J.M.; Grazu, V. *Nanobiotechnology, Inorganic Nanoparticles vs Organic Nanoparticles*; Frontiers of Nanoscience (Book Series); Elsevier: Amsterdam, The Netherlands, 2012; Volume 4.
184. Kozissnik, B.; Bohorquez, A.C.; Dobson, J.; Rinaldi, C. Magnetic fluid hyperthermia: Advances, challenges, and opportunity. *Int. J. Hyperth.* **2013**, *29*, 706–714. [CrossRef] [PubMed]
185. Andreu, I.; Natividad, E. Accuracy of available methods for quantifying the heat power generation of nanoparticles for magnetic hyperthermia. *Int. J. Hyperth.* **2013**, *29*, 739–751. [CrossRef]
186. Riedinger, A.; Guardia, P.; Curcio, A.; Garcia, M.A.; Cingolani, R.; Manna, L.; Pellegrino, T. Subnanometer local temperature probing and remotely controlled drug release based on azo-functionalized iron oxide nanoparticles. *Nano Lett.* **2013**, *13*, 2399–2406. [CrossRef]
187. Albarqi, H.A.; Wong, L.H.; Schumann, C.; Sabei, F.Y.; Korzun, T.; Li, X.; Hansen, M.N.; Dhagat, P.; Moses, A.S.; Taratula, O.; et al. Biocompatible Nanoclusters with High Heating Efficiency for Systemically Delivered Magnetic Hyperthermia. *ACS Nano* **2019**, *13*, 6383–6395. [CrossRef]
188. Ito, A.; Shinkai, M.; Honda, H.; Yoshikawa, K.; Saga, S.; Wakabayashi, T.; Yoshida, J.; Kobayashi, T. Heat shock protein 70 expression induces antitumor immunity during intracellular hyperthermia using magnetite nanoparticles. *Cancer Immunol. Immunother.* **2003**, *52*, 80–88. [CrossRef]
189. Yanase, M.; Shinkai, M.; Honda, H.; Wakabayashi, T.; Yoshida, J.; Kobayashi, T. Intracellular hyperthermia for cancer using magnetite cationic liposomes: Ex vivo study. *Jpn. J. Cancer Res.* **1997**, *88*, 630–632. [CrossRef]
190. Yanase, M.; Shinkai, M.; Honda, H.; Wakabayashi, T.; Yoshida, J.; Kobayashi, T. Antitumor immunity induction by intracellular hyperthermia using magnetite cationic liposomes. *Jpn. J. Cancer Res.* **1998**, *89*, 775–782. [CrossRef]
191. Rabias, I.; Tsitrouli, D.; Karakosta, E.; Kehagias, T.; Diamantopoulos, G.; Fardis, M.; Stamopoulos, D.; Maris, T.G.; Falaras, P.; Zouridakis, N.; et al. Rapid magnetic heating treatment by highly charged maghemite nanoparticles on Wistar rats exocranial glioma tumors atmicroliter volume. *Biomicrofluidics* **2010**, *4*, 024111. [CrossRef]
192. Ito, A.; Shinkai, M.; Honda, H.; Kobayashi, T. Heat-inducible TNF-alpha gene therapy combined with hyperthermia using magnetic nanoparticles as a novel tumor-targeted therapy. *Cancer Gene Ther.* **2001**, *8*, 649–654. [CrossRef]
193. van Landeghem, F.K.; Maier-Hauff, K.; Jordan, A.; Homann, K.T.; Gneveckow, U.; Scholz, R.; Thiesen, B.; Bruck, W.; von Deimling, A. Post-mortem studies in glioblastoma patients treated with thermotherapy using magnetic nanoparticles. *Biomaterials* **2009**, *30*, 52–57. [CrossRef]
194. Gupta, R.; Sharma, D. Biofunctionalization of magnetite nanoparticles with stevioside: Effect on the size and thermal behaviour for use in hyperthermia applications. *Int. J. Hyperth.* **2019**, *36*, 301–311. [CrossRef] [PubMed]
195. Maier-Hauff, K.; Rothe, R.; Scholz, R.; Gneveckow, U.; Wust, P.; Thiesen, B.; Feussner, A.; von Deimling, A.; Waldoefner, N.; Felix, R.; et al. Intracranial thermotherapy using magnetic nanoparticles combined with external beam radiotherapy: Results of a feasibility study on patients with glioblastoma multiforme. *J. Neuro -Oncol.* **2007**, *81*, 53–60. [CrossRef]
196. Maier-Hauff, K.; Ulrich, F.; Nestler, D.; Nieho, H.; Wust, P.; Thiesen, B.; Orawa, H.; Budach, V.; Jordan, A. Efficacy and safety of intratumoral thermotherapy using magnetic iron-oxide nanoparticles combined with external beam radiotherapy on patients with recurrent glioblastoma multiforme. *J. Neuro -Oncol.* **2011**, *103*, 317–324. [CrossRef] [PubMed]
197. Wismeth, C.; Dudel, C.; Pascher, C.; Ramm, P.; Pietsch, T.; Hirschmann, B.; Reinert, C.; Proescholdt, M.; Rummele, P.; Schuierer, G.; et al. Transcranial electro-hyperthermia combined with alkylating chemotherapy in patients with relapsed high-grade gliomas: Phase I clinical results. *J. Neuro -Oncol.* **2010**, *98*, 395–405. [CrossRef] [PubMed]
198. Kekalo, K.; Baker, I.; Meyers, R.; Shyong, J. Magnetic Nanoparticles with High Specific Absorption Rate at Low Alternating Magnetic Field. *Nano Life* **2015**, *5*, 1550002. [CrossRef]
199. Hanini, A.; Lartigue, L.; Gavard, J.; Schmitt, A.; Kacem, K.; Wilhelm, C.; Gazeau, F.; Chau, F.; Ammar, S. Thermosensitivity profile of malignant glioma U87-MG cells and human endothelial cells following γ-Fe_2O_3 NPs internalization and magnetic field application. *RSC Adv.* **2016**, *6*, 15415–15423. [CrossRef]
200. Hanini, A.; Lartigue, L.; Gavard, J.; Kacem, K.; Wilhelm, C.; Gazeau, F.; Chau, F.; Ammar, S. Zinc substituted ferrite nanoparticles with $Zn_{0.9}Fe_{2.1}O_4$ formula used as heating agents for in vitro hyperthermia assay on glioma cells. *J. Magn. Magn. Mater.* **2016**, *416*, 315–320. [CrossRef]
201. Shevtsov, M.A.; Multhoff, G. Recent developments of magnetic nanoparticles for theranostics of brain tumor. *Curr. Drug Metab.* **2016**, *17*, 737–744. [CrossRef]
202. Shinkai, M.; Yanase, M.; Honda, H.; Wakabayashi, T.; Yoshida, J.; Kobayashi, T. Intracellular Hyperthermia for Cancer Using Magnetite Cationic Liposomes: In vitro Study. *Jpn. J. Cancer Res.* **1996**, *87*, 1179–1183. [CrossRef] [PubMed]
203. Meenach, S.A.; Hilt, J.Z.; Anderson, K.W. Poly(ethylene glycol)-based magnetic hydrogel nanocomposites for hyperthermia cancer therapy. *Acta Biomater.* **2010**, *6*, 1039–1046. [CrossRef] [PubMed]

204. Altanerova, U.; Babincova, M.; Babinec, P.; Benejova, K.; Jakubechova, J.; Altanerova, V.; Zduriencikova, M.; Repiska, V.; Altaner, C. Human mesenchymal stem cell-derived iron oxide exosomes allow targeted ablation of tumor cells via magnetic hyperthermia. *Int. J. Nanomed.* **2017**, *12*, 7923–7936. [CrossRef]
205. Mannucci, S.; Tambalo, S.; Conti, G.; Ghin, L.; Milanese, A.; Carboncino, A.; Nicolato, E.; Marinozzi, M.R.; Benati, D.; Bassi, R.; et al. Magnetosomes Extracted from Magnetospirillum gryphiswaldense as Theranostic Agents in an Experimental Model of Glioblastoma. *Contrast Media Mol. Imaging* **2018**, *2018*, 2198703. [CrossRef] [PubMed]
206. Orlando, T.; Mannucci, S.; Fantechi, E.; Conti, G.; Tambalo, S.; Busato, A.; Innocenti, C.; Ghin, L.; Bassi, R.; Arosio, P.; et al. Characterization of magnetic nanoparticles from Magnetospirillum Gryphiswaldense as potential theranostics tools. *Contrast Media Mol. Imaging* **2016**, *11*, 139–145. [CrossRef]
207. Grapa, C.M.; Mocan, L.; Crisan, D.; Florea, M.; Mocan, T. Biomarkers in Pancreatic Cancer as Analytic Targets for Nanomediated Imaging and Therapy. *Materials* **2021**, *14*, 3083. [CrossRef] [PubMed]
208. Cheng, R.; Feng, F.; Meng, F.; Deng, C.; Feijen, J.; Zhong, Z. Glutathione-responsive nano-vehicles as a promising platform for targeted intracellular drug and gene delivery. *J. Control. Release* **2011**, *152*, 2–12. [CrossRef]
209. Kievit, F.M.; Stephen, Z.R.; Wang, K.; Dayringer, C.J.; Sham, J.G.; Ellenbogen, R.G.; Silber, J.R.; Zhang, M. Nanoparticle mediated silencing of DNA repair sensitizes pediatric brain tumor cells to γ-irradiation. *Mol. Oncol.* **2015**, *9*, 1071–1080. [CrossRef] [PubMed]
210. Revia, R.A.; Zhang, M. Magnetite nanoparticles for cancer diagnosis, treatment, and treatment monitoring: Recent advances. *Mater Today* **2016**, *19*, 157–168. [CrossRef] [PubMed]
211. Huber, D.L. Synthesis, properties, and applications of iron nanoparticles. *Small* **2005**, *1*, 482–501. [CrossRef]
212. Stockwell, B.R.; Friedmann Angeli, J.P.; Bayir, H.; Bush, A.I.; Conrad, M.; Dixon, S.J.; Fulda, S.; Gascón, S.; Hatzios, S.K.; Kagan, V.E.; et al. Ferroptosis: A Regulated Cell Death Nexus Linking Metabolism, Redox Biology, and Disease. *Cell* **2017**, *171*, 273–285. [CrossRef] [PubMed]
213. Chen, X.; Kang, R.; Kroemer, G.; Tang, D. Broadening horizons: The role of ferroptosis in cancer. *Nat. Rev. Clin. Oncol.* **2021**, *18*, 280–296. [CrossRef] [PubMed]
214. Battaglia, A.M.; Chirillo, R.; Aversa, I.; Sacco, A.; Costanzo, F.; Biamonte, F. Ferroptosis and Cancer: Mitochondria Meet the "Iron Maiden" Cell Death. *Cells* **2020**, *9*, 1505. [CrossRef] [PubMed]
215. Dixon, S.J.; Stockwell, B.R. The hallmarks of ferroptosis. *Annu. Rev. Cancer Biol.* **2019**, *3*, 35–54. [CrossRef]
216. Lu, B.; Chen, X.B.; Ying, M.D.; He, Q.J.; Cao, J.; Yang, B. The Role of Ferroptosis in Cancer Development and Treatment Response. *Front. Pharmacol.* **2018**, *8*, 992. [CrossRef] [PubMed]
217. Dixon, S.J.; Lemberg, K.M.; Lamprecht, M.R.; Skouta, R.; Zaitsev, E.M.; Gleason, C.E.; Patel, D.N.; Bauer, A.J.; Cantley, A.M.; Yang, W.S.; et al. Ferroptosis: An iron-dependent form of nonapoptotic cell death. *Cell* **2012**, *149*, 1060–1072. [CrossRef] [PubMed]
218. DeHart, D.N.; Fang, D.; Heslop, K.; Li, L.; Lemasters, J.J.; Maldonado, E.N. Opening of voltage dependent anion channels promotes reactive oxygen species generation, mitochondrial dysfunction and cell death in cancer cells. *Biochem. Pharmacol.* **2018**, *148*, 155–162. [CrossRef]
219. Agmon, E.; Solon, J.; Bassereau, P.; Stockwell, B.R. Modeling the effects of lipid peroxidation during ferroptosis on membrane properties. *Sci. Rep.* **2018**, *8*, 5155. [CrossRef] [PubMed]
220. Su, L.-J.; Zhang, J.-H.; Gomez, H.; Murugan, R.; Hong, X.; Xu, D.; Jiang, F.; Peng, Z.-Y. Review Article reactive oxygen species-induced lipid peroxidation in apoptosis, autophagy, and ferroptosis. *Oxidative Med. Cell. Longev.* **2019**, *2019*, 5080843. [CrossRef] [PubMed]
221. Xie, Y.; Hou, W.; Song, X.; Yu, Y.; Huang, J.; Sun, X.; Kang, R.; Tang, D. Ferroptosis: Process and function. *Cell Death Differ.* **2016**, *23*, 369–379. [CrossRef] [PubMed]
222. Mou, Y.; Wang, J.; Wu, J.; He, D.; Zhang, C.; Duan, C.; Li, B. Ferroptosis, a new form of cell death: Opportunities and challenges in cancer. *J. Hematol. Oncol.* **2019**, *12*, 34. [CrossRef]
223. Shen, Z.; Song, J.; Yung, B.C.; Zhou, Z.; Wu, A.; Chen, X. Emerging Strategies of Cancer Therapy Based on Ferroptosis. *Adv. Mater.* **2018**, *30*, 1704007. [CrossRef]
224. Ahmad, T.; Sarwar, R.; Iqbal, A.; Bashir, U.; Farooq, U.; Halim, S.A.; Khan, A.; Harassi, A. Recent advances in combinatorial cancer therapy via multifunctionalized gold nanoparticles. *Nanomedicine* **2020**, *15*, 1221–1237. [CrossRef]
225. Patil, R.M.; Thorat, N.D.; Shete, P.B.; Bedge, P.A.; Gavde, S.; Joshi, M.G.; Tofail, S.A.M.; Bohara, R.A. Comprehensive cytotoxicity studies of superparamagnetic iron oxide nanoparticles. *Biochem. Biophys. Rep.* **2018**, *13*, 63–72. [CrossRef]
226. Hassannia, B.; Vandenabeele, P.; Vanden Berghe, T. Targeting Ferroptosis to Iron Out Cancer. *Cancer Cell* **2019**, *35*, 830–849. [CrossRef]
227. Farzin, A.; Etesami, S.A.; Quint, J.; Memic, A.; Tamayol, A. Magnetic Nanoparticles in Cancer Therapy and Diagnosis. *Adv. Healthc. Mater.* **2020**, *9*, e1901058. [CrossRef]
228. Wu, K.; Su, D.; Liu, J.; Saha, R.; Wang, J.P. Magnetic nanoparticles in nanomedicine: A review of recent advances. *Nanotechnology* **2019**, *30*, 502003. [CrossRef] [PubMed]
229. Cavalu, S.; Fritea, L.; Brocks, M.; Barbaro, K.; Murvai, G.; Costea, T.O.; Antoniac, I.; Verona, C.; Romani, M.; Latini, A.; et al. Novel hybrid composites based on PVA/SeTiO_2 nanoparticles and natural hydroxyapatite for orthopedic applications: Correlations between structural, morphological and biocompatibility properties. *Materials* **2020**, *13*, 2077. [CrossRef]

230. Laptoiu, D.; Marinescu, R.; Balan, C.; Antoniac, I. Rheologic properties of some current hyaluronic acid products for viscosupplimentation-new trends for amelioration. *Mater. Plast.* **2015**, *52*, 500–503.
231. Petreus, T.; Stoica, B.A.; Petreus, O.; Goriuc, A.; Cotrutz, C.E.; Antoniac, I.V.; Barbu-Tudoran, L. Preparation and cytocompatibilityevaluation for hydrosolublephosphorous acid-derivatizedcellulose as tissue engineering scaffold material. *J. Mater. Sci. Mater. Med.* **2014**, *25*, 1115–1127. [CrossRef] [PubMed]
232. Cojocaru, F.D.; Balan, V.; Popa, M.I.; Lobiuc, A.; Antoniac, A.; Antoniac, I.V.; Verestiuc, L. Biopolymers - Calcium phosphates composites with inclusions of magnetic nanoparticles for bone tissue engineering. *Int. J. Biol. Macromol.* **2019**, *125*, 612–620. [CrossRef]
233. Guazzo, R.; Gardin, C.; Bellin, G.; Sbricoli, L.; Ferroni, L.; Ludovichetti, F.S.; Piattelli, A.; Antoniac, I.; Bressan, E.; Zavan, B. Graphene-based nanomaterials for tissue engineering in the dental field. *Nanomaterials* **2018**, *8*, 349. [CrossRef]
234. Sarosi, C.; Biris, A.R.; Antoniac, A.; Boboia, S.; Alb, C.; Antoniac, I.; Moldovan, M. The nanofiller effect on properties of experimental graphene dental nanocomposites. *J. Adhes. Sci. Technol.* **2016**, *30*, 1779–1794. [CrossRef]
235. Sinescu, C.; Marsavina, L.; Negrutiu, M.L.; Rusu, L.C.; Ardelean, L.; Rominu, M.; Antoniac, I.; Topala, F.I.; Podoleanu, A. New metallic nanoparticles modified adhesive used for time domain optical coherence tomography evaluation of class II direct composite restoration. *Revista Chimie* **2012**, *63*, 380–383.
236. Cavalu, S.; Kamel, E.; Laslo, V.; Fritea, L.; Costea, T.; Antoniac, I.V.; Vasile, E.; Antoniac, A.; Semenescu, A.; Mohan, A.; et al. Eco-friendly, facile and rapid way for synthesis of selenium nanoparticles production, structural and morphological characterization. *Revista Chimie* **2017**, *68*, 2963–2966. [CrossRef]
237. Cavalu, S.; Antoniac, I.V.; Mohan, A.; Bodog, F.; Doicin, C.; Mates, I.; Ulmeanu, M.; Murzac, R.; Semenescu, A. Nanoparticles and Nanostructured Surface Fabrication for Innovative Cranial and Maxillofacial Surgery. *Materials* **2020**, *13*, 5391. [CrossRef]
238. Cavalu, S.; Antoniac, I.V.; Fritea, L.; Mates, I.M.; Milea, C.; Laslo, V.; Vicas, S.; Mohan, A. Surface modifications of the titanium mesh for cranioplasty using selenium nanoparticles coating. *J. Adhes. Sci. Technol.* **2018**, *32*, 2509–2522. [CrossRef]
239. Szegedi, A.; Trendafilova, I.; Mihaly, J.; Lazar, K.; Nemeth, P.; Momekov, G.; Momekova, D.; Marinov, L.; Nikolova, I.; Popova, M. New insight on prednisolone polymorphs in mesoporous silica/maghemite nanocomposites. *J. Drug Deliv. Sci. Technol.* **2020**, *60*, 102092. [CrossRef]
240. Radeloff, K.; Ramos Tirado, M.; Haddad, D.; Breuer, K.; Müller, J.; Hochmuth, S.; Hackenberg, S.; Scherzad, A.; Kleinsasser, N.; Radeloff, A. Superparamagnetic Iron Oxide Particles (VSOPs) Show Genotoxic Effects but No Functional Impact on Human Adipose Tissue-Derived Stromal Cells (ASCs). *Materials* **2021**, *14*, 263. [CrossRef] [PubMed]
241. Ruiz-Pulido, G.; Medina, D.I.; Barani, M.; Rahdar, A.; Sargazi, G.; Baino, F.; Pandey, S. Nanomaterials for the Diagnosis and Treatment of Head and Neck Cancers: A Review. *Materials* **2021**, *14*, 3706. [CrossRef] [PubMed]

materials

MDPI

Article

Eco-Friendly Biosorbents Based on Microbial Biomass and Natural Polymers: Synthesis, Characterization and Application for the Removal of Drugs and Dyes from Aqueous Solutions

Lăcrămioara Rusu [1,*], Cristina-Gabriela Grigoraș [1,*], Elena Mirela Suceveanu [1], Andrei-Ionuț Simion [1], Andreea Veronica Dediu Botezatu [2], Bogdan Istrate [3] and Ioan Doroftei [3,*]

1 Department of Chemical and Food Engineering, Faculty of Engineering, "Vasile Alecsandri" University of Bacău, 157 Calea Mărășești, 600115 Bacău, Romania; mirela.suceveanu@ub.ro (E.M.S.); asimion@ub.ro (A.-I.S.)
2 Department of Chemistry, Physics and Environment, Faculty of Sciences and Environment, "Dunărea de Jos" University of Galați, 111 Domnească Street, 800201 Galați, Romania; andreea.botezatu@ugal.ro
3 Mechanical Engineering, Mechatronics and Robotics Department, Mechanical Engineering Faculty, "Gheorghe Asachi" Technical University of Iași, 43 Mangeron Blvd., 700050 Iași, Romania; bogdan_istrate1@yahoo.com
* Correspondence: lacraistrati04@yahoo.com (L.R.); cristina.grigoras@ub.ro (C.-G.G.); ioan.doroftei@academic.tuiasi.ro (I.D.)

Citation: Rusu, L.; Grigoraș, C.-G.; Suceveanu, E.M.; Simion, A.-I.; Dediu Botezatu, A.V.; Istrate, B.; Doroftei, I. Eco-Friendly Biosorbents Based on Microbial Biomass and Natural Polymers: Synthesis, Characterization and Application for the Removal of Drugs and Dyes from Aqueous Solutions. *Materials* **2021**, *14*, 4810. https://doi.org/10.3390/ma14174810

Academic Editor: Artur P. Terzyk

Received: 28 July 2021
Accepted: 23 August 2021
Published: 25 August 2021

Abstract: Pharmaceuticals and dyes are a very important part of the nonbiodegradable or hard biodegradable substances present in wastewater. Microorganisms are already known to be effective biosorbents, but the use of free microbial cells involves difficulties in their separation from effluents and limits their application in wastewater treatment. Thus, this study aimed to develop biosorbents by immobilizing *Saccharomyces cerevisiae*, *Saccharomyces pastorianus* and *Saccharomyces pastorianus* residual biomass on natural polymers (alginate and chitosan) and to evaluate the biosorptive potential for removal of pharmaceuticals and dyes from water. Six types of biosorbents were synthesized and characterized by Scanning Electron Microscopy and Fourier Transform Infrared Spectroscopy techniques and their biosorptive capacities for three drugs (cephalexin, rifampicin, ethacridine lactate) and two dyes (orange II and indigo carmine) were evaluated. The obtained results show that the removal efficiency depends on the polymer type used for the immobilization. In case of alginate the removal efficiency is between 40.05% and 96.41% for drugs and between 27.83% and 58.29% for dyes, while in the case of chitosan it is between 40.83% and 77.92% for drugs and between 17.17% and 44.77% for dyes. In general, the synthesized biosorbents proved to be promising for the removal of drugs and dyes from aqueous solutions.

Keywords: biosorption; *Saccharomyces cerevisiae*; *Saccharomyces pastorianus*; alginate; chitosan; drug; dye

1. Introduction

In the last few decades, due to the rapid industrialization and increasing of the world population, the ecosystems are seriously threatened due to the uncontrolled discharge of a large number of toxic organic or inorganic pollutants [1–4]. Among them, toxic metals, radionuclides as well as various organic pollutants, such as pesticides, pharmaceuticals, dyes, personal care products etc., can be listed [5–7].

Most of these inorganic and organic pollutants have accumulated in the environment due to their stability and difficult degradation and are detected in different environmental matrices: surface water, groundwater, soil and drinking water [2,3,8,9].

The World Health Organization states that water is an essential element for sustaining life, so water supply must be adequate, safe and accessible to all. Harmful effects on ecosystems and health hazards due to organic and inorganic pollutants present in water have been established over time, making it necessary to apply ever-higher standards for the detection and treatment of pollutants [10].

To remove pollutants from water, several physicochemical and biological processes have been developed such as adsorption, biosorption, coagulation, sedimentation, filtration, ion exchange, membrane technologies and biological treatments [11,12]. Some of these such as biological methods performed in wastewater treatment plants, are not effective in removing persistent organic pollutants [10].

Pharmaceuticals are a very important part of the nonbiodegradable or hard biodegradable substances present in wastewater and effluents from treatment plants. The annual production of pharmaceutical compounds has been estimated at thousands of tons worldwide [13]. Pharmaceutical residues have been detected in many environmental matrices around the world (e.g., in water, wastewater, sediment and sludge). The most important sources of pollution with these compounds are effluents from sewage treatment plants, hospitals and industrial units [14].

The presence of these pollutants in natural water poses a significant risk and some of them have already been declared as priority substances in water protection policies. Therefore, the development of innovative water treatment technologies is necessary to ensure water quality [15].

Lately, research has focused on removing persistent organic pollutants from water using as methods: adsorption/biosorption or advanced oxidation [11,16–20]. Biosorption is a relevant and increasingly studied process for the removal of pollutants. The literature shows that remarkable results have been obtained for the removal of persistent, inorganic and organic pollutants in low or medium concentrations in different types of aqueous effluents by using living and dead biomass, such as microorganisms, algae and agroindustrial waste [21,22].

This has been and remains a promising biotechnology for the removal of pollutants from aqueous solutions due to its simplicity, efficiency and to the availability of biomass and residual biomass. The biosorptive capacities of different types of biomasses in relation to the removal of different types of pollutants have been reported in numerous scientific papers [4,7,23–26].

Biosorption as a process is based on a variety of mechanisms including: adsorption, absorption, ion exchange, surface complexation and precipitation.

From the first research on the pollutants biosorption processes, the studies focused on obtaining efficient, viable, economical and easy-to-apply in wastewater treatment, biosorbents.

The literature presents numerous materials of biological origin that have been investigated to obtain biosorbents that include microbial biomass (bacteria, cyanobacteria, fungi, yeasts, microalgae), microbial residual biomass (resulting from fermentation processes), agricultural waste, food industry waste and other materials [10,23,27,28].

Both microorganisms and microbial residual biomass have been noted for good biosorptive capacity related to most pollutants tested. Still, they have a number of drawbacks, including the fact that free microbial cells are small in size and low in density, which involves difficulties in their separation from effluents and obviously, an increase in the cost of the biosorption process, which limits their application in wastewater treatment [10,23].

In this sense, the immobilization and encapsulation of microorganisms and residual biomass would be an alternative for obtaining biosorbents.

Different immobilization techniques and different supports were tested, among which we can mention polysulfone, polyvinyl alcohol, polyacrylamide, polyurethane, silica gel, carboxymethyl cellulose, brewer's spent grain, sodium alginate, chitosan [4,7,10,23,24].

Natural polymers such as chitosan and alginate have been used successfully in the processes of immobilization/encapsulation of enzymes and pharmaceuticals [29–31]. So, studies have expanded to obtain adsorbents/biosorbents by immobilizing microorganisms or other materials [4,7,24,26,32–41].

Chitosan has been demonstrated that it can be used in adsorption processes because it is an inert, biocompatible, biodegradable and nontoxic material [7,32,41].

Alginate is also an alternative for immobilization due to its good chemical stability, nontoxicity, incorporation efficiency and low cost [4,7,24,26].

Saccharomyces cerevisiae is one of the most researched species in biosorption processes. Both *Saccharomyces cerevisiae* and *Saccharomyces pastorianus* are used in brewing and are available in large quantities.

In this context, this paper focuses on obtaining eco-friendly biosorbents that can be used in the treatment of effluents containing persistent organic pollutants.

This study aimed to develop biosorbents by immobilizing *Saccharomyces cerevisiae*, *Saccharomyces pastorianus* and *Saccharomyces pastorianus* residual biomass on natural polymers (alginate and chitosan) and to evaluate the biosorptive potential for removal of pharmaceuticals and dyes from water.

Six types of biosorbents were synthesized and characterized by scanning electron microscopy (SEM) and Fourier transform infrared spectroscopy (FTIR) techniques and their biosorptive capacities for three drugs (cephalexin (CPX), rifampicin (RIF), ethacridine lactate (EL)) and two dyes (orange II (O II) and indigo carmine (IC)) were evaluated.

The pharmaceutical compounds were chosen as target molecules taking into account the following considerations:

- CPX is an antibiotic which is considered an emerging compound with a risk factor greater than 1 in a number of countries [42];
- RIF is an antibiotic which is a persistent compound in water matrices with high potential to induce antimicrobial resistance in bacterial cells, leading to mutagenic and carcinogenic effects [22];
- EL is an antiseptic and local analgesic used in large quantities worldwide which can be found in effluents of wastewater treatment plants. Therefore it represents a potential risk for aquatic and other environments [43].

As for the selected dyes, both IC and O II are used in various industrial sectors (food, cosmetics, pharmaceuticals, textiles etc.), discharged in considerable amounts in water and recognized as recalcitrant compounds in water policies, difficult to remove by conventional water treatment [7,44].

Our literature survey revealed that RIF, EL, IC and O II are less studied from the point of view of their removal by adsorptive/ biosorptive processes. Therefore, the present study aims to provide new insights into application of these types of biosorbents in the removal of drugs and dyes from water. To the best of our knowledge, this is the first study to evaluate the biosorptive capacity of *Saccharomyces pastorianus* residual biomass immobilized on natural polymers related to removal of drugs and dyes from aqueous solutions.

2. Materials and Methods

2.1. Reagents and Analytics

All reagents implied in the experimental step were of analytical grade and were used without further purification. Sodium alginate and chitosan were bought from Carl Roth, Karlsruhe, Germany; sodium hydroxide and calcium chloride were delivered by Chempur, Piekary Śląskie, Poland; hydrochloride acid, sodium chloride, acetic acid, di-sodium hydrogen phosphate, potassium di-hydrogen phosphate, bentonite and ethanol were purchased from Chemical Company, Iași, Romania; cephalexin was from Cayman Chemical, Tallinn, Estonia; orange II and indigo carmine green were procured from Merck, Darmstadt, Germany.

Dried yeast strains of *Saccharomyces cerevisiae* and *Saccharomyces pastorianus* and residual biomass of *Saccharomyces pastorianus* were provided by two local companies (Rompak, Pașcani, Romania and Albrau, Onești, Romania).

Stock solutions of drugs and dyes with concentrations of 500 mg/L were prepared by dissolving the reagents in distilled water and kept at 4 °C in closed recipients.

For the calibration curves, adequate volumes of stock solutions were placed in series of 10 mL volumetric flasks. Concentrations ranges were between 1 mg/L and 30 mg/L for CPX; 1 mg/L and 150 mg/L for EL; 1 mg/L and 50 mg/L for RIF; 1 mg/L and 15 mg/L for O II and between 1 mg/L and 20 mg/L for IC.

Samples absorbance were acquired with the help of a UV1280 spectrophotometer (Shimadzu, Tokyo, Japan) at 260 nm for CPX, 431 nm for EL, 475 nm for RIF, 485 nm for O II and 610 nm for IC. Calibration graphs (absorbance vs. concentration) served to recover linear regression equations.

The solutions' pH levels were adjusted with NaOH (0.1 M) or HCl (0.1 M).

2.2. Biosorbents Preparation

For *Saccharomyces cerevisiae*/calcium alginate (SC-A-5%) and *Saccharomyces pastorianus*/calcium alginate (SP-A-5%) beads preparation, aliquots of sodium alginate were placed in laboratory beakers containing hot water and mixed (Nahita heating magnetic plate (Auxilab, Beriain, Spain)) until complete dissolution. Dried yeasts strains were added to final concentrations of 5% on dry weight (d.w.) basis. The mixing process was continued until homogeneous suspensions were obtained. The resulting combinations were suspended in 2% calcium chloride solutions and kept at 4 °C.

Saccharomyces cerevisiae/chitosan (SC-C-2.5%) and *Saccharomyces pastorianus*/chitosan (SP-C-2.5%) beads were synthesized according to the method presented by Kaushal et al. [28] with some modifications. Briefly, homogeneous chitosan suspensions of 2.5% in 1% acetic acid (w/v) were prepared. In parallel, adequate amounts of dried yeasts and bentonite were dissolved into phosphate buffer (pH 7.5) and added to chitosan in order to reach final concentrations of 2.5% for yeasts strains (d.w.) and 0.05% for bentonite (d.w.). After complete homogeneity was attained, the mixtures were dripped in 2 M NaOH solutions, washed with phosphate buffer (pH 7.5) solution and stored at 4 °C.

For the biosorbents prepared with *Saccharomyces pastorianus* residual microbial biomass, the procedure used by de Rossi et al. [4] was considered. In a preliminary step, the biomass was submitted to multiple washing-decantation cycles. The recovered sediment was centrifuged for 20 min at 2500 rpm (Nahita 2615/1 centrifuge, Auxilab, Beriain, Spain) and then dried in a laboratory oven (AirPerformance AP60, Froilabo, Paris, France) for the determination of dried matter content, which was of 26.45%. Precise amounts of residual biomass (5% d.w. for SPRMB-A-5% biosorbent and 9% d.w. for SPRMB-A-9% biosorbent) were put in contact with 1% sodium alginate solutions in hot phosphate buffer (pH 7.0). The resulted homogeneous mixtures were suspended in 2% calcium solutions.

2.3. Biosorbents Characterization (Morphology, Functional Groups, Point of Zero Charge)

A Quanta 200 3D (FEI, Europe B.V., Eindhoven, The Netherlands) scanning electron microscope (SEM) with energy-dispersive X-ray system and a large field detector (LFD) was used for the surface morphology and elemental composition analyses of the synthesized biosorbents. The dried samples (50 °C, 2 h) were fixed to stubs with double adhesive carbon discs. Secondary electron mode (SE) in low vacuum was applied at an accelerating voltage of 20 kV, a working distance of 15 mm and a spot size of 5. The magnification range was between 1 mm and 10 μm.

Fourier transform infrared (FT-IR) spectra were registered between 4000 cm^{-1} and 400 cm^{-1} (32 scans coadded) with a resolution of 4 cm^{-1} on a Nicolet iS50 FT-IR spectrometer (Thermo Scientific, Dreiech, Germany) coupled with a built-in ATR accessory, DTGS detector and a KBr beam splitter. ATR ethanol cleaning was performed after each spectrum acquisition. The background spectrum reference with air was recorded and compared with the anterior one.

For the determination of biosorbents points of zero charge, the following modus operandi was applied. Volumes of 25 mL of 0.1 M NaCl solutions with initial pH values (pH_i) settled between 2 and 12 with HCl (0.1 M) or NaOH (0.1 M) and measured with a portable pH meter (Dostmann KLH9.1, 0–14 pH, Carl Roth, Karlsruhe, Germany) were stirred on magnetic plates with 0.5 g of biosorbent for 24 h, at room temperature. Then, the pH values (pH_f) were measured again. The pH_{pzc} were retrieved from curves $pH_f = f(pH_i)$.

2.4. Drugs and Dyes Biosorption

Samples of 25 mL of pharmaceuticals (CPX: 30 mg/L, pH 5; EL: 100 mg/L, pH 5; RIF: 50 mg/L, pH 6) and dyes (O II: 30 mg/L, pH 5; IC: 50 mg/L, pH 5) solutions were put in contact with 1 g of biosorbents beads for 12 h, at ambient temperature.

The residual concentrations were calculated by reading the samples absorbance at dedicated wavelength against the calibration curves.

The removal efficiency (*R*, %) and the biosorption capacity (*q*, mg/g) were determined with the Equations (1) and (2).

$$R = \frac{C_0 - C_f}{C_0} \cdot 100 \quad (1)$$

$$q = \frac{\left(C_0 - C_f\right) \cdot V}{m} \quad (2)$$

where C_0 and C_f are the initial and final concentrations (mg/L); *V* is the sample volume (L) and *m* is the biosorbent amount (g).

3. Results and Discussion

3.1. Biosorbents Synthesis

For the biosorbents synthesis, two types of natural polymers, sodium alginate and chitosan were tested.

Alginate (Figure 1A) is an alginic acid salt with a polymeric structure formed of blocks of α-L-guluronic acid and β-D-mannuronic acid linked by 1–4 glycosidic bonds [45] and which possess gelling properties. Various ions (hydrogen, calcium, aluminum, copper, zinc, iron, manganese etc.) can be responsible for alginate gelation but the most encountered one is the calcium cation. It interacts with guluronate units to form complexes which pair into dimers, which in turn associate to make multicomplexes [46]. Since this biopolymer is biodegradable, biocompatible and nontoxic, it is suitable for uses in processes such as formulation of pharmaceutical excipients [47], encapsulation of drugs or therapeutic agents [48,49], immobilization of different microorganisms [50,51], food packaging and stabilization [52] etc.

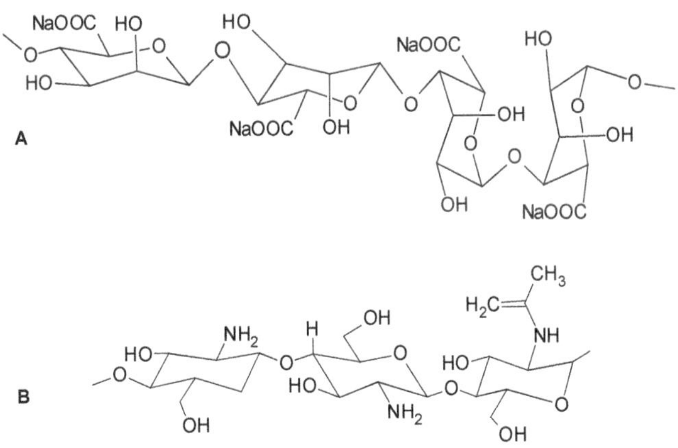

Figure 1. Molecular structure of sodium alginate (**A**) and chitosan (**B**) natural polymers.

Chitosan (Figure 1B) is a polysaccharide consisting of units of D-glucosamine deacetylated and *N*-acetyl-D-glucosamine linked β-1,4. It is biocompatible, biodegradable and easily transformable in hydrogel [53] by emulsion methods, coprecipitation and crosslinking, extrusion crosslinking etc. [54]. It finds applications in sectors as drug delivery [55], encapsulation [56] or immobilization [57] as well.

In our study, these polymers were used to immobilize two different dried yeasts strains, *Saccharomyces cerevisiae* and *Saccharomyces pastorianus*, and one residual microbial biomass of *Saccharomyces pastorianus* issued from beer production.

The exchange of cations existing in the polymers with those of the collecting solutions ensure the development of crosslinkages in matrices chains causing bead formation. The procedure is affected by working parameters including pH, temperature, mixing velocity, polymer solution concentration, collecting solution concentration etc. As general rule, the biomass was incorporated homogeneously and uniformly in the polymeric matrices. Sphericity was attained for all the biosorbents. In all cases, the granules had smooth surfaces and opaque appearance (Figure 2). White nuances were observed for the obtained biosorbents, excepting the SPRMB-5% and SPRMB-9% which had a darker color explained by its residual components.

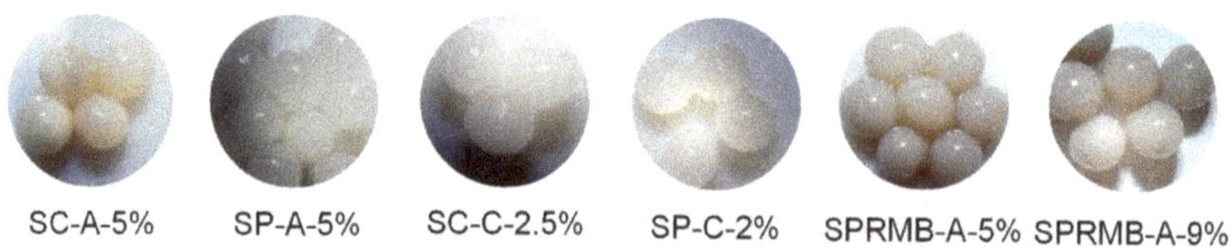

Figure 2. Biosorbents photographs.

Diameters of the adsorbent materials were comparable with average values of 3.064 ± 0.074 mm for SC-A-5%, 2.921 ± 0.198 mm for SP-A-5%, 3.543 ± 0.128 mm for SC-C-2.5%, 3.231 ± 0.130 mm for SP-C-2%, 3.169 ± 0.163 mm for SPRMB-A-5% and 3.534 ± 0.074 mm for SPRMB-A-9%.

3.2. *Biosorbents SEM Investigation*

Figures 3–5 present images recovered by SEM analyses of the synthesized biosorbents. A loss of sphericity can be detected. It is less accentuated for the materials obtained by immobilization of *Saccharomyces cerevisiae* and *Saccharomyces pastorianus* on alginate (Figure 3) than for those on chitosan (Figure 4).

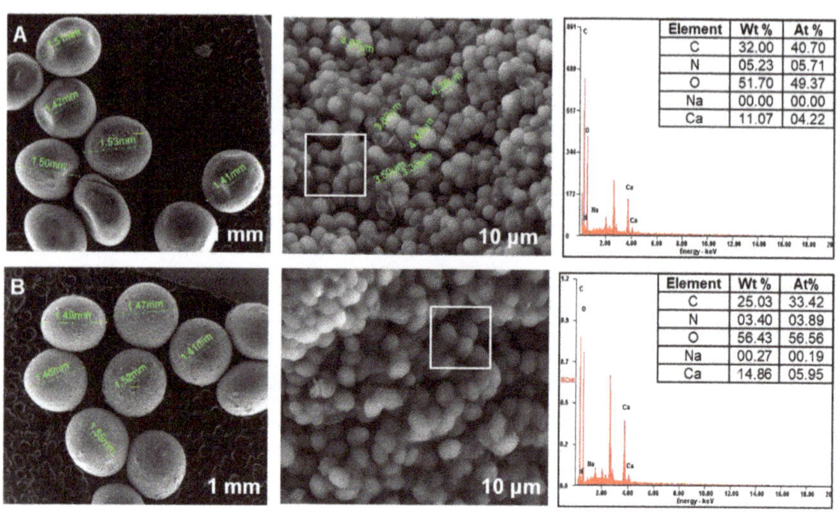

Figure 3. SEM images of biosorbents prepared by immobilization of yeast strains on calcium alginate polymer ((**A**): SC-A-5%; (**B**): SP-A-5%).

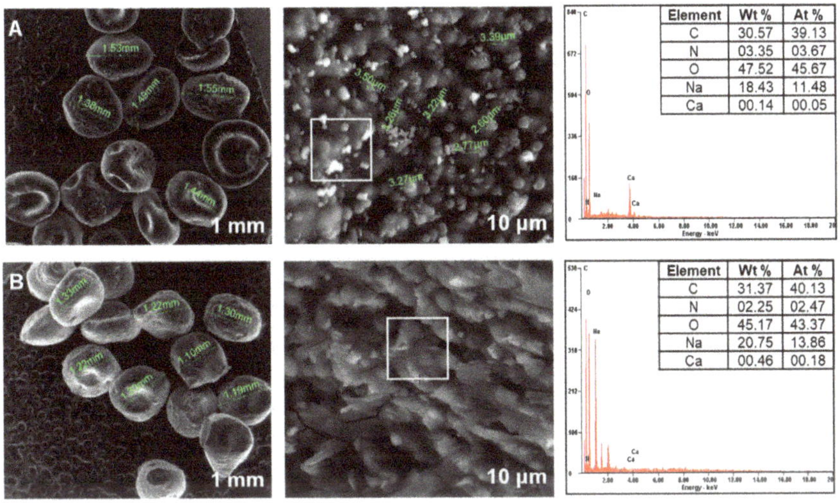

Figure 4. SEM images of biosorbents prepared by immobilization of yeast strains on chitosan ((**A**): SC-C-2.5%; (**B**): SP-C-2.5%).

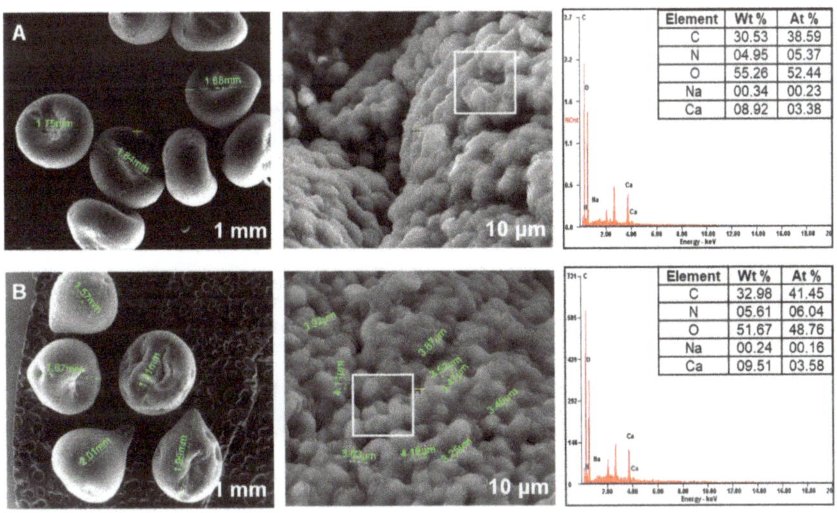

Figure 5. SEM images of biosorbents prepared by immobilization of residual microbial biomass on alginate ((**A**): SPRMB-A-5%; (**B**): SPRMB-A-9%).

As expected, this behavior is caused by the water removal on drying and it has been described in other studies. Gholamian et al. [58] prepared and characterized calcium alginate hydrogels in which they incorporated cumin essential oil. Their morphology tests reveal that freeze-drying affected the beads shapes causing contraction and polymer crease. A considerable shrinkage after beads drying was remarked also by Voo et al. [59] who obtained calcium alginate beads with promising properties for drugs or foods encapsulation. Drying is also responsible for residual crystals of bentonite that can be observed on the beads' surface of SC-C-2.5% (Figure 4A) and SP-C-2.5% (Figure 4B) materials.

In terms of elemental composition, similar proportions of carbon, nitrogen and oxygen exist in all the biosorbents. According to the preparation methods, differences in percentages of sodium and calcium are noted.

3.3. *FT-IR Analyses of the Biosorbents*

FT-IR spectroscopy is often used to characterize materials and qualitatively identify the presence of various functions and groups. FT-IR spectra of the biosorbents developed in this research are depicted in Figure 6.

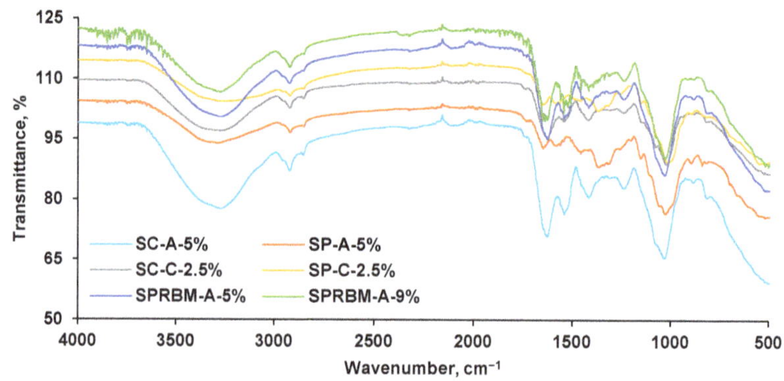

Figure 6. FT-IR spectra of the synthesized biosorbents.

In SC-A-5% and SP-A-5% biosorbents obtained by immobilization of microorganisms on alginate, a CH_2 bending vibration at approximatively 1020 cm^{-1} can be highlighted. At 1620 cm^{-1} and at 1418 cm^{-1}, signals of the asymmetric carboxylic group can be noted while the band of 3270 cm^{-1} is specific for hydroxyl from alginate [60].

For SC-C-2.5% and SP-C-2.5% adsorption bands can be attributed as follows: 2920 cm^{-1} to CH_3 symmetric stretch; 1646 cm^{-1} to N-H vibration (amide II) and 1576 cm^{-1} to C=O stretching (amide I) from chitosan but also specific for yeast cell walls constituents; 1363 cm^{-1} to CH_3 vibration [32,61].

In SPRMB-A-5% and SPRMB-A-9% at 1318 cm^{-1} and at 1238 cm^{-1} stretching for amide III (proteins) and for asymmetric and symmetric PO^{2-} (phosphorylated proteins and phospholipids) were detected and were caused by the increased amount of yeast.

3.4. *Biosorbents' pH_{PZC}*

Point of zero charge (pH_{pzc}), defined as the pH value at which the biosorbents' surface charge is null, i.e., the charge of the positive surface sites is the same as that of the negative ones, was determined with the drift method with sodium chloride as electrolyte. As shown in Figure 7, the shape is similar with an increase of pH_f with the augmentation of NaCl pH_i, with a plateau when the pH_i is between 4 and 10 and with a new increase step after pH_i 10. The same evolution was mentioned by Kragovic et al. [62] who studied among others the influence of alginate encapsulation on the point of zero charge of several zeolites.

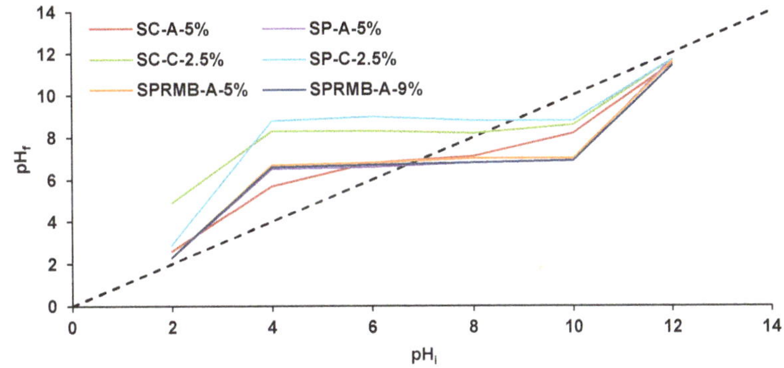

Figure 7. pH_{PZC} of the synthesized biosorbents.

In our case, the recorded results are 6.9 for SC-A-5%, 6.6 for SP-A-5%, 6.9 for SPRMB-A-5% and 6.8 for SPRBM-A-9%, 8.3 for SC-C-2.5% and 8.8 for SP-C-2.5%. It can be seen that the type of biomass did not affect the pH_{PZC} values but they are dependent on the natural polymer used for the immobilization. For biosorbents prepared with sodium alginate, pH_{PZC} values were lower than those collected for biosorbent prepared on chitosan-bentonite. At pH below pH_{PZC}, the biosorbents are considered positively charged while, when pH is superior to pH_{PZC}, the biosorbents possess negative charge.

3.5. *Biosorbents' Applications to Drug and Dye Removal*

In order to examine the viability of the obtained biosorbents, we decided to test their capacity to retain pollutants. Therefore, three pharmaceuticals and two dyes were chosen as target molecules (Figures 8 and 9).

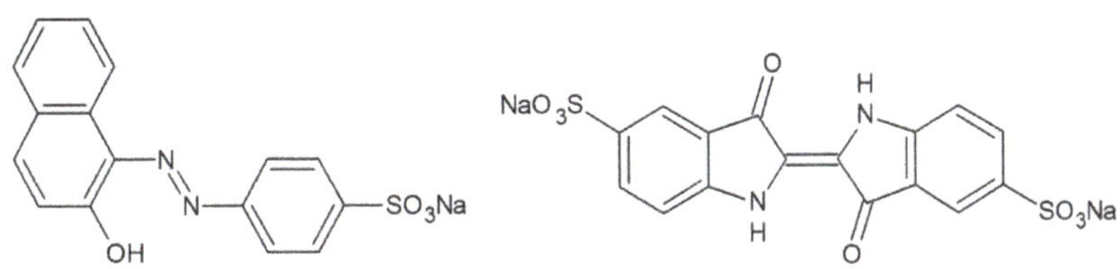

Orange II (O II) CAS: 633-96-5
MF: $C_{16}H_{11}N_2NaO_4S$ MW = 350.30 g/mol

Indigo carmine (IC) CAS: 860-22-0
MF: $C_{16}H_8N_2Na_2O_8S_2$ MW = 466.40 g/mol

Figure 8. Dyes tested on the obtained biosorbents.

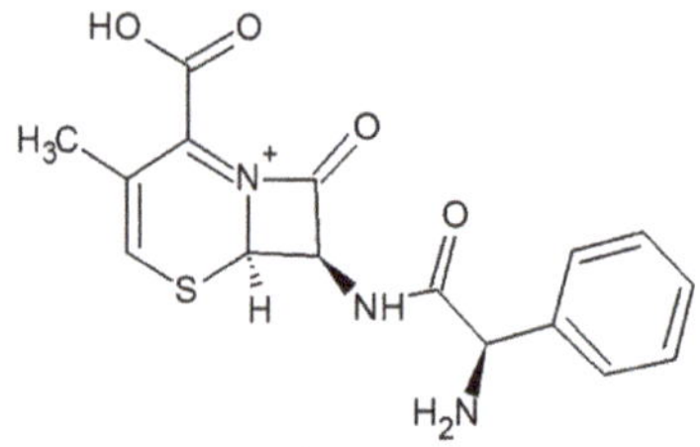

Cephalexin (CPX) CAS: 15686-71-2
MF: $C_{16}H_{17}N_3O_4S$ MW = 347.40 g/mol

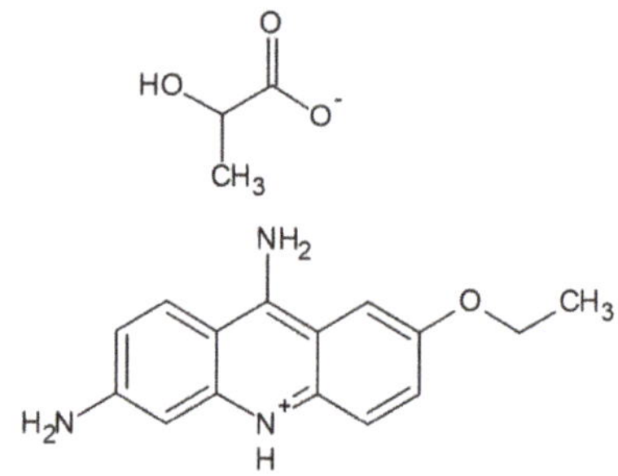

Ethacridine lactate (EL) CAS: 1837-57-6
MF: $C_{18}H_{21}N_3O_4$ MW = 343.40 g/mol

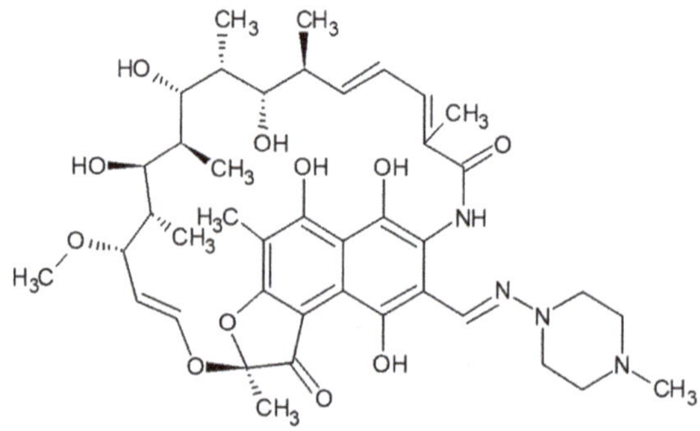

Rifampicin (RIF) CAS: 13292-46-1
MF: $C_{43}H_{58}N_4O_{12}$ MW = 822.90 g/mol

Figure 9. Drugs tested on the obtained biosorbents.

Aliquots of 1 g of each biosorbent were put in contact with volumes of 25 mL of aqueous contaminants solutions. The pH values were set taking into consideration the adsorbates chemical structures and the biosorbents characteristics, especially their pH_{PZC}. The calculated removal efficiencies are exhibited in Figure 10.

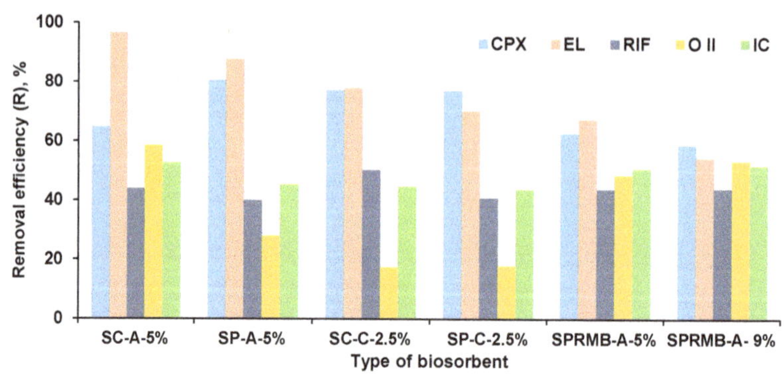

Figure 10. Removal efficiency of drugs and dyes on synthesized biosorbents.

For the antibiotic cephalexin, the best biosorption capacities were recorded for the biosorbents prepared on chitosan (22.78 mg/g for SC-C-2.5% and 28.42 mg/g for SP-C-2.5%) and the lowest ones for those prepared by the immobilization of residual microbial biomass of *Saccharomyces pastorianus* (4.79 mg/g for SPRMB-A-9% and 9.26 mg/g for SPRMB-A-5%).

Similar values were encountered when *Bacillus subtilis* bacterial biomass was used without treatment to remove CPX from aqueous systems [63,64]. As reported by Noman et al. [65], the retention mechanisms are electrostatic repulsion at pH below 4, electrostatic attraction, hydrophobic interactions, π-π bonds and Lewis acid-base interaction at pH between 4 and 6 and electrostatic repulsion and hydrogen bond at pH superior of 6. Moreover, between pKa values (2.56 and 6.88), the functional groups are negatively charged and CPX is in its molecular form fact that allows us to consider that at our working pH, the biosorption occurs by electrostatic attraction between the positive adsorbent surface and the zwitterionic form of CPX.

In the case of ethacridine lactate, the removal efficiencies were between 54.28% for SPRMB-A-5% and 96.40% for SC-A-5%. At present, very little information is available for the elimination of this molecule by adsorption. One of the publications dedicated to such study [43] reveals that the EL is well retained by activated carbons (in less than 60 min and at a concentration of 0.5 mM, more than 90% was eliminated from the solution).

Another of the tested molecules was the rifampicin, an antibiotic with a common structure consisting of a naphtohydroquinone chromophore spanned by an aliphatic ansa chain [66]. The recovered data show that the biosorption follows the same trend as in the case of CPX with the best values obtained for the biosorbents with chitosan as polymeric matrix (24.70 mg/g for SC-C-2.5% and 24.89 mg/g for SP-C-2.5%) while 1 g of SPRMB-A-5% was able to retain no more than 5.99 mg of pollutant. Electrostatic and π-π interactions and hydrogen bonding between RIF and the synthesized biosorbents may be responsible for the biosorption. Kais et al. [22] investigated RIF adsorption on a cocoa shell product and stated that the adsorption capacity was of 26.66 mg/g, while a hybrid nanomaterial prepared by Xu et al. [67] possessed an adsorption capacity of more than 90 mg/g for RIF found in solutions with concentrations between 3 mg/L and 20 mg/L.

Regarding the dyes' biosorption, from Figure 10, one can observe that the results are also satisfactory. For O II, the removal efficiencies were of approximatively 18% for the biosorbents prepared by yeast strains' immobilization on chitosan and significantly higher (more than 58% for SC-A-5%) for the other materials. For IC, the removal efficiencies present close and similar values for all the biosorbents (superior of 40% each time). The considered dyes are the subject of a wide variety of researches [7,44,68] which reveal that the adsorption process is due to electrostatic interactions and hydrogen bonding between the pollutants and the adsorbents.

In our tested conditions, the prepared biosorbents gave reasonable results in terms of the ability of removing drugs and dyes from aqueous solutions. However, it should be mentioned that in optimized settings, better uptake efficiencies could be expected.

4. Conclusions

The synthesized biosorbents exhibited similar characteristics, which proves that the techniques used for their synthesis are efficient.

The immobilization methods of microorganisms and residual biomass in the polymer matrix of alginate and chitosan were shown to be effective, the cells' immobilization being confirmed by SEM and FT-IR analyzes.

The points of zero charge were identified at 6.9 for SC-A-5%, 6.6 for SP-A-5%, 6.9 for SPRMB-A-5% and 6.8 for SPRBM-A-9%, 8.3 for SC-C-2.5% and 8.8 for SP-C-2%. The pH_{PZC} values depend on the polymer type used for the immobilization and not the type of biomass that was used.

The obtained results show that the removal efficiency depends on the type of biosorbent and the concentration of the pollutant. In the case of using the alginate for immobilization the removal efficiency is between 40.05% and 96.41% for drugs and between 27.83% and 58.29% for dyes, while in the case of chitosan it is between 40.83% and 77.92% for drugs and between 17.17% and 44.77% for dyes. The biosorption capacities obtained in the presented test conditions vary between 4.79 mg/g and 85.42 mg/g for drugs and between 4.34 mg/g and 26.87 mg/g for dyes. This indicates a good biosorption capacity if we take into account the amounts of pollutants present in different water matrices.

The mentioned results reveal that all the synthesized biosorbents can be useful for the removal of pharmaceuticals and dyes from aqueous solutions. Further, by using the immobilization technique on natural polymers, first of all, the disadvantages of using free cells are eliminated and secondly, eco-friendly materials are obtained.

Finally, additional work is needed to study the effect of process parameters (pH, initial concentration, biosorbent dose and temperature) in order to optimize biosorption capacity.

Author Contributions: Conceptualization, L.R. and C.-G.G.; methodology, L.R., E.M.S. and C.-G.G.; software, A.-I.S.; validation, L.R. and C.-G.G.; formal analysis, E.M.S., L.R., C.-G.G., A.V.D.B. and B.I.; investigation, E.M.S., A.-I.S., L.R. and C.-G.G.; resources, L.R.; writing—original draft preparation, L.R. and C.-G.G.; writing—review and editing, L.R., C.-G.G. and I.D.; supervision, L.R. and I.D.; project administration, L.R.; funding acquisition, L.R. All authors have read and agreed to the published version of the manuscript.

Funding: This research was funded by a grant of the Romanian Ministry of Research and Innovation, CCCDI-UEFISCDI, project number PN-III-P2-2.1-PED-2019-1063, within PNCDI III and the APC was funded by a grant of the Romanian Ministry of Research and Innovation, CCCDI-UEFISCDI, project number PN-III-P2-2.1-PED-2019-1063, within PNCDI III.

Institutional Review Board Statement: Not applicable.

Informed Consent Statement: Not applicable.

Data Availability Statement: All data produced in this study are presented in this paper.

Conflicts of Interest: The authors declare no conflict of interest. The funders had no role in the design of the study; in the collection, analyses, or interpretation of data; in the writing of the manuscript, or in the decision to publish the results.

References

1. Stawiński, W.; Węgrzyn, A.; Freitas, O.; Chmielarz, L.; Mordarski, G.; Figueiredo, S. Simultaneous removal of dyes and metal cations using an acid, acid-base and base modified vermiculite as a sustainable and recyclable adsorbent. *Sci. Total Environ.* **2017**, *576*, 398–408. [CrossRef] [PubMed]
2. Garrudo-Guirado, M.I.; Blanco-Flores, A.; Toledo-Jaldin, H.P.; Sánchez-Mendieta, V.; Vilchis-Néstor, A.R. Reuse of sustainable materials for xylenol orange dye and copper (II) ion ammoniacal removal. *J. Environ. Manag.* **2018**, *206*, 920–928. [CrossRef] [PubMed]
3. Visa, M.; Andronic, L.; Duta, A. Fly ash-TiO_2 nanocomposite material for multi-pollutants wastewater treatment. *J. Environ. Manag.* **2015**, *150*, 336–343. [CrossRef] [PubMed]
4. De Rossi, A.; Rigueto, C.V.T.; Dettmer, A.; Colla, L.M.; Piccin, J.S. Synthesis, characterization, and application of *Saccharomyces cerevisiae*/alginate composites beads for adsorption of heavy metals. *J. Environ. Chem. Eng.* **2020**, *8*, 104009. [CrossRef]
5. Fernández-Castro, P.; Vallejo, M.; San Román, M.F.; Ortiz, I. Insight on the fundamentals of advanced oxidation processes. Role and review of the determination methods of reactive oxygen species. *J. Chem. Technol. Biotechnol.* **2015**, *90*, 796–820. [CrossRef]
6. Anglada, Á.; Urtiaga, A.; Ortiz, I. Contributions of electrochemical oxidation to waste-water treatment: Fundamentals and review of applications. *J. Chem. Technol. Biotechnol.* **2009**, *84*, 1747–1755. [CrossRef]
7. Kekes, T.; Tzia, C. Adsorption of indigo carmine on functional chitosan and β-cyclodextrin/chitosan beads: Equilibrium, kinetics and mechanism studies. *J. Environ. Manag.* **2020**, *262*, 110372. [CrossRef]
8. Pal, A.; He, Y.; Jekel, M.; Reinhard, M.; Gin, K.Y.-H. Emerging contaminants of public health significance as water quality indicator compounds in the urban water cycle. *Environ. Int.* **2014**, *71*, 46–62. [CrossRef]
9. Rusu, L.; Suceveanu, M.; Șuteu, D.; Favier, L.; Harja, M. Assessment of groundwater and surface water contamination by landfill leachate: A case study in Neamț County, Romania. *Environ. Eng. Manag. J.* **2017**, *16*, 633–641. [CrossRef]
10. Fomina, M.; Gadd, G.M. Biosorption: Current perspectives on concept, definition and application. *Bioresour. Technol.* **2014**, *160*, 3–14. [CrossRef]
11. Suteu, D.; Coseri, S.; Rusu, L. Kinetics studies on the adsorption behaviour of Basic Blue 9 dye on macroporous ion exchanger resins. *Desalin. Water Treat.* **2016**, *57*, 14665–14673. [CrossRef]
12. Morosanu, I.; Teodosiu, C.; Fighir, D.; Paduraru, C. Simultaneous biosorption of micropollutants from aqueous effluents by rapeseed waste. *Process Saf. Environ. Prot.* **2019**, *132*, 231–239. [CrossRef]
13. Tran, N.H.; Urase, T.; Kusakabe, O. Biodegradation Characteristics of Pharmaceutical Substances by Whole Fungal Culture *Trametes versicolor* and its Laccase. *J. Water Environ. Technol.* **2010**, *8*, 125–140. [CrossRef]

14. Blanco, J.; Malato, S.; Fernández-Ibañez, P.; Alarcón, D.; Gernjak, W.; Maldonado, M.I. Review of feasible solar energy applications to water processes. *Renew. Sust. Energ. Rev.* **2009**, *13*, 1437–1445. [CrossRef]
15. Ribeiro, A.R.; Nunes, O.C.; Pereira, M.F.R.; Silva, A.M.T. An overview on the advanced oxidation processes applied for the treatment of water pollutants defined in the recently launched Directive 2013/39/EU. *Environ. Int.* **2015**, *75*, 33–51. [CrossRef] [PubMed]
16. Vrinceanu, N.; Hlihor, R.M.; Simion, A.I.; Rusu, L.; Fekete-Kertész, I.; Barka, N.; Favier, L. New Evidence of the Enhanced Elimination of a Persistent Drug Used as a Lipid Absorption Inhibitor by Advanced Oxidation with UV-A and Nanosized Catalysts. *Catalysts* **2019**, *9*, 761. [CrossRef]
17. Favier, L.; Rusu, L.; Simion, A.I.; Hlihor, R.M.; Păcală, M.L.; Augustyniak, A. Efficient degradation of clofibric acid by heterogeneous photocatalytic oxidation process. *Environ. Eng. Manag. J.* **2019**, *18*, 1683–1692. [CrossRef]
18. Şuteu, D.; Biliuţă, G.; Rusu, L.; Coseri, S.; Vial, C.; Nica, I. A regenerable microporous adsorbent based on microcrystalline cellulose for organic pollutants adsorption. *Desalin. Water Treat.* **2019**, *146*, 176–187. [CrossRef]
19. Smolin, S.; Kozyatnyk, I.; Klymenko, N. New approach for the assessment of the contribution of adsorption, biodegradation and self-bioregeneration in the dynamic process of biologically active carbon functioning. *Chemosphere* **2020**, *248*, 126022. [CrossRef]
20. Korotta-Gamage, S.M.; Sathasivan, A. A review: Potential and challenges of biologically activated carbon to remove natural organic matter in drinking water purification process. *Chemosphere* **2017**, *167*, 120–138. [CrossRef]
21. De Gisi, S.; Lofrano, G.; Grassi, M.; Notarnicola, M. Characteristics and adsorption capacities of low-cost sorbents for wastewater treatment: A review. *Sustain. Mater. Technol.* **2016**, *9*, 10–40. [CrossRef]
22. Kais, H.; Yeddou Mezenner, N.; Trari, M. Biosorption of rifampicin from wastewater using cocoa shells product. *Sep. Sci. Technol.* **2020**, *55*, 1984–1993. [CrossRef]
23. Kordialik-Bogacka, E. *Saccharomyces pastorianus* immobilized on brewer's spent grain in continuous system for lead ion biosorption. *Int. Biodeterior. Biodegrad.* **2014**, *96*, 191–197. [CrossRef]
24. De Araujo, L.G.; De Borba, T.R.; de Pádua Ferreira, R.V.; Canevesi, R.L.S.; Da Silva, E.A.; Dellamano, J.C.; Marumo, J.T. Use of calcium alginate beads and *Saccharomyces cerevisiae* for biosorption of 241Am. *J. Environ. Radioact.* **2020**, *223–224*, 106399. [CrossRef]
25. Huang, F.; Li, K.; Wu, R.-R.; Yan, Y.-J.; Xiao, R.-B. Insight into the Cd2+ biosorption by viable *Bacillus cereus* RC-1 immobilized on different biochars: Roles of bacterial cell and biochar matrix. *J. Clean. Prod.* **2020**, *272*, 122743. [CrossRef]
26. Horciu, L.I.; Zaharia, C.; Blaga, A.C.; Rusu, L.; Suteu, D. Brilliant Red HE-3B Dye Biosorption by Immobilized Residual Consortium *Bacillus* sp. Biomass: Fixed-Bed Column Studies. *Appl. Sci.* **2021**, *11*, 4498. [CrossRef]
27. Park, D.; Yun, Y.-S.; Park, J.M. The past, present, and future trends of biosorption. *Biotechnol. Bioprocess Eng.* **2010**, *15*, 86–102. [CrossRef]
28. Dhankhar, R.; Hooda, A. Fungal biosorption—An alternative to meet the challenges of heavy metal pollution in aqueous solutions. *Environ. Technol.* **2011**, *32*, 467–491. [CrossRef]
29. Wahba, M.I. Porous chitosan beads of superior mechanical properties for the covalent immobilization of enzymes. *Int. J. Biol. Macromol.* **2017**, *105*, 894–904. [CrossRef] [PubMed]
30. Kaushal, J.; Seema; Singh, G.; Arya, S.K. Immobilization of catalase onto chitosan and chitosan–bentonite complex: A comparative study. *Biotechnol. Rep.* **2018**, *18*, e00258.
31. Singh, R.S.; Singh, R.P.; Kennedy, J.F. Immobilization of yeast inulinase on chitosan beads for the hydrolysis of inulin in a batch system. *Int. J. Biol. Macromol.* **2017**, *95*, 87–93. [CrossRef]
32. Dilarri, G.; Corso, C.R. *Saccharomyces cerevisiae* immobilized onto cross-linked chitosan beads: Application of a novel material for the removal of dye toxicity. *Environ. Technol.* **2018**, *39*, 1851–1867. [CrossRef] [PubMed]
33. Sun, X.; Hu, D.; Yang, L.Y.; Wang, N.; Wang, Y.-G.; Ouyang, X.K. Efficient adsorption of Levofloxacin from aqueous solution using calcium alginate/metal organic frameworks composite beads. *J. Sol-Gel. Sci. Technol.* **2019**, *91*, 353–363. [CrossRef]
34. Afzal, M.Z.; Sun, X.-F.; Liu, J.; Song, C.; Wang, S.-G.; Javed, A. Enhancement of ciprofloxacin sorption on chitosan/biochar hydrogel beads. *Sci. Total Environ.* **2018**, *639*, 560–569. [CrossRef]
35. Vakili, M.; Qiu, W.; Cagnetta, G.; Huang, J.; Yu, G. Mechanochemically oxidized chitosan-based adsorbents with outstanding Penicillin G adsorption capacity. *J. Environ. Chem. Eng.* **2021**, *9*, 105454. [CrossRef]
36. Kekes, T.; Kolliopoulos, G.; Tzia, C. Hexavalent chromium adsorption onto crosslinked chitosan and chitosan/β-cyclodextrin beads: Novel materials for water decontamination. *J. Environ. Chem. Eng.* **2021**, *9*, 105581. [CrossRef]
37. Danalıoğlu, S.T.; Bayazit, Ş.S.; Kerkez Kuyumcu, Ö.; Salam, M.A. Efficient removal of antibiotics by a novel magnetic adsorbent: Magnetic activated carbon/chitosan (MACC) nanocomposite. *J. Mol. Liq.* **2017**, *240*, 589–596. [CrossRef]
38. Ranjbari, S.; Tanhaei, B.; Ayati, A.; Khadempir, S.; Sillanpää, M. Efficient tetracycline adsorptive removal using tricaprylmethylammonium chloride conjugated chitosan hydrogel beads: Mechanism, kinetic, isotherms and thermodynamic study. *Int. J. Biol. Macromol.* **2020**, *155*, 421–429. [CrossRef] [PubMed]
39. Sadiq, A.C.; Rahim, N.Y.; Suah, F.B.M. Adsorption and desorption of malachite green by using chitosan-deep eutectic solvents beads. *Int. J. Biol. Macromol.* **2020**, *164*, 3965–3973. [CrossRef] [PubMed]
40. Tong, K. Preparation and biosorption evaluation of *Bacillus subtilis*/alginate-chitosan microcapsule. *Nanotechnol. Sci. Appl.* **2017**, *10*, 35–43. [CrossRef]

41. Yaqubi, O.; Tai, M.H.; Mitra, D.; Gerente, C.; Neoh, K.G.; Wang, C.-H.; Andres, Y. Adsorptive removal of tetracycline and amoxicillin from aqueous solution by leached carbon black waste and chitosan-carbon composite beads. *J. Environ. Chem. Eng.* **2021**, *9*, 104988. [CrossRef]
42. Meng, L.; Wang, J.; Li, X.; Cui, F. Insight into effect of high-level cephalexin on fate and driver mechanism of antibiotics resistance genes in antibiotic wastewater treatment system. *Ecotoxicol. Environ. Saf.* **2020**, *201*, 110739. [CrossRef] [PubMed]
43. Talman, R.Y.; Gokturk, S.; Bastug, A.S.; Caliskan, E. Adsorption of rivanol on activated carbon from aqueous solutions. *EMChiE 2010 Conf. Proc.* **2010**, *2*, 815–821.
44. Jin, X.; Yu, B.; Chen, Z.; Arocena, J.M.; Thring, R.W. Adsorption of Orange II dye in aqueous solution onto surfactant-coated zeolite: Characterization, kinetic and thermodynamic studies. *J. Colloid Interface Sci.* **2014**, *435*, 15–20. [CrossRef] [PubMed]
45. Mrudulakumari Vasudevan, U.; Lee, O.K.; Lee, E.Y. Alginate derived functional oligosaccharides: Recent developments, barriers, and future outlooks. *Carbohydr. Polym.* **2021**, *267*, 118158. [CrossRef] [PubMed]
46. Hu, C.; Lu, W.; Mata, A.; Nishinari, K.; Fang, Y. Ions-induced gelation of alginate: Mechanisms and applications. *Int. J. Biol. Macromol.* **2021**, *177*, 578–588. [CrossRef]
47. Sanchez-Ballester, N.M.; Bataille, B.; Soulairol, I. Sodium alginate and alginic acid as pharmaceutical excipients for tablet formulation: Structure-function relationship. *Carbohydr. Polym.* **2021**, *270*, 118399. [CrossRef]
48. Lakkakula, J.R.; Gujarathi, P.; Pansare, P.; Tripathi, S. A comprehensive review on alginate-based delivery systems for the delivery of chemotherapeutic agent: Doxorubicin. *Carbohydr. Polym.* **2021**, *259*, 117696. [CrossRef]
49. Ahmad Raus, R.; Wan Nawawi, W.M.F.; Nasaruddin, R.R. Alginate and alginate composites for biomedical applications. *Asian J. Pharm. Sci.* **2021**, *16*, 280–306. [CrossRef]
50. López-Menchero, J.R.; Ogawa, M.; Mauricio, J.C.; Moreno, J.; Moreno-García, J. Effect of calcium alginate coating on the cell retention and fermentation of a fungus-yeast immobilization system. *LWT* **2021**, *144*, 111250. [CrossRef]
51. Ercole, A.; Raganati, F.; Salatino, P.; Marzocchella, A. Continuous succinic acid production by immobilized cells of Actinobacillus succinogenes in a fluidized bed reactor: Entrapment in alginate beads. *Biochem. Eng. J.* **2021**, *169*, 107968. [CrossRef]
52. de Oliveira Filho, J.G.; Rodrigues, J.M.; Valadares, A.C.F.; de Almeida, A.B.; de Lima, T.M.; Takeuchi, K.P.; Alves, C.C.F.; de Figueiredo Sousa, H.A.; da Silva, E.R.; Dyszy, F.H.; et al. Active food packaging: Alginate films with cottonseed protein hydrolysates. *Food Hydrocoll.* **2019**, *92*, 267–275. [CrossRef]
53. Azmana, M.; Mahmood, S.; Hilles, A.R.; Rahman, A.; Arifin, M.A.B.; Ahmed, S. A review on chitosan and chitosan-based bionanocomposites: Promising material for combatting global issues and its applications. *Int. J. Biol. Macromol.* **2021**, *185*, 832–848. [CrossRef] [PubMed]
54. Qu, B.; Luo, Y. Chitosan-based hydrogel beads: Preparations, modifications and applications in food and agriculture sectors—A review. *Int. J. Biol. Macromol.* **2020**, *152*, 437–448. [CrossRef]
55. Shi, W.; Ching, Y.C.; Chuah, C.H. Preparation of aerogel beads and microspheres based on chitosan and cellulose for drug delivery: A review. *Int. J. Biol. Macromol.* **2021**, *170*, 751–767. [CrossRef]
56. Karthikeyan, P.; Meenakshi, S. Fabrication of hybrid chitosan encapsulated magnetic-kaolin beads for adsorption of phosphate and nitrate ions from aqueous solutions. *Int. J. Biol. Macromol.* **2021**, *168*, 750–759. [CrossRef]
57. Costa-Silva, T.A.; Carvalho, A.K.F.; Souza, C.R.F.; De Castro, H.F.; Bachmann, L.; Said, S.; Oliveira, W.P. Enhancement lipase activity via immobilization onto chitosan beads used as seed particles during fluidized bed drying: Application in butyl butyrate production. *Appl. Catal. A Gen.* **2021**, *622*, 118217. [CrossRef]
58. Gholamian, S.; Nourani, M.; Bakhshi, N. Formation and characterization of calcium alginate hydrogel beads filled with cumin seeds essential oil. *Food Chem.* **2021**, *338*, 128143. [CrossRef] [PubMed]
59. Voo, W.-P.; Ooi, C.-W.; Islam, A.; Tey, B.-T.; Chan, E.-S. Calcium alginate hydrogel beads with high stiffness and extended dissolution behaviour. *Eur. Polym. J.* **2016**, *75*, 343–353. [CrossRef]
60. Das, T.K.; Scott, Q.; Bezbaruah, A.N. Montmorillonite-iron crosslinked alginate beads for aqueous phosphate removal. *Chemosphere* **2021**, *281*, 130837. [CrossRef] [PubMed]
61. Abdelkrim, S.; Mokhtar, A.; Djelad, A.; Bennabi, F.; Souna, A.; Bengueddach, A.; Sassi, M. Chitosan/Ag-Bentonite Nanocomposites: Preparation, Characterization, Swelling and Biological Properties. *J. Inorg. Organomet. Polym. Mater.* **2020**, *30*, 831–840. [CrossRef]
62. Kragović, M.; Stojmenović, M.; Petrović, J.; Loredo, J.; Pašalić, S.; Nedeljković, A.; Ristović, I. Influence of Alginate Encapsulation on Point of Zero Charge (pHpzc) and Thermodynamic Properties of the Natural and Fe(III)—Modified Zeolite. *Procedia Manuf.* **2019**, *32*, 286–293. [CrossRef]
63. Al-Gheethi, A.A.S.; Ismail, N. Biodegradation of Pharmaceutical Wastes in Treated Sewage Effluents by *Bacillus subtilis* 1556WTNC. *Environ. Process.* **2014**, *1*, 459–481. [CrossRef]
64. Adel, A.S.; Lalung, J.; Efaq, A.N.; Ismail, N. Removal of cephalexin antibiotic and heavy metals from pharmaceutical effluents using *Bacillus subtilis* strain. *Expert Opin. Environ. Biol.* **2015**, *4*, 1000117.
65. Ali Noman, E.; Al-Gheethi, A.; Saphira Radin Mohamed, R.M.; Talip, B.A.; Hossain, M.S.; Ali Hamood Altowayti, W.; Ismail, N. Sustainable approaches for removal of cephalexin antibiotic from non-clinical environments: A critical review. *J. Hazard. Mater.* **2021**, *417*, 126040. [CrossRef]
66. Yulug, B.; Hanoglu, L.; Kilic, E.; Schabitz, W.R. RIFAMPICIN: An antibiotic with brain protective function. *Brain Res. Bull.* **2014**, *107*, 37–42. [CrossRef]

67. Xu, Q.; Owens, G.; Chen, Z. Adsorption and catalytic reduction of rifampicin in wastewaters using hybrid rGO@Fe/Pd nanoparticles. *J. Clean. Prod.* **2020**, *264*, 121617. [CrossRef]
68. Abramian, L.; El-Rassy, H. Adsorption kinetics and thermodynamics of azo-dye Orange II onto highly porous titania aerogel. *Chem. Eng. J.* **2009**, *150*, 403–410. [CrossRef]

materials

MDPI

Article

The Influence of the Environment for Glass-Reinforced Plastic Composite Material Used for Ground Water Transport Pipes

Ana Diana Ancaș [1], Corneliu Munteanu [2], Bogdan Istrate [2,*], Mihai Profire [1] and Florin-Emilian Țurcanu [1,*]

[1] Building Services Department, Faculty of Civil Engineering and Building Services, Gheorghe Asachi Technical University, 700050 Iaşi, Romania; ancas05@yahoo.com (A.D.A.); profiremihai@yahoo.com (M.P.)
[2] Mechanical Engineering, Mechatronics and Robotics Department, Faculty of Mechanical Engineering, Gheorghe Asachi Technical University, 700050 Iaşi, Romania; cornelmun@gmail.com
* Correspondence: bogdan_istrate1@yahoo.com (B.I.); emilian-florin.turcanu@tuiasi.ro (F.-E.Ț.)

Abstract: Glass-reinforced plastic (GRP) composite materials are mainly used in the construction of pipes due to the wide range of sizes, ease of installation, adaptability to the specific situation in the field and, last but not least, the more competitive price as the nominal diameter increases. Their wide range of applications: drinking and raw water transport, sewerage, industrial waters, desalination plants, mining, etc., has led to the need to tailor the behaviour of the composite material to different fields, with pH values that are not neutral. Based on the experimental data, we aimed to study the change in the structure of the composite material as influenced by the soil characteristics: neutral, basic and acidic. In addition, starting with the pH of the three types of soil—basic, acidic and neutral—which significantly affect GRP composite materials, we calculated the pipe damage index and the Pearson correlation coefficients for axial tension. The results highlight the significant influence of the soil pH on the behaviour over time of the buried GRP pipes. Thus, laying the pipe in acidic soil significantly reduces its life, which should be taken into consideration during the design phase.

Keywords: glass-reinforced plastic composite material; ground water transport pipes; pipe damage index; axial tension

Citation: Ancaș, A.D.; Munteanu, C.; Istrate, B.; Profire, M.; Țurcanu, F.-E. The Influence of the Environment for Glass-Reinforced Plastic Composite Material Used for Ground Water Transport Pipes. *Materials* **2021**, *14*, 3160. https://doi.org/10.3390/ma14123160

Academic Editor: Aleksander Muc

Received: 20 April 2021
Accepted: 4 June 2021
Published: 8 June 2021

1. Introduction

Water transport pipes constructed from composite materials are relatively new. Thus, it is advantageous to extrapolate their behaviour throughout the long operating time to establish the trends that these pipelines follow in service. Behaviour comparisons between pipes constructed from composite materials and those constructed from classical materials (cast iron, concrete, steel) are also extremely useful in design in order to choose the most suitable solution for different situations [1,2].

Thus, it is necessary to conduct new studies to evaluate other aspects of the behaviour of composite pipes in different situations (effects generated by the nature and properties of the site, the impact of natural hazards such as landslides, floods, earthquakes, shocks, explosions, etc.) [3].

Glass-reinforced plastic materials are composites produced by creating alternate layers of fibreglass, resin, sand, quartz, dye, etc. They are used in various applications and fields, such as water and sewerage pipes, tanks, the automotive industry, aeronautics, petrochemicals, energy, sports products, construction, etc. The chemical composition of the material is at least 98% silicon dioxide (quartz sand), calcium alumina silicates (fibreglass) and thermally hardened resins [4].

The manufacturing technology consists of winding and centrifugation. The centrifugation technology uses only chopped glass wires, randomly arranged in the mass of the material, having the advantage of a higher density. Modern technology has enabled the highly compact lamination of the three materials, namely glass, sand and resin, thus optimizing the composition of the pipe according to the actual requirements in the field—both a circumferentially arranged continuous wire and a fibreglass shredder are used for increased axial strength. The addition of quartz sand increases the pipe wall thickness, thus removing the material from the neutral axis to increase the annular rigidity of the pipe. Therefore, three layers are created: the outer structural layer, the core and the lower structural layer [5,6].

For pressurised pipes or for those laid in trenches, the main challenge is the circumference, so the use of continuous wire rods in this context has unique advantages in operation. The main element of the technology is a rotating mould, with a length of 6 m; the length of each section is also limited [7,8]. Through a feed arm moving along the mould, it is subjected to various predetermined amounts (depending on the characteristics of the product pipe) of fibreglass shredder, resin and quartz sand while rotating at a relatively low speed. The quantities of the three components and the sequence of their delivery in the mould are controlled by a computer. After the final quantities are delivered to the mould, the rotational speed increases significantly in order to achieve the centrifugal force needed to ensure the desired degree of compression and ensure the dispersion of the materials according to the requirements. After cooling and stopping the rotational movement, the pipe is released from the mould, and the machined ends and the plug are added. The resistance structure consists of layers of resin and fibreglass, and the middle is filled with quartz sand mixed with resin. The application of these materials is performed continuously and successively, followed by heat treatment [9,10].

The standards for pipes constructed from fibreglass-reinforced polyester resins consider that the stresses to which the pipes are subjected are reflected in their mechanical deformation. Their projected lifespan is 50 years. Over the years, test results have demonstrated that their stability is much better than expected, leading to the notion that their lifespan can be extended to 150 years (in the case of Flowtite Gray Pipes, for example) [11,12].

Pipes constructed from composite materials, such as GRP, can be found in a wide range of corrosive environments, which can cause changes in the properties of the material.

The mass of the composite, due to the action of the underlying medium, and the phenomenon of modification of the structure of the material lead to its mechanical degradation [13–17].

Taking the above into consideration, it is necessary to adapt the material to the requirements of the particular environment in order to determine the type and the amount of fibreglass and resin to be used.

2. Materials and Methods

The resistance structure of GRP pipes consists of layers of resin and fibreglass, and the middle is filled with quartz sand mixed with resin. The application of these materials is performed continuously, in successive layers. After the formation of the pipeline, the heat treatment is applied, during which all the chemical processes are completed, and the length of the pipeline is established. The end is calibrated, and the plug is mounted.

Based on prior experience with water pipes, some damage has been attributed to errors in installation and problems relating to the minimum characteristics required for the filling material. Indeed, the pH of the soil also has an essential effect on the destruction of the pipe by degrading the mechanical characteristics of the GRP composite material. The experimental studies performed so far on GRP pipes have mainly focused on their behaviour when filled with substances, and less on the soil in which they are located. Regarding the influence of the soil on the composite material, many variables are involved; consequently, we used the Pearson correlation coefficient, developed by the English mathematician Karl Pearson at the beginning of the 20th century [18].

One of the previously completed studies [19,20] proposed a formula for quantifying the damage to the composite pipe applied to the flow load (the value of the force from which the behaviour of the pipe changes from elastic to rigid):

$$D = 1 - \frac{F_c^*}{F_c} \quad (1)$$

where: D—the pipe damage parameter; F_c^*—break strength in basic and acidic soil; F_c—break strength in neutral ground.

The choice of the flow load for measuring the impact of pH on the axial behaviour of the composite material should take into consideration that, in practice, the load values will not equal the breaking load (due to the joining with plugs, possibly resulting in a plug-in in the case of a large axial load). In blocked or in laminated joints, although less commonly used (less than 3% of GRP pipes worldwide), the breaking load could be relevant in the short term [21,22].

We conducted the experimental study in acidic and basic soils based on these data. Neutral soil was used as control ground.

To highlight the influence of the terrain on the GRP pipes, we buried three sections of length 150 cm, DN250, wall thickness 6.2 mm, yield point 25 MPa, in three experimental locations.

The percentage composition was determined after calcination (JEPE) using the following values:

- Resin 26.41%;
- Fibreglass ($CaAlSiO_8$)—38.28%;
- Sand (SiO_2)—35.31%.

The first sample was taken from a petroleum field (the first experimental location), in the Zemeș area, Romania, which also contains water from a section of the main pipeline. The second sample was taken from a salt field (the second experimental location) in the Amara area of Romania, which was a basic soil.

For the third experiment, the control sample (neutral soil) was taken from an area where GRP pipes had been previously laid, which did not present problems in operation over time (Bacău area, Romania—the third experimental location).

These samples were analysed at the OSPA laboratory, Iași, Romania.

The method used was soil reaction, which is a measure of soil acidity or alkalinity. The soil pH (Table 1) was defined as the logarithm in base 10 of the activities of hydrogen ions (H^+) or of hydronium ions (H_3O^+) in a solution. For simplification and standardization, a numerical scale from 0 to 14 was adopted. This size was considered the main variable of the soil. Moreover, its influence on the behaviour of GRP pipes was treated as a priority in this material. The principle of determining the soil reaction consists of extracting soluble forms from the soil in distilled water.

Table 1. pH variation intervals depending on soil reaction.

Soil Reaction	pH Variation Intervals	Experimental Soil Type Perimeter
Strongly acidic	≤5	-
Moderately acidic	5.01–5.80	5.40
Weakly acidic	5.81–6.80	-
Neutral	6.81–7.20	6.90
Weakly basic	7.21–8.40	8.10
Moderately, strongly basic	≥8.40	-

The humus and organic carbon content, the carbonate content and the soil solution composition were also determined. The method used was analysis by wet combustion of the soil using chromic acid.

Sample pipes constructed from GRP composite material used for buried water transport pipes were also buried in the three experimental locations. After 24 months, the samples were recovered; 3 specimens were extracted and evaluated, from a structural and morphological point of view, by electron microscopy and X-ray diffraction.

Morphological SEM (scanning electron microscopy) analyses were performed using the SEM FEI Quanta 200 3D (Brno, Czech Republic) equipment. Samples were scanned using an LFD (large field detector) detector with 20 kV high voltage at a working distance of 15 mm. XRD (X-ray diffraction) analysis was performed on the Xpert PRO MPD 3060 equipment, Panalytical (Almelo, The Netherlands), using a copper X-ray tube ($K\alpha = 1.54051°$), 2θ: 10°–80°.

An INSTRON 4400 universal test machine (Techincal University of Gheorghe Asachi from Jassy, Jassy, Romania) was used to determine the axial stress. A sample cut from the pipes was subjected to an increasing tensile force until it broke. The deformation and the respective force were measured and, taking into account the section, the mechanical and elastic characteristics were determined.

Based on the results obtained, we calculated the pipe damage index for axial tension and the Pearson coefficients for axial tension.

Soil Characteristics Analysis

The results obtained from the tests performed at OSPA Iași, Romania are presented in Tables 2–4.

Table 2. Acidic field results.

Soil Type	pH	P	$P_{corrected}$	K	$CaCO_3$	Humus	N_t	Na^+ tot	Na^+ sol	Na^+ comp	Ca^{2+}	Mg^{2+}
		ppm	ppm	ppm	%	%	%	me/100 g	me/100 g	me/100 g	me/100 g	me/100 g
acid	5.94	38.0	21.0	183.0	7.84	4.06	0.211	1.09	0.217	0.873	18.0	6.08

Table 3. Basic field results.

Soil Type	pH	P	$P_{corrected}$	K	$CaCO_3$	Humus	N_t	Na^+ tot	Na^+ sol	Na^+ comp	Ca^{2+}	Mg^{2+}
		ppm	ppm	ppm	%	%	%	me/100 g	me/100 g	me/100 g	me/100 g	me/100 g
basic	8.19	44.0	31.0	177.0	1.67	7.56	0.371	7.82	2.8	5.02	46.0	20.67

Table 4. Neutral field results (control field).

Soil Type	pH	P	$P_{corrected}$	K	$CaCO_3$	Humus	N_t	Na^+ tot	Na^+ sol	Na^+ comp	Ca^{2+}	Mg^{2+}
		ppm	ppm	ppm	%	%	%	me/100 g	me/100 g	me/100 g	me/100 g	me/100 g
neutral	7.03	21.0	10.0	157.0	3.50	1.20	0.068	0.87	0.87	0.217	20.0	6.08

3. Results and Discussion

Figure 1 shows two photos taken at different locations along the pipeline: basic ground and acidic ground. In the case of the pipe buried in basic soil, exfiltration phenomena can be observed visually through the points of rust that appear at the level of the pipe, indicating the complete destruction of the pipe in the case of acidic ground.

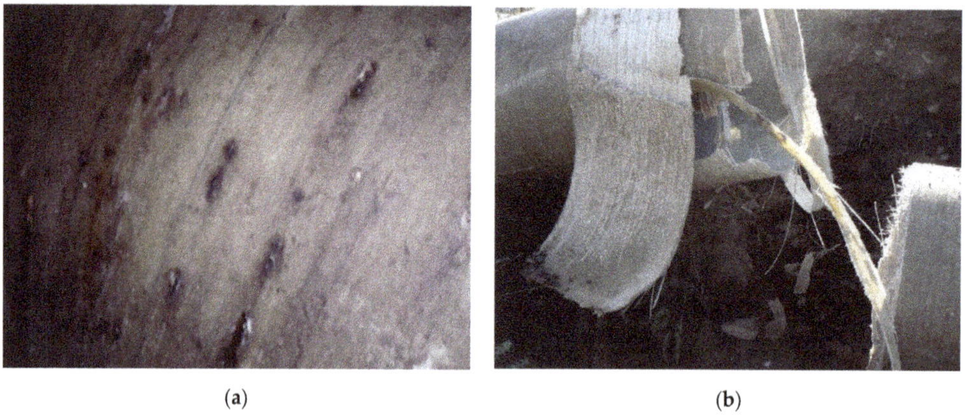

Figure 1. Damage to the glass-reinforced plastic pipeline in (**a**) basic soil and (**b**) acidic soil.

In the case of acidic soil, the damage was manifested in a remarkable way. By changing the mechanical characteristics of acid-contaminated soils, their bearing capacity decreases. There is also a drastic alteration of the pipe material due to the significant effects on the glass fibre.

In the case of basic ground, leaks and colour changes could be observed in the pipe, without significant damage to the fibreglass.

3.1. SEM Analysis of GRP Materials

The evaluation of the modification of the GRP material structure was performed by analysing the areas related to the three characteristic layers of the pipe material: the outer structural layer, the core and the lower structural layer.

For structural analysis using scanning electron microscopy, samples were taken according to Figure 2, and the specific analytical areas are highlighted by section in the longitudinal area of the pipe.

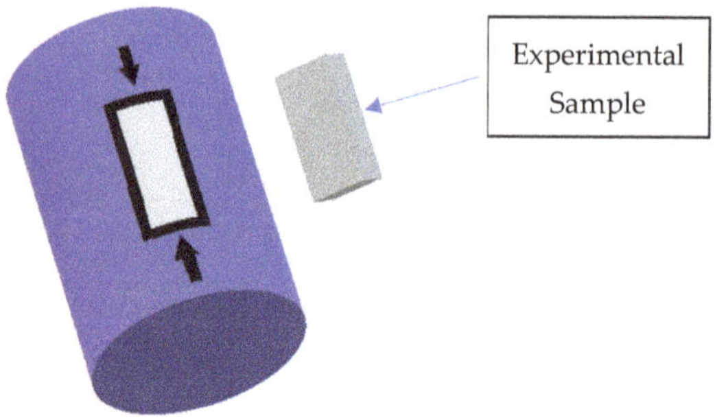

Figure 2. Sampling area for structural analysis.

3.1.1. Analysis of Changes in the Core of the Material over Time

In Figure 3a–c, we have highlighted the structures in the core area of the GRP material at a magnification power of 100× for all three types of material samples collected from the three types of terrain. There is a structure with granulation and uneven distribution of glass particles and fibres, which is similar to the samples collected from neutral and basic terrain. In the acid field test, the structurally highlighted fibres were much smaller and thinner, which confirms the degradation of the glass fibres in the composite mass and thus the plasticization of the core structure in the acidic environment.

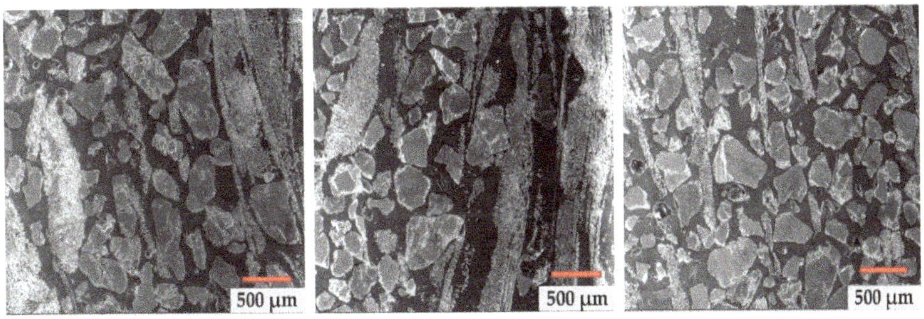

(**a**) (**b**) (**c**)

Figure 3. Morphological structure of the material core: (**a**) neutral soil, (**b**) basic soil, (**c**) acidic soil.

Figure 3 shows the morphology of the component particles in the pipe in the specific medium (neutral, acidic and basic). There is a more accentuated fragmentation for samples a and c, respectively, in neutral and acidic environments, and a lower fragmentation for sample b, due to the presence of continuous fibres arranged in the longitudinal direction. The average dimensions of the particles have the following values (Figure 4): (a) neutral: 339.50 µm ± 146.63 µm, (b) basic: 321.58 µm ± 124.15 µm and (c) acid: 295.78 µm ± 111.58 µm.

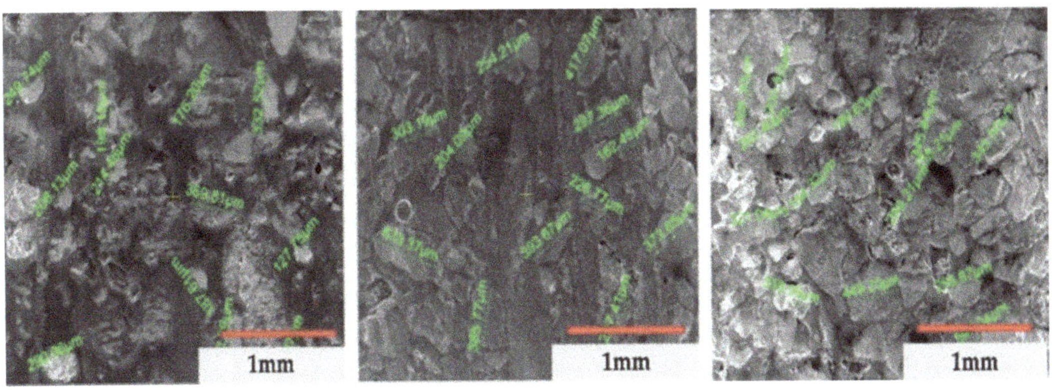

(**a**) (**b**) (**c**)

Figure 4. Average values of the particles: (**a**) neutral soil, (**b**) basic soil, (**c**) acidic soil.

3.1.2. Analysis of Changes in the Outer Layer Thickness over Time

From the analysis of the outer layer for the three cases—shown in Figure 5a–c—the fibrous structure obtained through the arrangement of glass fibres with different thicknesses is highlighted, with an average value of 4.608 mm for the neutral ground sample; an average value for 2.368 mm for the primary field sample and an average value of 1.34 mm for the acidic field sample. There is a reactive activity of the environment on the thickness of the outer layer, with a significant reduction in thickness by 50% in the case of basic soil and by 70% in the case of acidic soil.

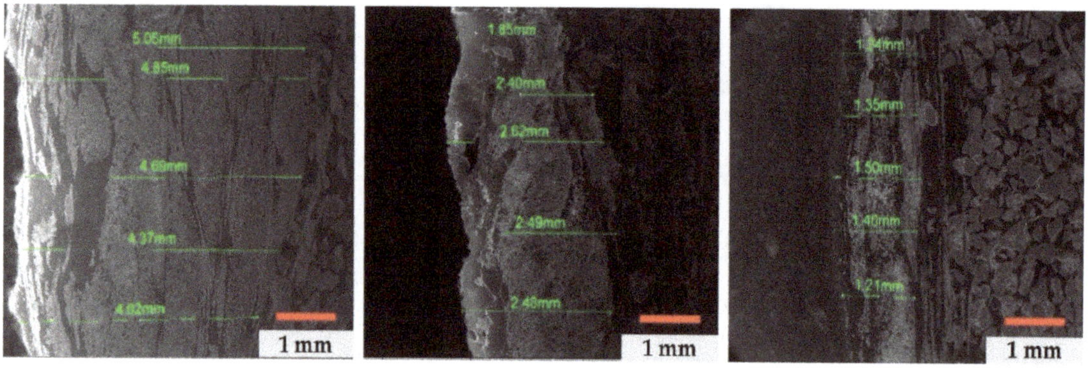

(**a**) (**b**) (**c**)

Figure 5. Dimensional variation in the outer layer: (**a**) neutral soil, (**b**) basic soil, (**c**) acidic soil.

Structural analysis of external areas—as shown in Figure 6a–c—highlights fibrous structures with different orientations with the same calcium aluminosilicate morphology.

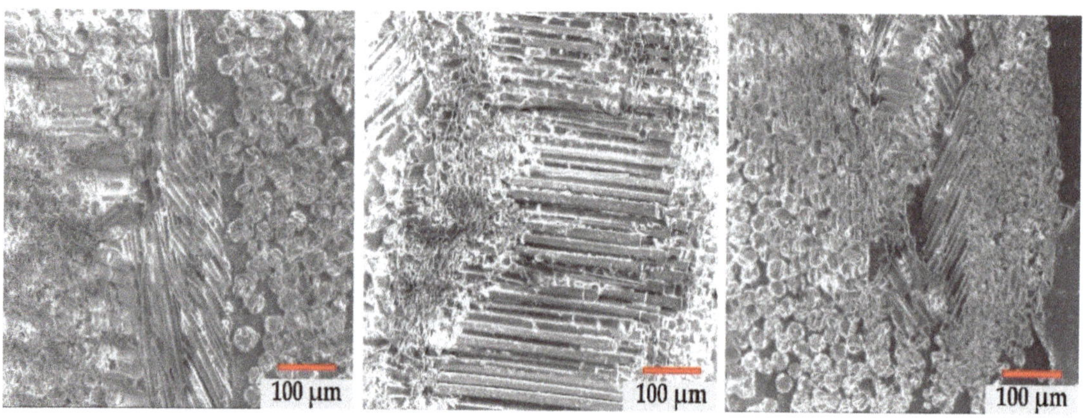

(**a**) (**b**) (**c**)

Figure 6. Morphological structure of the outer layer: (**a**) neutral soil, (**b**) basic soil, (**c**) acidic soil.

Figure 6 confirms the continuity of the fibres in the structure of the pipes used in a neutral environment, having a parallel and substantially equal fibre structure in different areas of the material and with different orientations, without a preferential orientation. We also highlighted the size of the component particles by extending the SEM analysis (Figure 4).

The mechanism of degradation of the pipe material differs from acidic soils with respect to neutral basic soils. In particular, in acid soils, there is a degradation of the outer layer by 70% as a result of increasing roughness, aging of the material, exfoliation, cracks and damage to the pipe.

The core of the material is negatively influenced by the pH of the soil, destroying the glass fibres and leading to plasticization of the material.

Fibreglass tends to undergo accelerated aging when exposed to an acidic solution; the fibres break, causing an overload of the rest of the fibres, thus causing major ruptures.

3.2. XRD Analysis of GRP Materials

Figure 7 shows the X-ray diffraction analysis results for the core of each of the three samples. The figure highlights the specific diffraction peaks of the two main components (SiO_2 and $CaAl_2Si_2O_8$).

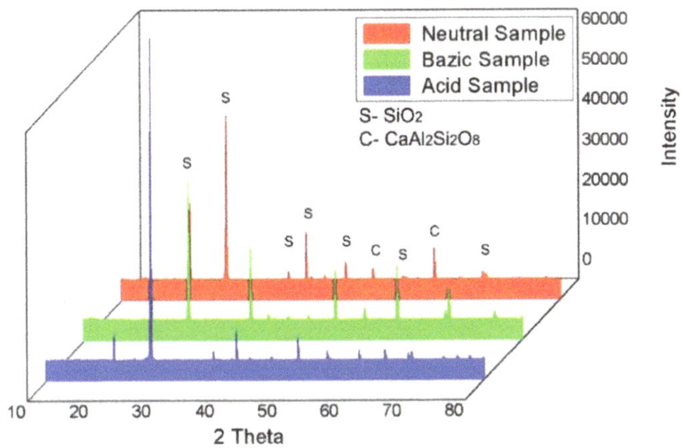

Figure 7. XRD analysis of 3 experimental samples.

Figure 7 shows the diffraction peaks for the component phases of the pipes used in the three different media. The figure presents the three cases in different colours and provides the related explanations.

EDS analysis of the component particles and base material was an important aspect of this study. The results are justified by the XRD analysis, in terms of the identified compounds. The diffraction peaks did not show displacements in terms of the 2θ angle, but there were clear differences in intensity for the acidic sample, which explains the higher crystallinity, Figure 8. The EDS results collected in three different areas are presented in Table 5.

Figure 8. SEM images for different areas used for EDS chemical analysis. (1) SEM images for General aspect area, (2) SEM images for Continous Fibres area, (3) SEM images for Particles area.

Table 5. EDS chemical analysis.

EDS Chemical Analysis		%wt.C	%wt.O	%wt.Al	%wt.Si	%wt.Ca
General aspect	Neutral	28.89	37.41	0.61	28.08	5.01
	Basic	32.62	39.03	1.8	24.82	1.74
	Acidic	26.25	39.20	1.15	31.97	1.41
Continuous fibres	Neutral	34.37	32.74	3.47	18.46	10.96
	Basic	25.94	34.85	5.58	23.56	10.06
	Acidic	34.85	32.59	4.71	20.20	7.64
Particles	Neutral	8.58	47.89	0.42	43.60	0.50
	Basic	6.14	47.64	0.57	45.25	0.41
	Acidic	7.73	47.95	0.64	43.33	0.35

3.3. Calculation of the Pipe Damage Index for Axial Tension

The values of the force from which the behaviour of the sample changes from elastic to rigid are as follows:

F_c^* = 4.31 KN for basic soil;
F_c^* = 3.99 KN for acidic soil;
F_c^* = 4.95 KN for neutral soil.

The corresponding result of the pipe damage index is:

$D = 1 - \frac{F_c^*}{F_c} = 1 - \frac{4.31}{4.95} = 0.13 = 13\%$ for acidic soil;
$D = 1 - \frac{F_c^*}{F_c} = 1 - \frac{3.99}{4.95} = 0.13 = 13\%$ for basic soil.

3.4. Calculation of Pearson Correlation Coefficients for Axial Tension

The average reference value for axial tension and standard deviations are presented in Table 6.

Table 6. Mean values of axial tension.

Variable	Neutral Soil	Standard Deviations	Basic Soil	Standard Deviations	Acidic Soil	Standard Deviations
Δ (%)	6.60	±0.07	4.70	±0.06	4.70	±0.06
F (KN)	4.95	±0.03	3.99	±0.06	4.31	±0.05
E (MPa)	672.00	±3.26	852.00	±4.66	844.00	±4.05
σ_r (MPa)	46.38	±0.08	40.50	±0.14	39.45	±0.07
σ_{max} (MPa)	46.40	±0.11	40.50	±0.16	39.45	±0.09

Δ—maximum deformation; F—break strength; E—Young's modulus; σ_r—failure strain; σ_{max}—maximum stress.

The results of the axial tensile test are presented in Table 7 for the correlation coefficients of pH and axial stress.

Table 7. pH–axial stress correlation coefficients (Pearson correlation matrix).

Variable	pH	Δ	F	E	σ_r	σ_{max}
pH	1	−0.015	−0.342	0.055	0.125	0.125
Δ	−0.015	1	0.945	−0.999	0.990	0.990
F	−0.342	0.945	1	−0.957	0.890	0.890
E	0.055	−0.999	−0.957	1	−0.984	−0.984
σ_r	0.125	0.990	0.890	−0.984	1	1000
σ_{max}	0.125	0.990	0.890	−0.984	1000	1

Δ—maximum deformation; F—break strength; E—Young's modulus; σ_r—failure strain; σ_{max}—maximum stress.

The measured values of Na (in me/100 g) related to the three fields, calculated using the Pearson correlation coefficients, yielded the following Pearson correlation matrix results (Table 8).

Table 8. Na^+ correlation coefficients–axial effort (Pearson correlation matrix).

Variable	Na^+	Δ	F	E	σ_r	σ_{max}
Na^+ (me/100 g)	1	−0.524	−0.774	0.557	−0.399	−0.399
Δ	−0.524	1	0.945	−0.999	0.990	0.990
F	−0.774	0.945	1	−0.957	0.890	0.890
E	0.557	−0.999	−0.957	1	−0.984	−0.984
σ_r	−0.399	0.990	0.890	−0.984	1	1000
σ_{max}	−0.399	0.990	0.890	−0.984	1000	1

Δ—maximum deformation; F—break strength; E—Young's modulus; σ_r—failure strain; σ_{max}—maximum stress.

The low values of the pH correlation index (all other sizes) show a moderate influence of the pH variation on the mechanical characteristics. The only significant value (−0.342) shows the tendency for the load value at which the plastic deformation decreases slightly.

Positive correlation index values were calculated only between the concentration of Na ions and Young's modulus (0.557), indicating the composite material's tendency to plasticise. The relatively high value (−0.774) of the correlation index between the concentration of Na ions and the load at flow shows a general tendency to decrease the mechanical resistance in the axial direction with an increase in the basic content of the soil. Taken together, these two values present image of an aging phenomenon of the composite material, reflected especially by the change in the behaviour from semi-elastic to plastic.

4. Discussion

The evaluation of the structure of glass-reinforced plastic material in the outer layer and in the core indicates the degradation of the glass fibre in an acidic environment. The most used fibreglass is type E, with a high boron content, which gives it special mechanical properties but is affected by an acidic environment.

Scanning electron analysis presents a structure with granulation and uneven distribution of glass particles and fibres, a problem that is similar to the structures obtained from neutral and basic terrain. In the acid field test, fibres were found to be much smaller and thinner, which confirms the degradation of the glass fibres in the composite mass and thus the plasticization of the core structure in the acidic environment.

From the XRD point of view, the predominant peak is identified at the 2θ angle of 26.64° and is formed of silicon dioxide (COD: 96-710-3015) and has a crystalline structure of the hexagonal type. Secondary peaks consist of silicon dioxide and calcium aluminosilicate (COD: 96-154-0253), identified at various 2θ angles of 51.42° and 60.31°, having a crystalline structure of the tetragonal type. These results are in accordance with the structural and morphological analysis obtained by SEM analysis.

In the case of basic soil, the increase in the Young's modulus in the axial direction is highlighted and the ability to work with the ground in the vicinity of the pipeline disappears; thus, the pipeline operates independently of the ground. However, due to the use of plugs for joining, which can be assimilated with elastic joints, the loss of elasticity in the axial direction is compensated.

From the point of view of the morphological structure of the material (Figure 2), it can be observed that it is more affected in the case of acidic soil than basic soil.

5. Conclusions

The results of our experimental study show that a unit deviation in pH in the acidic spectrum generates a decrease in the GRP pipe's lifespan by 10 years. Thus, if the typical lifespan is 50 years for pipes buried in the ground with a neutral pH (pH = 7), at a pH value of 5, the lifespan will decrease to 40 years, and a pH value of 9 would further reduce the lifespan, from the perspective of stiffness depreciation.

In the case of the basic ground, although the morphological structure of the material is affected, the pipe does not lose its elastic character in the axial direction due to the joints, meaning that the service life is not affected.

In order to avoid the deterioration of the behaviour of the buried GRP pipes over time, a series of measures can be taken during the design phase; in particular, a petrochemical study is necessary to determine the pH of the soil solution at the required depth, as well as the presence of Na^+, oil residues or other forms of pollution (anthropogenic or not).

In addition, the choice of a higher class of pressure and rigidity than the one obtained from the static calculation will lead to sufficient and long-term values of the mechanical characteristics for safety in operation, despite the steep regression curve of these characteristics over time.

From the microstructural analysis, it could be concluded that the samples presented a structure with granulation and uneven distribution of glass particles and fibres. In the acid field test, the microstructure displayed fibres that were much smaller and thinner than the other two media. During the XRD analysis, the predominant peaks were identified at the 2θ angle of 26.64° and were formed of silicon dioxide with a crystalline structure of the hexagonal type. Secondary peaks consisted of silicon dioxide and calcium aluminosilicate, identified at various 2θ angles of 51.42° and 60.31°, with a crystalline structure of the tetragonal type.

It would be advantageous to develop the concept of residual chemical stress, to introduce a factor for quantifying the structural degradation caused by chemical aggression on pipes constructed from composite materials.

6. Patents

This section is not mandatory but may be added if there are patents resulting from the work reported in this manuscript.

Author Contributions: Conceptualization, A.D.A. and F.-E.Ţ.; methodology, A.D.A., B.I. and C.M.; investigation, A.D.A. and F.-E.Ţ.; resources, M.P. and C.M.; data curation, A.D.A.; writing—original draft preparation, A.D.A. and B.I.; writing—review and editing, B.I. and F.-E.Ţ.; visualization, M.P.; supervision, M.P. and C.M. All authors have read and agreed to the published version of the manuscript.

Funding: This research received no external funding.

Institutional Review Board Statement: Not applicable.

Informed Consent Statement: Not applicable.

Data Availability Statement: The data underlying this article will be shared on reasonable request from the corresponding author.

Conflicts of Interest: The authors declare no conflict of interest.

References

1. Faria, H.; Guedesb, R.M. Long-term behaviour of GFRP pipes: Reducing the prediction test duration. *Polym. Test.* **2010**, *29*, 337–345. [CrossRef]
2. Rafiee, R. Experimental and theoretical investigations on the failure of filament wound GRP pipes. *Compos. Part B* **2013**, *45*, 257–267. [CrossRef]
3. Stamenović, M.; Putić, S.; Rakin, M.; Medjo, B.; Čikara, D. Effect of alkaline and acidic solutions on the tensile properties of glass–polyester pipes. *Mater. Des.* **2011**, *32*, 2456–2461. [CrossRef]
4. Ancas, A.D.; Profire, M. Effects of soil pH on mechanical properties of glass-reinforced plastic pipes. *Environ. Eng. Manag. J.* **2018**, *17*, 1879–1885. [CrossRef]
5. Ancas, A.D.; Cirstolovean, I.L.; Mihai, M.P. Influence of soil pollution on glass-reinforced plastic water pipes. *J. Environ. Prot. Ecol.* **2018**, *19*, 721–726.
6. Hogg, P.J. Factors affecting the stress corrosion of GRP in acid environments. *Composites* **1983**, *14*, 254–261. [CrossRef]
7. Banna, M.H.; Shirokoff, J.; Molgaard, J. Effects of two aqueous acidic solutions on polyester and bisphenol A epoxy vinyl ester resins. *Mater. Sci. Eng. A* **2011**, *528*, 2137–2142. [CrossRef]

8. Shin, S.; Lee, S.; Judi, D.R.; Parvania, M.; Goharian, E.; McPherson, T.; Burian, S.J. A Systematic Review of Quantitative Resilience Measures for Water Infrastructure Systems. *Water* **2018**, *10*, 164. [CrossRef]
9. Marzouk, M.; Hamid, S.A.; El-Said, M. A methodology for prioritizing water mains rehabilitation in Egypt. *HBRC J.* **2015**, *11*, 114–128. [CrossRef]
10. Akimwumi, I.I.; Diwa, D. Effects of crude oil contamination on the index properties, strength and permeability of lateritic clay. *Int. J. Appl. Sci. Eng. Res.* **2014**, *3*, 816–824.
11. Mugume, S.N.; Diao, K.; Astaraie-Imani, M.; Fu, G.; Farmani, R.; Butler, D. Enhancing resilience in urban water systems for future cities. *Water Sci. Technol. Water Supply* **2015**, *15*, 1343–1352. [CrossRef]
12. Aşchilean, I.; Badea, G.; Giurca, I.; Naghiu, G.S.; Iloaie, F.G. Determining Priorities Concerning Water Distribution Network Rehabilitation. *Energy Procedia* **2017**, *112*, 27–34. [CrossRef]
13. Adewumi, J.R.; Ilemobade, A.A.; Van Zyl, J.E. Treated wastewater reuse in South Africa: Overview, potential and challenges. *Resour. Conserv. Recycl.* **2010**, *55*, 221–231. [CrossRef]
14. Aşchilean, I.; Iliescu, M.; Ciont, N.; Giurca, I. The Unfavourable Impact of Street Traffic on Water Distribution Pipelines. *Water* **2018**, *10*, 1086. [CrossRef]
15. Cawley, P.; Adams, R.D. Defect types and non-destructive testing techniques for composites and bonded joints. *Mater. Sci. Technol.* **1989**, *5*, 413–425. [CrossRef]
16. Li, S.; Wang, R.; Wu, W.; Sun, J.; Jing, Y. Non-hydraulic Factors Analysis of Pipe Burst in Water Distribution Systems. *Procedia Eng.* **2015**, *119*, 53–62. [CrossRef]
17. Chaallal, O.; Arockiasamy, M.; Godat, A. Field test performance of buried flexible pipes under live truck loads. *J. Perform. Constr. Facil.* **2015**, *29*, 04014124. [CrossRef]
18. Ancas, A.D.; Profire, M.; Cojocaru, G. Experimental evaluation a tensile strength of PAFSIN pipes in different types of land. *J. Appl.* **2017**, *7*, 11–16. [CrossRef]
19. Reis, P.N.B.; Neto, M.A.; Amaro, A.M. Effect of hostile solutions on stress relaxation of carbon/epoxy composites. *Polym. Degrad. Stab.* **2019**, *165*, 60–67. [CrossRef]
20. Amaro, A.M.; Reis, P.N.B.; Neto, M.A.; Louro, C. Effects of alkaline and acid solutions on glass/epoxy composites. *Polym. Degrad. Stab.* **2013**, *98*, 853–862. [CrossRef]
21. Profire, M.; Ancaş, A.-D.; Toma, D.; Cojocaru, G. Calculation Of Underground GRP Pipes. *Bull. Polytech. Inst. Iasi Constr. Archit. Sect.* **2018**, *68*, 73–82.
22. Watkins, R.K.; Anderson, L.R. *Structural Mechanics of Buried Pipes*; CRC Press: Boca Raton, FL, USA, 1999; ISBN 9781420049572.

materials

Article

Determination of Physicomechanical Characteristics of the Cement Mortar with Added Fiberglass Waste Treated with Hydrogen Plasma

Marius Lucian Lupu [1], Dorina Nicolina Isopescu [1,*], Ioan Tuns [2], Ioana-Roxana Baciu [1,*] and Sebastian George Maxineasa [1,*]

1 Department of Civil and Industrial Engineering, Gheorghe Asachi Technical University of Iasi-Romania, Blvd. Mangeron, No. 1, 700050 Iasi, Romania; mariuslucianlupu@gmail.com
2 Department of Construction, The Transilvania University of Brasov-Romania, Tower Street, No. 5, 500152 Brașov, Romania; ioan.tuns@unitbv.ro
* Correspondence: isopescu@tuiasi.ro (D.N.I.); ioanaroxana.baciu@yahoo.com (I.-R.B.); sebastian.maxineasa@tuiasi.ro (S.G.M.)

Abstract: Solving the environmental problems and the economic aspects of the construction sector represent a global priority. The considerable quantities of raw materials and the energy consumed by this sector make it one of the most polluting economic activities. Fiberglass in various forms is widely used in the construction sector. In the manufacturing process and during the usage of fiberglass products, a significant amount of indestructible waste results, negatively impacting the environment. An innovative solution for utilizing this type of waste is the treatment with hydrogen plasma. This process results in two products: the first in the gaseous state used to obtain synthetic fuel and the second in solid-state, named slag. The composition of solid waste contains chemical compounds that can increase their strength if used as additives in mortars or concretes. This study presents the laboratory tests on mortars, in which a part of the cement amount was replaced with the solid component resulting from the plasma treatment of glass fiber waste. The results showed that replacing a part of the cement with these materials is a solution that minimizes the ecological footprint of the buildings.

Keywords: fiberglass; hydrogen; plasma; cement; mortar; vitrification

Citation: Lupu, M.L.; Isopescu, D.N.; Tuns, I.; Baciu, I.-R.; Maxineasa, S.G. Determination of Physicomechanical Characteristics of the Cement Mortar with Added Fiberglass Waste Treated with Hydrogen Plasma. *Materials* **2021**, *14*, 1718. https://doi.org/10.3390/ma14071718

Academic Editors: Corneliu Munteanu and Dolores Eliche Quesada

Received: 4 February 2021
Accepted: 29 March 2021
Published: 31 March 2021

1. Introduction

In the current circumstances, where there is an increasing question of the negative influence humanity has on the environment, effective municipal waste management is a hotly debated topic, internationally. It is known that the construction sector is responsible for the consumption of a significant amount of raw materials and non-renewable energy, which positions it as one of the most polluting economic activities globally. The construction sector has a high impact on a community, ecologically, socially, and economically [1,2]. Since the construction materials industry is of great interest in using waste in producing new materials, the new categories of materials are developing in line with sustainability policies [3,4].

One of the most used materials in the construction sector is concrete. Concrete is a mixture of natural aggregates, cement, and water, making it easy to produce anywhere. Concrete currently contains approximately 12% cement, 8% water, and 80% aggregate in mass. Aggregates and water come from natural resources, while cement is a manufactured product. Its production process has a significant negative impact on the environment (for each kilogram of cement produced, approximately one kilogram of CO_2 equivalent is released into the atmosphere) [5,6].

In line with the steps to protect the environment and reduce pollution globally, various replacement materials, or the reduction of aggregate and cement use, have been studied over time. In the specialty literature, several studies were conducted that were mainly aimed at the use of various wastes in the construction sector, thus noting the study on cement mortar containing oil-contaminated sand, which presents the characteristics, strength development, and microstructure of this type of mortar [7].

Another commonly used material in the construction sector is fiberglass (Figure 1a). Fiberglass is predominantly used in the form of glass wool (as a heat-insulating and pho-no-insulating material), or fiberglass mesh (for reinforcement of sheets or plasters). Even though fiberglass has many benefits, it does not biodegrade, and recycling solutions are needed. When fiberglass products are manufactured, and after their use, a large amount of waste is not biodegradable and has a considerable impact on the environment [8].

(a)

(b)

Figure 1. (**a**) Fiberglass waste; (**b**) fiberglass waste vitrified with hydrogen plasma.

The waste management process, in general, involves the treatment of waste in different forms of aggregation. The selection of treatment has, as essential input data, the characteristics of waste biodegradability. A variety of waste recycling solutions can be used today in this process, these being extended on a large scale by the circular economy principles implementation. In the construction sector, waste is primarily solid and non-biodegradable, having a significant impact on the environment. Thus, waste management processes and methods are needed to minimize the environmental impact. One of the most recent and most effective methods of waste treatment is hydrogen plasma conversion.

The plasma conversion system transforms any waste using a neutral gas, hydrogen in our case, through which a strong electromagnetic field passes until it becomes ionized, increasing its electrical conductivity [9]. It is essential to note that it is not about incineration but dissociation at a molecular level in this system. When a substance is subjected to molecular dissociation, it is not only changing its condition, but disintegrates. At this point, it is no longer "a substance"; all that remains are the component atoms and the decimated molecules. When the molecules are subjected to intense energies (plasma torch), the bonds that hold the molecules are broken [10,11]. What remains are the component elements of the molecules. Dangerous substances such as cyanide, for example, results in carbon and nitrogen atoms. Organic compounds are volatile and converted into synthetic gas (syngas), which can be used as a fuel source if purified. Inorganic compounds are melted and converted into a vitrified rock containing metal in addition to other inorganic substances, such as slag [9–11].

Depending on the type of waste material, the plasma treatment can consist of two phenomena: gasification and vitrification [12,13]. Gasification is a process that transforms carbon-containing materials such as coal, petroleum coke, municipal solid waste, or biomass into a synthesis gas (syngas), composed mainly of carbon monoxide and hydrogen [13,14]. Vitrification is the process by which silicates, or a mixture of silicates, are transformed into a glassy amorphous mass following exposure to high temperatures, resulting in melting followed by solidification [15,16].

Research has led to the use of grinding, incineration, and pyrolysis for the recycling of fiberglass. Fiberglass finds its way into different industries and can be used in various end products [10]. For example, fiberglass reinforcement effectively reduces concrete shrinkage, thereby increasing its durability [17]. This concrete can best be used in frozen temperate areas for floors, sidewalks, and concrete curbs [18–20].

In the context of the sustainable development of the construction sector, this research studies the use of hydrogen plasma-treated fiberglass waste (Figure 1b) and its incorporation into cement mortars, particularly looking at the influence that these additional materials have on the mechanical characteristics of mortar.

This study presents the laboratory tests on mortars in which a part of the cement amount was replaced with the solid component resulting from the plasma treatment of glass fiber waste. In the research regarding the replacement of cement with other substances to diminish the negative effect of this material on the environment, a limitation of 20–25 wt %, is accepted. Considering the novelty of the research presented in this article, it was initially aimed at replacing cement in smaller percentages, namely 3 wt %, 6 wt %, and 10 wt %. For the beginning, these percentages were chosen, and if the results were favorable, the research would have continued, taking into account higher percentages.

The obtained results showed replacing a part of the cement with these material additions is a solution that minimizes the ecological footprint of the buildings in terms of the environmental impact. The results obtained by replacing a small amount of cement (3 wt %; 6 wt %; 10 wt %) highlighted that the solid component obtained from the plasma treatment of glass fiber waste, which has in the chemical component Si and Ca, has a positive influence on the mortar resistance. These chemical compounds are complementary to the formation of hydration products of cement, respectively, for the calcium silicates.

2. Materials and Methods

2.1. Analysis of Fiberglass Waste (before and after the Hydrogen Plasma Treatment)—Scanning Electron Microscopy (SEM) and Energy Dispersive X-Ray (EDX) Analysis

Determination of fiberglass waste characteristics before and after the hydrogen plasma process was made with the QUANTA 200 3D electron microscope [21]. First of all, a scanning electron microscope (SEM) (FEI, Brno, Czech Republic) analysis was conducted, which presents a focused electron beam over a surface to create an image. The electrons in the beam interact with the sample, producing various signals that can be used to obtain information about the surface topography and composition. Secondly, there were carried out energy dispersive X-ray analysis (EDX) (FEI, Brno, Czech Republic), an X-ray technique used to identify materials' elemental composition. This analysis is applicable in materials and product research [22–24].

In the SEM analysis of the fiberglass waste can be observed the dimensions of the fibers between 11.9–13.12 µm, and for the fiberglass waste treated with hydrogen plasma, the analysis presents glomerules and irregularities resulting from the treatment with hydrogen plasma (Figure 2).

(**a**) (**b**)

Figure 2. SEM analysis for (**a**) fiberglass waste; (**b**) fiberglass waste treated with hydrogen plasma.

EDX analysis indicated the presence of a silicon (Si) peak at 1.9 keV for both fiberglass waste and fiberglass waste treated with hydrogen plasma. The difference is the value of the silicon peak, noting that in the case of the treated fiberglass waste, the value is approximately 5.1 KCnt. In contrast, untreated fiberglass waste is about 1.9 KCnt (Figure 3). It can also be observed that in the case of untreated fiberglass waste, carbon (C) and potassium (K) are present. After the hydrogen plasma treatment, these two chemical elements disappear, being replaced by sodium (Na).

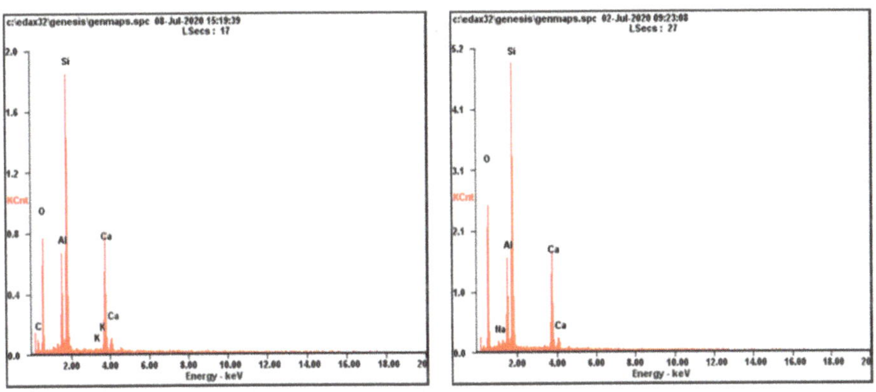

(**a**) (**b**)

Figure 3. EDX analysis for (**a**) fiberglass waste; (**b**) fiberglass waste treated with hydrogen plasma.

Recent studies show that sodium silicate permeates into concrete, which favors the reaction with portlandite in the cement matrix to yield calcium–silicate hydrates. This reaction blocks the concrete pores, increases its surface hardness, and structural impermeability, and durability. The investigations based on these substances did not reveal the exact mechanism by which the concrete performance was improved [25–28].

Further X-ray diffraction (XRD) (PANalytical, Almelo, Netherlands) analysis on the fiberglass waste treated with hydrogen plasma sample revealed SiO_2 peaks (COD 96-153-8065), the major peak 26.64°, and the minor peak at 51.15°, having a hexagonal crystalline structure. The XRD analysis results (Figure 4) are consistent with EDX analysis, in which the main elements are Si and O.

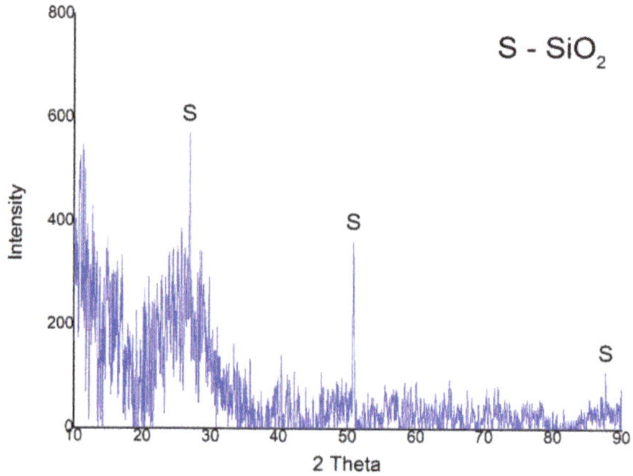

Figure 4. XRD pattern of fiberglass waste treated with hydrogen plasma.

2.2. Experiment Description—Determination of Flexural and Compressive Strength

Flexural and compressive strength were determined under SR EN 196-1 with prismatic test samples with a size of 40 mm × 40 mm × 160 mm. The sand used is standardized C.E.N., according to EN 196-1, each prepackaged sandbag weighing 1350 ± 5 g [29]. The cement used is CEM II/B-LL 42.5 N. Tap water from the public network was used to carry out the samples.

The proportions by mass for the control samples (PM) are one part cement, three parts standardized sand C.E.N. and half part water, water/cement ratio of 0.5 [29].

Each mixture of three control samples (PM) consists of [29]:

450 ± 2 g cement
1350 ± 5 g standardized sand CEN
225 ± 1 g water

Starting from the above proportions, various percentages of cement were replaced to develop new materials that incorporate products resulting from the hydrogen plasma conversion of fiberglass waste as follows (recipes are for three samples type):

- PS 3 wt %—436.5 ± 1 g cement

  13.5 ± 0.5 g hydrogen plasma vitrified fiberglass
  1350 ± 5 g standardized sand CEN
  225 ± 1 g water

- PS 6 wt %—423 ± 1 g cement

  27 ± 0.5 g hydrogen plasma vitrified fiberglass
  1350 ± 5 g standardized sand CEN
  225 ± 1 g water

- PS 10 wt %—405 ± 1 g cement

  45 ± 0.5 g hydrogen plasma vitrified fiberglass
  1350 ± 5 g standardized sand CEN
  225 ± 1 g water

The mortar was prepared by mechanical mixing and compacted into a mold. The molded samples were stored in a humid atmosphere (relative humidity of 70%) for 24 h, then demolded to be immersed in water until resistance testing. At the optimum time, samples were taken from the water, subjected to bending until breaking, resulting in two halves, and half being tested at compression [29]. The machine used to determine flexural and compressive strength is a hydraulic universal testing machine, type WAW-600E.

The testing term of the prisms has been calculated from the moment they were molded with the following limits [29]:

- 2 days ± 30 min.
- 7 days ± 30 min.
- 28 days ± 30 min.

All test procedures complied with the provisions of SR EN 196-1 [29].

Flexural strength, R_f, was determined according to Equation (1) [29]:

$$R_f = 1.5 \times F_f \times l/b^3 \quad (1)$$

where R_f represents the flexural strength (MPa), associated with the side of the square section of prism b (mm), the load applied to the middle of the prism at fracture F_f (N), and the distance between the supports, l (mm) [29].

Compressive strength, R_c, was evaluated according to Equation (2) [29]:

$$R_c = F_c/A_c \quad (2)$$

where R_c represents the compressive strength (MPa), associated with the maximum load at fracture F_c (N), and the cross-sectional area of the specimen on which the compressive force acts (1600 mm^2) [29].

3. Results

3.1. Determination of Flexural Strength

The flexural strength test's average value was calculated as the arithmetic means of the three individual results [29].

Table 1 presents the control sample's flexural strength results and the three hydrogen plasma vitrified fiberglass samples tested (PM, P.S.3 wt %, P.S.6 wt %, P.S.10 wt %), their fracture load, and the standard deviation for all three ages of curing (2, 7, and 28 days).

Table 1. Flexural strength (R_f), fracture load (F_f), and the standard deviation (*SD*) results for the control sample and the hydrogen plasma vitrified fiberglass samples tested (PM, PS 3 wt %, PS 6 wt %, and PS 10 wt %).

	Samples	Age of Curing					
		2 Days	SD	7 Days	SD	28 Days	SD
Fracture load—F_f [kN]	PM	1.31	-	2.33	-	2.63	-
	PS 3 wt %	1.40	-	2.40	-	2.82	-
	PS 6 wt %	1.36	-	2.08	-	2.65	-
	PS 10 wt %	1.21	-	2.03	-	2.44	-
Flexural strength—R_f [MPa]	PM	3.1	0.27	5.5	0.05	6.3	0.41
	PS 3 wt %	3.3	0.12	5.6	0.56	6.6	0.21
	PS 6 wt %	3.2	0.23	4.9	0.37	6.2	0.18
	PS 10 wt %	2.8	0.21	4.8	0.37	5.6	0.49

In Figure 5 are presented the diagrams of the flexural strength results for samples with hydrogen plasma vitrified fiberglass (PS 3 wt %, PS 6 wt %, and PS 10 wt %), and the control sample for the three tested periods (2 days, 7 days, and 28 days), highlighting the behavior of each batch of samples and their evolution over time.

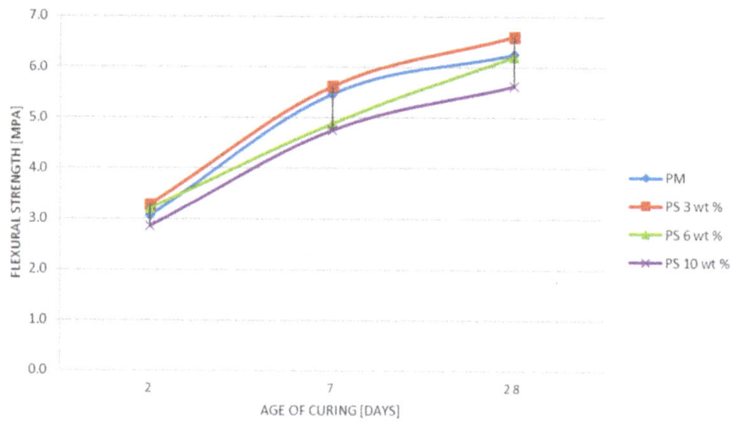

Figure 5. The diagrams of flexural strength results for samples with hydrogen plasma vitrified fiberglass (PS 3 wt %, PS 6 wt %, and PS 10 wt %) and the control samples (PM).

3.2. Determination of Compression Strength

Compression strength test results were calculated as the arithmetic means of the six individual results [29].

Table 2 presents the control sample's compression strength results and the six hydrogen plasma vitrified fiberglass samples tested (PM, PS 3 wt %, PS 6 wt %, PS 10 wt %), their fracture load, and the standard deviation for all three ages of curing (2, 7, and 28 days). The compression strength results for samples with hydrogen plasma vitrified fiberglass (PS 3 wt %, PS 6 wt %, and PS 10 wt %) and the control sample for the three test periods (2 days, 7 days, and 28 days), highlighting the average behavior of each batch of samples and their evolution over time are presented in Figure 6.

Table 2. Compression strength (R_c), fracture load (F_f), and the standard deviation (SD) results for the control sample and the hydrogen plasma vitrified fiberglass samples tested (PM, PS 3 wt %, PS 6 wt %, and PS 10 wt %).

	Samples	Age of Curing					
		2 Days	SD	7 Days	SD	28 Days	SD
Fracture load—F_c [kN]	PM	11.08	-	20.48	-	30.84	-
	PS 3 wt %	16.01	-	33.55	-	49.95	-
	PS 6 wt %	15.95	-	33.67	-	43.64	-
	PS 10 wt %	17.54	-	33.58	-	43.25	-
Compression strength—R_c [MPa]	PM	6.90	0.52	12.8	0.47	19.0	0.91
	PS 3 wt %	10.0	0.65	21.0	0.94	31.2	0.98
	PS 6 wt %	10.0	0.32	21.0	0.84	27.3	0.80
	PS 10 wt %	11.0	0.57	21.0	0.76	26.8	0.67

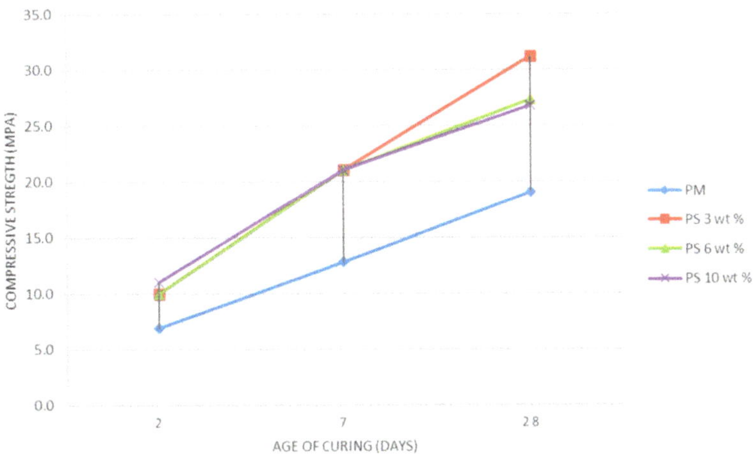

Figure 6. The diagrams of compression strength results for samples with hydrogen plasma vitrified fiberglass (PS 3 wt %, PS 6 wt %, and PS 10 wt %) and the control sample (PM).

4. Discussion

The presented research's primary goals are to analyze, develop, and optimize different solutions to significantly reduce the construction sector's negative influence on the natural environment. At the same time, new plasma technologies can turn waste into raw materials, contributing to environmental protection in large urban areas. In the technical literature, the research on reducing cement consumption in mortars and concretes by replacing a part of it with by-products (ultrafine silica, power plant ash, slag, etc.) has experienced a growing trend. The studies have been carried out to ensure that new products developed (mortars or concretes that comply with standardized strength class with different composition) have strengths and rigidity characteristics applicable in the construction sector [30–34].

Following the tests carried out, the average values obtained in the case of the flexural strength of the mortar specimens underlined the fact that the reduction of cement by 3 wt % and its replacement with products resulting from fiberglass waste treated with hydrogen plasma improves the flexural strength. The values obtained for the PS 3 wt % specimens show an improvement, with approximately 5% compared to the control specimens, PM, for all tested curing times, given by standards [29].

The average values of flexural strength, in the cases of reductions of 6 wt % and 10 wt %, the PS 6 wt % and PS 10 wt % samples, are lower than those of the control specimens; therefore, it can be stated that, after these tests, there is no certainty of obtaining mortars comparable, in terms of flexural strength, with control mortars.

For the mortars' compressive strengths, the results obtained in all studied cases reveal that the average values are higher than the average value of the control samples' compressive strengths. The highest increase is also obtained in the case of PS 3 wt % samples.

5. Conclusions

The research presented in the study showed that a quantity of cement could be replaced with products resulting from fiberglass waste treated with hydrogen plasma. In construction, structural elements are rarely subjected to compression forces only. Most often, they are subjected to bending or, to bending and compression. Therefore, a limit in the amount of replaced cement by fiberglass waste treated with hydrogen plasma should be considered 3 wt %.

The experiment results showed that the particles of fiberglass vitrified with plasma hydrogen increased the viscosity, and slightly improved the mortars' strength, due to their pozzolanic activity. The added particles led to the quantitative increase of the hydration products, and thus enhanced the value of the compressive strength of the mortars due to the supplementary formation of a calcium-silicon hydrated compound, which strengthened the microstructure of the mortars. In addition, in the case of compressive strength of mortars, a positive influence was due to the larger particles added that consolidated the mortar's structure.

The tests for flexural strengths have shown that the addition of fiberglass waste treated with hydrogen plasma in percentages higher than 3 wt % can decrease the resistances due to the granularity of the mineral addition. Even if the mortar's compactness is better, the volume of included air increased, thus causing a decrease in the mortar's tensile strength.

By replacing a quantity of cement with fiberglass waste treated with hydrogen plasma, in terms of mechanical characteristics, mortars similar to the usual ones can be obtained, having advantages from an economic and ecological perspective. The results obtained are promising and underline the necessity for continuing the research regarding the strength and rigidity characteristics in conditions of the static and dynamic loads, and life-cycle assessment analysis of these materials. It is observed that the replacement of cement with the complementary product analyzed in the paper, within certain limits, leads to improvement of resistances. Based on the research carried out and presented in this article, a new research direction can be established using different additives to increase the workability and reduce the water/cement ratio.

Author Contributions: M.L.L., D.N.I., and S.G.M. designed the experiments; M.L.L. and I.-R.B. performed the experiments; M.L.L., D.N.I., I.-R.B., and S.G.M. analyzed the data; M.L.L., D.N.I., and I.T. carried out the investigations; D.N.I. and I.T. found resources; M.L.L., D.N.I., and S.G.M. wrote, reviewed and edited the paper. All authors have read and agreed to the published version of the manuscript.

Funding: This paper was elaborated with the support of the "Ecoinnovative Products and Technologies for Energy Efficiency in Constructions—EFECON" research grant, project ID P_40_295/105524, a program co-financed by the European Regional Development Fund through Operational Program Competitiveness 2014–2020.

Institutional Review Board Statement: Not applicable.

Informed Consent Statement: Not applicable.

Data Availability Statement: Data sharing not applicable.

Conflicts of Interest: The authors declare no conflict of interest.

References

1. Lupu, M.L.; Isopescu, D.N.; Cucoş, I.; Antonescu, I.; Maxineasa, S.G.; Baciu, I.-R. Researches on energy conversion of municipal waste by plasma decomposition for energy-efficiency in civil engineering. *Iop Conf. Ser. Mater. Sci. Eng.* **2020**, *789*, 012035. [CrossRef]
2. Umberto, A. Process and technological aspects of municipal solid waste gasification. A review. *Waste Manag.* **2012**, *32*, 625–639.
3. Leal-Quirós, E. Plasma Processing of Municipal Solid Waste. *Braz. J. Phys.* **2004**, *34*, 1587–1593. [CrossRef]
4. Lal, P.M.; Singh, C.J. Plasma Gasification: A Sustainable Solution for the Municipal Solid Waste Management in the State of Madhya Pradesh. *Int. J. Environ. Sci.* **2012**, *3*, 306.
5. Maxineasa, S.G.; Taranu, N.; Bejan, L.; Isopescu, D.; Banu, O.M. Environmental impact of carbon fibre-reinforced polymer flexural strengthening solutions of reinforced concrete beams. *Int. J. Life Cycle Assess.* **2015**, *20*, 1343–1358. [CrossRef]
6. Miller, A.; Ip, K. Sustainable construction materials. In *Design and Management of Sustainable Built Environments*; Yao, R., Ed.; Springer: London, UK, 2013; pp. 341–358.
7. Abousnina, R.; Manalo, A.; Ferdous, W.; Lokuge, W.; Benabed, B.; Al-Jabri, K.S. Characteristics, strength development and microstructure of cement mortar containing oil-contaminated sand. *Constr. Build. Mater.* **2020**, *252*, 119155. [CrossRef]
8. Cameron, N.N.; Rapp, C.F. Fiberglass. *Encycl. Mater. Sci. Technol.* **2001**, *2*, 3142–3146.
9. Linhart, J.G. *Plasma Physycs*; EURATOM: Brussels, Belgium, 1969.

10. Carpinlioglu, M.O.; Sanlisoy, A. Performance assessment of plasma gasification for waste to energy conversion: A methodology for thermodynamic analysis. *Int. J. Hydrogen Energy* **2018**, *43*, 11493–11504. [CrossRef]
11. Morozov, A.I. *Introduction to Plasma Dynamics*; CRC Press: Boca Raton, FL, USA, 2012.
12. Anyaegbunam, F.N.C. Thermal Plasma Process for Hazardous Waste Treatment. *Int. J. Eng. Res. Technol.* **2014**, *3*, 1769–1773.
13. Moustakas, K.; Fatta, D.; Malamis, S.; Haralambous, K.; Loizidou, M. Demonstration plasma gasification/vitrification system for effective hazardous waste treatment. *J. Hazard. Mater.* **2005**, *123*, 120–126. [CrossRef]
14. University of Leeds. *Report FLEXI-PYROCAT—Flexible Pyrolysis-Catalysis Processing of Waste Plastics, Supported by the Research and Innovation Staff Exchange Project—Grant Number: 643322-FLEXI-PYROCAT*; University of Leeds: Leeds, UK, 2016.
15. Academia Română. *Micul Dicționar Academic*; Univers Enciclopedic: Bucharest, Romania, 2010; Volume 2.
16. US Environmental Protection Agency. *Handbook: Vitrification Technologies for Treatment of Hazardous and Radioactive Waste*; US Environmental Protection Agency: Washington, DC, USA, 2013.
17. Chawla, K.K. Glass Fibers. *Encycl. Mater. Sci. Technol.* **2001**, *2*, 3541–3545.
18. Siddika, A.; Mamun, A.; Ferdous, W.; Saha, A.K.; Alyousef, R. 3D-printed concrete: Applications, performance, and challenges. *J. Sustain. Cem. Based Mater.* **2019**, *9*, 127–164. [CrossRef]
19. Khazanov, V.E.; Kolesov, Y.I.; Trofimov, N.N. *Glass Fibres, Fibre Science and Technology. Soviet Advanced Composites Technology Series*; Springer: Dordrecht, The Netherlands, 1995; Volume 5, pp. 15–230.
20. Johnson, T. Recycling Composite Materials. ThoughtCo. Available online: thoughtco.com/recycling-composite-materials-820337 (accessed on 27 January 2021).
21. The Quanta 200 3D. Simplified Operation Manual. Available online: mri.psu.edu/sites/default/files/file_attach/fib_sop.pdf (accessed on 26 January 2021).
22. Silva, M.A.P.; Ferri, F.A. *Nanocharacterization Techniques*; William Andrew Publishing: Oxford, UK, 2017; pp. 1–35.
23. Orasugh, J.T.O.; Ghosh, S.K.; Chattopadhyay, D. *Fiber-Reinforced Nanocomposites: Fundamentals and Applications*; Elsevier: Amsterdam, The Netherlands, 2020; pp. 199–233.
24. Patel, J.P.; Parsania, P.H. *Biodegradable and Biocompatible Polymer Composites*; Woodhead Publishing: Duxford, UK, 2018; pp. 55–79.
25. Franzoni, E.; Pigino, B.; Pistolesi, C. Ethyl silicate for surface protection of concrete: Performance in comparison with other inorganic surface treatments. *Cem. Concr. Compos.* **2013**, *44*, 69–76. [CrossRef]
26. Baltazar, L.; Santana, J.; Lopes, B.; Rodrigues, M.P.; Correia, J.R. Surface skin protection of concrete with silicate-based impregnations: Influence of the substrate roughness and moisture. *Constr. Build. Mater.* **2014**, *70*, 191–200. [CrossRef]
27. LaRosa Thompson, J.L.; Silsbee, M.R.; Gill, P.M.; Scheetz, B.E. Characterization of silicate sealers on concrete. *Cem. Concr. Res.* **1997**, *27*, 1561–1567. [CrossRef]
28. Pigino, B.; Leemann, A.; Franzoni, E.; Lura, P. Ethyl silicate for surface treatment of concrete—Part II: Characteristics and performance. *Cem. Concr. Compos.* **2012**, *34*, 313–321. [CrossRef]
29. Asociația de Standardizare din România. *SR EN 196-1:2016—Methods of Testing Cement—Part 1: Determination of Strength*; Asociația de Standardizare din România: Bucharest, Romania, 2016.
30. Abdel-Gawwad, H.A.; Metwally, K.A.; Tawfik, T.A.; Mohammed, M.S.; Hassan, H.S.; Heikal, M.; El-Kattan, I.M. Evaluating the performance of high volume fly ash-blended-cement mortar individually containing nano- and ultrafine micro-magnesia. *J. Build. Eng.* **2021**, *36*, 102129. [CrossRef]
31. Yang, J.; Hu, H.; He, X.; Su, Y.; Wang, Y.; Tan, H.; Pan, H. Effect of steam curing on compressive strength and microstructure of high volume ultrafine fly ash cement mortar. *Constr. Build. Mater.* **2021**, *266*, 120894. [CrossRef]
32. Sáez del Bosque, I.F.; Medina, J.M.; Frías, M.; Sánchez de Rojas, M.I.; Medina, C. Use of biomass-fired power plant bottom ash as an addition in new blended cements: Effect on the structure of the C-S-H gel formed during hydration. *Constr. Build. Mater.* **2019**, *228*, 117081. [CrossRef]
33. Teixeira, E.R.; Camões, A.; Branco, F.G.; Matos, J.C. Effect of Biomass Fly Ash on Fresh and Hardened Properties of High Volume Fly Ash Mortars. *Crystals* **2021**, *11*, 233. [CrossRef]
34. Deboucha, W.; Alachek, I.; Plassiard, J.-P.; Plé, O. New Composite Material for Masonry Repair: Mortar Formulations and Experimental Studies. *Materials* **2021**, *14*, 912. [CrossRef] [PubMed]

MDPI AG
Grosspeteranlage 5
4052 Basel
Switzerland
Tel.: +41 61 683 77 34

Materials Editorial Office
E-mail: materials@mdpi.com
www.mdpi.com/journal/materials

www.ingramcontent.com/pod-product-compliance
Lightning Source LLC
Chambersburg PA
CBHW041941260926
51850CB00018B/393
9783725815401